Progress in Gauge Field Theory

NATO ASI Series

Advanced Science Institutes Series

A series presenting the results of activities sponsored by the NATO Science Committee, which aims at the dissemination of advanced scientific and technological knowledge, with a view to strengthening links between scientific communities.

The series is published by an international board of publishers in conjunction with the NATO Scientific Affairs Division

A Life Sciences **B Physics**	Plenum Publishing Corporation New York and London
C Mathematical and Physical Sciences	D. Reidel Publishing Company Dordrecht , Boston, and Lancaster
D Behavioral and Social Sciences **E Engineering and Materials Sciences**	Martinus Nijhoff Publishers The Hague, Boston, and Lancaster
F Computer and Systems Sciences **G Ecological Sciences**	Springer-Verlag Berlin, Heidelberg, New York, and Tokyo

Recent Volumes in this Series

Volume 111—Monopole '83
edited by James L. Stone

Volume 112—Condensed Matter Research Using Neutrons: Today and Tomorrow
edited by Stephen W. Lovesey and Reinhard Scherm

Volume 113—The Electronic Structure of Complex Systems
edited by P. Phariseau and W. M. Temmerman

Volume 114—Energy Transfer Processes in Condensed Matter
edited by Baldassare Di Bartolo

Volume 115—Progress in Gauge Field Theory
edited by G.'t Hooft, A. Jaffe, H. Lehmann, P. K. Mitter, I. M. Singer, and R. Stora

Volume 116—Nonequilibrium Cooperative Phenomena in Physics and Related Fields
edited by Manuel G. Velarde

Volume 117—Moment Formation in Solids
edited by W. J. L. Buyers

Series B: Physics

Progress in Gauge Field Theory

Edited by

G. 't Hooft
Institute for Theoretical Physics
Utrecht, The Netherlands

A. Jaffe
Harvard University
Cambridge, Massachusetts

H. Lehmann
University of Hamburg
Hamburg, Federal Republic of Germany

P. K. Mitter
University of Paris VI
Paris, France

I. M. Singer
University of California
Berkeley, California

and

R. Stora
Laboratory of Particle Physics
Annecy, France

Plenum Press
New York and London
Published in cooperation with NATO Scientific Affairs Division

Proceedings of a NATO Advanced Study Institute on
Progress in Gauge Field Theory,
held September 1–15, 1983, in Cargèse, Corsica, France

ISBN 978-1-4757-0282-8 ISBN 978-1-4757-0280-4 (eBook)
DOI 10.1007/978-1-4757-0280-4

Library of Congress Cataloging in Publication Data

NATO Advanced Study Institute on Progress in Gauge Field Theory (1983: Cargèse, Corsica)
Progress in gauge field theory.

(NATO ASI series. Series B, Physics; v. 115)
"Published in cooperation with NATO Scientific Affairs Division."
"Proceedings of a NATO Advanced Study Institute on Progress in Gauge Field Theory, held September 1–15, 1983, in Cargèse, Corsica, France"—Verso t.p.
Includes bibliographical references and index.
1. Gauge fields (Physics)—Congresses. I. 't Hooft, G. II. Title. III. Series.
QC793.3.F5N37P 1983 530.1′43 84-17869
ISBN 978-1-4757-0282-8

Softcover reprint of the hardcover 1st edition 1984
A Division of Plenum Publishing Corporation
233 Spring Street, New York, N.Y. 10013

To the memory of Kurt Symanzik

Kurt Symanzik (1923-1983)

PREFACE

The importance of gauge theory for elementary particle physics is by now firmly established. Recent experiments have yielded convincing evidence for the existence of intermediate bosons, the carriers of the electroweak gauge force, as well as for the presence of gluons, the carriers of the strong gauge force, in hadronic interactions. For the gauge theory of strong interactions, however, a number of important theoretical problems remain to be definitely resolved. They include the quark confinement problem, the quantitative study of the hadron mass spectrum as well as the role of topology in quantum gauge field theory. These problems require for their solution the development and application of non-perturbative methods in quantum gauge field theory. These problems, and their non-perturbative analysis, formed the central interest of the 1983 Cargèse summer institute on "Progress in Gauge Field Theory." In this sense it was a natural sequel to the 1979 Cargèse summer institute on "Recent Developments in Gauge Theories."

Lattice gauge theory provides a systematic framework for the investigation of non-perturbative quantum effects. Accordingly, a large number of lectures dealt with lattice gauge theory. Following a systematic introduction to the subject, the renormalization group method was developed both as a rigorous tool for fundamental questions, and in the block-spin formulation, the computations by Monte Carlo programs. A detailed analysis was presented of the problems encountered in computer simulations. Results obtained by this method on the mass spectrum were reviewed. In addition lecture series were devoted to a new approach to the mass spectrum based on a perturbative scheme and asymptotic freedom, and to the mathematical study of asymptotically free renormalisable field theories in the planar diagram approximation. Recent work in constructive field theory based on random walk methods, the random surface approach to gauge field theory, and random lattice theories were reviewed. Another series of lectures were devoted to monopoles, the calculus of variations and topology, as well as recent works on anomalies and supersymmetry and index theory. These principal lectures were supplemented by a seminar program with invited speakers.

A notable absence in the school was Kurt Symanzik who was ailing at the time. Kurt Symanzik was to be a principal lecturer at this school, and his work was of extreme relevance to the central interests of the school. It was this with profound sorrow that we learned of his subsequent death. His death is a great loss for Theoretical Physics and also for the Cargèse summer institute with which he had been associated for many years. He lectured successively at Cargèse in 1970, 1973, 1976 and 1979 and his lectures had a seminal impact on developments in quantum field theory and statistical mechanics. With deep sorrow we dedicate these proceedings to his memory.

We wish to express our gratitude to NATO whose generous financial contribution made it possible to organize this school. We thank Maurice Lévy, the Director of the Institut d'Etudes Scientifiques de Cargèse, as well as the University of Nice, for making available to us the facilities of the Institut. Grateful thanks are due to Marie-France Hanseler, for much help with the material aspects of the organization. Last but not least we thank the lecturers and the participants for their enthusiastic involvement which contributed much to the success of the school.

G. 't Hooft

A. Jaffe

H. Lehmann

P.K. Mitter

I.M. Singer

R. Stora

CONTENTS

An Introduction to Gravitational Anomalies 1
L. Alvarez-Gaumé

Dirac Monopoles, from d = 2 to d = 5, Lecture I . . . 23
F.A. Bais

Charge-Pole Dynamics, Lecture II 51
F.A. Bais

Exact Renormalization Group for Gauge Theories 79
T. Balaban, J. Imbrie and A. Jaffe

The Spectrum in Lattice Gauge Theories 105
B. Berg

Lattices, Demons and the Microcanonical Ensemble . . . 155
G. Bhanot

Quantum Field Theory in Terms of Random Walks and Random Surfaces 169
J. Fröhlich

Rigorous Renormalization Group and Large N 235
K. Gawedzki and A. Kupiainen

On Kähler's Geometric Description of Dirac Fields . . 247
M. Göckeler and H. Joos

Planar Diagram Field Theories 271
G. 't Hooft

Fields on a Random Lattice 337
C. Itzykson

Defect Mediated Phase Transitions in Superfluids, Solids, and their Relation to Lattice Gauge Theories . . 373
H. Kleinert

Lattice Gauge Theory 403
C.P. Korthals Altes

Constructive Theory of Critical Phenomena 435
A. Kupiainen and K. Glawedzki

On a Relation Between Finite Size Effects and Elastic Scattering Processes 451
M. Lüscher

Renormalization Group and Mayer Expansions 473
G. Mack

Gauge Invariant Frequency Splitting in Non Abelian Lattice Gauge Theory 497
P.K. Mitter

Prolegomena to any Future Computer Evaluation of the QCD Mass Spectrum 531
G. Parisi

Algebraic Structure and Topological Origin of Anomalies . 543
R. Stora

Morse Theory and Monopoles: Topology in Long Range Forces 563
C.H. Taubes

Monte Carlo Renormalization Group and the Three Dimensional Ising Model 589
K.G. Wilson

Index . . . .

AN INTRODUCTION TO GRAVITATIONAL ANOMALIES

Luis Alvarez-Gaumé

Lyman Laboratory of Physics
Harvard University
Cambridge, MA 02138

INTRODUCTION

The study of anomalies in global and gauge currents in Quantum Field Theory has had a remarkable number of important applications during the 70's. In the original version of the anomaly[1] one considers a massless fermion triangle diagram with one axial current and two vector currents. Requiring the vector currents to be conserved, one finds that the axial current is not conserved therefore leading to a breakdown of chiral symmetry in the presence of gauge fields coupled to conserved vector currents. This breakdown of chiral symmetry led to the understanding of π^0 decay and to the resolution of the u(1) problem.[2] The anomaly has also been instrumental in posing constraints to insure the mathematical consistency of gauge theories coupled to chiral currents. If one considers a theory with gauge fields coupled for instance to left handed currents, one must look at a fermion triangle diagram with V-A currents at each vertex. Again, this diagram is anomalous, and unless the anomalies cancel when summing over all the fermion species running around the loop, one finds that the V-A currents are not conserved, implying that gauge invariance is broken and thus the anomaly renders the theory inconsistent. The anomaly cancellation condition has proven to be very useful in constraining the particle content of unified gauge theories.[3] More recently[4], the anomaly has also been shown to be useful in analyzing the spectrum of massless fermions in confining theories. In the context of low energy chiral theories, the Wess-Zumino lagrangian[5] has recently played a central role in showing that the soliton solutions of certain models[6] can be identified with baryons.[7] This recent development has in turn shed new light into our understanding of chiral anomalies.

Due to the renewed interest in Kaluza-Klein theories, some authors[8] have recently discussed the question of anomalies in space-times of dimensions higher than four. In any even number of dimensions there is an anomaly analogous to the triangle anomaly. In 2n dimensions one has to consider polygon diagrams with n+1 vertices and fermions running around the loop. These diagrams are again anomalous[8], and the anomaly cancellation condition provides severe restrictions on possible Kaluza-Klein theories. The main part of these lectures will be concentrated on unraveling another generalization of the standard chiral anomalies, namely the possibility of having anomalies under general coordinate transformations.[9] The existence of anomalies with respect to coordinates changes implies that the energy momentum tensor is not conserved and that the coupling to gravitation is inconsistent. It will be shown that in the context of higher dimensional theories the requirement of anomaly cancellations for these type of transformations is very hard to meet.

Finally, as explained in Professor Stora's lectures, it has been shown by some formal manipulations that the full structure of the non-abelian anomaly in 2n dimensions can be obtained in terms of the expression for the abelian chiral anomaly in two higher dimensions.[10] This is interesting because it will ultimately lead to a purely topological understanding of the non-abelian anomaly which was lacking until now. This line of thinking is now being actively pursued by both mathematicians[11] and physicists.[12]

The outline of these lectures is as follows: We will first analyze the abelian anomaly from the point of view of the Atiyah-Singer index theorem. This is clearly not the first time that this analysis has been carried out, but it will give us a chance of introducing a general method of computing anomalies based on supersymmetric quantum mechanics. Then we will present the general strategy for identifying and computing the anomalies in the energy-momentum tensor and what can be learned from them.

Since the literature on anomalies has been rather prolific, it was necessary to make a selection of references. This selection however is based on historical or pedagogical reasons, and I do apologize to those whose work has not been included in the reference list.

ANOMALIES AND INDEX THEORY

The close relation between the anomaly in axial u(1) currents and the Atiyah-Singer index theories has been known since the mid 70's.[13] We would like to present here a different way of looking at this connection. To fix ideas we will concentrate on the gravitational contribution to the axial anomaly for the time being.

After the discovery of the anomalous triangle diagram, it was soon realized that one could also consider triangle diagrams with an axial current and two energy momentum tensors. This diagram is again anomalous, and leads to the celebrated breakdown of chirality conservation in the presence of an external gravitational field which has been extensively analyzed in the literature.[14]

The connection between this anomaly and the index theorem for the Dirac operator is best understood if one computes the anomaly through a method introduced by Fujikawa.[15] However, before we do that, it is necessary to explain why all of this is in any way related to supersymmetric quantum mechanics.

In [16], E. Witten introduced a quantity $\Delta = \mathrm{Tr}(-1)^F$ ($(-1)^F$ in the fermion number operator) which "counts" the number of bosonic states minus the number of fermionic states. In a supersymmetric theory, the hamiltonian can be written as the square of a fermionic charge S (the supersymmetric charge), $H = S^2$, thus for any bosonic state of energy $E \neq 0$, $|E\rangle$, there is a fermionic state with the same energy $S|E\rangle$, so that for non-zero energy states, their contribution to Δ cancel in pairs leading to the conclusion that Δ is uniquely determined by the zero energy states of the system under considerations (in order to avoid confusion it should be mentioned that these arguments are carried out in a box with finite volume V, where the concept of several zero energy states makes sense). In other words, Δ is equal to the number of zero energy bosonic states minus the number of zero energy fermionic states. As E. Witten showed, Δ is extraordinarily useful in determining whether certain theories (among them supersymmetric Yang-Mills theories) will undergo dynamical supersymmetry breaking. Moreover, if we restrict our considerations to the zero momentum sector, we have that in any supersymmetric theory $\{Q,Q^*\} = H$, $S = Q+Q^*$ and $Q^2 = 0$, $Q^{*2} = 0$; which immediately implies that Δ = index Q thus for field theories in 0+1 dimensions, i.e. quantum mechanical theories, H is simply a 2nd order elliptic operator and Q and Q^* are first order elliptic operators which define a square root of H, and it was soon realized that $\mathrm{Tr}(-1)^F$ could be used to generate a proof of the Atiyah-Singer index theorem for judiciously chosen supersymmetric systems whose supercharge coincides with differential operators whose index is relevant both in geometry and physics.[17,18,19,20] For example, T. Parker[17] used $\mathrm{Tr}(-1)^F$ to derive the Gauss-Bonnet formula for the Euler character, and to study some properties of heat kernel expansions, D. Friedan and P. Widney[18] used Δ to derive the index formula for the Dirac operator, and the present author[19,20] carried out a more exhaustive search of supersymmetric quantum mechanical systems which are related to interesting index problems as well as their corresponding G-index generalizations (generalized fixed point theorems of M. Atiyah and R. Bott.[20] The outline presented in [19] for a proof of the Atiyah-Singer index theorem was turned into a rigorous proof by E. Getzler.[22]

Since we will only be interested in anomalies generated by fermion fields, let us quickly derive the index formula for the Dirac operator in the presence of both external gravitational and gauge fields. The starting point consists of writing a supersymmetric system whose hamiltonian is given by the square of the Dirac operator in the presence of the external field of interest. For example if we only consider an external gravitational field, $g_{ij}(x)$, which represents the metric of a compact manifold M_{2n} of dimension 2n, the following lagrangian is easily shown to be supersymmetric:

$$L = \frac{1}{2} g_{ij}(x) \frac{dx^i}{dt} \frac{dx^j}{dt} + \frac{i}{2} \psi^a \left(\frac{d}{dt} \psi^a + \frac{dx^i}{dt} \omega_i{}^a{}_b \psi^b \right) \tag{1}$$

and the supercharge is given by:

$$S = g_{ij} \psi^i \frac{dx^j}{dt}, \qquad \psi^i = e^i{}_a \psi^a, \qquad g^{ij} = e^i{}_a e^j{}_b \delta^{ab}. \tag{2}$$

In (1) the x^i's represent the coordinates on the manifold M_{2n}, $i = 1, \ldots, 2n$, the $\psi^a(t)$'s are 2n real antcommutating grassman functions, $e^i{}_a(x)$ is the vierbein field and $\omega_i{}^a{}_b$ is the vierbein or spin connection defined by the torsion free condition of riemannian geometry:

$$\partial_i e^a_j + \Gamma^k_{ij} e^a_k + \omega^a_{ib} e^b{}_j = 0. \tag{3}$$

$\Gamma^i{}_{jk}$ stands for the standard Christoffel symbol. If we canonically quantize the theory defined by (1), we find that the fermions satisfy the following anticommutation relations: $\{\psi^a, \psi^b\} = \delta^{ab}$, so that a convenient way of realizing the fermionic operators in the theory is $\psi^a = \psi^a/\sqrt{2}$, where the γ^a's are the Dirac matrices in 2n dimensions. The canonical conjugate momenta associated to the x^i's are:

$$p_i = g_{ij}(x) \frac{dx^j}{dt} + \frac{i}{4} \omega_{iab} [\psi^a, \psi^b] = + \frac{1}{i} \frac{\partial}{\partial x^i} \tag{4}$$

where we have used the fact that $\omega_{iab} = -\omega_{iba}$. Hence the supercharge s becomes

$$S = -i\psi^i \left(\frac{\partial}{\partial x^i} + \frac{1}{4} \omega_{iab} [\psi^a, \psi^b] \right) = - \frac{i}{\sqrt{2}} \left(\frac{\partial}{\partial x^i} + \frac{1}{2} \omega_{iab} \sigma^{ab} \right), \tag{5}$$

$$\sigma^{ab} = \frac{1}{4} [\gamma^a, \gamma^b], \quad \gamma^i = e^i{}_a \gamma^a,$$

i.e. S becomes the Dirac equation, and $H = (i\not{D})^2/2$. Since fermion number is an operator which anticommutes with all the fermion operators in the theory, it follows that $(-1)^F$ is $\Gamma_{2n+1} = i^n \gamma^1\gamma^2\cdots\gamma^{2n}$, and that:

$$\mathrm{Tr}(-1)^F e^{-\beta H} = \mathrm{Tr}\,\Gamma_{2n+1} \exp-\beta(i\not{D})^2/2 = \mathrm{index}\,\not{D} \tag{6}$$

the computation of $\mathrm{ind}(\not{D})$ is carried out by first writing down a path integral representation of (6). Since (6) is just the partition function for an ensemble at temperature "β^{-1}" described by the density matrix $\rho = (-1)^F e^{-\beta H}$, is follows that

$$\mathrm{Tr}(-1)\ e^{-\beta H} = \int_{\mathrm{P.B.C.}} [dx(t)d\psi(t)]e^{-S_E(x,\psi)} \tag{7}$$

where P.B.C. stands for periodic boundary conditions. The presence of $(-1)^F$ in the trace implies that both bosonic and fermionic fields must be integrated over with periodic boundary conditions: $x(\beta)=x(0)$, $\psi(\beta) = \psi(0)$. In (7) S_E is the Wick rotated version of (1). Since we are only interested in the $\beta\to 0$ limit, we can Fourier expand $x(t)$ and $\psi(t)$. As $\beta\to 0$ the leading contribution would correspond to purely constant configurations. However, this contribution vanishes because the constant configurations are zero action solutions of the classical equations of motion, therefore the first non-vanishing contribution comes from the gaussian fluctuations around the constant configurations. Carrying out the expansion of (1) around the constant configurations (x_0,ψ_0) to second order using riemannian normal coordinates, we obtain

$$L^{(2)} = \frac{1}{2} g_{ij}(x_0)\dot{\xi}^i\dot{\xi}^j + \frac{1}{4} R_{ijk\ell}(x_0)\psi_0^k\psi_0^\ell\, \xi^i\dot{\xi}^j + \frac{i}{2} x^a \frac{d}{dt} x^a . \tag{8}$$

ξ^i and x^j are the bosonic and fermionic fluctuations respectively. Wick rotating (8) and performing the Gaussian approximation to (7), we find after some straightforward computation that:

$$\mathrm{Tr}(-1)^F e^{-\beta H} = \frac{1}{(2\pi)^n} \int d\,\mathrm{Vol}\int (d\psi_0) \prod_{\alpha=1}^{n} \frac{x_\alpha/2}{\sinh x_\alpha/2} \tag{9}$$

where the x_α's are the skew eigenvalues of the antisymmetric matrix $R_{abcd}\psi_0^c\psi_0^d/2$. Even though the integrand in (9) looks non-polynomial, if we expand the integrand around $x_\alpha = 0$ we obtain a finite polynomial because each x_α is a bilinear in the Grassmann numbers ψ_0^a, and the highest non-zero polynomials in the ψ_0^a's that can be formed is of order 2n. If we restrict (9) to the case of four dimensional space-time, we find

$$\mathrm{Ind}(i\not{D}) = \frac{1}{384\pi^2}\int (dx)\, R^a{}_{bcd} R^b{}_{aef}\, \varepsilon^{cdef} \tag{10}$$

The connection between the previous computations and the gravitational contribution to the axial anomaly follows from a generalization of a procedure first introduced by Fujikawa.[15] The basic idea of Fujikawa's procedure is to notice that the symmetries of the classical action are not necessarily symmetries of the measure in the functional integral which defines the quantum theory, therefore symmetries of the classical theory may cease to be symmetries at the quantum level. Let us consider a Dirac fermion $\psi(x)$ coupled to an external gravitational field $g_{\mu\nu}$ with euclidean signature. The coupling of ψ to $g_{\mu\nu}$ is given by the action:

$$S = \int e\,\bar{\psi}(x)\, i\gamma^\mu \left(\partial_\mu + \frac{1}{2}\omega_{\mu ab}\sigma^{ab}\right)\psi\, d^4x$$

$$e = \det(e^a{}_i) \tag{10}$$

using the same notation as before. This theory is invariant under global axial rotations:

$$\psi(x) \to e^{i\alpha\gamma_5}\psi(x) \tag{11}$$

under an infinitesimal space-time dependent chiral rotation (11), the change in the action is:

$$\delta S = \int d^4x\, e\, \alpha(x)\, D_\mu(\bar{\psi}\gamma^\mu\gamma_5\psi) \tag{12}$$

where in parenthesis is the axial current. In order to check whether the axial current is still conserved at the quantum level, we consider the effective action:

$$e^{-\Gamma(g)} = \int (d\bar{\psi}d\psi)\, e^{-S(e,\psi,\bar{\psi})} \tag{13}$$

and make a change of variables $\psi \to \psi' = \psi + i\alpha(x)\gamma_5\psi$. As pointed out in [15], the only term which could lead to anomalous contributions to the Ward identities generated by the axial symmetry is the measure in (13). The easiest way to define the measure in general, is to expand ψ and $\bar{\psi}$ in terms of the eigenfunctions of the Dirac operator:

$$i\not{D}\psi_n = \lambda_n\psi_n \qquad \psi = \sum_n a_n\psi_n, \quad \bar{\psi} = \sum_n \psi_n^\dagger(x)\bar{b}_n \tag{14}$$

and the measure becomes

$$\prod_n d\bar{b}_n da_n .$$

Under an infinitesimal axial rotation, the measure changes by a jacobian factor:

$$\prod_{n,m} d\bar{b}_n da_n \to \left(\prod_{n,m} d\bar{b}_n da_m\right) \exp -2i \int dx\, e \sum_n \psi_n^\dagger(x)\alpha(x)\gamma_5\psi(x) \tag{15}$$

the minus sign in the exponent is due to Fermi statistics. If for simplicity we choose α to be constant, the computation of the anomaly is reduced to computing the trace in (15), or rather its regularized form

$$\lim_{\beta\to 0} \int (dx) \sum_n \psi_n^\dagger(x)\gamma_5\psi_n(x) e^{-\beta\lambda_n^2} = \lim_{\beta\to 0} \mathrm{Tr}\, \gamma_5 e^{-\beta(i\not{D})^2} \tag{16}$$

which in terms of our previous arguments is nothing but $\mathrm{Tr}(-1)^F e^{-\beta H}$ for the theory defined by (1), since we showed before that $\mathrm{Tr}(-1)^F e^{-\beta H}$ is the index of the Dirac operator, we immediately obtain the anomaly equation:

$$D_\mu \langle J^{\mu 5}\rangle = -\frac{2i}{384\pi^2} R\tilde{R} \tag{17}$$

the factor of two appears because we have considered a Dirac fermion rather than a Weyl fermion. The generalization of (17) to higher dimensional space-times is straightforward and is left as an exercise to the reader. Finally, if we want to include contributions to (17) from external gauge fields we have to extend (1) as to generate a theory whose hamiltonian is the square of the Dirac operator in the presence of external gauge and gravitational fields. Let G be the gauge group, and assume that the fermions transform according to the representation of G generated by the matrices $T^\alpha{}_{AB}$, $A,B = 1,2\ldots,N$, $(T^\alpha)^\dagger = T^\alpha$. For each index A, B, we introduce a pair of fermion operators satisfying the standard anticommutation relations $\{c_A, c_B^*\} = \delta_{AB}$. Calling L_0 to the lagrangian defined in (1), it is easy to check that:

$$L = L_0 + ic_A^*\left(\frac{d}{dt} c_A + \frac{dx^i}{dt} iA_i^\alpha T^\alpha_{AB} c_B\right) - \frac{i}{2}\psi^a\psi^b F^\alpha_{ab} c_A^* T_{AB} c_B \tag{18}$$

(where A_i^α is the gauge field and F_{ij}^α its gauge field strength) is a theory whose hamiltonian is equal to $(i\not{D})^2/2$ and:

$$\not{D} = \gamma^i\left(\partial_i + \frac{1}{2}\,\omega_{iab}\sigma^{ab} + iA_i^\alpha T^\alpha\right) \tag{19}$$

and compute Tr $\Gamma_{2n+1}\, e^{-\beta(i\not{D})^2}$ is equivalent to computing (7) where $(-1)^F$ is the fermion number operator for the ψ^a fermion, thus the c-fermions must be integrated over with antiperiodic boundary conditions. Before writing down the final answer for $D_i\langle J^{i5}\rangle$ in four dimensions, it should be mentioned that the path integral over the c-fermions has to be restricted to the set of one-particle states, because we want to calculate the index of $i\not{D}$ (or the anomaly) for fermions in the representation (T^α) and not in any of its tensor products, and clearly the only set of states in the Hilbert space generated by the c-fermions which generate this representation are the one-particle states. We evaluate Tr $\Gamma_{2n+1}\exp-\beta(i\not{D})^2$ by following the same steps as in the purely gravitational case. In the $\beta\to 0$ limit, we can expand around the constant configurations. Since the c's are integrated over with antiperiodic boundary conditions, the only constant configuration for the c's is c=0, thus (18) is already second order in c and c^*. The expansion of x(t) and $\psi(t)$ is carried out as before. Computing now the Gaussian fluctuations around $(x_0, \psi_0, c=0)$ and projecting onto the one particle states for c and c^*, we obtain[20]:

$$\text{ind}\,\not{D} = \frac{1}{(2\pi)^n}\int d\,\text{Vol}\int (d\psi_0)(\text{Tr}\ e^F)\ \prod_\alpha \frac{x_\alpha/2}{\sinh x_\alpha/2} \tag{20}$$

$$F = \frac{1}{2}\,F^\alpha_{ab}T^\alpha\psi_0^a\psi_0^b$$

and the x_α's are defined as in Equation (9). Restricting (20) to the four dimensional case, we obtain for Dirac fermions:

$$D_\mu\langle J^{\mu 5}\rangle = -\frac{\dim T}{192\pi^2}\,R\tilde{R} + \frac{1}{16\pi^2}\,\text{Tr}\ F\tilde{F} \tag{21}$$

as should be.[23]

Similar arguments can be carried out for the case of the spin 3/2 field, and in general for the computation of virtually any chiral anomaly for global currents. If we compute the anomaly using Fujikawa's procedure, we only have to compute the anomalous variation of the measure. This in turn reduces to the computation of a trace

$$\sum_n \psi_n^\dagger\gamma_5\, L\,\psi_n$$

over the spectrum of a suitable Dirac operator, where L is some local operator which implements the infinitesimal transformation on the fields of the theory, but as shown before this trace can be represented as a Green's function for a supersymmetric quantum mechanical system: $\mathrm{Tr}(-1)^F e^{-\beta H_L}(x,p)$ in the $\beta\to 0$ limit. This procedure can for instance be applied to the gravitational anomalies to be discussed below.

GRAVITATIONAL ANOMALIES

In order to identify theories which may contain anomalies in the energy-momentum tensor, I would like to first present a general criterion which makes rather easy to identify those cases which could be potentially troublesome. Consider first the standard non-abelian anomaly for Weyl fermions coupled to an external gauge field. In euclidean space the action is

$$S = \int d^4x\, \bar{x}\, i\not{D}\left(\frac{1-\gamma_5}{2}\right)\lambda$$

$$\not{D} = \gamma^\mu(\partial_\mu + A_\mu^a T^a), \quad T^a = -T^{a+} \tag{21}$$

for some gauge group G, and the left and right handed spinors λ, $\bar{\chi}$ are completely independent. The question one is interested in is whether the effective action $\Gamma(A)$ for fermion in the representation T is gauge invariant. Under a gauge transformation, $A_\mu \to A_\mu - D_\mu\varepsilon$:

$$\Gamma(A-D\varepsilon) = \Gamma(A) + \int d^4x\, \varepsilon^a(x)\, D_\mu \frac{\delta\Gamma}{\delta A_\mu^a} \tag{22}$$

so that $D_\mu\, \delta/\delta A_\mu^a$ is the generator of gauge transformations for the effective action. If one computes the anomalous variation by Feynman diagrams, the result is[1,3]

$$D_\mu \frac{\delta\Gamma(A)}{\delta A_\mu^a} = \frac{i}{24\pi^2}\, \varepsilon^{\mu\nu\alpha\beta}\, \mathrm{Tr}\, T^a \partial_\mu\left(A_\nu\partial_\alpha A_\beta + \frac{1}{2} A_\nu A_\alpha A_\beta\right) \tag{23}$$

One important feature of (23) is the overall factor of i. The anomaly is always purely imaginary in euclidean space. This can be argued in several ways. Since the anomaly is parity violating, the presence of the ε symbol requires the imaginous unit so that after Wick rotating back to Minkowski space we obtain a real answer. More importantly however the factor of i in (23) implies that only the imaginary part of the effective action can pick up an anomaly. To prove this, let $\Gamma(A,T)$ be the effective action of a Weyl fermion transforming under a representation T of the gauge group G, and $\Gamma(A,T^*)$ the effective action in the complex conjugate representation

T^*. Since the action for T^* is basically the complex conjugate of (21), apart from trivial factors we get that $\Gamma(A,T^*) = \Gamma(A,T)^*$, hence $2\,\mathrm{Re}\,\Gamma(A,T) = \Gamma(A,T) + \Gamma(A,T^*)$. But $\Gamma(A,T) + \Gamma(A,T^*)$ is the same as the effective action for a fermion transforming on the $T \oplus T^*$ representation which is real, and for real representations of the gauge group one can always introduce bare masses, which amounts to saying that $\Gamma(A,T \oplus T^*)$ can be regularized in a gauge invariant way using Pauli-Villars regulators. Thus $2\,\mathrm{Re}\,\Gamma(A,T)$ is gauge invaraint and as claimed the anomaly only shows up in the real part of the effective action. We conclude therefore that if the fermions transform under a real representation of G, the effective action is gauge invariant. If the representation is pseudoreal, then $T \oplus T$ is real which implies that $e^{-2\Gamma(A,T)}$ is gauge invariant; however, in order to get $e^{-\Gamma(A,T)}$ we have to extract the square root whose sign may not be globally defined as in the case of the SU(2) anomaly.[24] Thus, modulo anomalies with respect to non-trivial finite gauge transformations, we only need to worry about theories with Weyl fermions transforming according to complex representations of the gauge group. The question of gauge invariance of the effective action is equivalent to current conservation, because

$$\delta\Gamma/\delta A_\mu^a = \langle\bar{\chi}\gamma_\mu T^a(1-\gamma_5)/2\lambda\rangle^A$$

the induced current in the presence of the gauge field A. Hence

$$D_\mu \frac{\delta\Gamma}{\delta A_\mu^a} = D_\mu \langle J^{\mu a}\rangle.$$

The extension of the preceding argument to the gravitational case is straightforward. Now the action in 2n dimensions is:

$$S = \int (d^{2n}x)\, e\, \bar{\chi}\, i\gamma^\mu\left(\partial_\mu + \frac{1}{2}\,\omega_{\mu ab}\sigma^{ab}\right)\left(\frac{1-\gamma_5}{2}\right)\lambda \tag{24}$$

the σ^{ab}'s are the generators of the spinor representation of SO(2n). Since the only term in (24) which could generate a complex action is the term containing the σ^{ab}'s, the question of whether there will be gravitational anomalies is reduced to the question of which spinor representations of SO(K) are complex.[25] If K is odd, SO(K) has a single spinor representation and it cannot be complex. This is equivalent to saying that in space-times of odd dimensionality there is no chirality. If K is even, SO(2n) has two spinor representations, and two cases can be considered. If n is even i.e. SO(4K), both spinor representations are either real or pseudoreal. Only when n is odd i.e. SO(4K+2) the spinor representations are complex and anomalies could arise. A trivial corollary of these remarks is that in four dimensions the energy momentum tensor has no purely gravitational anomalies. This is probably the reason why this type of anomalies were not considered before. The effective action one obtains after integrating out the fermions in (24) will only depend on the gravitational field $\Gamma(g)$. Under an infinitesimal general coordinate transformation

$$g_{ij} \to g_{ij} - D_i\xi_j - D_j\xi_i$$

and

$$\Gamma(g_{ij} - D_i\xi_j - D_j\xi_i) - \Gamma(g) = \int (dx)\, \xi^i D^j \frac{\delta\Gamma}{\delta g^{ij}} \qquad (25)$$

since $\delta\Gamma/\delta g^{ij} = \langle 0|T_{ij}|0\rangle$ (energy-momentum tensor) we see that invariance of $\Gamma(g)$ under infinitesimal coordinate changes in equivalent to the conservation of the induced energy momentum tensor as claimed in the introduction. Since a consistent coupling to gravity requires a conserved energy momentum tensor it follows that unless (25) vanishes the theory under consideration is mathematically inconsistent.

Which complex representation of SO(4K+2) are relevant? For fermions we may consider spin 1/2 or spin 3/2 fields of definite chirality. Each of these fields gives rise to one loop anomalies. In addition, it seems that there is one type of bose field that suffers from an anomaly. The simplest complex bosonic representation of SO(4K+2) is an antisymmetric tensor $F_{\mu_2 \cdots \mu_{2K+1}}$ that obeys a self-duality condition

$$F_{\mu_1 \cdots \mu_{2K+1}} = \pm i/(2K+1)!\ \ \varepsilon_{\mu_1 \cdots \mu_{2K+1}\,\nu_1 \cdots \nu_{2K+1}} F^{\nu_1 \cdots \nu_{2K+1}} .$$

Certain supergravity theories contain an antisymmetric tensor field $A_{\mu_1 \cdots \mu_{2K}}$ whose curl is constrained to obey such a condition. In fact, such fields also generate one loop anomalies.

We will now present a diagrammatic evaluation of the gravitational anomaly in 4K+2 dimensions for a spin 1/2 Weyl fermion. For the spin 3/2 field or the antisymmetric tensor field only the final result will be presented. Further details for these fields can be found in [9].

In 4K+2 dimensions we consider a one-loop diagram with n external graviton lines. The amplitude will depend on the momenta $p^{(i)}_\mu$ i=1,...,4K+2 restricted to $\Sigma\, p^{(i)}_\mu = 0$ and the graviton polarization tensors $\varepsilon^{(i)}_{\mu\nu}$. By our general arguments, only the parity violating amplitudes are anomalous. A parity violating amplitude is necessarily proportional to the Levi-Civita symbol

$\varepsilon_{\mu_1 \cdots \mu_{4K+2}}$. The ε-symbol, being antisymmetric can be contracted at most with n polarization tensors and n-1 momenta $p^{(i)}_{\mu}$. Thus $n+n-1 \geq 4K+2$ or $n \geq 2K+2$. Thus in 4K+2 dimensions, diagrams with less than 2K+2 external gravitons are not anomalous. We will evaluate the Feynman diagrams with 2K+2 external gravitons. Graphs with more than 2K+2 gravitons also have anomalies, and their contribution could probably be determined by consistency conditions of the Wess-Zumino type.[26]

The action for a Weyl fermion in 4K+2 dimensions is:

$$S = \int dx \det e\, e^{\mu a} \frac{i}{2} \bar{\psi}\, i\, \gamma_a D_\mu \left(\frac{1-\gamma_5}{2}\right) \psi$$

$$D_\mu \psi = \left(\partial_\mu + \frac{1}{2} \omega_{\mu ab} \sigma^{ab}\right) \psi \tag{26}$$

so $S = S_1 + S_2$:

$$S_1 = \frac{i}{2} \int dx \det e\, e^{\mu a}\, \bar{\psi}\, \gamma_a \partial_\mu \left(\frac{1-\gamma_5}{2}\right) \psi$$

$$\begin{aligned} S_2 &= \frac{i}{2} \int dx \det e\, e^{\mu a} \omega_\mu{}^{cd}\, \bar{\psi} \left\{\gamma_a, \sigma_{cd}/2\right\} \left(\frac{1-\gamma_5}{2}\right) 4 \\ &= \frac{i}{4} \int dx \det e\, e^{\mu a} \omega_\mu{}^{cd}\, \bar{\psi}\, \Gamma_{acd} \frac{1-\gamma_5}{2} \psi \end{aligned} \tag{27}$$

where Γ_{acd} is the antisymmetric product of three gamma matrices. We will choose ψ to be a Weyl fermion i.e. $\gamma_5 \psi = -\psi$. Since we will be doing perturbation theory, we will choose $g_{\mu\nu} = \eta_{\mu\nu} + h_{\mu\nu}$, and work in the gauge $e_{\mu a} = \eta_{\mu a} + 1/2\, h_{\mu a}$. The restriction to this gauge means that we will only consider anomalies with respect to the coordinate changes but not in local Lorentz rotations, although the latter can be computed by similar methods. Expanding to lowest order in h, we get:

$$S_1 = -\frac{i}{4} \int dx\, h^{\mu\nu}\, \bar{\psi}\, \gamma_\mu \partial_\nu \left(\frac{1-\gamma_5}{2}\right) \psi + O(h^2)$$

$$S_2 = -\frac{i}{16} \int dx\, (h_{\lambda\alpha} \partial_\mu h_{\nu\alpha}) \bar{\psi}\, \Gamma^{\mu\lambda\nu} \frac{1-\gamma_5}{2} \psi + O(H^3) \tag{28}$$

in S_2 there is also a term linear in h which cancels due to the antisymmetry of $\Gamma^{\mu\lambda\nu}$. Thus vertices originating from S_1 contain at least one graviton line whereas vertices originating from S_2 contain at least two graviton lines. Parity violating terms only appear due to the presence of $1-\gamma_5/2$ factors in the vertices, and a trace containing γ_5 will vanish unless at least 4K+2 γ-matrices are present. In fact as we will show in detail later, after contracting a graviton line with its momentum to test for conservation we always loose two γ-matrices, hence only diagrams containing at least 4K+4 γ-matrices will be anomalous. Consider a one loop diagram with n_1 vertices coming from S_1, and n_2 vertices coming from S_2. Such a diagram contains n_1+n_2 internal lines and therefore n_1+n_2 γ-matrices coming from the fermion propagators. Also we have one γ matrix for each S_1 vertex, and three for each S_2 vertex. The total number of γ-matrices being $2(n_1+2n_2) \geq 4K+4$ or $n_1 + 2n_2 \geq 2K+2$. We can also get an inequality in the other direction if we recall that we only want to compute graphs with 2K+2 external gravitons. Since there is at least one graviton for each S_1 vertex and there are at least three for each S_2 vertex, we see that $2K+2 \geq n_1+2n_2$. Combining these inequalities we see that the anomalous diagrams with 2K+2 gravitons have $n_1+2n_2 = 2K+2$ with exactly one (two) graviton line attached at each vertex originating in $S_1(S_2)$, and we can neglect the $O(h^2)$, $O(h^3)$ terms in (28) in deriving the Feynman rules.

Finally, we must specify how the diagrams will be regulated. We will use a procedure originally due to Adler. Although the one loop diagrams have bose symmetry in the external lines, the simplest method for extracting the anomaly is to insert a single factor of $(1-\gamma_5)/2$ at one vertex and none at the others, and then regularized à la Pauli-Villars by subtracting the same diagram for a fermion of mass M. The amplitude constructed in this way will be anomalous only in the channel where $(1-\gamma_5)/2$ is inserted. By bose symmetrizing at the end, the amplitude which satisfied bose symmetry will have an anomaly in each channel equal to 1/(2K+2) times the anomaly we will calculate in one channel. We now indicate how the diagrammatic computation proceeds. The Feynman rules for (28) can be read off in Figure 1, and the diagram we will be considering is drawn in Figure 2.

In the dangerous channel contain the factor of $(1-\gamma_5)/2$, we take the graviton momentum to be p_μ and its polarization vector to be $i(p_\mu\varepsilon_\nu+p_\nu\varepsilon_\mu)$ which is equivalent to a coordinate transformation $x^\mu \to x^\mu+\varepsilon^\mu$. Thus if the energy-momentum tensor is conserved, an amplitude with such a polarization vector should vanish. In order to keep the algebra as simple as possible, we will choose the polarization vectors in the remaining 2K+1 vertices to be of the form $\varepsilon^{(i)}_{\mu\nu} = \varepsilon^{(i)}_\mu\varepsilon^{(i)}_\nu$. In the regularization method that we have selected, the anomaly appears only because for the massive regulator fermion λ, the energy momentum tensor with an insertion of $(1-\gamma_5)/2$ is not conserved even formally, and it is only the regulator diagram that contributes to the anomaly. In fact, for a regulator field satisfying $(i\not{D}-M)\lambda=0$, it easily follows that $D_\nu(\bar\lambda\,\gamma^\mu D^\nu(1-\gamma_5)/2\lambda)=0$ and $D_\mu\,\bar\lambda\gamma^\mu\,D_\nu\,1/2(1-\gamma_5)\lambda = -iM\,\bar\lambda\,D_\nu\gamma_5\lambda$. Consequently, in the dangerous channel we may replace $i(p_\mu\varepsilon_\nu+p_\nu\varepsilon_\mu)\cdot(-i/4)\bar\lambda\,\gamma^\mu D^\nu\,\frac{1}{2}(1-\gamma_5)\lambda$

by $-1/4\, M\varepsilon^{\nu}\bar{\lambda}D_{\nu}\gamma_5\lambda$, the reader may notice that a γ-matrix has disappeared in this manipulation as claimed above.

In order to carry out the computation of the amplitude in Figure 2, we will proceed in steps. First we do the Dirac algebra. After rationalizing the propagators $i(\not{p}+M)/(p^2-M^2)$ the diagram contains exactly one factor at γ_5 at the anomalous vertex Q. Remembering that the polarization tensors are $\varepsilon^{(i)}_{\mu\nu} = \varepsilon^{(i)}_{\mu}\varepsilon^{(i)}_{\nu}$, we have at each non-anomalous vertex a factor of $\not{\varepsilon}^{(i)}$ so that the Dirac trace is:

$$A = \mathrm{Tr}\ \gamma_5(\not{k}+M)\not{\varepsilon}^{(1)}(\not{k}-\not{p}^{(1)}+M)\not{\varepsilon}^{(2)}(\not{k}-\not{p}^{(1)}-\not{p}^{(2)}+M)\ldots$$
$$\not{\varepsilon}^{(2K+1)}(\not{k}-\not{p}^{(1)}-\not{p}^{(2)}-\ldots-\not{p}^{(2K+1)}+M)$$

since there are at most 4K+3 γ-matrices in A, a non-zero trace requires that we must pick out terms that are precisely in M. This finally brings down the original number of 4K+4 γ-matrices down to 4K+2. Using:

$$\mathrm{Tr}\ \gamma_5\gamma_{\mu_1}\gamma_{\mu_2}\cdots\gamma_{4K+2} = -2^{2K+1}\varepsilon_{\mu_1\mu_2\cdots\mu_{4K+2}}$$

we obtain

$$A = 2^{2K+1}\, M\, R(\varepsilon^{(i)},p^{(i)})$$

where

$$R(\varepsilon^{(i)},p^{(i)}) = -\varepsilon_{\mu_1\mu_2\cdots\mu_{4K+2}}\, p^{(1)}_{\mu_1}\varepsilon^{(2)}_{\mu_2}p^{(2)}_{\mu_3}\varepsilon^{(2)}_{\mu_4}$$
$$\cdots\ p^{(2K+1)}_{\mu_{4K+1}}\varepsilon^{(2K+1)}_{\mu_{4K+2}}.$$

After eliminating the Dirac algebra we are left with the following "effective" Feynman rules: the propagators become $i/(p^2-M^2)$, the propagator for a massive scalar of mass M. At the ith vertex there is a factor

$$-\frac{i}{4}\,\varepsilon^{(i)}_{\mu}(p+p')^{\mu}$$

where p and p' are the incoming and outgoing momenta of the scalar particles, and at the anomalous vertex we have a factor of

$$(-i/4)\varepsilon^{(0)}_{\mu}(p^{\mu}+p'^{\mu}).$$

These Feynman rules have a simple interpretation. The factors

$$-\frac{i}{4}\,\varepsilon^{(i)}_{\mu}\,(p+p')^{\mu}$$

is the amplitude for absorbing a photon of polarization $\varepsilon^{(i)}_{\mu}$ and momentum $(p'-p)^{\mu}$ by a charged particle of charge 1/4. The reader may be wondering that in scalar electrodynamics there are also seagull diagrams with two photons of polarizations $\varepsilon^{(1)}_{\mu}$, $\varepsilon^{(2)}_{\mu}$ being absorved with amplitude $2i\,\varepsilon^{(1)}\cdot\,\varepsilon^{(2)}$. In our problem, however there are gravitational seagull diagrams coming from the S_2 vertices after factoring out the Dirac algebra. The gravitational seagulls become electromagnetic seagulls in the analog problem. Putting all this together, the anomalous diverge of the amplitude with 2K+2 gravitons is

$$I_{1/2} = 2^{2K+1}\, iM^{2}\; R(\varepsilon,p)Z(\varepsilon^{(i)},p^{(j)}) \tag{29}$$

where Z is the amplitude for a charged particle of mass M and charge 1/4 interacting with 2K+2 photons of momentum and polarization $p^{(j)}$, $\varepsilon^{(i)}$ $i,j=0,1,2,\ldots,2K+1$. Since we are only interested in $I_{1/2}$ in the limit when $M\to\infty$ the computation is simpler than it looks. First of all, Z is gauge invariant and must be constructed exclusively in terms of the field strength $F_{\mu\nu}$ and possibly its derivatives, but Z is of order 2K+2 in $F_{\mu\nu}$, so by dimensional analysis Z vanishes at least as $1/M^2$ as $M\to\infty$; and terms in Z including derivatives of F will vanish faster than 1/M giving a vanishing contribution to $I_{1/2}$. This means that we can regard Z as the amplitude for a scalar propagating in the presence of a constant electromagnetic field

$$F_{\mu\nu} = -i \sum_{j=0}^{2K+1} \left(p^{(j)}_{\mu}\varepsilon^{(j)}_{\nu} - p^{(j)}_{\nu}\varepsilon^{(j)}_{\mu}\right).$$

This amplitude is to be evaluated and expanded in $\varepsilon^{(i)}$, $p^{(j)}$ to extract the term linear in each $p^{(i)}$ and $\varepsilon^{(j)}$

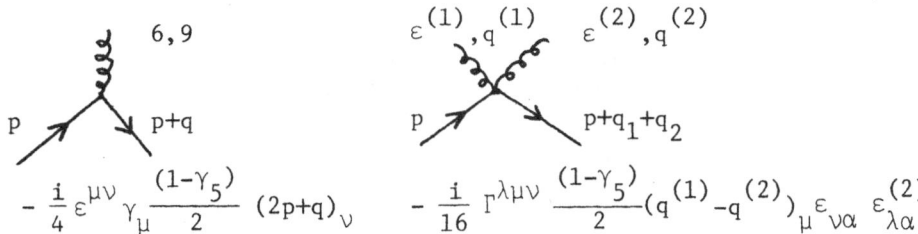

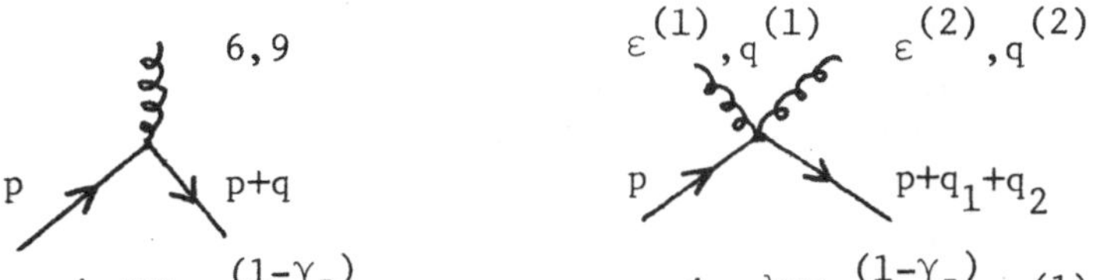

$$-\frac{i}{4}\,\varepsilon^{\mu\nu}\,\gamma_{\mu}\frac{(1-\gamma_5)}{2}\,(2p+q)_{\nu} \qquad -\frac{i}{16}\,\Gamma^{\lambda\mu\nu}\,\frac{(1-\gamma_5)}{2}(q^{(1)}-q^{(2)})_{\mu}\,\varepsilon_{\nu\alpha}\,\varepsilon^{(2)}_{\lambda\alpha}$$

Figure 1

Feynman Rules for S_1 and S_2

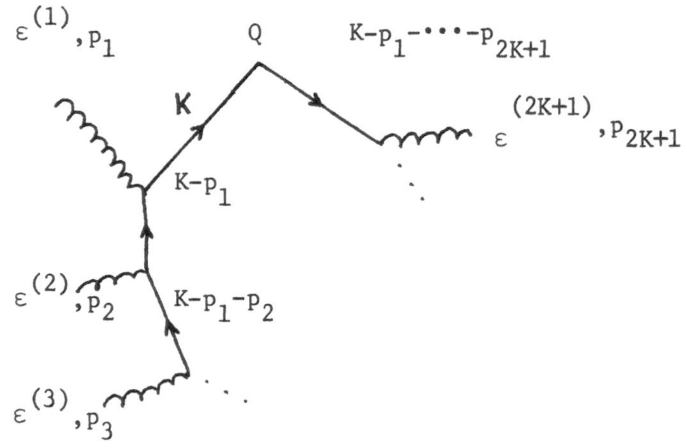

Figure 2

Anomalous Diagram for a Spin 1/2 Fermion

We can now evaluate Z by a method due to Schwinger.[27] In euclidean space, the effective action density for a charged scalar with mass M propagating in the presence of a constant electromagnetic field is

$$Z = \frac{1}{\mathrm{Vol}} \ln \det(-D_\mu D^\mu + M^2)$$
$$= -\frac{1}{\mathrm{Vol}} \int_0^\infty \frac{ds}{s} \operatorname{Tr} e^{sD_\mu D^\mu} e^{-SM^2} \tag{30}$$

bring F to canonical form:

$$F_{\mu\nu} = 2\begin{pmatrix} & x_1 & & \\ -x_1 & & & \\ & & & x_{2K+1} \\ & & -x_{2K+1} & \end{pmatrix} \tag{31}$$

with skew eigenvalues $x_1, x_2, \ldots, x_{2K+1}$. For a particle of charge e interacting with a magnetic field of strength B in two dimensions

$$\frac{1}{\mathrm{Vol}} \operatorname{Tr} e^{sD_\mu D^\mu} = \frac{1}{4\pi} \frac{eB}{\sinh eBs} \tag{32}$$

with F brought to canonical form, $H = -D_\mu D^\mu$ is a sum of commuting two-dimensional operators so that

$$Z = -\int_0^\infty \frac{ds}{s} \left(\prod_{i=1}^{2K+1} \frac{x_i/2}{4\pi \sinh s x_i/2} \right) e^{-sM^2} \tag{33}$$

where the term in parenthesis is to be expanded to order 2K+2 in the x_i's. Then (33) reduces to:

$$Z = -\frac{1}{(4\pi)^{2K+1}} \frac{1}{M^2} \prod_{i=1}^{2K+1} \frac{x_i/2}{\sinh x_i/2} \tag{34}$$

where it is understood that the right hand side of (34) is to be expanded in powers of the x_i, and only the terms of order 2K+2 should be kept. We thus get:

$$I_{i/2} = -\frac{i}{(2\pi)^{2K+1}} R(\varepsilon^{(i)}, p^{(j)}) \prod_{i=1}^{2K+1} \frac{x_i/2}{\sinh x_i/2} \tag{35}$$

Similar methods can be applied to the spin 3/2 and anti-symmetric self-dual tensor field.[9] What one finds is that the purely kinematical factor is the same and only the function of the eigenvalues of F changes. To summarize, if we extract the factor of $-iR(\varepsilon,p)/(2n)^{2K+1}$, we obtain

$$I_{1/2} = \prod_{i=1}^{2K+1} \frac{x_i/2}{\sinh x_i/2}$$

$$I_{3/2} = \left(\prod_{i=1}^{2K+1} \frac{x_i/2}{\sinh x_i/2} \right) \left(-1+2 \sum_{ij=1}^{2K+1} \cosh x_j \right) \tag{36}$$

$$I_A = -\frac{1}{8} \prod_{i=1}^{2K+1} \frac{x_i}{\mathrm{tauh}\, x_i}.$$

By $I_{1/2}$, $I_{3/2}$ we mean the anomalies of positive chirality complex Weyl fields. I_A is the anomaly for a real self-dual real antisymmetric tensor field; and by a self-dual tensor we mean the representation that arises in combining two positive chirality spinors.

We next analyze field multiplets which produce an anomaly cancellation in two, six and ten dimensions. The formulae (36) are all symmetric under permutations of the x_i and under $x_i \to -x_i$ for any i. So each term in their power series expansion has the same symmetries. To count the independent tensor structures appearing in the anomalies, it is enough to express $I_{1/2}$, $I_{3/2}$, I_A in terms of the symmetric polynomials

$$p_1 = \Sigma x_i^2, \quad p_2 = \sum_{i<j} x_i^2 x_j^2, \quad p_3 = \sum_{i<j<k} x_i^2 x_j^2 x_k^2,$$

etc. In 4K+2 dimensions we have to consider symmetric homogeneous polynomials of order 2K+2. This gives a finite number of tensor structures for each dimension.

In two dimensions:

$$\begin{aligned} I_{1/2} &= -\frac{1}{24} p_1 \\ I_{3/2} &= \frac{23}{24} p_1 \\ I_A &= -\frac{1}{24} p_1 \end{aligned} \tag{37}$$

and the anomalies can be cancelled in a variety of ways. The fact that $I_{1/2} = I_A$ effect the fact that after standard bosonization a positive chirality spinor is equivalent to a right moving scalar (a real self-dual antisymmetric tensor gauge field in two dimensions).

For Kaluza-Klein theories our interest is in six dimensions or more. In six dimensions:

$$\begin{aligned} I_{1/2} &= \frac{1}{5760} (7 p_1^2 - 4 p_2) \\ I_{3/2} &= \frac{1}{5760} (275 p_1^2 - 980 p_2) \\ I_A &= \frac{1}{5760} (16 p_1^2 - 112 p_2). \end{aligned} \tag{38}$$

It may be seen that any two of these expressions are linearly independent, so that anomaly cancellation is possible only if all three spins are present. However with three fields and only two independent anomalies (p_1^2 and p_2), there is a linear combination of these expressions that vanishes. The simplest nontrivial solution is $21\, I_{1/2} - I_{3/2} + 8\, I_A = 0$. There are six dimensional supergravity theories with this field content as shown recently by P. Townsend.[28]

Turning now to ten dimensions, we find three tensor structures for the anomaly: $p_1^3, p_1 p_2, p_3$ and

$$I_{1/2} = \frac{1}{967680} \left(-31 p_1^3 + 44 p_1 p_2 - 16 p_3\right)$$

$$I_{3/2} = \frac{1}{967680} \left(225p_1^3 - 1620p_1p_2 + 7920p_3\right)$$

$$I_A = \frac{1}{967680} \left(-256p_1^3 + 1664p_1p_2 - 7963p_3\right) \qquad (39)$$

Thus N=1 supergravity in ten dimensions is anomalous because it contains a single left handed Weyl gravitino. If we consider N=2 supergravity in ten dimensions, one may consider two cases. The N=2 supergravity obtained from dimensionally reducing the N=1 eleven dimensional supergravity[29] is parity symmetric. This theory is anomaly free because it is parity invariant. The second case is the chiral N=2 supergravity theory where both gravitinos have the same chirality.[30] This is the naive low energy limit of one of the ten dimensional supersymmetric string theories.[31] Here we meet a real surprise. The expressions for $I_{1/2}$, $I_{3/2}$ and I_A are linearly dependent. In addition the minimal solution to the anomaly cancellation condition is remarkably simple:

$$-I_{1/2} + I_{3/2} + I_A = 0,$$

and modulo fields which do not contribute to the anomaly, this is precisely the field content of the ten dimensional chiral supergravity theory. It is now natural to ask whether similar cancellations are possible beyond the dimensions. The answer is no because in fourteen dimensions the anomaly depends on five different tensor structures

$$p_1^4,\ p_1^2,\ p_2,\ p_1p_3,\ p_2^2,\ p_4,$$

and it is easy to check that the expressions for $I_{1/2}$, $I_{3/2}$, I_A are linearly independent. Since there is no miracle in fourteen dimensions, there is no miracle in 14+4K dimensions. Consider a 14+4K dimensional manifold of topology $M^{14}\mathrm{x}B$, M^{14} being fourteen dimensional Minkowski's space and B a compact manifold of dimension 4n on which the Dirac, Rarita-Schwinger, and antisymmetric tensor equations have non-zero index. Thus an arbitrary chiral theory in 14+4K dimensions will reduce on $M^{14}\mathrm{x}B$ to a fourteen dimensional chiral theory. Since this fourteen dimensional chiral theory is anomalous, it follows that any chiral theory that might have been considered in 14+4n dimensions also has anomalies.

In conclusion, we have shown that the unique chiral theory with non-trivial anomaly cancellation is given by the chiral N=2 supergravity in ten dimensions. In six dimensions, the anomaly cancellation is also possible but it requires a fairly elaborate field content.

To finish, it should be said that we could extend our computations to include mixed gravitational and gauge anomalies in any even number of dimensions although they will not be explored here (the interested reader may find the details of the computation in [9] section 9).

ACKNOWLEDGEMENTS

I would like to thank the organizers of the school for giving me the opportunity of presenting this material in such a stimulating environment. I would also like to thank Professor R. Stora for very enlightening discussions on the general theory of anomalies.

This research is supported in part by the National Science Foundation under Grant No. PHY-82-15249.

REFERENCES

1. S. Adler, Phys. Rev. 177 (1969) 2426, and in "Lectures in Elementary Particles and Quantum Field Theory", ed. S. Deser et al. (M.I.T. Press, 1970); J. Bell and R. Jackiw, Nuovo Cimento 60A (1969) 47; R. Jackiw, in "Lectures on Current Algebra and its Applications" (Princeton University Press, 1972); S. L. Adler and W. Bardeen, Phys. Rev. 182 (1969) 1517; W. A. Bardeen, Phys. Rev. 184 (1969) 1848; R. W. Brown, C. C. Shi, and B. L. Young, Phys. Rev. 186 (1969) 1491; J. Wess and B. Zumino, Phys. Lett. 37B (1971) 95; A. Zee, Phys. Rev. Lett. 29 (1972) 1198.
2. J. Steinberg, Phys. Rev. 76 (1949) 1180; J. Schwinger, Phys. Rev. 82 (1951) 664; L. Rosenberg, Phys. Rev. 129 (1963) 7786; R. Jackiw and K. Johnson, Phys. Rev. 182 (1969) 1459; S. Adler and D. G. Boulware, Phys. Rev. 184 (1969), 1740; S. L. Adler, B. W. Lee, S. B. Treiman, and A. Zee, Phys. Rev. D4 (1971) 3497; R. Aviv and A. Zee, Phys. Rev. D5 (1972) 2372; M. V. Terentiev, J.E.T.P. Letters 14 (1971) 140; A. M. Belevin, A. M. Polyakov, A. S. Schwarz and Yn. S. Tyupkin, Phys. Lett. 59B (1975) 85; G. 't Hooft, Phys. Rev. Lett. 37 (1976) 8, Phys. Rev. D14 (1976) 3432; C. Callan, R. Dashen, and D. J. Gross, Phys. Lett. 63B (1976) 334; R. Jackiw and C. Rebbi, Phys. Rev. Lett. 37 (1976) 172.
3. P. J. Gross and R. Jackiw, Phys. Rev. D6 (1972) 477; C. Bouchiat, J. Iliopoulos and Ph. Meyer, Phys. Lett. 38B (1972) 519; H. Georgi and S. L. Glashow, Phys. Rev. D6 (1972) 429.
4. G. 't Hooft, in "Recent Developments in Gauge Theories", G. 't Hooft et al. eds. (Plenum Press, New York, 1980); A. A. Ansel'm, J.E.T.P. Lett. 32 (1980) 138; A. Zee, Phys. Lett.

95B (1980) 290; Y. Frishman, A. Schwimmer, T. Banks, and S. S. Yankielowicz, Nucl. Phys. B177 (1981) 157; S. Colemand and B. Grossman, Nucl. Phys. B203 (1982) 205; G. R. Farrar, Phys. Lett. 96B (1980) 273; S. Weinberg, Phys. Lett. 102B (1981) 401; C. H. Albright, Phys. Rev. D24 (1981) 1969; I. Bars, Phys. Lett. 109B (1982) 73; T. Banks, S. Yankielowicz, and A. Schwimmer, Phys. Lett. 96B (1980) 67; A. Schwimmer, Nucl. Phys. B198 (1982) 269.

5. J. Wess and B. Zumino, Phys. Lett. 37B (1971) 95.
6. J. H. R. Skyrme, Proc. Roy. Soc. (London) A260 (1961) 127.
7. A. D. Balachandran, Y. P. Nair, S. G. Raicev, and A. Stern, Phys. Rev. Lett. 49 (1982) 1182 and Syracuse University Preprint (1982). E. Witten, "Global Aspects of Current Algebra" and Current Algebra, Baryons and Quark Confinement", Princeton Preprints.
8. P. H. Frampton and T. W. Kephart, Phys. Rev. Lett. 50 (1983) 1343, 1347; P. K. Townsend and G. Sierra (L.P.T.E.N.S. Preprint, 1983), Y. Matsuki and A. Hill (OSV Preprints, 1983).
9. L. Alvarez-Gaumé and E. Witten, Harvard Prepirnt HUTP-83/A039.
10. B. Zumino, W. Y. Shi and A. Zee (Univ. of Washington Preprint, 1983); B. Zumino, Lectures at the Les Houches Summer School, August 1983; R. Stora and B. Zumino, in preparation, and R. Stora, Lectures at Les Houches Summer School.
11. M. F. Atiyah and I. M. Singer, paper in preparation. Quillen in preparation.
12. O. Alvarez, I. M. Singer, and B. Zumino, in preparation, L. Alvarez-Gaumé and P. Ginsparg, in preparation.
13. R. Jackiw and C. Rebbi, Phys. Rev. D14 (1976) 517; N. K. Nielsen, H. Römer, and B. Schroer, Phys. Lett. 70B (1977) 445.
14. R. Delbourgo and A. Salam, Phys. Lett. 40B (1972) 381; T. Eguchi and P. Freund, Phys. Rev. Lett. 37 (1976) 1251; S. W. Hawking and C. N. Pope, Nucl. Phys. B146 (1978) 381; M. J. Perry, Nucl. Phys. B143 (1978) 114; N. K. Nielsen, H. Römer, and B. Shroer, Nucl. Phys. B136 (1978) 475; N. K. Nielsen, M. T. Grisaru, H. Römer, and P. V. Nieuwenhuisen, Nucl. Phys. B140 (1978) 477; A. J. Hanson and H. Römer, Phys. Lett. 80B (1978) 58; R. Critchley, Phys. Lett. 78B (1978) 410; S. M. Christensen and M. J. Duff, Phys. Lett. 76B (1978) 571, Nucl. Phys. B154 (1979) 301; T. Eguchi, P. B. Gilkey, and A. J. Hanson, Phys. Rep. 66 (1980) 213.
15. K. Fujikawa, Phys. Rev. Lett. 42 (1979) 1195, 44 (1980) 1733, Phys. Rev. D21 (1980) 2848; D22 (1980) 1499 (E); D23 (1981) 2262; M. B. Einhorn and D. R. T. Jones, U.M.-Th. 83-3 Preprint; A. Balachandran et al., Phys. Rev. D25 (1982) 2713.
16. E. Witten, Nucl. Phys. B202 (1982) 253.
17. T. Parker, unpublished and private communication.
18. D. Friedan and P. Windey, paper in preparation.
19. L. Alvarez-Gaumé, Comm. Math. Phys. 90 (1983) 161.
20. L. Alvarez-Gaumé, Harvard Preprint HUTP-83/A035.

21. M. F. Atiyah and R. Bott, Ann. Math. 86 (1967) 374; 88 (1968) 451.
22. E. Getzler, Harvard Preprint.
23. T. Eguchi, P. B. Gilkey, and A. J. Hanson, Phys. Rep. 66 (1980) 213.
24. E. Witten, Phys. Lett. 117B (1982) 324.
25. See, for example, H. Georgi, "Lie Algebras and Particle Physics", (Benjamin Publ., 1981).
26. R. Stora and T. Schücker have outlined a general method to derive the Wess-Zumino consistency conditions in this case (Private communication).
27. J. S. Schwinger, Phys. Rev. 82 (1951) 664.
28. P. K. Townsend, University of Texas (Austin) Preprint, 1983.
29. E. Cremmer, B. Julia, and J. Scherk, Phys. Lett. 76B (1978) 409; E. Cremmer and B. Julia, Nucl. Phys. B159 (1979) 141.
30. W. Nahm, Nucl. Phys. B135 (1978) 149; M. B. Green and J. H. Schwartz, Phys. Lett. 109B (1982) 444, 122B (1983) 143; J. H. Schwartz and P. West, Phys. Lett. 126B (1983) 301; P. S. Howe and P. West, King's College preprint (1983).
31. M. B. Green and J. H. Schwartz, Nucl. Phys. B181 502, B198 (1982) 252, 441; Phys. Lett. 109B (1982) 444; J. H. Schwartz, Phys. Reports 89 (1982) 223.

DIRAC MONOPOLES, FROM d = 2 TO d = 5, LECTURE I

F. Alexander Bais

Institute for Theoretical Physics
Princetonplein 5, P.O. Box 80.006
3508 TA Utrecht, The Netherlands

ABSTRACT

We review abelian (Dirac type) monopole solutions in an increasing number of dimensions. In doing so, we tie together three remarkable ideas, all of which date back to the twenties and thirties; the Dirac monopole (1931), the Hopf map (bundle) (1931) and the Kaluza-Klein dimensional reduction (compactification) scheme. Starting from Maxwell's equations on the two sphere, we arrive via euclidean selfdual Einstein spaces, at the recently discovered Kaluza-Klein monopoles. These are regular time independent solutions to the five-dimensional theory of general relativity corresponding to monopoles in the Kaluza-Klein frame-work. Various properties of these solutions are briefly discussed.

I.1. INTRODUCTION

We discuss various manifestations of the Dirac monopole in 2-, 3-, 4- and 5-dimensional spaces. Though few new results will be obtained during this geometric excursion, we will explicitly demonstrate the intimate relationship between three profound physical and mathematical ideas, all of which originated way back in the roaring twenties or shortly thereafter. Firstly there is the fundamental 1931 paper of Dirac on magnetic monopoles[1]. Secondly, and this is the mathematical counter part of Dirac's monopole, Hopf's observation (also in 1931) that S^3 exemplifies a nontrivial U_1 bundle over S^2 [2]. Finally the Kaluza-Klein idea (1921/26) of dimensional reduction or compactification as a means of unifying the fundamental gauge interactions with gravity[3]. In its most rudimentary form the Kaluza-Klein formalism provides exactly the *trait d'union* between the Dirac monopoles and their fibre bundle description. Alternatively one may say that Kaluza-Klein theories provide the fibre bundle description of topological excitations in gauge theories with a direct physical interpretation.

In this lecture we will avoid formal problems and mathematical finetuning, instead, rather elementary examples will be worked out in some detail to convey the underlying structure. This lecture is composed as follows. We start by introducing basic notions and notations from geometry by describing S^2. Considering the Maxwell equations in S^2 leads to the essential Dirac monopole as a solution with nontrivial topology. We then proceed to Hopf's observation that S^3 constitutes a U_1 principle fibre bundle over S^2. These two objects become one and the same thing in the Kaluza-Klein frame-work, that is, there is a natural (U_1 invariant) metric on S^3, for which S^3 (in a differential geometrical sense) is the Kaluza-Klein description of the (fundamental) Dirac monopole on S^2 [4]. The next question we ask is whether this solution can be extended to a four-dimensional space which asymptotically approaches $R^3 \times S^1$ (locally, in the sense that $S^3 = S^2 \times S^1$ locally). Here one encounters a pleasant surprise; a large class of the wanted solutions are known exactly and moreover they are completely regular; they have neither a horizon nor any physical singularity! Most of them correspond to the well known selfdual Einstein spaces in four euclidean dimensions [5], whose general form has been given by Hawking and Gibbons (1978) [6]. Because these are empty space solutions, one may add the time component of the metric in a trivial way as to obtain what are called the Kaluza-Klein monopoles, i.e. regular time independent (soliton like) solutions to the five-dimensional Kaluza-Klein theory [7,8]. We spend one section on some of the remarkable properties of these solutions, such as their mass, 5-dimensional geodesics, fermion dynamics and multi monopole solutions. The paper concludes with a perspective on various conceivable generalizations to higher dimensions, including non-abelian analogies.

I.2. S^2

We establish a notation for the basic geometrical objects we will be dealing with throughout. Rather than using the metric $g_{\mu\nu}$ and the connection coefficients (Christoffel symbols) $\Gamma^{\mu}_{\nu\sigma}$, we use Cartan's formulation of Riemannian geometry employing the *frame field (or vielbein)* e^a and the *spin connection one form* $\omega^a{}_b$. Apart from calculational simplicity it furnishes a direct analogy between gravity and ordinary gauge theories.

Recall that the metric may be written as

$$g_{\mu\nu} = e^a_\mu e^b_\nu \eta_{ab} \tag{2.1}$$

where we have introduced the *frame field* $e^a(x)$

$$e^a = e^a_\mu dx^\mu \tag{2.2}$$

which defines an orthonormal basis for the cotangent space of the manifold at the point x. A distinction is made between curved (greek) and flat (latin) indices which are raised and lowered by g and the flat metric η (with appropriate signature) respectively. The inverse of e^a_μ is defined through

$$e^\mu_a = \eta_{ab} g^{\mu\nu} e^b_\nu \ . \tag{2.3}$$

For the standard line element on S^2,

$$ds^2 = d\theta^2 + \sin^2\theta \, d\varphi^2 \tag{2.4}$$

a possible choice for the basis one forms is

$$e^\theta = d\theta \ , \qquad e^\varphi = \sin\theta \, d\varphi \ . \tag{2.5}$$

Note that these forms are not necessarily exact, hence not closed:

$$de^\varphi = \cos\theta \, d\theta \wedge d\varphi \neq o \ .$$

Next we define the (Levi-Cevita) *spin connection one form* $\omega^a{}_b$ and its relation with the connection coefficients $\Gamma^\mu_{\nu\sigma}$. Their relation is expressed through the (tetrad) postulate stating that the total covariant derivative of the vielbein vanishes:

$$\partial_\mu e^a_\nu - \Gamma^\alpha_{\mu\nu} e^a_\alpha + \omega^a{}_{b\mu} e^b_\nu = o \ . \tag{2.6}$$

The appearance of ω in this formula clearly exhibits its role of connection (gauge potential) for the local orthogonal (Lorentz) transformations. If one adds the assumption of vanishing torsion (i.e. the vanishing of the antisymmetric part of the Christoffel symbols) then (2.6) yields

$$de^a + \omega^a{}_b \wedge e^b = o \ . \tag{2.7}$$

The Levi-Cevita connection is defined through the condition that the metric be covariantly constant, which translates with help of (2.6) into the condition that ω be antisymmetric

$$\omega_{ab} = -\omega_{ba} \ . \tag{2.8}$$

With these assumptions, (2.7) and (2.8) allow one to solve ω algebraically in terms of the vielbein and its inverse.

A trivial calculation shows that for the two sphere the only non-vanishing spin connection is

$$\omega_{\theta\varphi} = -\omega_{\varphi\theta} = -\cos\theta\, d\varphi \; . \qquad (2.9)$$

Finally the *curvature two form* $R^a{}_b$

$$R^a{}_b \equiv R^a{}_{b\mu\nu}\, dx^\mu \wedge dx^\nu \qquad (2.10)$$

is defined in terms of the connection in the usual way

$$R^a{}_b = d\omega^a{}_b + \omega^a{}_c \wedge \omega^c{}_b \; . \qquad (2.11)$$

Because in the case of S^2 the local SO_2 invariance is an abelian symmetry one simply has that,

$$R^\theta{}_\varphi = d\omega^\theta{}_\varphi = e^\theta \wedge e^\varphi \; . \qquad (2.12)$$

The spheres S^n are maximally symmetric spaces which admit $\frac{1}{2}n(n+1)$ Killing vectors corresponding to the generators of the isometry group SO_{n+1}. For a maximally symmetric space the curvature tensor is of the form

$$R_{ab} = \tfrac{1}{2}\left(\delta_{ac}\delta_{ba} - \delta_{ad}\delta_{bc}\right) e^c \wedge e^d \; . \qquad (2.13)$$

The components of the Riemann curvature tensor are related with those of the two form as

$$R^a{}_{b\mu\nu} = R^\alpha{}_{\tau\mu\nu}\, e^a_\alpha e^\tau_b \; . \qquad (2.14)$$

The scalar curvature for S^2 is a positive constant

$$R^{ab}{}_{ab} = 2 \qquad (2.15)$$

as it should.

I.3 THE ESSENTIAL DIRAC MONOPOLE

We now consider the electromagnetic field on the two sphere. In particular we are interested in the monopole potential proposed by Dirac:

$$eA \equiv eA_\mu dx^\mu = \tfrac{1}{2}(1+\cos\theta)d\varphi \; . \qquad (1.3)$$

[From now on we will absorb e into the definition of A.]

The field strength is defined as

$$F = dA \ , \tag{3.2}$$

and naively calculated to equal $F = -\frac{1}{2} e^{\theta} \wedge e^{\varphi}$. The Bianchi identity

$$ddA = 0 \tag{3.3}$$

and the Maxwell equations

$$d{*}F = d{*}dA = 0 \tag{3.4}$$

are satisfied. [The dual or Hodge * of a p-form on a n-dimensional manifold, is in the basis we adopted defined as

$${*}\left(e^{a_1} \wedge e^{a_2} \dots \wedge e^{a_p}\right) \equiv \frac{1}{(n-p)!}\, \varepsilon^{a_1 \dots a_p}{}_{a_{p+1} \dots a_n} \left(e^{a_{p+1}} \wedge \dots \wedge e^{a_n}\right).]$$

We overlooked a profound subtlety. The potential (3.1) is singular for $\theta = 0$ (in the sense that if we write $A = A_{\hat{\varphi}} e^{\varphi}$ then $A_{\hat{\varphi}} = \frac{1}{2}(1+\cos\theta)/\sin\theta$), which makes the statements which follow ill defined. The basic problem is the following. If one insists that (3.2) holds everywhere then Stokes' theorem tells you that the total flux (charge) vanishes:

$$\int_{S^2} F = 0 \ . \tag{3.5}$$

Divide S^2 into two parts H_I and H_{II} with as boundary an oriented closed curve C(-C), then

$$\int_{S^2} F = \int_{H_I} F + \int_{H_{II}} F = \oint_C A - \oint_C A = 0 \ . \tag{3.6}$$

Our naive calculation of dA following (3.2) is wrong as it produces a total flux equal to (2π) thereby violating the Bianchi identity (3.3). The curl receives in fact a singular contribution at $\theta = 0$:

$$dA = \left(-\tfrac{1}{2}\sin\theta + \delta(\theta)\right) d\theta \wedge d\varphi \ , \tag{3.7}$$

regaining consistency with (3.3). This is the well known fact that Dirac's potential describes an infinitely thin magnetic vortex along the positive Z-axis.

Dirac's original resolution to this problem was to decree that the definition of the field strength be changed (at least on one point on S^2) as to obtain the true monopole field

$$F = dA - G \tag{3.8}$$

where G stands for the "Dirac string", i.e. the second term in (3.7).
The crucial question was then under what circumstances such a radical modification could be justified from a physical point of view. As charged matter fields couple to the gauge potential A rather then to the field strength F the condition had to be that charged fields should not be able to detect the string singularity in A. This requirement leads to the famous Dirac quantization condition

$$eg = 2\pi n \ . \tag{3.9}$$

The existence of a single monopole would thus imply that all electric charges be integral multiples of some fundamental charge e_0. And indeed as Dirac puts it "Under these circumstances one would be surprised if nature had made no use of it".
An alternative cure to the problem, proposed by Wu and Yang[9] (1975) is more in line with the contemporary mathematical point of view. Here the basic observation is that there is no need to describe a monopole field with a single vector potential everywhere. It would be very much like trying to cover S^2 with a single coordinate system, which necessarily leads to coordinate singularities which have nothing to do with the underlying space which is smooth everywhere. The remedy for the gauge singularity is entirely analogous to that for the coordinate singularity: introduce various overlapping regions (coordinate patches), give a regular gauge potential in each of them and demand that in the overlap regions the gauge potentials differ by a well defined gauge transformation (i.e. that F is unique). So on S^2 we define two patches

$$\begin{aligned} I &= \left\{ o \leqslant \theta \leqslant \pi/2 + \varepsilon \ , \ o \leqslant \varphi < 2\pi \right\} \\ II &= \left\{ \frac{\pi}{2} - \varepsilon \leqslant \theta \leqslant \pi \ , \ o \leqslant \varphi < 2\pi \right\} . \end{aligned} \tag{3.10}$$

In region II a good potential is provided by (3.1), whereas in region I we may take

$$A^I = \tfrac{1}{2}(\cos\theta - 1)\,d\varphi \tag{3.11}$$

with a string for $\theta = \pi$. The two potentials differ by a gauge transformation

$$A^I - A^{II} = -d\varphi$$

which is well defined in the overlap region, as the corresponding gauge (transition) function $\Omega \in U_1$ is

$$\Omega = e^{-i\varphi} \tag{3.12}$$

which is indeed single valued. Note that the fact that the overlap region is not simply connected is important, if we would have given the potentials (3.1) and (3.11) an arbitrary coefficient μ ($\equiv$ eg) then we would have found a transformation $\Omega = \exp i\mu\varphi$ leading to the Dirac quantization condition (3.9). The field strength is defined by (3.2) everywhere, but expressed in the appropriate gauge potential. The topology of the monopole is manifest in this formulation and carried by the transition function Ω; if φ runs from o to 2π then Ω goes around the group $U_1 = S^1$ n times. This winding number of the transition function Ω: $S^1 \to S^1$ is associated with an element of the first homotopy group $\pi_1(U_1)$, and completely characterizes the topology of the solution.

The extension of this solution of the Maxwell equations to R^3 is trivial. One chooses the basis vectors

$$e^r = dr \qquad e^\theta = rd\theta \qquad e^\varphi = r\sin\theta\, d\varphi \tag{3.13}$$

and the solution scales trivially

$$A = \frac{1}{2r}(\mp 1 + \cos\theta)d\varphi \tag{3.14}$$

leading to the point singular magnetic field

$$B_r = F_{\theta\varphi} = -\frac{1}{2r^2}\ . \tag{3.15}$$

I.4. SPINORS ON S^2

We make a digression to the problem of spinors on S^2. It is not essential in order to understand the rest of this lecture but it will serve some purpose when we are to discuss fermion monopole dynamics in higher dimensions. For a two component spinor we introduce the γ matrices

$$\gamma^1 = \sigma_1\ , \quad \gamma^2 = \sigma_2\ , \quad \gamma^5 = -i\sigma_1\sigma_2 = \sigma_3\ . \tag{4.1}$$

Local rotations are generated by $\frac{1}{8}[\gamma^u,\gamma^b]$, and as the spin connection is the associated gauge potential, we have the covariant derivative

$$\nabla_\mu = \partial_\mu + \tfrac{1}{8}[\gamma^a,\gamma^b]\omega_{ab\mu}\ . \tag{4.2}$$

Substituting (2.9) yields

$$\nabla_\mu = \partial_\mu - \delta_{\mu\varphi} \frac{i\sigma_3}{2} \cos\theta \,. \tag{4.3}$$

A first question which arises is whether there exist a covariantly constant operator on S^2, i.e. a solution of

$$\nabla_\mu \psi = o \,. \tag{4.4}$$

The answer is easily found, by checking the integrability of (4.4):

$$\left[\nabla_\mu, \nabla_\nu\right]\psi = R_{\mu\nu}\psi = \varepsilon_{\mu\nu} \frac{i\sigma_3}{2} \sin\theta\psi$$

which is clearly nonvanishing for any $\psi \neq o$. Hence, no covariantly constant spinor on S^2. There is in fact a more powerful theorem due to Lichnerowitz[10], stating that on any positive curvature euclidean space the Dirac operator does not have any zero modes. This situation changes if one introduces the (fundamental) Dirac monopole. We have to consider the gauge covariant derivative

$$D_\mu = \nabla_\mu - iA_\mu \tag{4.5}$$

and now the integrability condition reads

$$\left[D_\mu, D_\nu\right]\psi = \left(R_{\mu\nu} - iF_{\mu\nu}\right)\psi = o \tag{4.6}$$

or (with $\frac{eg}{4\pi} = \mp \frac{1}{2}$)

$$\frac{i}{2} \varepsilon_{\mu\nu} \sin\theta \, (\sigma_3 \pm 1)\psi = o \tag{4.7}$$

so there exists a nontrivial covariantly constant spinor in this case. This implies a zero mode of the Dirac equation in the field of the fundamental (anti)pole

$$\not{D}\psi = \gamma^\mu D_\mu = \gamma^a e_a^\mu \, D_\mu \psi = o \,. \tag{4.8}$$

For multiple charged poles again no covariantly constant spinors exist, and to determine the number of zero modes of the Dirac operator one may use the Atiyah-Singer index theorem. One observes that

$$\{\not{D}, \sigma_3\} = o \,, \tag{4.9}$$

implying that if

$$\not{D}\psi_\lambda = \lambda\psi_\lambda \,, \tag{4.10}$$

then it follows that

$$\not{D}(\sigma_3\psi_\lambda) = -\lambda(\sigma_3\psi_\lambda) \tag{4.11}$$

i.e. the eigenspinors come in pairs with opposite eigenvalues. However, if $\lambda = o$ then this is no longer necessary, and one can choose the modes to have a definite chirality

$$\sigma_3\psi_0 = \pm\, \psi_0 \ . \tag{4.12}$$

The index theorem is the statement that the number of zero modes with positive chirality minus the number of zero modes with negative chirality of the Dirac operator on some compact manifold equals the value of some topological invariant of the gauge field on the manifold. If we define $\not{D}^\pm$ as

$$\not{D}^\pm = \not{D}\ \tfrac{1}{2}(1\pm\sigma_3) \tag{4.13}$$

then in our elementary case the index theorem would read

$$\dim(\mathrm{Ker}\ \not{D}^+) - \dim(\mathrm{Ker}\ \not{D}^-) = \frac{1}{2\pi}\int F = n\ . \tag{4.14}$$

One can in fact show that for $|n| = \pm n$ only $\not{D}_\pm$ has zero modes (the other operator being positive definite). This is related to an important property in the four-dimensional case, where the zero mode in the lowest possible state of total angular momentum $j = |\frac{n}{2}| - \frac{1}{2}$ has a $2j+1 = |n|$ fold degeneracy.

I.5. THE HOPF BUNDLE

At the one hand as an introduction to the ultimate subject of this lecture, at the other as to possibly deepen our insight into the geometrical nature of the Dirac monopole, we give a compact exposé of its manifestation as a principle fibre bundle[4]. In this context the basic observation is that Dirac monopoles are nontrivial U_1 bundles over S^2. To get to grips with this brainy statement we consider the case of the fundamental $n = 1$ monopole in some detail.

A principle fibre bundle is a collection of objects: a *bundle space* E (here S^3), a *base manifold* M ($\simeq S^2$) and a *group* G ($\simeq U_1 = S^1$) together with a *projection* π: E $\to$ M. The original $\pi^{-1}(x)$ for any point $x \in M$, is called a *fibre* and isomorphic to G. The manifold E has a local direct product structure $E = M \times G$, but its topology may be more complicated globally in which case one speaks of a *nontrivial* bundle. The topology of the principle G bundle is characterized by some topological invariants usually of the homotopy type. For the case at hand, i.e. Dirac monopoles on S^2, the invariant corresponds to an element of

$$\pi_{d-1}(G) = \pi_1(S^1) = Z \tag{5.1}$$

where Z is the group of integers under addition and d = dim M = 2. Apparently there are an infinite number of topologically inequivalent S^1 bundles over S^2. Before turning to the differential geometry (i.e. questions related to connection, metric etc.) of the bundles, let us illustrate what we have said so far for the simplest nontrivial case where $E = S^3$.

In order to exhibit S^3 as an S^1 bundle over S^2 it is convenient to introduce complex coordinates

$$z_0 = x_1 + ix_2 \ , \qquad z_1 = x_3 + ix_4 \ . \tag{5.2}$$

S^3 is then parametrized by

$$z_0\bar{z}_0 + z_1\bar{z}_1 = 1 \tag{5.3}$$

The projection π: $S^3 \to S^2$ is the famous Hopf map[2] alluded to before:

$$(z_0,z_1) \to z_1/z_0 \in \mathbb{C}P^1 \simeq S^2 \ . \tag{5.4}$$

Complex projective space $\mathbb{C}P^1$ is the collection of lines in $\mathbb{C}^2$ and homeomorphic to S^2. Note that the point at infinity i.e. $z_0 = o$ has to be included, it corresponds to the z_1-axis and the south pole of S^2. Explicitly we parametrize S^3 as

$$\begin{aligned} z_0 &= \cos\,\theta/2 \exp\, i(\alpha+\varphi/2) \\ z_1 &= \sin\,\theta/2 \exp\, i(\alpha-\varphi/2) \end{aligned} \tag{5.5}$$

with

$$o \leqslant \theta \leqslant \pi \ , \qquad o \leqslant \varphi < 2\pi \ , \qquad o \leqslant \alpha < 2\pi \ .$$

Indeed z_1/z_0 does parametrize a two sphere. The action of $G = U_1$ on $E = S^3$ is given by

$$(z_0,z_1)' = (z_0g,z_1g) \ , \qquad g \in U_1 \ . \tag{5.6}$$

The fibres $(\pi^{-1}(\theta,\varphi))$ are circles parametrized by α.

In order to establish the relation between the gauge theory on M and the geometry of the bundle we need to introduce a connection and a metric on E.

A connection on E can be introduced in a number of ways which are in fact uniquely related. We will refrain here from spelling out its important geometrical interpretation as the agent which allows one to move sections in the bundle, and limit ourselves to a plain definition. A connection on E is a Lie algebra valued one form ω whose component along the fibre equals the Maurer Cartan form $-ig^{-1}dg = d\alpha$ (the point being that it should not vanish in the vertical direction as will be pointed out later on). So with coordinates $y = (x,g) \in E$ such that $\pi(y) = x$ one may write

$$\omega = g^{-1}\tilde{A}_\mu(x)dx^\mu g - ig^{-1}dg = d\alpha + \tilde{A}_\mu dx^\mu \ . \tag{5.7}$$

We remark that the connection defined here has nothing to do with a (Levi-Cevita) spin connection as defined in (2.7). Under a gauge transformation

$$g \rightarrow gg_0 \tag{5.8}$$

one has the transformation property

$$\omega \rightarrow g_0^{-1}\omega g_0 - ig_0^{-1}dg_0 \ . \tag{5.9}$$

Note however that $\tilde{A}_\mu(x)dx^\mu$ remains *invariant*, indicating that our notation has been treacherous because apparently $\tilde{A}_\mu(x)dx^\mu$ is not the gauge potential on M. To obtain the gauge potential on M (i.e. a Lie algebra valued one form) is slightly involved. It amounts to choosing a *section* f

$$f: M \rightarrow E \tag{5.10}$$

such that πf is the identity map on M. In other words it corresponds to fixing a gauge by giving $g = g(x)$. The good old gauge potentials (in a particular gauge) are now obtained as the *pull back* $f^*\omega$ of ω, i.e.

$$A = f^*\omega = (g^{-1}\tilde{A}_\mu g + g^{-1}\partial_\mu g)dx^\mu \ . \tag{5.11}$$

The fun of a nontrivial bundle is that one cannot choose a section globally, so one has to choose overlapping coordinate patches on M and define different local sections $g^{I}(x)$, $g^{II}(x)$, etc. The sections in the overlap regions should be related by well defined transition functions Φ

$$g^{I}(x) = g^{II}(x)\Phi_{II,I}(x) \qquad x \in I \cap II \ . \tag{5.12}$$

Consequently the gauge potentials A_I and A_{II} differ in the overlap region only by a gauge transformation $\Phi_{II,I}$. (Indeed this is exactly the origin of the Wu-Yang prescription discussed in section 3.) We emphasize that ω itself is insensitive to the choice of sections, it generates through the choice of different sections the gauge potentials on M in all possible gauges. At this point the precise form of $\tilde{A}_\mu$ in (5.7) is immaterial as long as it is compatible with the correct topology. In particular there is nothing whatsoever that tells us that we have to take an $\tilde{A}_\mu$ which satisfies Maxwell's equations. As we shall see in the next section, the missing link is provided by the Kaluza-Klein idea. In anticipation thereof and also to complete our lightning review of the Hopf bundle we will construct a metric on E which is suitable for our purposes.

A metric γ on E, i.e. a bilinear form in the cotangent space T^*_{Ey} at every point $y \in E$ which works on vectors in the tangent space T_{Ey}. The tangent space can be split into a horizontal and vertical component using the connection ω and π respectively,

$$T_{Ey} = H_y + V_y \; . \tag{5.13}$$

The metric γ is orthogonal in H and V, i.e.

$$\gamma = \gamma_H + \gamma_V \; . \tag{5.14}$$

The horizontal subspace H is defined as the Kernel of ω;

$$\omega(H) = o \; . \tag{5.14}$$

This property allows us to construct γ_V as

$$\gamma_V(Y,Z) = -\chi k\Big(\omega(Y),\omega(Z)\Big) \qquad Y,Z \in T_E \; . \tag{5.15}$$

The following comments are in order. Because $\omega(Y)$ is a Lie algebra valued function, we use the Killing metric k on the group to obtain a scalar. For semi simple G the negative definite metric is defined in terms of the structure constants

$$k_{ab} = C^c_{ad} C^d_{bc} \; . \tag{5.16}$$

And for U_1 we take $k = -1$. χ is an at this point arbitrary positive function $\chi = \chi(x)$.

The vertical subspace V is defined through π. Since $\pi\colon E \to M$ (say $x = x(y)$), its derivative $d\pi$ provides us with an onto mapping $d\pi\colon T^*_{Ey} \to T^*_{M_x}$. The Kernel of $d\pi$ is the vertical subspace (for the obvious reason that $d\pi$ annihilates the components along the fiber. With

$$d\pi(V) = o \tag{5.17}$$

it follows that $d\pi$ restricted to H has vanishing Kernel, implying that H_y is isomorphic to T_{M_x}. Hence, for γ_H we can just take the metric g on M:

$$\gamma_H(Y,Z) = g\Big(d\pi(Y),d\pi(Z)\Big) \; . \tag{5.18}$$

This completes our construction of a natural metric on E which reads

$$\gamma(Y,Z) = g\Big(d\pi(Y), d\pi(Z)\Big) - \chi k\Big(\omega(Y), \omega(Z)\Big) \tag{5.19}$$

Let us now return to our favourite example. The metric on S^3 of the desired form is just the natural metric in terms of the coordinates (5.6)

$$\begin{aligned} ds^2 &= 4(d\bar{z}_0 dz_0 + d\bar{z}_1 dz_1) = \\ &= d\theta^2 + \sin^2\theta\, d\varphi^2 + (2d\alpha + \cos\theta\, d\varphi)^2 \, . \end{aligned} \tag{5.20}$$

Comparing with (5.19) yields

$$\begin{aligned} \chi &= 4 \\ \omega &= d\alpha + \tfrac{1}{2}\cos\theta\, d\varphi \, . \end{aligned} \tag{5.21}$$

A natural orthonormal basis for T^*_E is

$$e^\theta = d\theta \qquad e^\varphi = \sin\theta\, d\varphi \qquad e^\alpha = 2d\alpha + \cos\theta\, d\varphi$$

and its dual for T_E:

$$e_\theta = \partial_\theta \, , \; e_\varphi = \frac{1}{\sin\theta}\left(\partial_\varphi - \frac{\cos\theta}{2}\,\partial_\alpha\right) , \; e_\alpha = \tfrac{1}{2}\partial_\alpha \, .$$

Clearly H is spanned by e_θ and e_φ and V by e_α.
Going back to the definition of the gauge potentials on M in terms of ω as given by (5.11-12) and choosing $\alpha = \varphi$ and $\alpha = -\varphi$ in the regions I and II of (3.10), one finds from (5.21) exactly the Wu-Yang potentials for the fundamental monopole. Again we emphasize that this is somewhat miraculous. What is so special about S^3 that it generates exactly the correct monopole potential? The answer was given by Kaluza and Klein sixty years ago!

I.6. KALUZA-KLEIN

During the nineteen twenties already a remarkable attempt to unify electromagnetism and gravity was made by Kaluza and Klein[3] The idea was to achieve this unification through a dimensional reduction (or compactification) of the theory of general relativity in five dimensions. Rather than five-dimensional Minkowski space, the groundstate was assumed to be the direct product of four-dimensional Minkowski space and a circle S^1. This "internal" space S^1 would be experimentally not detectable provided its radius would be small enough. Clearly imposing dimensional reduction on the solutions of the Einstein equations, that is restricting the metric $\gamma_{\Sigma\Lambda}$ to be independent of x^5, reduces the symmetry of the residual system considerably. Five-dimensional coordinate invariance is broken down to a product of four-dimensional general coordinate invariance (yielding the graviton associated with $\gamma_{\mu\nu}$) and a local U_1 (yielding the photon associated with $\gamma_{\mu 5}$). Finally because scale invariance is broken by the fixed radius of the internal circle, one expects a massless goldstone particle (dilaton)(associated with γ_{55}) in the resulting theory. The appearance of this massless scalar particle which couples to matter with the same strength as gravity, is what makes this theory unrealistic. The reader who beared with us through the previous section should now have an "Aha Erlebniss". (S)he is invited to observe that indeed *Kaluza-Klein spaces are nothing but U_1 principle bundles over four-dimensional manifolds.* But apparently a severe restriction has been added: *the bundle manifold E should be an Einstein space.* The metric on E should satisfy the Einstein equations with or without cosmological constant. Evidently all we have to do is to plug the canonical bundle metric (5.19) into the Einstein Lagrangian and turn the crank. For the U_1 bundle we had

$$\gamma_{\Sigma\Lambda} = \begin{pmatrix} g_{\mu\nu}+A_\mu A_\nu \chi & A_\mu \chi \\ A_\nu \chi & \chi \end{pmatrix} , \tag{6.1}$$

or

$$e^N_\Sigma = \begin{pmatrix} e^n_\mu & o \\ \chi^{\frac{1}{2}} A_\mu & \chi^{\frac{1}{2}} \end{pmatrix} . \tag{6.2}$$

With inverses

$$\gamma^{\Sigma\Lambda} = \begin{pmatrix} g^{\mu\nu} & -A^\mu \\ -A^\nu & \frac{1}{\chi} + A_\mu A^\mu \end{pmatrix} , \tag{6.3}$$

and

$$e^\Sigma_N = \begin{pmatrix} e^\mu_n & o \\ -A_n & \chi^{-\frac{1}{2}} \end{pmatrix} . \tag{6.4}$$

We do the reduction in several steps starting from a (d+1)-dimensional action

$$S = - \frac{1}{16\pi\tilde{G}} \int dx^{d+1} \sqrt{-\gamma}\, \mathbb{R}^{(d+1)}(\gamma) . \tag{6.5}$$

First we make a conformal rescaling of γ by setting

$$\gamma_{\Sigma\Lambda} = \chi\bar{\gamma}_{\Sigma\Lambda} \tag{6.6}$$

which yields

$$\mathbb{R}^{(n)}(\gamma) = \frac{1}{\chi}\left[\mathbb{R}^{(n)}(\bar{\gamma}) + 2(1-n)\bar{\gamma}^{\mu\nu}\chi_{;\mu\nu} - (n-1)(n-2)\bar{\gamma}^{\mu\nu}\chi_{,\mu}\chi_{,\nu}\right] . \qquad (6.7)$$

The field λ is defined through

$$\chi = e^{2\lambda} . \qquad (6.8)$$

Dimensional reduction of $\mathbb{R}^{(n)}(\bar{\gamma})$ gives the fundamental relation between Maxwell's and Einstein's theory:

$$\mathbb{R}^{(d+1)}(\bar{\gamma}) = \mathbb{R}^{(d)}(\bar{g}) - \tfrac{1}{4}F_{\mu\nu}F_{\sigma\tau}\bar{g}^{\mu\sigma}\bar{g}^{\nu\tau} . \qquad (6.9)$$

Given that under conformal rescaling

$$g^{\mu\nu}\chi_{;\mu\nu} = \frac{1}{\chi}\left[\bar{g}^{\mu\nu}\chi_{;\mu\nu} + (n-2)\bar{g}^{\mu\nu}\chi_{,\mu}\chi_{,\nu}\right] \qquad (6.10)$$

one obtains from (6.7-11) the final result

$$\mathbb{R}^{(d+1)}(\gamma) = \mathbb{R}^{(d)}(g) - \frac{\chi}{4}F_{\mu\nu}F_{\sigma\tau}g^{\mu\sigma}g^{\nu\tau} - 2g^{\mu\nu}\chi_{;\mu\nu} - 2g^{\mu\nu}\chi_{,\mu}\lambda_{,\nu} . \qquad (6.11)$$

It is evident from this expression that if χ is given a constant value (which is undetermined classically) then the reduced theory is indeed the coupled Maxwell-Einstein system in d-dimensions.

At this point it may be instructive to look back at the Hopf bundle of the previous section. The conclusion is simple: S^3 is indeed an Einstein space (of constant curvature) which allows a Kaluza-Klein interpretation as a solution of the coupled Einstein-Maxwell system in two dimensions, this solution being a Dirac monopole on S^2. One may wonder whether there exist maybe other solutions where $\chi \neq 4$. The answer is negative, there is no such thing as a squashed three sphere (this in contrast with S^7 as an S^3 bundle over S^4 where one has two solutions).

I.7. SELF DUAL EINSTEIN SPACES

At the end of section 3 we mentioned that the essential monopole solution on S^2 could be trivially extended to a solution on R^3 by adding the radial coordinate in a trivial way. The field strength necessarily becomes singular at the origin consistent with the presence of a magnetic point charge. The corresponding solution to the coupled Einstein-Maxwell system is the Reissner-Nordström magnetically charged black hole solution. In the previous sections we have shown that the monopole solution on S^2 may be considered as an S^3 itself which suggests the question whether it is possible to extend this solution to a euclidean four-dimensional manifold by adding the radial coordinate. One would like to exploit the possibility of local compactification (i.e. where the radius of S^1 would become r dependent) to possibly eliminate the singularity, as to obtain a regular (source free) solution to the euclidean 4-dimensional Einstein equations. Though the question was never phrased this way a large class of solutions of desired type were constructed some years ago, These are the self dual euclidean Einstein spaces[5].

(Anti) self dual spaces have by definition a Riemann tensor which satisfies

$$R_{abcd} = \pm \tfrac{1}{2}\varepsilon_{abef} R_{efcd} \; . \qquad (7.1)$$

That these satisfy the empty space Einstein equation (with vanishing cosmological constant) is implied by the cyclic identity

$$\varepsilon_{abcd} R_{ebcd} = o = \mp 2\left(\mathbf{R}_{ae} - \delta_{ae}\mathbb{R}\right) . \qquad (7.2)$$

At first sight, the self duality condition (7.1) only relates various components of the Riemann tensor and does not seem to lower the order of the differential equations. That this is not true follows from the following observation. If the (Levi-Cevita) spin connection itself is (anti) self dual, i.e.

$$\omega_{ab} = \pm \varepsilon_{abcd}\omega_{cd} \qquad (7.3)$$

then it is easily verified that R_{abcd} is (anti) self dual. However, since (7.3) is not covariant under local O_4 transformations (ω transforms like a connection), and (7.1) is covariant, clearly not all self dual Riemann tensors come from self dual connections. Nevertheless one can show that any connection ω giving rise to a self dual space, can itself be made self dual by a local O_4 transformation, so indeed it suffices to restrict one self to solutions of (7.3).

We briefly describe the Hawking and Gibbons multicentered metrics, which contain the gravitational instantons which were obtained earlier, like the Eguchi-Hanson and Taub-NUT solutions as special cases. They are usually given in the form

$$ds^2 = V(\vec{x})d\vec{x}^2 + \frac{1}{V(\vec{x})}\,(d\tau + \vec{\tilde{A}}.d\vec{x})^2 \qquad (7.4)$$

and are clearly of the U_1 bundle type (5.19). Imposing (anti) self duality on the spin connection one obtains that the following equations have to be satisfied

$$\vec{\nabla}V(\vec{x}) = \pm (\vec{\nabla}x\vec{\tilde{A}}(\vec{x}) . \qquad (7.5)$$

V is like the magnetic scalar potential satisfying the Laplace equation

$$\vec{\nabla}^2 V = o \qquad (7.6)$$

with solutions

$$V = \varepsilon + \sum_i \frac{2m_i}{|\vec{x}-\vec{x}_i|} \qquad (7.7)$$

where ε = o, 1 (scaled to unity).

At this point we restrict ourselves to the asymptotically flat (ε= 1) Taub-NUT solution with

$$V = 1 + \frac{2m}{x} , \qquad A = 2m \cos\theta \, d\varphi . \qquad (7.8)$$

For x >> m it describes a monopole field on R^3. Comparing with (5.19) one finds that $\chi = 16m^2$ (i.e. the radius of S^1 tends to R_∞ = 4m) and for the fundamental monopole $|eg| = 2\pi$, $\frac{\tau}{4m}$ should have period 2π.

There is some reason to get worried about the singularity at x = o. The surprising thing is that this so called NUT singularity can be removed by a coordinate transformation provided certain conditions are met. Introducing new coordinates

$$x = \frac{1}{8m}\rho^2 \qquad \sigma = \frac{\tau}{2m} \qquad (7.9)$$

one finds that for x → o the metric takes the form

$$ds^2 \underset{\rho\to o}{\simeq} d\rho^2 + \frac{\rho^2}{4}\left[d\Omega^2_{(2)} + (d\sigma + \cos\theta \, d\varphi)^2\right] . \qquad (7.10)$$

This is exactly the line element of R^4 in (four-dimensional) polar coordinates if σ has a period 4π. This is a remarkable result: the condition that the singularity for r = o is removable corresponds to the Dirac quantization condition for the monopole field at infinity! In spite of being regular, the origin remains a peculiar point in the sense that the radius $R = \sqrt{\chi}$ goes to zero there.

I.8. THE d = 5 KALUZA-KLEIN MONOPOLE

Self dual spaces satisfy the empty space Einstein equation (with vanishing cosmological constant). Adding the time coordinate in a trivial way i.e. setting $g_{oo} = -1$, $g_{oa} = o$, turns the euclidean solution into a regular, time independent soliton type solution of the original 5-dimensional Kaluza-Klein theory. Asymptotically it corresponds to a monopole field on flat Minkowski space. This solution was first discussed by Sorkin[7], and Perry and Gross[8]. Though the Kaluza-Klein unification of electromagnetism with gravity differs essentially from the familiar gauge unification (say GUTs), both approaches share the intrinsic quantization of electric charge (in the sense of a compact U_1) as well as the existence of magnetic monopoles as regular solutions to the field equations. It is worthwhile to briefly discuss some of the physical properties of the Kaluza-Klein monopoles. The remainder of this section is divided into subsections devoted to various aspects, most of which were covered in the articles[7,8] mentioned before.

I.8.A. FUNDAMENTAL CONSTANTS

Taking the five-dimensional theory as a starting point we now want to identify the physical constants in the effective four-dimensional theory.

$$\begin{aligned} S &= \frac{1}{16\pi\tilde{G}} \int d^5x \sqrt{\gamma}\, \mathbb{R}^{(5)} \\ &= \frac{2\pi R_\infty}{16\pi\tilde{G}} \int d^4x \sqrt{g} \left(\frac{R}{R_\infty}\right) \left[\mathbb{R}^{(4)} - \tfrac{1}{4}\left(\frac{R}{R_\infty}\right)^2 R_\infty^2 F^2 + \ldots\right] . \end{aligned} \tag{8.1}$$

Asymptotically where $R(\infty) = R_\infty$ we obtain the four-dimensional gravitational constant G as

$$G = \frac{\tilde{G}}{2\pi R_\infty} . \tag{8.2}$$

The coefficient in front of F^2 should equal $\frac{1}{e^2}$ which yields for the fine structure constant

$$\alpha = \frac{4G}{R_\infty^2} . \tag{8.3}$$

In other words the radius R_∞ of the internal space is fixed by the ratio of the gravitational constant and the fine structure constant

$$R_\infty = 2\sqrt{\frac{G}{\alpha}} \simeq 10^{-32} \text{ cm} , \tag{8.4}$$

exceedingly small indeed.

I.8.B. MASSES

In general relativity only the covariant derivative of the

stress energy tensor vanishes, this causes a problem if one wants to define a *conserved* (angular) momentum. In an asymptotically flat space time it is however possible to derive conserved quantities from an appropriately defined pseudo tensor[11]. As the name indicates, this pseudo tensor is not generally covariant but it *is* covariant with respect to the linear transformations in particular the Lorentz group. In five dimensions the conserved momentum of the gravitational field may be expressed in terms of the metric as

$$P^{\Sigma} = \frac{1}{16\pi\tilde{G}} \int dx^5 \int d^3x \; \partial_i \partial_\Lambda \left[-\gamma \left(\gamma^{\Sigma o} \gamma^{i\Lambda} - \gamma^{\Sigma i} \gamma^{o\Lambda} \right) \right] . \tag{8.5}$$

With the Kaluza-Klein metric (6.1) one obtains for the fifth component

$$P^5 = \frac{1}{16\pi G} \int d^3x \; \vec{\nabla} . \vec{E} \tag{8.6}$$

corresponding to the total electric charge of the solution, which in the present case happens to vanish. As we are in the rest frame of the monopole one finds for the mass

$$P^{\mu} \equiv M\delta^{\mu o} ,$$

a value

$$M = \frac{m}{2G} = \frac{R_\infty}{8G} = \frac{m_{p\ell}}{4\sqrt{\alpha}} . \tag{8.7}$$

The expression is characteristic for a soliton mass; the ratio of the typical mass scale in the theory divided by a dimensionless coupling constant (compare with the GUT monopole mass $M = M_x/\alpha$). The mass calculated through the energy momentum tensor is by definition the *intertial mass*. In four-dimensional theories this inertial mass by the principle of equivalence is exactly equal to the *gravitational mass* (say the one that appears in Newton's law). At this point we are in for a surprise. To find out what the gravitational mass is it suffices to recall that effective Newtonian potential U appears in γ_{oo} as

$$\gamma_{oo} \simeq -1 + 2U \tag{8.8}$$

(for example in the conventional Schwarzschild solution $g_{oo} = -(1 - 2m/r)$. However, all five-dimensional metrics obtained as trivial extensions of d = 4 self dual spaces have $\gamma_{oo} = -1$, hence vanishing gravitational mass!
Apparently five-dimensional general covariance allows for such bizar phenomena like having gravitational objects (localized

lumps of energy) and yet a test particle does not fall towards them. In fact going back to the multicentered metrics (7.4) one realizes that they generate stationary multi monopole solutions implying that also Kaluza-Klein monopoles with charges of the same sign do not interact classically. This reminds us of the 't Hooft-Polyakov monopoles which have the same property in the Bogomolny - Prasad-Sommerfield limit. In that case the phenomenon is well understood and due to the long range force between the monopoles caused by a massless scalar (Higgs) field which exactly cancels the interaction due to the massless gauge bosons. Here a similar phenomenon takes place, the attractive gravitational potential is cancelled by a repulsive potential due to the massless dilaton field. It calls for a closer look at the geodesics, to which we now turn.

I.8.C. GEODESICS

Consider the geodesic Lagrangian for a unit mass test particle

$$\mathcal{L} = \tfrac{1}{2}\gamma_{\Sigma\Lambda}\dot{x}^{\Sigma}\dot{x}^{\Lambda} \tag{8.9}$$

where the dot refers to differentiation with respect to proper time. It is convenient to introduce a new radial coordinate r which is related to the coordinate x of section 7 through

$$r^2 = (x+m)^2 - m^2 \tag{8.10}$$

with the attractive feature that it gives the standard r^2 coefficient in front of the line element of S^2 (θ,φ) in the metric, so that

$$\mathcal{L} = \tfrac{1}{2}\left[-\dot{t}^2+A\dot{r}^2+r^2\left(\dot{\theta}^2+\sin^2\theta\dot{\varphi}^2\right)+C\left(\dot{\alpha}+\tfrac{1}{2}\cos\theta\dot{\varphi}\right)^2\right]. \tag{8.11a}$$

A and C are the functions

$$A = \left(1+\frac{m}{\rho}\right)^2 , \qquad C = \left(\frac{4mr}{\rho+m}\right)^2 , \tag{8.11b}$$

where

$$\rho^2 = r^2 + m^2 . \tag{8.11c}$$

The conserved total angular momentum $\vec{J}$ can be expressed as

$$\vec{J} = \vec{L} - S\hat{r} \tag{8.12}$$

where $\vec{L}$ is the orbital angular momentum

$$\vec{L} = r^2(\dot{\theta}\hat{\varphi} - \sin\theta\dot{\varphi}\hat{\theta}) \ , \tag{8.13}$$

and S the equivalent of the angular momentum contribution of the electromagnetic field $\frac{qg}{4\pi}$

$$S = - C(\dot{\alpha} + \tfrac{1}{2}\cos\theta\dot{\varphi}) \ . \tag{8.14}$$

Because $\vec{L}$ and $S\hat{r}$ are orthogonal we have that both $|\vec{L}|$ and S are conserved (related to the Killing vectors ∂_φ and ∂_α respectively) which fixes the motion of the particle to lie on a cone with opening angle

$$B = \text{arctg}\,\frac{|\vec{L}|}{S} \ . \tag{8.15}$$

Up to this point the situation is completely identical to that of the classical motion of a charge in the field of a Dirac monopole.

The radial motion is determined in terms of $|\vec{L}|$, S and the energy ε (i.e. the conserved quantity related to the Killing vector ∂_t) through

$$A\dot{r}^2 + \frac{L^2}{r^2} + \frac{S^2}{C} = \varepsilon > o \ . \tag{8.16}$$

This corresponds to the motion of a particle with r dependent mass 2A and energy ε in a potential

$$V(r) = \frac{L^2}{r^2} + \frac{S^2}{C} \ . \tag{8.17}$$

It is here that a notable difference with the conventional problem shows up. In the conventional situation the term S^2/C in (8.17) is absent, leading in that case to a well known pathology. A radially in falling charged particle would not feel any force and hence go straight through the monopole, *en passant* violating the conservation of the total angular momentum $\vec{J}$ (because $\hat{r}$ flips direction).

The situation in the case of the Kaluza-Klein monopole with the potential (8.17), is rather healthy indeed. Even if $|\vec{L}| = o$, a repulsive $\frac{1}{r^2}$ core due to the S^2/C term prevents the particle from reaching the origin. The minimal distance of approach r_0 is determined by the condition

$$C(r_0) = \varepsilon/S^2 \ . \tag{8.18}$$

Angular momentum conservation is saved.

It is amusing to see what is happening in the five-dimensional setting. Let the particle fall in along the z-axis, r = z and both θ and φ disappear altogether. The motion is in the (x_5,z) space, which is a curved conical surface obtained by rotating the curve $\sqrt{C(z)}$

around the z-axis. The geodesic spirals in around the cone until the turning point is reached, where the z-component of the velocity reverses sign and it starts spiraling out. This so called *magnetic mirror effect* is very similar to what happens in the scattering of a charged, $\vec{L} \neq o$, particle of a Dirac monopole.

What about the peculiar circumstance that the Kaluza-Klein monopole has vanishing gravitational mass alluded to before? This stipulates the following amusing observation in connection with the geodesics. Indeed, a particular solution to the equations is obtained by having all $\dot{x}^\Sigma = o$, in other words if $|\vec{L}| = S = \varepsilon = o$ then $\dot{r} = o$. A neutral observer is allowed to sit right next to It, stare at It forever, without feeling any attraction.

I.8.D. FERMION DYNAMICS

We conclude our discussion of the Kaluza-Klein monopole with an analysis of the Dirac equation in the corresponding 5-dimensional manifold[12]. Our main purpose is to perform the dimensional reduction and subsequently compare the result with the conventional four-dimensional Dirac equation in the field of the abelian monopole. The latter problem has been studied at various occasions, most thoroughly by Goldhaber, Kazama and Yang[13]. The subject of charge-pole dynamics including fermions is discussed at length in a following lecture to which we refer the interested reader who wants to refresh his memory of many facts which will be stated without explanation in this section.

The 5-dimensional Dirac equation, covariant with respect to general coordinate and local Lorentz transformations reads (see (4.2))

$$\not{D}\psi = e^{\Sigma}{}_{A}\, \gamma^{A}\left(\partial^{\Sigma} + \omega_{\Sigma AB}\, \frac{1}{8}\left[\gamma^{A}, \gamma^{B}\right]\right)\tilde{\psi} = o \ . \qquad (8.19)$$

The five-dimensional Dirac matrices γ^A (flat indices) are just the usual 4 x 4 matrices satisfying

$$\left\{\gamma^{A}, \gamma^{B}\right\} = 2\eta^{AB} \ . \qquad (8.20)$$

We will work in the coordinate system which was employed in the previous section on geodesics. To achieve the dimensional reduction of (8.19), we make an harmonic analysis of the spinor wave function on the internal space. This amounts to writing ψ as a sum of Fourier components on S^1 with the coordinate α $(o \leqslant \alpha \leqslant 2\pi)$:

$$\tilde{\psi} = \sum_{n=-\infty}^{\infty} \tilde{\psi}_n(x^\mu)e^{in\alpha} \; . \tag{8.21}$$

Substitution in (8.19) of the above expression as well as the general form of the Kaluza-Klein metric (6.1), where we have set $\chi = R^2$ (R = radius of S^1), one obtains the following set of equations:

$$\Big\{\gamma^\mu\Big(\partial_\mu - i\,n A_\mu + \frac{1}{8}\Big[\gamma^a,\gamma^b\Big]\omega_{\mu ab} + \frac{\partial_\mu R}{2R}\Big) - \\ + m - \frac{n}{|n|}\,\frac{i}{16}\Big[\gamma^\mu,\gamma^\nu\Big]F_{\mu\nu}R\Big\}\Psi_n(x^\mu) = o \; . \tag{8.22}$$

Here some comments should be made.

i) The matrices γ^μ with curved indices are defined in the four-dimensional context:

$$\gamma^\mu \equiv e^\mu{}_a\,\gamma^a \; . \tag{8.23}$$

This in contradiction with their five-dimensional analogs

$$\Gamma^\Sigma = e^\Sigma{}_A\,\gamma^A .$$

ii) The chiraly rotated field Ψ_n is defined in terms of Ψ_n as

$$\Psi_n = \tfrac{1}{2}(1-\gamma_5)\tilde{\psi}_n \pm \frac{i}{2}(1+\gamma_5)\tilde{\Psi}_n \quad \text{for} \quad n = \pm\,|n| \; .$$

This transformation is applied to obtain the conventional positive mass term in (8.22) where

$$m = \frac{|n|}{R} \; . \tag{8.24}$$

iii) From the coupling to the photon in (8.22) we verify that the charge of $\tilde{\Psi}_n$ is ne as expected, from (8.24) we see that the mass is proportional to the charge, though not necessarily constant ($R = R(r)$). This implies that by finetuning the mass in the five-dimensional theory, we can at most arrange two state of the infinite tower of states Ψ_n, to be massless in $d = 4$. Hence for each low mass particle ($m \ll m_{p\ell}$) in the $d = 4$ theory we have to introduce a different field in $d = 5$ theory.

iv) The last term in the equation, describes the effect of an anomalous magnetic moment, corresponding to the so called

'Pauli term' in the nonrelativistic theory. Comparing the usual expression

$$\Delta H = \frac{i\kappa\, n}{m}\, F_{\mu\nu}\, \frac{1}{8}\left[\gamma^{\mu},\gamma^{\nu}\right] \tag{8.25}$$

with (8.22) we find that in the present case the anomalous magnetic moment κ has the particular value

$$\kappa = -\tfrac{1}{2} \tag{8.26}$$

For the Kaluza-Klein monopole we know $\gamma^{\mu}, \omega_{\mu ab}$, $F_{\mu\nu}$ and A_{μ} explicitly. Substitution of these, followed by a suitable transformation to eliminate the angular components of the spin connection leads to an equation which is rather similar to the usual 4-dimensional equation in the field of a magnetic monopole:

$$\Big\{-\frac{1}{\Lambda}(\vec{\alpha}.\hat{r})\left(-i\partial_r + \frac{i}{2r}(\Lambda-1)\right) + \frac{1}{r}(\vec{\alpha}.\hat{\theta})(-i\partial_{\theta})$$
$$+ \frac{1}{r\sin\theta}(\vec{\alpha}.\hat{\varphi})(-i\partial_{\varphi} - \frac{n}{2}\cos\theta) + \beta m - \frac{1}{8}\frac{n}{|n|}\frac{R}{r^2}\beta(\vec{\sigma}.\hat{r}) = E\Psi_n. \tag{8.27}$$

The functions $A = \sqrt{(A(r)}$ and $R = \sqrt{C(r)}$ are defined in (8.11b). The angular part of the equation is exactly the same as in the conventional case and the same analysis goes through. The conserved angular momentum is

$$\vec{J} = \vec{r}\times\vec{\pi} - \mu\hat{r} + \tfrac{1}{2}\vec{\sigma} \tag{8.28}$$

where $\mu = \frac{eg}{4\pi} = -\frac{n}{2}$. It is most convenient to simultaneously diagonalize the operators H, J^2, J_z and $(\vec{\sigma}.\hat{r})$ with eigenvalues E, $j(j+1)$, m and s respectively. We will restrict ourselves to the most interesting mode, which corresponds to the lowest angular momentum state with

$$j = |\mu| - \tfrac{1}{2}\ . \tag{8.29}$$

The remarkable fact is that $\vec{\sigma}.\hat{r}$ can only have a single eigenvalue $s = \frac{\mu}{|\mu|}$. Choosing the eigenfunctions

$$(\vec{\sigma}.\hat{r})\eta_m = \frac{\mu}{|\mu|}\eta_m \tag{8.30}$$

one may show that

$$(\vec{\sigma}.\vec{\pi})f\eta_m = -i\frac{\mu}{|\mu|}\left(\frac{1}{\Lambda}\partial_r + \frac{1}{2r} + \frac{1}{2\Lambda r}\right)f\eta_m \tag{8.31}$$

Setting

$$\Psi = \begin{pmatrix} f\eta_m \\ g\eta_m \end{pmatrix} \tag{8.32}$$

we finally arrive at the radial equations

$$\begin{aligned} \left[(E-m) - \frac{1}{8}\frac{R}{r^2}\right] f + i\frac{\mu}{|\mu|}\left(\frac{1}{\Lambda}\partial_r + \frac{1}{2r} + \frac{1}{2\Lambda r}\right) g = o \\ \left[(E+m) + \frac{1}{8}\frac{R}{r^2}\right] g + i\frac{\mu}{|\mu|}\left(\frac{1}{\Lambda}\partial_r + \frac{1}{2r} + \frac{1}{2\Lambda r}\right) f = o \end{aligned} \tag{8.33}$$

These equations are exactly equal to the conventional ones (with anomalous magnetic moment $\kappa = -\frac{1}{2}$) if $\Lambda = 1$ and $R = R_\infty$, i.e. in the asymptotic region where $r > R_\infty$. In other words the asymptotic solution should be the conventional one which for the case of scattering means that

$$\Psi(r \geqslant R_\infty) \sim \begin{pmatrix} \frac{\mu}{|\mu|}\frac{1}{r} \sin\,(kr+\delta)\eta_m \\ \frac{1}{ir}\frac{k}{E+m} \cos\,(kr+\delta)\eta_m \end{pmatrix} \tag{8.34}$$

where $k^2 = E^2 - m^2 > o$. The phase shift will ultimately be fixed through the boundary conditions on Ψ at the origin. The small r behaviour turns out to be different from the conventional one. This is caused by the radial dependence of Λ and R. Defining

$$G = ig \quad , \quad F = \frac{\mu}{|\mu|} f, \tag{8.35}$$

one finds that the leading behaviour near $r = o$ is determined by

$$\begin{aligned} \left(\partial_r + \frac{3}{2r}\right) G + \frac{1}{r}\left(|n| + \tfrac{1}{2}\right) F = o \\ \left(\partial_r + \frac{3}{2r}\right) F + \frac{1}{r}\left(|n| + \tfrac{1}{2}\right) G = o \end{aligned} \tag{8.36}$$

yielding that,

$$G(r \to o) = -F(r \to o) \sim r^{|n|-1} .$$

Consequently Ψ behaves like

$$\Psi(r \to o) \sim r^{|n|-1} . \tag{8.38}$$

which does not vanish for $|n| = 1$. i.e. in the $\vec{J} = o$ channel. At first sight this implies that we encounter the Lipkin-Weisberger-Peshkin problem, namely that the Jacobi identity of the conjugate momentum operators is not satisfied on the physical states. The

L problem is a consequence of the fact that in the prescence of a singular monopole

$$\tfrac{1}{2}\varepsilon_{ijk}\left[\pi_i[\pi_j,\pi_k]\right] = e\vec{\nabla}.\vec{B} = \mu\delta^3(\vec{r}) \tag{8.39}$$

In the four-dimensional theory this problem disappears only if $\Psi(o) = o$. However in our case the condition is less stringent because the metric contains an extra factor R in the numerator. Since $R(r \to o) = 2r$ we only have to require that

$$\Psi(r \to o) \sim r^{-\frac{1}{2}+|\varepsilon|} . \tag{8.40}$$

This condition is met for all n in our case, hence the LWP problem is evaded. We should not be surprised with this result, it could be anticipated from the fact that also at the classical level the analogous problem of a charged particle moving through the monopole does not occur, as we pointed out in the previous section. Anyway the existence of a normalizable solution with the correct behaviour at $r = o$ implies that the minimal angular momentum amplitude describes pure helicity flip. No surprise either; the natural presence of the anomalous magnetic moment term in (8.22) breaks helicity conservation. This in turn is related to the fact that in the massless five-dimensional Dirac equation there is in general no chiral symmetry. The bound state problem will be considered elsewhere.

I.9. GENERALIZATIONS AND PERSPECTIVES

In this lecture we have described the Kaluza-Klein monopole and its roots in considerable detail. One may speculate on conceivable generalizations. One important generalization namely to the multimonopole solution was briefly mentioned, they are obtained by replacing the Taub-NUT by the Gibbons-Hawking multicentered metrics[6]. Remarkably enough there are also exact solutions which contain both monopoles and antimonopoles[8], the simplest one, a dipole, corresponds to the (non selfdual) euclidean Kerr (and Schwarzschild) solutions.

Another interesting question is the generalization from monopoles to dyons i.e. solutions with both electric and magnetic charge.

We have emphasized, that the topology of the Kaluza-Klein monopole is intimately connected with its magnetic charge, hence with the spatial boundary of the manifold being S^3 (a nontrivial S^1 bundle over S^2). This suggest that the occurrence of K-K monopoles depends crucially on whether the theory allows nontrivial bundles over S^2 at the boundary. A class of such (associated vector) bundles is furnished by S^n bundles over S^2. The topological invariant in question would be given by $\pi_1(G)$ where $G \simeq SO_{n+1}$; the isotropy group of S^n. Because

$$\pi_1\left(SO_{n+1}\right) = Z_2 \qquad n \geqslant 2$$

there are reasons to expect interesting non-abelian monopoles in theories, where one has dimensional reduction from d = n+4 to 4 dimensions over S^n. Because S^n is a space of positive curvature, one needs an extra ingredient to make the compactification possible in the first place. The only way known to me is to invoke the Freund-Rubin device[14]. One introduces a higher rank totally antisymmetric gauge field A, whose field tensor with (d-4) or 4 indices is given a vacuum expectation value

$$F_{\mu\nu\sigma\varepsilon\ldots} = \varepsilon_{\mu\nu\sigma\tau\ldots}$$

Depending on whether one introduces a (fine tuned) cosmological constant one obtains a four-dimensional Minkowski or anti De Sitter space. Though this procedure seems extremely ad hoc one should realize that in supergravity theories such higher rank gauge fields are naturally present. It is also in those theories that soliton like solutions are to be expected[15].

To those, whose aspirations are even less bound by the quest for realistic models, yet more esoteric schemes may appeal, where the compactification happens in steps. Topological excitations which live (partially) in the internal space are possible. For example abelian theories based on U_1 bundles over $\mathbb{C}P^n$. The simplest nontrivial bundles in such a case correspond to the spheres S^{2n+1} [4]. One may also think of instanton bundles with nonabelian groups as fibres. The simplest case is the also due to Hopf; S^7 as S^3 bundle over S^4. We mention en passant that the octionic bundles (Hopf), the simplest one being S^{15} as S^7 bundle over S^8, have not yet found their vocation in the arena of theoretical physics. Some of the questions touched upon here are presently under investigation.

ACKNOWLEDGEMENTS

I'd like to thank P. Batenburg for carrying out various calculations, and P. van Baal and other participants of the Cargèse school for interesting conversations.

REFERENCES

1. P.A.M. Dirac, Proc. Roy. Soc. A133 (1931) 60.
2. H. Hopf, Math. Ann. 104 (1931) 637.

3. Th. Kaluza, Sitzungsber. Preuss. Akad. Wiss., Berlin, Math. Phys. Kl (1921) 966.
O. Klein, Z. Phys. 37 (1926) 895.
4. A. Trautman, Int. Journ. Theor. Phys. 16 (1977) 561.
5. T. Eguchi, P.B. Gilkey and A.J. Hanson, Phys. Rep. 66 (1980) 215.
6. G.W. Gibbons and S.W. Hawking, Phys. Lett. 78B (1978) 430.
7. R.D. Sorkin, Phys. Rev. Lett. 51 (1983) 87.
8. D.J. Gross and M.J. Perry, Nucl. Phys. B226 (1983) 29.
9. T.T. Wu and C.N. Yang, Phys. Rev. D12 (1975) 3845.
10. A. Lichnerowitz, C.R. Acad. Sci. Ser. A257 (1963) 7.
11. For example: *L.D. Landau and E.M. Lifshitz, The Classical Theory of Fields*, Pergamon Press (1971).
12. P. Batenburg, private communication (to be published).
13. Y. Kazama, C.N. Yang and A.S. Goldhaber, Phys. Rev. D15 (1977) 2287; Y. Kazama and C.N. Yang, Phys. Rev. D15 (1977) 2300.
14. P. Freund and M. Rubin, Phys. Lett. 97B (1980) 233.
15. P. van Baal, F.A. Bais and P. van Nieuwenhuizen, Nucl. Phys. B (to be published); P. van Baal and F.A. Bais, Phys. Lett. (to be published).

CHARGE-POLE DYNAMICS, LECTURE II

F. Alexander Bais

Institute for Theoretical Physics
Princetonplein 5, P.O. Box 80.006
3508 TA Utrecht, The Netherlands

ABSTRACT

We review the characteristic properties of the interactions between charges and magnetic monopoles in abelian and non-abelian gauge theories. Aspects which are suspected to underly the baryon decay catalysis phenomenon in GUTs are emphasized.

1. INTRODUCTION

Dirac firmly introduced the notion of magnetic monopoles into physics as the fundamental explanation for the observed quantization of electric charge[1]. The rather drastic developments in our understanding of the fundamental interactions, which have taken place since, have only strengthened and deepened his basic argument. Indeed, the only way to quantize charge we know of is through unification with other interactions, being either the strong and weak interactions (GUTs), or (and) gravitation (Kaluza-Klein). Both approaches share the necessary existence of magnetic monopoles[2,3]. Ever since its inception the theory of magnetic monopoles has continued to be as profound as surprising. The monopole is certainly an "enfant terrible" of physics as it withstands any treatment by naive arguments or routine, usually forcing us to make subtle exceptions to cherished rules. The most recent challenge to our understanding is certainly the catalysis of baryon number violating processes by monopoles in Grand Unified Theories (GUTs) as predicted by Rubakov[4] and Callan[5]. This remarkable phenomenon requires and deserves several lectures by itself.

In this pedagogical lecture, we restrict ourselves to the

well established physics of charges and monopoles, which, though indispensable for a full appreciation of the catalysis effect, does not add up to more than circumstantial evidence for it.

The following basic subjects will be covered. In Section 2 we consider the classical, static, charge-pole system and recall the fundamental fact that the electro-magnetic field of the configuration carries a net angular momentum. How this affects the classical dynamics is reviewed in Section 3. In Section 4 we examine the Schrödinger equation, stating the pathology of the dynamical Lie algebra known as the Lipkin-Weisberger-Peshkin (LWP) problem[6]. This problem becomes acute in the study of the Dirac equation of a charge in the field of a magnetic monopole, which is taken up in Section 5. We summarize the most outstanding results of Goldhaber, Kazama and Yang[7,8], notably the occurrence helicity-flip amplitudes and zero energy boundstates. In Section 6 we extend this analysis to the case of an SU_2 doublet of fermions in the field of an 't Hooft-Polyakov monopole, where now instead of helicity-flip, charge exchange is found to occur. Even in the abelian limit the LWP problem is evaded. At this point it becomes clear that for a better understanding of fermion-monopole dynamics, the back reaction of the monopole has to be taken into account. Simple but with a limited scope, is the kinematical approach where one considers the conservation laws in the system of relevant degrees of freedom[9]. Interestingly enough this procedure uniquely relates the incoming and outgoing states and produces the selection rules obtained by Rubakov and Callan. This is the subject of Section 7. More ambitous and convincing are the dynamical field theoretical calculations of Rubakov and Callan, which we do not discuss here.

2. THE STATIC CHARGE-POLE SYSTEM

The calculation of the angular momentum in the field of a static configuration of a magnetic monopole of strength g and an electric charge e, was apparently suggested as an exercise by J. J. Thomson as early as 1904. At any given point $\vec{r}$, $\vec{E}$ and $\vec{B}$ are nonvanishing and nonparallel so that the Poynting vector $\vec{S} = \vec{E} \times \vec{B}$ is nonvanishing. Clearly $\vec{S}$ is tangent to the circle which is obtained by totating $\vec{r}$ around the axis through e and g. This means that the contribution of all the points on the circle to the total field momentum $P = \int \vec{S} d^3x$ vanishes, but they do yield a nonvanishing contribution to the total angular momentum $\vec{L} = \int (\vec{r} \times \vec{S}) d^3x$. After performing the integral one obtains that

$$\vec{L} = \frac{eg}{4\pi c}\,\hat{n}\ , \tag{2.1}$$

directed from the (positive) charge to the (positive) monopole. The result is independent of the pole-charge separation.

Though not justified (as will become clear in the next section), one may invoke the Bohr-Sommerfeld quantization rule for angular momenta:

$$L_z = \frac{n}{2}\hbar \; , \tag{2.2}$$

as a cheap way to obtain the charge quantization condition of Dirac

$$\mu = \frac{eg}{4\pi} = \frac{n}{2}\hbar c \; . \tag{2.3}$$

More spectacular, and at this point certainly less justified, is the conjecture burried in the Dirac quantization condition (2.3) that fermions could be made out of bosons. Imagine a boundstate (say an s-state) of two bosons, one with minimal electric charge e_o and one with minimal magnetic charge, then the total angular momentum of the state would be $\frac{1}{2}\hbar c$. The question whether such a so-called "dyon" obeys Fermi-Dirac statistics is a separate one. Both the spin[10] and the statistics[11] conjecture have been established to be correct. It constitutes one of the more astounding features of monopole physics.

3. CLASSICAL CHARGE-POLE DYNAMICS

The motion of a charge e in the field of a monopole g located at the origin is determined by the Lorentz force law

$$\vec{F} = \frac{\mu}{r^3}(\vec{v} \times \vec{r}) \; . \tag{3.1}$$

With respect to the solutions of this equation, the following observations were apparently first made by Poincaré in 1896[12]:

i) Because $\vec{F}.\vec{v} = o$ the kinetic energy $\frac{1}{2}mv^2$ is a constant of the motion.

ii) The orbital angular momentum

$$\vec{l} \equiv \vec{r} \times m\vec{v} = -\mu m r^3 \vec{F} \; . \tag{3.2}$$

The torque $\vec{\tau} \equiv \vec{r} \times \vec{F} = \partial_t \vec{l}$ has the property that $\vec{\tau}.\vec{l} = o$, hence $|\vec{l}|$ is a constant of the motion.

iii) From (3.1) one further more obtains directly that

$$\vec{\tau} = \partial_t \vec{L}_{em} \tag{3.3}$$

where $\vec{L}_{em}$ is the field angular momentum (2.1)

$$\vec{L}_{em} = -\mu\hat{r} \; . \tag{3.4}$$

iv) Define the total angular momentum $\vec{L}$

$$\vec{L} = \vec{I} + \vec{L}_{em} \,, \tag{3.5}$$

it is clearly conserved because $\partial_t \vec{L} = \vec{\tau} - \vec{\tau} = o$.

Given the constants μ, $|\vec{I}|$ and $|\vec{v}|$ the motion is completely fixed, it corresponds to a geodesic motion on a cone with axis $\vec{L}$ and opening angle α determined through

$$\text{arctg } \alpha = |\vec{I}|/\mu \,. \tag{3.6}$$

The minimal distance b at which the charge approaches the pole is given by

$$b = |\vec{I}|/m|\vec{v}| \,. \tag{3.7}$$

Consider now the case where $\vec{I} = b = o$. The charge comes in radially and because the Lorentz force vanishes, the particle just passes through the monopole. Here one encounters a pathology. Because $\vec{L} = \mu\hat{r}$ in this particular case, it will not be conserved, as the direction of $\hat{r}$ changes on passing through the origin. Indeed, $b = o$ is too close for comfort. This is the classical manifestation of the Lipkin-Weisberger-Peshkin problem to which we return in the next section. It is amusing to note that if one replaces the point charge by a little charged sphere, then in the $|\vec{I}| = o$ case it will start spinning at exactly the rate necessary to compensate the angular momentum loss. This invites a speculation on the fate of a spin $s = \frac{1}{2}$ particle. Assume that the total angular momentum is $\vec{J} = \vec{L} + \vec{s}$. If $\mu = \frac{1}{2}$ it is possible to have $\vec{J} = o$ (corresponding to $\vec{I} = o$). Clearly $\hat{r}.\vec{J} = o$ implies that the radial component of the spin equals $\hat{r}.\vec{s} = \mu$. This corresponds to a negative helicity particle moving in, and a positive helicity particle moving out. This suggests helicity flip to occur; a possibility which will be considered in detail when we examine the Dirac equation in Section 5.

There is one other remark concerning a conceivable escape from this humiliation of angular momentum. One may imagine the details of the dynamical model to be such that it allows for an electric charge 2e to be deposited on the monopole. Then upon passing through the origin both μ and $\hat{r}$ would change sign, saving angular momentum.

4. THE SCHRÖDINGER PROBLEM

We consider the Hamiltonian

$$\mathcal{H} = \frac{\vec{\pi}^2}{2m} \tag{4.1}$$

where the conjugate momentum $\vec{\pi}$ equals

$$\vec{\pi} = \vec{p}-e\vec{A} \ . \tag{4.2}$$

In the presence of a monopole (say at $\vec{r} = o$), one finds that the Jacobi identity for these momentum operators fails to be satisfied

$$\tfrac{1}{2}\varepsilon_{ijk}\left[\pi_i,[\pi_j,\pi_k]\right] = e\vec{\nabla}.\vec{B} = \mu\ \delta^3(\vec{r}) \ . \tag{4.3}$$

This is the well-known Lipkin-Weisberger-Peshkin problem alluded to before[6]. It suggests that the theory is inconsistent, except if all the wave functions vanish at $\vec{r} = o$.

The naive candidate for the angular momentum operator,

$$\vec{I} = \vec{r} \times \vec{\pi} \tag{4.4}$$

satisfies the correct commutation relations with $\vec{\pi}$ and $\vec{r}$, however it does not satisfy the standard SO_3 commutation relations as

$$\left[I_i,I_j\right] = i\varepsilon_{ijk}\left[I_k+\mu\hat{r}_k\right] \ . \tag{4.5}$$

The classical analysis of the previous section where the same problem manifested itself, suggests the cure. Instead of (4.4) one should take the operators

$$\vec{L} = \vec{I}-\mu\hat{r} \tag{4.6}$$

which do indeed form a representation of the SO_3 algebra. The dynamical Lie algebra has the following form

$$\begin{aligned} \left[L_i,L_j\right] &= i\varepsilon_{ijk}L_k \\ \left[L_i,r_j\right] &= i\varepsilon_{ijk}r_k \\ \left[r_i,r_j\right] &= o \ . \end{aligned} \tag{4.7}$$

This is precisely the algebra E_3 of the Euclidean group in three dimensions. There are two quadratic Casimir invariants

$$\begin{aligned} C_1 &= L^2 \\ C_2 &= \vec{r}.\vec{L} \ . \end{aligned} \tag{4.8}$$

The solution of the Schrödinger problem, amounts to the construction of the irreducible representations of E_3, which can be done by a standard technique[6]. The key result is that unitary representations only exist if the Dirac quantization is met, i.e. $\mu = \frac{eq}{4\pi} = \frac{n}{2}$ (n integer). The smallest eigenvalue for L equals

$\ell = |\mu|$. The angular wavefunctions $Y_{\mu,\ell,m}$ (often called monopole harmonics) are defined by

$$L^2 Y_{\mu,\ell,m} = \ell(\ell+1) Y_{\mu,\ell,m} \qquad \ell = \mu, \mu+1, \ldots$$
$$L_z Y_{\mu,\ell,m} = m\, Y_{\mu,\ell,m} . \tag{4.9}$$

They differ from the ordinary spherical harmonics because of the different definition of $\vec{L}$ in (4.6). Up to a phase they can be constructed starting from the eigenfunction of highest weight satiffying $L_+ Y_{\mu,\ell,m} = o$, and then repeatedly applying L_-. The raising and lowering operators are

$$L_\pm = e^{\pm i\varphi} [\pm\partial_\theta + i\cot\theta\,\partial_\varphi - \mu\cot\theta/2] \tag{4.10}$$

their action on the $Y_{\mu,\ell,m}$ is the usual one:

$$L_\pm\, Y_{\mu,\ell,m} = \sqrt{(\ell\mp m)\;(\ell\pm m+1)}\; Y_{\mu,\ell,m\pm 1} . \tag{4.11}$$

This way one obtains for the normalized angular wave functions

$$Y_{\mu,\ell,m}(\theta,\varphi) = \sqrt{\frac{2j+1}{4\pi}}\, d^j_{m,-\mu}(\theta)\, e^{i(m+\mu)\varphi} . \tag{4.12}$$

The Hamiltonian decomposes as

$$= \frac{1}{2m}\left(p_r^2 - \frac{\vec{I}^2}{r^2}\right) \tag{4.13}$$

which now yields the radial equation

$$\left(\partial_r^2 + \frac{2}{r}\,\partial_r - \frac{\ell(\ell+1)-\mu^2}{r^2}\right) R = 2mER. \tag{4.14}$$

The solutions correspond to spherical Bessel functions (4.15)

$$R_i = j_i\ (kr) \tag{4.16}$$

where $i(i+1)$ is the eigenvalue of $\vec{I}^2$,

$$i = \sqrt{(\ell+\tfrac{1}{2})^2-\mu^2} - \tfrac{1}{2} \tag{4.17}$$

and

$$k = \sqrt{2mE} . \tag{4.18}$$

For small r one has that $j_i(kr) \sim r^i$, recalling that $\ell \geqslant |\mu|$ one has that $i > o$ if $|\mu| > o$. Hence, all wavefunctions vanish at the origin, eliminating the LWP problem.

5. THE DIRAC EQUATION

The Dirac equation in an external magnetic monopole field has been studied thoroughly[7,8,13]. We briefly review some of the outstanding physical aspects, notably the helicity-flip amplitudes and the existence of zero energy bound states. This exposé is based on the extensive work by Goldhaber, Kazama and Yang.

The Dirac equation

$$(\not{D}-im)\psi = o \qquad (5.1)$$

in the standard (γ^0 diagonal) representation reads

$$(E\mp m)u_{\pm} = \vec{\sigma}.\vec{\pi}u_{\mp} \qquad (5.2)$$

where $u_{\pm}$ are two compact spinors defined through

$$\psi = \begin{pmatrix} u_+ \\ u_- \end{pmatrix} . \qquad (5.3)$$

A complete set of mutually commuting operators is furnished by $\mathcal{H}$, J^2, J_z and $\vec{\sigma}.\hat{r}$. The total angular momentum is given by

$$\vec{J} = \vec{L} + \tfrac{1}{2}\vec{\sum} = \vec{r} \times \vec{\pi} - \mu\hat{r} + \tfrac{1}{2}\vec{\sigma} , \qquad (5.4)$$

with eigenvalues,

$$\begin{aligned} j &= \ell+\tfrac{1}{2} = |\mu| + \tfrac{1}{2},\ |\mu| + {}^3/_2 ,\ldots \\ j &= \ell-\tfrac{1}{2} = |\mu| - \tfrac{1}{2},\ |\mu| + \tfrac{1}{2} \ ,\ldots \end{aligned} \qquad (5.5)$$

The fate of helicity in this system, deserves a short digression[8]. There is the following paradox. Consider the operator

$$h = \vec{\sigma}.\vec{\pi} \qquad (5.6)$$

in an $A_0 = o$ gauge we have

$$\partial_t h = e\vec{\sigma}.\vec{E} \qquad (5.7)$$

where $\vec{E}$ is the external electric field. The magnitude of h is given by

$$h^2 = \pi^2 - e\vec{\sigma}.\vec{B} . \qquad (5.8)$$

Equations (5.7-8) establish the well known fact that the helicity $h/|h|$ is conserved in a static purely magnetic field $\vec{E} = \partial_t\vec{B} = \vec{o}$. Let us now turn to the lowest available angular momentum state where $j = |\mu| - \frac{1}{2}$ (i.e. $\ell = \mu$). First note that

$$|\hat{r}.\vec{J}| = |\tfrac{1}{2}(\vec{\sigma}.\hat{r})-\mu| \leqslant |\mu| - \tfrac{1}{2} \tag{5.9}$$

which in conjunction with the fact that $(\vec{\sigma}.\hat{r})^2 = 1$ yields the important relation

$$(\vec{\sigma}.\hat{r})\ \eta_m(\mu) = \frac{\mu}{|\mu|}\ \eta_m(\mu) \tag{5.10}$$

for the eigenvalue of $\vec{\sigma}.\hat{r}$ on the lowest angular momentum eigenvectors. In other words, the conserved radial component of the spin can only take on a *single* value! This fact in turn implies that helicity has to flip in the minimal angular momentum channel where the particle motion is radially in and out (recall our comments in Section 2). The paradox arose basically because we were somewhat cavalier in dealing with the momentum operator $\vec{\pi}$, which as we emphasized before, is plagued by the LWP syndrome. As we will see the helicity operator is not well defined at $\vec{r} = o$. The argument necessitates a further investigation of the $j = |\mu| - \frac{1}{2}$ channel. As it happens to be the case that this is the only channel where the LWP problem rears its head; we will limit ourselves to that particular case in the remainder of this section.

To find out what the operator $\vec{\sigma}.\vec{\pi}$ does on the eigen spinors $\eta_m(\mu)$ of $(\vec{\sigma}.\hat{r})$ we exploit the useful identity

$$\begin{aligned}(\vec{\sigma}.\hat{r})(\vec{\sigma}.\vec{\pi}) &= \hat{r}.\vec{\pi} + i\vec{\sigma}.(\hat{r}x\vec{\pi}) \\ &= -i\partial_r + \frac{i}{r}\ (\vec{J}^2-\vec{L}^2-3/4) + \frac{i}{r}\ |\mu| \\ &= -i\left(\partial_r + \frac{1}{r}\right)\end{aligned} \tag{5.11}$$

where we used that $j = |\mu| - \frac{1}{2}$ and $\ell = |\mu|$. From (5.11) we infer that

$$(\vec{\sigma}.\vec{\pi})h(r)\eta_m(\mu) = -i\ \frac{\mu}{|\mu|}\left(\partial_r + \frac{1}{r}\right) h(r)\eta_m(\mu) \tag{5.12}$$

which implies an enormous simplification as the result only depends on the sign of μ. Clearly both u^+ and u^- contain $\eta(\mu)$ (we omit the index m) and writing

$$u_\pm = h_\pm(r)\eta(\mu)\ , \tag{5.13}$$

one obtains the radial equations

$$(E\mp m)h_\pm(r) = -i\ \frac{\mu}{|\mu|}\left(\partial_r + \frac{1}{r}\right) h_\mp(r)\ . \tag{5.14}$$

Defining

$$h_{+} = \frac{1}{r} \frac{\mu}{|\mu|} F$$
$$h_{-} = - \frac{i}{r} G \qquad (5.15)$$

(5.14) reduces to the second order system

$$\left(\partial_r^2 + k^2\right) F = o$$
$$G = - \frac{1}{(E+m)} \partial_r F \qquad (5.16)$$

where $k^2 = E^2 - m^2$.
First we give the solutions of (5.16) and subsequently analyse to what extent they are acceptable in view of the physical requirements that the Hamiltonian be well defined everywhere.

i) Scattering solutions

The normalized solution witk $k > o$ is of the form[7]

$$\psi = \begin{pmatrix} \frac{\mu}{|\mu|} \frac{1}{r} \sin(kr+\delta)\ \eta(\mu) \\ \frac{-ik}{E+m} \frac{1}{r} \cos(kr+\delta)\ \eta(\mu) \end{pmatrix} \qquad (5.17)$$

and contains an arbitrary phase δ. Clearly ψ does not vanish at the origin implying that we will have to deal with the LWP problem. Before doing so, we first like to discuss some general requirements on the Hamiltonian.

For the Hamiltonian to be acceptable as a physical observable it has to be self adjoint (hermitean) everywhere, so that it defines a complete set of states. For the case at hand we may consider the radial Hamiltonian H_r

$$H_r = -i \frac{\mu}{|\mu|} \partial_r \sigma_1 + m\sigma_3 \qquad (5.18a)$$

working on the column vector

$$\chi(r) = \begin{pmatrix} rh_{+} \\ rh_{-} \end{pmatrix} .$$

The requirement of self adjointness translates into the condition

$$o = \int \chi_1^\dagger (H_r \chi_2) dr - \int (H_r \chi_1^\dagger) \chi_2 dr$$

$$= -i \frac{\mu}{|\mu|} [\chi_1^\dagger(o) \sigma_1 \chi_2(o)] \tag{5.20}$$

Where χ_1 and χ_2 represent normalizable solutions which vanish sufficiently rapidly at $r \to \infty$. It is crucial to observe that since the radial coordinate is only defined on a half line, the self adjointness of the Hamiltonian depends on the boundary condition of the wave function at the origin. For wave functions which do vanish at $\vec{r} = o$ the condition is trivially met. Generally equation (5.20) imposes that

$$\left({}^{h_{1+}(o)}\!/_{h_{1-}(o)} \right)^* = - {}^{h_{2+}(o)}\!/_{h_{2-}(o)} \tag{5.21}$$

In particular by choosing $h_2 = h_1$ one obtains that the ratio $h_{1+}(o)/h_{1-}(o)$ must be purely imaginary. Taking the solution (5.17) one finds that this condition is met provided δ is real:

$$h_+(o)/h_-(o) = i \frac{\mu}{|\mu|} \frac{E+m}{k} \, tg\delta \tag{5.22}$$

Apparently for all δ, the Hamiltonian is well defined and acceptable as an observable. It is easy to verify that the helicity operator h of (5.6) is not well defined, for it to be self adjoint one has to impose

$$h_+^*(o) \, h_+(o) + h_-^*(o) \, h_-(o) = o \tag{5.23}$$

that is $h_+(o) = h_-(o) = o$. This condition is not met by the solution (5.17), henceforth there is no reason for h to be conserved in the lowest angular momentum channel, resolving our initial paradox. It also provides an explanation for the fact that in the higher j channels, where $h_\pm(o) = o$, helicity conservation is obeyed.

Let us now return to (5.17) and explicitly verify that helicity flip does occur for the simplest case where $|\mu| = \frac{1}{2}$, i.e. scattering of a fundamental monopole in the j = o channel. To analyse the physical content of the solution one has to match the asymptotic large r behaviour with solutions of the free Dirac equation. We recall the explicit form of the radially incoming/outgoing *plane wave* solutions with positive energy ($E > o$).

A radially incoming particle with negative/positive helicity:

$$\psi(-k\hat{r},\vec{\sigma}.\hat{r} = \pm) = \sqrt{\frac{E+m}{2E}} \begin{pmatrix} \xi^{\pm} \\ \frac{\mp k}{E+m}\xi^{\pm} \end{pmatrix} e^{-ikr-iEt} \tag{5.24a}$$

A radially outgoing particle with positive/negative helicity:

$$\psi(k\hat{r},\vec{\sigma}.\hat{r} = \pm) = \sqrt{\frac{E+m}{2E}} \begin{pmatrix} \xi^{\pm} \\ \frac{\pm k}{E+m}\xi^{\pm} \end{pmatrix} e^{ikr-iEt} \tag{5.24b}$$

A radially incoming antiparticle with negative/positive helicity:

$$\psi(-k\hat{r},\vec{\sigma}.\hat{r} = \pm) = \sqrt{\frac{E+m}{2E}} \begin{pmatrix} \frac{\mp k}{E+m}\xi^{\pm} \\ \xi^{\pm} \end{pmatrix} e^{+ikr+iEt} \tag{5.24c}$$

A radially outgoing antiparticle with positive/negative helicity:

$$\psi(k\hat{r},\vec{\sigma}.\hat{r} = \pm) = \sqrt{\frac{E+m}{2E}} \begin{pmatrix} \frac{\pm k}{E+m}\xi^{\pm} \\ \xi^{\pm} \end{pmatrix} e^{-ikr+iEt} \tag{5.24d}$$

Here the two component spinors $\xi^{\pm}$ are the eigenspinors of $(\vec{\sigma}.\hat{r})$ associated with the free Dirac equation, given by

$$\xi^{+} = \frac{1}{\sqrt{4\pi}} \begin{pmatrix} \cos\theta/_2\, e^{-i\varphi} \\ \sin\theta/_2 \end{pmatrix}, \quad \xi^{-} = \frac{1}{\sqrt{4\pi}} \begin{pmatrix} \sin\theta/_2 \\ -\cos\theta/_2\, e^{i\varphi} \end{pmatrix} \tag{5.24e}$$

Observing that

$$\eta(\mu = \pm\tfrac{1}{2}) = e^{\pm i\varphi}\xi^{\pm} \tag{5.25}$$

the solution (5.17) is rewritten in an obvious fashion as

$$\psi_{scat}(\mu = \pm\tfrac{1}{2}) = \pm\frac{1}{2ir}\, e^{\pm i\varphi+i\delta} \begin{pmatrix} \xi^{\pm} \\ \frac{\pm k}{E+m}\xi^{\pm} \end{pmatrix} e^{ikr}$$

$$\mp \frac{1}{2ir} e^{\pm i\varphi - i\delta} \begin{pmatrix} \xi^{\pm} \\ \frac{\mp k}{E+m} \xi^{\pm} \end{pmatrix} e^{-ikr}. \tag{5.26}$$

So for positive E we see that the scattering solution describes a radially incoming particle with helicity $-\mu/|\mu|$ and a radially outgoing particle with helicity $+\mu/|\mu|$. Similarly for negative E, the solution describes a radially incoming $\mu/|\mu|$ helicity antiparticle and a radially outgoing $-\mu/|\mu|$ helicity antiparticle. Evidently, in the j = o channel we obtain pure helicity-flip amplitudes.

We mentioned before that there exists a one parameter (δ) family of self adjoint extensions of the Hamiltonian. Goldhaber[8] showed that δ can be fixed (up to a sign) by imposing further discrete symmetries on the theory. In the presence of a monopole one has that the standard C, P and T invariances are only respected if one simultaneously reverses the sign of the magnetic charge. This implies that with μ fixed one may impose the combined symmetries CP, PT etc. One may verify that under a CP inversion H_r transforms as

$$H_r \to \sigma_1 H_r^* \sigma_1 = -H_r \tag{5.27}$$

In other words for CP symmetry to be realised one has to impose an invariant boundary condition

$$\begin{pmatrix} h_+(o) \\ h_-(o) \end{pmatrix} \to \sigma_1 \begin{pmatrix} h_+(o) \\ h_-(o) \end{pmatrix}^* = \begin{pmatrix} h_-^*(o) \\ h_+^*(o) \end{pmatrix} \tag{5.28}$$

or in view of (5.22)

$$|h_+(o)|^2 / |h_-(o)|^2 = \left(\frac{E+m}{k}\right)^2 tg^2 \delta = 1. \tag{5.29}$$

Consequently δ is fixed by CP-invariance

$$\delta = \pm \operatorname{arctg} \frac{k}{E+m} \tag{5.30}$$

This result indicates that δ may be related to the CP violating angle θ, that this is actually the case has been shown explicitly by Yamagishi[14].

We conclude our discussion of the scattering solutions with a closer look at the LWP problem. We emphasized before that not withstanding the fact that the Hamiltonian is well defined, the

LWP problem persists due to the nonvanishing of the wavefunction at the origin. In the work of Goldhaber, Kazama and Yang this problem was resolved by adding an (infinitesimal) anomalous magnetic moment term to the Hamiltonian

$$\Delta H = -\frac{\kappa\mu}{2m}\gamma^{o}\vec{\Sigma}\cdot\frac{\hat{r}}{r^2} \tag{5.31}$$

This term drastically changes the behaviour near the origin

$$F(r \to o) = -G(r \to o) \sim \exp\left[-\frac{|\kappa\mu|}{2mr}\right] \tag{5.32}$$

i.e. for any $\kappa \neq o$, $\psi(r \to o)$ vanishes faster then any power of r, and the LWP problem is clearly (over)killed. We note that the boundary condition in (5.32) satisfies (5.29) as should be expected from the CP-invariant perturbation (5.31). We agree that at the present level of analysis (i.e. not doing quantum field theory) adding an anomalous magnetic moment term is an entirely legitimate, even realistic thing to do. From the formal point of view the LWP problem persist in the $\kappa = o$ case, where it in our opinion should be interpreted as an essential incompleteness of the theory. As we shall see in the following section where we consider the abelian limit of the 't Hooft-Polyakov monopole; the LWP problem can be evaded even if the wavefunction does not vanish at $\vec{r} = o$ by imposing a boundary condition which leads to charge exchange.

ii) Boundstates

Equations (5.10) have a normalizable zero energy boundstate

$$F = -G = e^{-mr} \tag{5.33}$$

It satisfies the CP-invariant boundary condition. It is a peculiar boundstate because although it carries electric charge it is exactly degenerate with the monopole. According to the spin-statistics theorem this state is a fermion or a boson depending on wether μ is integer or half integer. A zero energy boundstate exists for each $j = |\mu|-\frac{1}{2}, |\mu|+\frac{1}{2}, \ldots$ but we have to keep in mind what we have said about the LWP problem before in the case where $\kappa = o$. The existence of these tightly bound zero energy states suggest the formation of a particle-antiparticle condensate around the monopole[7]. The total energy it costs to create a pair in such a state would be +2m (paircreation) -2m (binding energy) $-\sum$ (electric Coulomb energy of the pair), which adds up to a negative quantity.

6. THE J = 0 MODES OF A FERMION DOUBLET IN THE FIELD OF THE 't HOOFT-POLYAKOV MONOPOLE

In the previous section we showed that a charged fermion coupled to the abelian Dirac monopole exhibits some interesting dynamics in the lowest angular momentum channel. Helicity turned out to be not conserved which could be traced back to the fact that the condition that the Hamiltonian be self adjoint (and CP invariant) imposed some nontrivial boundary condition on Ψ at the origin which was incompatible with the selfadjointness of the helicity operator.

It is well known that spontaneously broken gauge theories whose residual gauge group contains a separate U_1 factor feature 't Hooft-Polyakov magnetic monopoles[2]. These are regular solutions to the field equations with a topologically conserved charge. It is this regularity (i.e. the vanishing of the external fields) at the origin, which implies that there is no such thing as the LWP problem. As was pointed out by Goldhaber[8], in this case both the Hamiltonian and the helicity operator will be well defined everywhere. *Helicity must be conserved.* And we are left with the question what actually will happen to a radially infalling fermion. Here we recall a remark made in Section 2. It was pointed out that an alternative way to save angular momentum conservation is to have processes where charge is exchanged between the fermion and the monopole. Such a possibility can in principle be realized in non abelian gauge theories, because a single fermion multiplet contains states with different charges. The remainder of this section will be devoted to an analysis of this effect in the simplest model.

We consider the original 't Hooft-Polyakov monopole, in the SO_3 gauge theory broken down to U_1 by a scalar triplet Φ and couple a doublet of Dirac fermions to it. Fermions in the field of a non abelian monopole have been studied extensively[15]. Recently Marciano and Muzinich solved the SU_2 doublet problem exactly in the Bogomolny-Prasad-Sommerfield limit[16]. For the case at hand we do expect that all the interesting phenomena will occur in the J = o channel, we will be content with a simple direct analysis partially following arguments presented by Kazama[17].

After we have presented the relevant equation in its general form, we first consider the "abelian limit". It is the limit where the mass of the charged guage bosons become infinite and the non abelian monopole core shrinks to zero, leading again to a singular abelian monopole solution embedded in the non-abelian group. Subsequently we determine the all important boundary conditions at the origin from the small r approximation of the non-abelian solution. It is then shown that this indeed leads to charge exchange scattering. Various comments as to the implications of this result to the SU_5 model are made.

The Dirac equation for a fermion doublet coupled to the SU_2 gauge potential $A_\mu = A_\mu{}^a \tau_a/2$ and the scalar triplet $\Phi = \Phi^a \tau_a/2$ reads:

$$(i\gamma^\mu \partial_\mu + \gamma^\mu A_\mu - G\Phi)\psi = o \qquad (6.1)$$

G is the Yukawa coupling Constant. The spinor wavefunction carries both a Lorentz index α ($\alpha = 1,..,4$) and an SU_2 index i ($i = 1,2$), which we will denote as a tensor product:

$$\psi \simeq \psi_{\alpha i} = (u_\alpha \otimes \xi_i) \qquad (6.2)$$

Using a similar decomposition of the Dirac operator, (6.1) takes the form

$$[i(\gamma^\mu \otimes 1)\partial_\mu + \tfrac{1}{2}(\gamma^\mu \otimes \tau^a)\, A^a_\mu + \tfrac{1}{2}(1 \otimes \tau^a)\, G\Phi^a]\; (u \otimes \xi) = o \qquad (6.3)$$

We recall that for the 't Hooft-Polyakov solution one has

$$\vec{A} = \left(\frac{1-k}{r}\right) \tfrac{1}{2}(\hat{r}x\vec{\tau})$$

$$\Phi = \tfrac{1}{2}\phi(\hat{r}.\vec{\tau}) \qquad (6.4)$$

Where the radial functions $k = k(r)$ and $\phi = \phi(r)$ have the following asymptotic behaviour

$$k(o) = 1,\; k(\infty) = o;$$

$$\phi(o) = o,\; \phi(\infty) = \phi_o. \qquad (6.5)$$

The conserved angular momentum is of the form

$$\vec{J} = \vec{r} \times \vec{\pi} + \tfrac{1}{2}\vec{\sigma} + \tfrac{1}{2}\vec{\tau} \qquad (6.6)$$

It should be noted that spin ($\vec{\sigma}$) and isospin ($\vec{\tau}$) appear on equal footing, which is the clue to the peculiar spin-statistics properties of the charge monopole boundstates. Let us now restrict ourselves to the abelian limit, i.e. $k = k(\infty) = o$ and $\phi = \phi(\infty) = \phi_o$. A suitable set of mutually commuting operators consists as in the previous section, of H, J^2, J_z, $\vec{\sigma}.\hat{r}$ but in addition $\vec{\tau}.\hat{r}$. Diagonalizing the latter amounts to expanding ξ in terms of the eigenspinors $\eta_\pm$ of $(\vec{\tau}.\hat{r})$

$$(\vec{\tau}.\hat{r})\; \eta_\pm\; (\theta,\varphi) = \pm\eta_\pm\; (\theta,\varphi) \qquad (6.7a)$$

where

$$\eta_+ = \begin{pmatrix} \cos\theta/_2 \\ e^{i\varphi}\sin\theta/_2 \end{pmatrix}, \quad \eta_- = \begin{pmatrix} e^{-i\varphi}\sin\theta/_2 \\ -\cos\theta/_2 \end{pmatrix} \tag{6.7b}$$

As we limit our analysis to the J = o channel we have the important relation

$$o = \hat{r}.\vec{J} = \tfrac{1}{2}(\vec{\sigma}.\hat{r}) + \tfrac{1}{2}(\vec{\tau}.\hat{r}). \tag{6.8}$$

This implies that we may decompose (6.2) as,

$$\psi = u_- \otimes \eta_+ + u_+ \otimes \eta_-, \tag{6.9}$$

where because of (6.8),

$$(\vec{\sigma}.\hat{r})\, u_\pm = \pm u_\pm \tag{6.10}$$

or in view of (6.7)

$$u_\pm = \begin{pmatrix} f_\pm(r)\ \eta_\pm \\ g_\pm(r)\ \eta_\pm \end{pmatrix} \tag{6.11}$$

We have arrived at the folowing form for the J = o wavefunction

$$\psi = \begin{pmatrix} f_-\eta_- \otimes \eta_+ + f_+\eta_+ \otimes \eta_- \\ g_-\eta_- \otimes \eta_+ + g_+\eta_+ \otimes \eta_- \end{pmatrix} \tag{6.12}$$

Which may be written as the following 2x2 matrix:

$$\psi = \tfrac{1}{2}i \begin{pmatrix} [(f_+-f_-) + (f_++f_-)\ (\vec{\tau}.\hat{r})]\ \tau_2 \\ [(g_+-g_-) + (g_++g_-)\ (\vec{\tau}.\hat{r})]\ \tau_2 \end{pmatrix} \tag{6.13}$$

This is exactly the expression obtained by Jackiw and Rebbi[15].

At this point it is convenient to apply the 't Hooft gauge transformation[2] which rotates $(\vec{\tau}.\hat{r})$ into τ_3 etc., to make the correspondence with the abelian theory even more explicit. The gauge fields take the form

$$A_\mu = A_\mu{}^D \tau_3\ , \quad \Phi = \phi_o \tau_3/D \tag{6.14}$$

where A_μ^D is the usual Dirac monopole potential

$$\vec{A}^D = \frac{1}{2}\,\frac{(1-\cos\theta)}{\sin\theta}\,\hat{\varphi},\quad A_o = o \tag{6.15}$$

(the overall strength of the term is correct, the charges of the doublet are $\pm\, e/2$, but the magnetic charge is $g = 4\pi/e$. In this gauge (6.3) takes the form

$$\{[\gamma^\mu(i\partial_\mu + A_\mu^D) - m]\; u_-\} \otimes \eta_+{}' \; +$$

$$\{[\gamma^\mu(i\partial_\mu - A_\mu^D) + m]\; u_+\} \otimes \eta_-{}' = o \tag{6.16}$$

where $m = G\phi_o/2$, and $\eta_\pm{}'$ are the constant eigenspinors of τ_3. The result is clear, the Hamiltonian splits into two uncoupled abelian Hamiltonians describing two oppositely charged fermions in the field of the fundamental Dirac monopole. We are back to the problem which we tackled in the previous section, were it not for the all important boundary condition at the origin which has to be destilled from the small r behaviour of the full non-abelian equation (6.1). In view of (6.5) this amounts to solving the free Dirac equation

$$[i(\gamma^\mu \otimes 1)\partial_\mu - m(1\otimes \vec{\tau}.\hat{r})]\;\psi = o \tag{6.17}$$

Once more we substitute the J = o wave function (6.13) and use $(\vec{\tau}.\vec{\nabla})\;(\vec{\tau}.\hat{r}) = 2/r$ to obtain:

$$\begin{aligned}
&(E-m)\;(f_+-f_-) + i\partial_r(g_++g_-) + \frac{2i}{r}\,(g_++g_-) = o\\
&(E+m)\;(g_+-g_-) + i\partial_r(f_++f_-) + \frac{2i}{r}\,(f_++f_-) = o\\
&(E+m)\;(g_++g_-) + i\partial_r(f_+-f_-) = o\\
&(E-m)\;(f_++f_-) + i\partial_r(g_+-g_-) = o
\end{aligned} \tag{6.18}$$

The last term of the first two equations would cause non integrable $1/r^2$ singularities in the wavefunction which better be absent. Hence at r = o we have to impose the boundary condition

$$(f_+ + f_-)|_{r=o} = o$$
$$(g_+ + g_-)|_{r=o} = o \qquad (6.19)$$

One finds that the boundary condition at the origin does indeed connect the two different isospin components, thereby constituting a vital ingredient in the Rubakov-Callen effect. The solution to (6.16) satisfying the boundary condition (6.19) have another important property, namely that the LWP problem does not occur even though the wavefunction does not vanish at r = o. We recall that this was left somewhat unspecified in the previous section (assuming that there was no animation magnetic moment term). In the present situation with the anstatz (6.12) we have

$$P = \tfrac{1}{2}\varepsilon_{ijk}\left[\pi_i,[\pi_j,\pi_k]\right] = \tfrac{1}{2}(\vec{\tau}.\hat{r})\delta^3(\vec{r})$$

So that:

$$\int \psi^\dagger P \Psi d^3x = \int [(f_-^2+g_-^2) - (f_+^2+g_+^2)]\ \delta(r) r^2 dr = o,$$

independently of the behaviour of f and g at r = o merely because of the boundary condition (6.19).

We complete our review with a discussion of the solutions to the abelian system (6.16) obeying the boundary conditions (6.19). First we introduce

$$f_\pm = \pm\frac{1}{r}F_\pm \qquad g_\pm = \frac{1}{i}\frac{1}{r}G_\pm \qquad (6.20)$$

to obtain

$$(\partial_r^{\ 2}+k^2)F_- = o \qquad G_- = \frac{1}{(E+m)}\partial_r F_-$$
$$(\partial_r^{\ 2}+k^2)G_+ = o \qquad G_+ = \frac{1}{E-m}\partial_r F_+ \qquad (6.21)$$

There is a zero energy boundstate solution with the correct boundary condition

$$F_- = F_+ = G_+ = -G_- = e^{-mr} \qquad (6.22)$$

It corresponds to the abelian limit of the Jackiw-Rebbi zero mode[15]. Note however that this solution is only possible if the mass term for the fermions is generated through the Higgs mechanism as in (6.1).

In our discussion of the scattering solutions we limit ourselves for simplicity to the massless case m = o (the extension to finite mass is straightforward). From (6.20-21) one obtains

$$
\begin{aligned}
f_- &= -\frac{1}{r}\left(Ae^{ikr} + Be^{-ikr}\right) \\
g_- &= \frac{1}{r}\left(Ae^{ikr} - Be^{-ikr}\right) \\
f_+ &= \frac{1}{r}\left(Ce^{ikr} + De^{-ikr}\right) \\
g_+ &= \frac{1}{r}\left(Ce^{ikr} - De^{-ikr}\right)
\end{aligned}
\qquad (6.23)
$$

with A, B, C and D arbitrary constants up to the normalization condition. The boundary condition (6.19) imposes A = D and C = B. For E > o one easily verifies that our discussion in the previous section (in particular (5.24)). that for example A corresponds to a radially outgoing negative helicity particle with positive isospin. Similarly one finds that D corresponds to a radially incoming negative helicity particle with negative isospin. Consequently the boundary condition A = D implies a process in which helicity is conserved and charge is not. One may go through the other states including the ones with negative energy, and obtain the grand total as displayed in the following Table.

Motion	$\vec{\tau}.\hat{r}$	hel.	Fermion #	
↓ In	+	−	+	B
Out	−	+	+	C
↓ In	−	+	+	D
Out	+	−	+	A
↑ Out	+	−	−	B
In	−	+	−	C
↑ Out	−	+	−	D
In	+	−	−	A

J = o states for a massless fermion doublet

It is worthwile to make some remarks at this point.

i) The results given in the Table are in agreement with the exact results for the exact non abelian equations in the BPS-limit of reference 16. Continuing the analysis one finds that the cross section for charge exchange scattering becomes of geometrical size $\sigma \sim \pi/k^2$, determined by the only dimensional parameter in the problem. This is very much larger than

π/M^2 which would be the size of the non abelian core.

ii) Including a mass term for the fermions would also introduce a helicity flipping amplitude proportional to m.

iii) What does the obtained result teach us about a maybe realistic model like the minimal SU_5 GUT[18]. For the fundamental monopole in SU_5 with one family of fermions one has the following four left handed doublets:

$$\begin{pmatrix} u_2^{\,c} \\ u_1 \end{pmatrix}_L , \begin{pmatrix} -u_1^{\,c} \\ u_2 \end{pmatrix}_L , \begin{pmatrix} d_3 \\ e^+ \end{pmatrix}_L , \begin{pmatrix} -e^- \\ d_3^{\,c} \end{pmatrix}_L \tag{6.24}$$

where the numerial subscripts denote the color indices. One easily verifies from the Table that processes like

$$e^+_L + \text{Mon} \rightarrow d_{3_L} + \text{Mon (?)} \tag{6.25}$$

are allowed. Such a process does not just violate Baryon number, but also charge and color conservation. Clearly (6.25) and therefore the analysis we have done so far can not be the whole story. Indeed from the work of Rubakov[4] and Callan[5] we do know that a lot is missing.

iv) The most upsetting in the naive picture above is the breaking of electric charge conservation. Even black holes respect local symmetrics. Where did the charge go? A natural possibility is that the lost charge is actually deposited on the monopole turning it into a dyon. This idea is made more precise in the next Section. This may be expected to cost a lot of energy, because the mass of a dyon is roughly (exact in the BPS limit):

$$M_{dyon} = M_{mon} \sqrt{1 + q^2/g^2} . \tag{6.26a}$$

The difference between the dyon and monopole mass can be understood as being due to the Coulomb energy of a charge q inside the core of radion $r_o = 1/M_x$:

$$\Delta M \simeq \int_{r>r_o} \frac{E^2}{2} d^3x = 4\pi \int_{r_o}^{\infty} \tfrac{1}{2} \left(\frac{q}{4\pi r^2}\right)^2 r^2 dr = \tfrac{1}{2} \frac{q^2}{g^2} M_{mon} . \tag{6.26b}$$

To continue our line of thought, let us assume that the monopole is converted into a dyon in (6.25). Then the following question innediately poses itself. Is the dyon stable in the presence of light (or massless) charged fermions? This question has been answered in detail by Blair, Christ and Tang[19], who showed that the dyon is unstable indeed. It would lead us to far to give a detailed account of their work here, it runs in many ways parallel to the analysis we presented here for the charge exchange (for a dyon rather than a monopole though). The main result is that a positively charged dyon emits fermion ($\vec{\tau}.\hat{r} = +\frac{1}{2}$, i.e. $q = \frac{1}{2}e$) anti fermion ($\vec{\tau}.\hat{r} = -\frac{1}{2}$, $q = \frac{1}{2}e$) pairs, thereby lowering its charge. Again this calculation is limited by the same problem we encountered before, namely that the back reaction of the dyon is not taken into account. Still, assuming that both charge exchange and the dyon instability do occur, it is easy to concoct baryon number violating processes which do obey electric charge conservation. For example the following "two step" process

$$\begin{aligned} & d_{3_R} + \text{Mon.}(Q=o) \rightarrow e^+{}_R + \text{Dyon } (Q=1) \rightarrow \\ & \rightarrow e^+_R + u_1{}^c{}_L + u_2{}^c{}_L + \text{Mon. } (Q=o) \end{aligned} \tag{6.27}$$

effects one of the transitions considered by Rubakov and Callan. We emphasize that at this point we have put in crucial ingredients by hand, and there is no hope to obtain a quantitative result for a process like (6.27).

Summarizing we may say that the study of fermions in the fixed monopole (dyon) background strongly suggests that peculiar things may happen, but is certainly not conclusive about what the precise nature of the processes is. Can one do better? The answer is wellknown to be yes. There are various ways possible however. The simplist way to improve our results is to consider the symmetries and their corresponding conserved charges in the model[9]. This we will review in the next section, the main conclusion being that surprisingly enough *the conservation laws determine the selection rules in the J = o channel completely*! The best way to go is find some suitable approximation scheme to calculate the dynamics in the J = o channel, as Rubakov and also Callan did. Indeed for massless fermions the J = o dynamics is described by a variant the exactly solvable two dimensional (massless) Schwinger model.

We conclude this section with some qualitative remarks on an intermediate approach, namely the semiclassical approximation which was so succesfull in understanding the physics of the anomaly in the vacuum sector[20]. Instanton configurations are solutions to the unbroken Euclidean Yang-Mills equations with an integer topological charge.

$$\frac{1}{32\pi^2} \int d^4 x \, F\widetilde{F} = \nu \tag{7.28}$$

and minimal action

$$S_o = \frac{8\pi^2}{e^2}\,\nu. \qquad (6.29)$$

These instantons describe tunneling between topologically distinct degenerate components of the vacuum. This leads to the introduction of the θ-vacua which are coherent superpositions of the previously mentioned components

$$|\theta\rangle = \sum_n e^{i\theta n}|n\rangle. \qquad (6.30)$$

The angle θ is a free (CP violating) parameter in the theory. In the euclidean path integral formalism the anomaly manifests itself through the occurence of fermionic zero modes in the instanton field. The correct interpretation of the naively vanishing fermionic determinant det ($\not{D}$), is that the θ-vacuum is characterized by some "condensate" which breaks the axial U_1 symmetry i.e.

$$<\theta|\prod_{i=1}^{N_f}\bar{\psi}_i\gamma_5\psi_i|\theta> \neq o \qquad (6.31)$$

where N_f corresponds to the number of flavors. The product of all different fermions is needed to absorb all the zero modes so that a nonvanishing result obtains. The actual magnitude of (6.31) is exponentially small i.e. of order exp $(-8\pi^2/e_2)$, and clearly nonperturbation in nature. 't Hooft pointed out that in the θ-vacua baryon number is no longer conserved implying that also in the standard $SU_3 \times SU_2 \times U_1$ model the proton would decay at an exceedingly small rate determined by the same small exponential. One may envisage to do a similar calculation but now in the monopole - rather than in the vacuum sector. Rubakov showed that indeed configurations with nontrivial topological charge (6.28) do ixist, but these have *vanishing* action. This causes the semiclassical approximation to break down because higher critical points are not suppressed. Nevertheless it may be possible to extract a qualitative picture (say the selection rules) of the baryon number violation through an analysis of the fermionic zeromodes.

7. CONSERVATION LAWS IN THE FERMION-MONOPOLE SYSTEM

In this section we consider the symmetries and corresponding conserved charges in the monopole-massless fermion system[9]. The J = o condition adds constraints on the charges of the incoming and outgoing states which together with the conservation laws completely determine the finalstate for a given initial state. This is an elegant and elementary way to obtain the selection rules for the Rubakov-Callan effect. These considerations also clarify the role of the anomaly, in particular it will become clear that baryon number violating processes are allowed which do not involve the anomaly[22].

For symplicity we restrict ourselves to a fundamental SU_5 monopole with one generation of fermions, i.e. we have two massless doublets (or four Weyl doublets) coupled to the monopole.

$$\psi_1 = \begin{pmatrix} \psi_{1-} \\ \psi_{1+} \end{pmatrix} = \begin{pmatrix} d_3 \\ e^+ \end{pmatrix} \qquad \psi_2 = \begin{pmatrix} \psi_{2-} \\ \psi_{2+} \end{pmatrix} = \begin{pmatrix} u_2^c \\ u_1 \end{pmatrix} \tag{7.1}$$

The notation $\psi_{i\pm}$ is in line with the notation of the previous section and facilitates the use of the results derived there. We may define eight charges which completely specify the fermion-number and chirality content of the state as

$$Q_{i\pm} = \int \bar{\psi}_{i\pm} \gamma^o \psi_{i\pm} d^3x \tag{7.2a}$$

$$Q^5_{i\pm} = \int \bar{\psi}_{i\pm} \gamma^o \gamma^5 \psi_{i\pm} d^3x \tag{7.2b}$$

In the abelian limit the (non)conservation of these charges reads

$$\partial_t Q = \int (\vec{\nabla}.\vec{j})\, d^3x = \left[r^2 \int j^r d^2\Omega \right]_{r=o}^{r=\infty} \tag{7.3}$$

Again, for the $j \neq o$ modes, the wave functions vanish both at $r = \infty$ and $r = o$ fast enough so that the r.h.s. of (7.3) does indeed vanish. However in the $j = o$ mode a contribution of the origin may survive, and we obtain ($\dot{Q} = \partial_t Q$)

$$\dot{Q} = \lim_{r \to o} (-4\pi r^2 j^r). \tag{7.4}$$

For the radial current we can write:

$$j^r = \bar{\psi}(\vec{\gamma}.\hat{r})\psi = \psi^\dagger \begin{pmatrix} o & \vec{\tau}.\hat{r} \\ \vec{\tau}.\hat{r} & o \end{pmatrix} \psi = \frac{\mu}{|\mu|} (\psi^\dagger \gamma^5 \psi) \tag{7.5}$$

This provides the following relations

$$j^r_{i\pm} = \pm j^{t\,5}_{i\pm} = \pm u^\dagger_{i\pm} \gamma^5 u_{i\pm} = \pm \frac{1}{4\pi} (f^\dagger_{i\pm} g_{i\pm} + g^\dagger_{i\pm} f_{i\pm}) \tag{7.6a}$$

and

$$j^{r\,5}_{i\pm} = \pm j^t_{i\pm} = \pm u^\dagger_{i\pm} u_{i\pm} = \pm \frac{1}{4\pi} (f^\dagger_{i\pm} f_{i\pm} + g^\dagger_{i\pm} g_{i\pm}), \tag{7.6b}$$

where we have substituted the j = o spinor doublet decomposition (6.9-11). Equation (7.4) states that the charges are conserved except for a flow into the core (origin) of the monopole, which in its turn is completely determined by the r = o boundary conditions (6.19). We also recall the important relation (6.8) for j = o fermions which states that for in (out)going states the [isospin] charge equals plus (minus) the helicity. In other words for outgoing states we have

$$Q^5_{i\pm} \mp Q_{i\pm} = o \qquad (7.7a)$$

whereas for ingoing states the condition is the

$$Q^5_{i\pm} \pm Q_{i\pm} = o. \qquad (7.7b)$$

In the previous section it was shown that the j = o solution satisfying (7.7) is given by (6.23), where the boundary condition (6.19) leads to A = D and C = B. Using these results in (7.4-6) one obtains simply

$$\begin{aligned} \dot{Q}_{i+} &= -\dot{Q}_{i-} = 2(A_i^2 - B_i^2) \\ \dot{Q}^5_{i+} &= -\dot{Q}^5_{i-} = -2(A_i^2 + B_i^2) \end{aligned} \qquad (7.8)$$

Henceforth the following charges are strictly conserved

$$\begin{aligned} S_1 &= Q_{1+} + Q_{1-} \\ S_2 &= Q_{2+} + Q_{2-} \\ S_3 &= Q^5_{1+} + Q^5_{1-} - Q^5_{2+} - Q^5_{2-} \end{aligned} \qquad (7.9)$$

Formally also the total axial charge

$$A = Q^5_{1+} + Q^5_{1-} + Q^5_{2+} + Q^5_{2-} \qquad (7.10)$$

is conserved as far as the boundary term is conserved, but it is well known that it is violated locally by the Adler-Bell-Jackiw anomaly. The charge corresponding to the unbroken U_1 gauge group

$$Q = Q_{1-} - Q_{1+} + Q_{2-} - Q_{2+} , \qquad (7.11)$$

is strictly speaking conserved only up to the boundary term

$$\dot{Q} = 4(B_1^2 - A_1^2 + B_2^2 - A_2^2). \qquad (7.12)$$

In other words (7.12) gives us the charge deposited on the monopole core which is not necessarily zero. From dynamical considerations, namely the suppression of processes where charge would be deposited on the monopole due to the electric energy, it is reasonable to assume that (7.12) has to vanish, so that only those processes occur which conserve the total electric charge of the fermions. We arrive at the following picture: the eight charges (7.2) satisfying four constraints (7.7) leave us with four independent quantities both for the ingoing and the outgoing states, these quantities are uniquely related by the conservation of the four charges S_1, S_2, S_3 and Q. Let us consider some examples. We take incoming d_{3R} ($Q^5_{1-} = \frac{1}{2}$, $Q_{1-} = \frac{1}{2}$) and e^-_R ($Q_{1+} = -\frac{1}{2}$, $Q^5_{1+} = +\frac{1}{2}$) then since $S_1 = S_2 = o$, $S_3 = 1$, and $Q = 1$, the only outgoing sulution is u^c_{2L} ($Q_{2-} = -Q^5_{2-} = \frac{1}{2}$) and u^c_{1L} ($Q_{2+} = -\frac{1}{2}Q^5_{2+} = -\frac{1}{2}$). In short we have obtained that a variant of (6.27) is allowed:

$$d_{3R} + e^-_R + \text{Mon.} \rightarrow u_{2_L}^c + u_{1_L}^c + \text{Mon.} \tag{7.13}$$

A similar reasoning leads to the following process:

$$d_{3R} + u_{1L} + \text{Mon} \rightarrow e^+_R + u^c_{2L} + \text{Mon} \tag{7.14}$$

where the in- and outgoing states are characterized by the charges $S_1 = S_2 = \frac{1}{2}$, $S_3 = 1$ and $Q = o$. Both processes violate baryon number B according to $\Delta B = -1$, there is however an important difference between the two which concerns the role of the anomaly[22]. For the first process (7.13) the axial charge is not conserved $\Delta A = -2$; this process apparently involves the anomaly. In the second process (7.14) the axial charge is actually conserved and it constitutes an example of a baryon number violating process which does not proceed through the anomaly; but rather through the nontrivial boundary condition at the core. Let us finally consider the process

$$d_{3R} + u_1{}^c{}_R + \text{Mon} \rightarrow d_{3L} + u_1{}^c{}_L + \text{Mon} \tag{7.15}$$

it obviously conserves S_1, S_2, S_3 and Q, but not only that, it also conserves baryon number. Only chirality is violated: $\Delta A = -2$, which is due to the anomaly. This process is somewhat surprising in view of our results in the previous section, where chirality flip dit not seem to occur, indeed, it can only happen because of the anomaly. This transition is consistent with one of the $B = o$ condensates calculated by Rubakov[4]. It accounts for the fact that even with more generations of fermions present still $\Delta B = 1$ processes are possible[4,21].

Summarizing we may say that from studying the conservation laws in the SU_5 theory one obtains in a straight forward way all the selection rules which follow from the fermion condensates given by Rubakov and Callan. This analysis can easily be generalized to the non-minimal monopoles in SU_5.

8. CONCLUSIONS

In the foregoing pages we have reviewed the most interesting aspects of the dynamics of charged particles in the field of abelian and non-abelian monopoles. We have not reviewed the original Rubakov calculation where the $j = o$ dynamics for massless fermions is solved exactly through a reduction to the massless Schwinger model in two dimensions, nor the calculations of Callan based on bosonisation techniques. Nevertheless we gave a rather detailed exposé of the underlying physics including a discussion of the Rubakov-Callan selection rules for monopole induced baryon number violating processes. We hope this will enable the reader to catch on with the current work in this exciting new branch of monopole physics.

Acknowledgements: I thank G. 't Hooft for interesting discussions.

REFERENCES

For an introduction to the subject of abelian and non-abelian monopoles we may refer the reader to some of the existing review articles:

S. Coleman, Lectures at the International School of Subnuclear Physics, Ettore Majorana, Erice, Sicily both in 1975 and 1981.

P. Goddard and D. Olive, Rep. Prog. Phys. 41 (1978) 1357.

F.A. Bais, Proceedings of Meeting "Geometric Techniques in Gauge Theories", Lecture Notes in Mathematics 926, Springer (1981).

1. P.A.M. Dirac, Proc. Roy. Soc. (London), Series A, 133 (1931) 60.

2. G. 't Hooft, Nucl. Phys. B79 (1974) 276;
   A.M. Polyakov, JETP Lett. 20 (1974) 194.

3. R.D. Sorkin, Phys. Rev. Lett. 51 (1983) 87;
   D.J. Gross and M.J. Perry, Nucl. Phys. B226 (1983) 29.

4. V.A. Rubakov, JETP Lett. 33 (1981) 644;
   Nucl. Phys. B203 (1982) 311.

5. C.G. Callan, Phys. Rev. D25 (1982) 2141;
   Phys. Rev. D26 (1982) 2058; Nucl. Phys. B212 (1983) 391.

6. H.J. Lipkin, W.I. Weisberger and M. Peshkin, Ann. Phys. (New York) 53 (1969) 203.

7. Y. Kazama, C.N. Yang and A.S. Goldhaber, Phys. Rev. D15 (1977) 2287;
   Y. Kazama and C.N. Yang, Phys. Rev. D15 (1977) 2300.

8. A.S. Goldhaber, Phys. Rev. D16 (1977) 1815.

9. A. Sen, Phys. Rev. D28 (1983) 876.

10. P. Hasenfratz and G. 't Hooft, Phys. Rev. Lett. 36 (1976) 1119;
    R. Jackiw and C. Rebbi, Phys. Rev. Lett 36 (1976) 1116.

11. A.S. Goldhaber, Phys. Rev. Lett. 36 (1976) 1122.

12. H. Poincaré, Comtes Rendus 123 (1896) 530.

13. P.B. Banderet, Helv. Phys. Acta 19 (1946) 503;
    Harish-Chandra, Phys. Rev. 74 (1948) 883;
    A.S. Goldhaber, Phys. Rev. 140 (1965) B1407;

J. Schwinger et al., Ann. Phys. (New York) 101 (1976) 451;
D. Boulware et al., Phys. Rev. D14 (1976) 2708.

14. H. Yamagishi, Phys. Rev. D27 (1983) 2383.

15. T. Dereli, J.H. Swank and L.J. Swank, Phys. Rev. D11 (1975) 3541; Phys. Rev. D12 (1975) 1096;
D. Boulware et al. Phys. Rev. D14 (1976) 2708;
R. Jackiw and C. Rebbi, Phys. Rev. D13 (1976) 3398;
C. Callias, Phys. Rev. D16 (1977) 3068;
P. Cox and A. Yildiz, Phys. Rev. D18 (1978) 1211;
F.A. Bais and W. Troost, Nucl. Phys. B178 (1981) 125;
Ju-Fei Tang, Phys. Rev. D26 (1982) 510.

16. M. Prasad and C. Sommerfield, Phys. Rev. Lett. 35 (1975) 760;
E. Bogomolny, Sov. J. Nucl. Phys. 24 (1976) 449.
W. Marciano and I. Muzinich, Phys. Rev. Lett. 50 (1983) 1035.

17. Y. Kazama, Proceedings of the workshop "Proton Decay and Monopoles", Oct. 18-20, 1982, Kamioka, Japan.

18. H. Georgi and S.L. Glashow, Phys. Rev. Lett. 32 (1974) 438.

19. A. Blaer, N. Christ and J-F. Tang, Phys. Rev. Lett 47 (1981) 1364; Phys. Rev. D25 (1982) 2128.

20. G. 't Hooft, Phys. Rev. Lett. 37 (1976) 8; Phys. Rev. D14 (1976) 3432.

21. F.A. Bais, J. Ellis, D.V. Nanopoulos and K.A. Olive, Nucl. Phys. B219 (1983) 189.

22. K. Seo, Phys. Lett. B. (to be published).

EXACT RENORMALIZATION GROUP FOR GAUGE THEORIES[+]

Tadeusz Balaban, John Imbrie,* and Arthur Jaffe

Lyman Laboratory of Physics
Harvard University
Cambridge, MA 02138 USA

1. INTRODUCTION

Renormalization group ideas have been extremely important to progress in our understanding of gauge field theory. Particularly the idea of asymptotic freedom leads us to hope that nonabelian gauge theories exist in four dimensions and yet are capable of producing the physics we observe--quarks confined in meson and baryon states. For a thorough understanding of the ultraviolet behavior of gauge theories, we need to go beyond the approximation of the theory at some momentum scale by theories with one or a small number of coupling constants. In other words, we need a method of performing exact renormalization group transformations, keeping control of higher order effects, nonlocal effects, and large field effects that are usually ignored.

Rigorous renormalization group methods have been described or proposed in the lectures of Gawedzki, Kupiainen, Mack, and Mitter. Earlier work of Glimm and Jaffe[1] and Gallavotti et al.[2] on the ϕ^4 model in three dimensions were quite important to later developments in this area.

We present here a block spin procedure which works for gauge theories, at least in the superrenormalizable case. It should be enlightening for the reader to compare the various methods described in these proceedings--especially from the point of view of how each method is suited to the physics of the problem it is used to study.

---

*Junior Fellow, Harvard Society of Fellows. [+]Supported in part by the National Science Foundation under grant no. PHY-82-03669.

We believe that our approach is advantageous for the study of gauge theories.

The problem we present in some detail is the abelian Higgs model in two or three dimensions, with a "wine bottle" potential for the Higgs field. The continuum action density is

$$\frac{1}{2}\,|(\partial_\mu - ieA_\mu)\phi|^2 + \frac{1}{4}F_{\mu\nu}^2 + \lambda|\phi|^4 - \frac{1}{4}m^2|\phi|^2 + \text{counterterms},$$

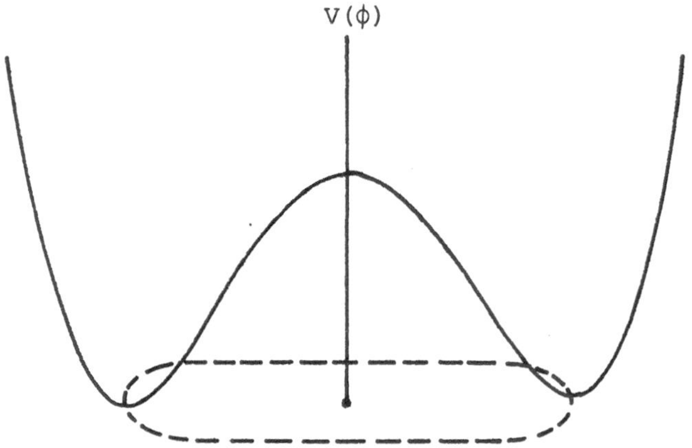

where ϕ is a complex scalar field. We use a lattice approximation and renormalization transformations in the form of block spins to study two problems.

1. Ultraviolet stability. This phrase describes the bounds necessary to control the continuum limit, i.e. estimates uniform in the lattice spacing ε for finite volume partition functions and correlation functions.

2. The Higgs mechanism. Here infrared problems complementary to ultraviolet stability are considered. We seek bounds on correlations that have exponential decay in the separation between observables. The rate of decay should be uniform in the lattice spacing and in the volume, and it should be independent of the observables which are considered. Such a uniform decay rate ensures a gap in the mass spectrum of the model, and this gap is also known as the Higgs effect. The mass gap at first appears surprising, since one might expect a zero-mass scalar boson (Higgs particle) to arise from transverse motion in the wine bottle potential. One might also expect a massless gauge particle (photon) because the gauge field has no explicit mass term.

The uniformity of the estimates as $\varepsilon \to 0$ and as the volume goes to infinity means that bounds carry over to the corresponding limits, if they exist. The exponential decay should allow us to take the infinite volume limit, because distant regions contribute only exponentially small effects to a given correlation function. The $\varepsilon \to 0$ limit is more difficult, but we expect that our methods will apply.

The standard heuristic physics explanation of the Higgs mechanism arises from consideration of the model in the "unitary" gauge. In this point of view, the scalar field is written in polar form, $\phi = re^{i\theta}$, with $r \geqslant 0$, and one chooses the gauge $\theta = 0$. In this manner, the transverse (angular) fluctuations of ϕ are eliminated, while radial fluctuations in a neighborhood of the potential minimum at $r = r_0 = m(8\lambda)^{-1/2}$ are characterized by a mass m. (In the semiclassical limit, the mass squared is twice the sectional curvature of the potential.) Hence one expects a Higgs particle with a mass m_H which is close to m. We choose $m = O(1)$.

Furthermore, in the unitary gauge $1/2|(\partial_\mu - ieA_\mu)\phi|^2 = 1/2|\partial_\mu\phi|^2 + 1/2\, e^2A_\mu^2|\phi|^2$. Assuming that $\phi - r_0$ is small, this yields a semiclassical photon mass equal to $er_0 = me/(8\lambda)^{1/2}$. Thus in order to obtain semiclassical masses which are $O(1)$, we also also require that $e^2/\lambda = O(1)$. The quantum effects for small e,λ will modify the Higgs mass m_H and the photon mass m_{ph} from their classical values,

$$m_H = m(1 + O(e,\lambda)), \qquad m_{ph} = me(8\lambda)^{-1/2}(1 + O(e,\lambda)) .$$

The modifications are expected to be small for $0 \leqslant e^2, \lambda \ll 1$, with m, $e^2/\lambda = O(1)$, and this is the region in which we work.

What is wrong with this picture of mass generation and the Higgs effect? The problem is that the unitary $(\theta = 0)$ gauge used above is a nonrenormalizable gauge. Propagators which occur in perturbation theory are very badly behaved for large momenta, i.e. they have bad local regularity properties for small ε. This points to the fact that the $\theta = 0$ gauge is physically a poor way of understanding the Higgs model. It intertwines the infrared and the ultraviolet problems, making each less clear. Furthermore, there are unanswered questions about the role of configurations where $\phi \approx 0$, namely singular gauge configurations or vortex configurations.

Our answer to these difficulties is to treat high-momentum degrees of freedom differently from low-momentum degrees of freedom. The high-momentum degrees of freedom are studied using gauges which have well-behaved ultraviolet properties. The function space integral over these degrees of freedom can be carried out and analyzed. The result is an effective low momentum theory. For this resulting model there are no problems with the $\theta = 0$ gauge, and we use this unitary gauge to exhibit the generation of a positive mass.

The first problem is to find an appropriate gauge-invariant splitting of the model into its high and low momentum parts. We choose a block spin method to achieve this splitting. Thus while the Higgs effect is not generally considered as a renormalization group problem, we do so and perform the momentum space splitting in a series of steps. Each step will transform a model on a lattice with spacing δ into another model on a lattice with spacing δL, where L is a small positive integer. After a finite number $k = k(\varepsilon)$ steps, $\varepsilon L^k = O(1)$, we say that we have arrived at the "unit lattice" model. We analyze this model using the unitary gauge and find a mass gap $\bar{m} > 0$. Here $\bar{m}$ is uniform in ε and $|\Lambda|$ as $\varepsilon \to 0$ and $|\Lambda| \to \infty$.

In order to control the errors at each stage of renormalization, it is necessary to eliminate portions of the effective action which couple distant space-time regions with very small probability; in other words we "localize" terms in the effective action. Furthermore we stop the renormalization transformation procedure in spacial neighborhoods of places where the fields produce a large action. In such regions we find it impossible to define a useful definition of background field configurations and fluctuations about such configurations. However these events have a small probability and contribute little to expectations. All these complications are handled at each length scale, and hence lead us to a "multi-scale cluster expansion." This is the technical tool we use for estimates.

This work on Higgs models is built upon years of experience with superrenormalizable quantum field theories--many models with bosons and fermions in two and three dimensions have been constructed and properties such as particle structure and phase structure have been analyzed--see Glimm and Jaffe's book[3] for references. The most popular approach (which we pursue here) has been to work with Euclidean functional integrals, verifying the Osterwalder-Schrader axioms[4] which guarantee the existence of a Minkowski theory satisfying the basic axioms of quantum field theory.

The work on gauge theories to date is more limited. The most complete construction is of the two dimensional abelian Higgs model in two dimensions, by Brydges, Fröhlich, and Seiler[5]. For this model all the Osterwalder-Schrader axioms were verified except clustering. In particular, the existence of a mass gap is open, even for the two dimensional model. More limited results were obtained by Challifour and Weingarten[6] and by Ito[7] for two dimensional quantum electrodynamics, see also Seiler.[8]

The work described here is partly published and partly in progress. Balaban's papers[9] prove ultraviolet stability for the somewhat simpler case of the abelian Higgs model with a massive vector field. Our work on the Higgs mechanism for the model without an explicit mass for the gauge field (Wilson action) is not yet complete, but is sufficiently far along for us to present the main ideas here. Balaban is extending the stability result to the nonabelian case in two and three dimensions--see Ref. 10 and subsequent papers in preparation. Readers particularly interested in the nonabelian model should consult these papers as they appear, see also Ref. 11. However, the overall strategy and many of the details are common to the abelian and nonabelian case, so our discussion here of the abelian model should serve as an introduction to the methods used in all of these works.

Of course, the most interesting case is four dimensional nonabelian Yang Mills. We are still far from understanding this model (we will see below some features of the method which seem to be particularly troublesome for four dimensions). However, an understanding of how to do exact renormalization group theory for nonabelian models in three dimensions is surely a prerequisite to results in four dimensions. We refer the reader to the Cargèse lectures for the year 2003 for a construction of $(YM)_4$. (We don't know the authors yet)!

2. BLOCK SPINS FOR GAUGE THEORIES

2.1 The Lattice Approximation

We work with a lattice regularization because it is extremely important to preserve gauge invariance--not just at the start but also for all effective theories obtained after applying some number of renormalization transformations. Of course there are disadvantages of the lattice approximation: loss of Euclidean invariance, for example, or loss of an obvious geometric interpretation. However, for us (and many other authors) the advantages outweigh the disadvantages.

We also find it convenient to work with a compact action (the Wilson action). We hope that noncompact actions, or actions where the gauge field has a larger period than the Higgs field can be treated with some additional work. One can convert a noncompact model to a compact one at the price of introducing vortices. This device was used in Ref. 12 to study the noncompact model with a fixed (not arbitrarily small) lattice spacing.

On the ε-lattice we use fields $u_b \in U(1)$, $\phi(x) \in \mathbb{C}$, where b is a bond and x is a site on a periodic lattice T_ε. If we write

$u_{\langle x,x+\varepsilon e_\mu\rangle} = e^{ie\varepsilon A_\mu(x)}$, then $A_\mu(x)$ corresponds to the usual gauge field for continuum formulations. By convention, bonds are oriented, and $u_b = u_{-b}^{-1}$, where $-b$ denotes the reverse of b. If Γ is any oriented contour composed of bonds, then we define $u(\Gamma) = \Pi_{b\in\Gamma}\, u_b$. In particular, $u(p)$ is the usual plaquette variable formed by taking the product of u_b's around the plaquette p. We define the covariant derivative of ϕ as $(D_u\phi)_b = \varepsilon^{-1}(u_b\phi_{b_+} - \phi_{b_-})$, where b_+, b_- are the forward, backward sites of b.

The action on the ε-lattice is

$$S^\varepsilon(u,\phi) = \sum_p \varepsilon^d \frac{1}{e^2\varepsilon^4}(1-\mathrm{Re}\, u(p)) + \frac{1}{2}\sum_b \varepsilon^d |(D_u^\varepsilon\phi)_b|^2$$

$$+ \sum_x \varepsilon^d(\lambda|\phi(x)|^4 - \frac{1}{4}m^2|\phi(x)|^2 - \frac{1}{2}\delta m^2|\phi(x)|^2) + E\ .$$

Here $\delta m^2 = \delta m^2(e,\lambda,\varepsilon)$ is a mass counterterm for the Higgs field. It is given by a diagrammatic expansion to some order in coupling constants, and is divergent as $\varepsilon\to 0$. The constant E contains vacuum energy subtractions, including divergent counterterms, again given by a diagrammatic expansion. From this action we obtain the partition function

$$Z = \int du d\phi\ e^{-S^\varepsilon(u,\phi)}$$

and expectations

$$\langle F\rangle = Z^{-1}\int du d\phi\ Fe^{-S^\varepsilon(u,\phi)}\quad .$$

We consider only gauge invariant observables. (Non-invariant observables like $\phi(x)$ are easily seen to have vanishing expectation--no gauge fix is needed above because of the compactness of the u-integration.) The gauge transformations for these fields are

$$\phi(x) \to \phi(x)e^{i\lambda(x)}$$

$$u_b \to u_b e^{-i(\lambda(b_+)-\lambda(b_-))}$$

where λ is a real-valued function on sites. The action is easily seen to be invariant, as are observables such as $|\phi(x)|^2$, $\phi(b_-)u_b\phi(b_+)$, $u(p)$.

The goal of this work is to prove bounds on expectation of appropriately renormalized observables that are uniform in the lattice spacing and in the volume. Furthermore, we wish to take the infinite volume limit and establish exponential clustering of truncated correlations $\langle F_1F_2\rangle - \langle F_1\rangle\langle F_2\rangle$ at a rate uniform in ε (the Higgs mechanism). The existence and Euclidean in variance of the limit $\varepsilon \to 0$ is open for the moment. We expect that our methods will eventually be used to understand these questions, thereby yielding a complete construction of the model and verification of the axioms. Partial results in this direction have been obtained by C. King.

2.2 Rescaling to the Unit Lattice

It is convenient to work with unit lattice spacing for each effective theory. Thus we rescale the ε-lattice to the unit lattice, performing canonical scalings on the fields $(\phi^{\varepsilon} = \varepsilon^{-(d-2)/2}\phi^1$, etc.). The partition function becomes

$$\int du d\phi \; e^{-S(u,\phi)}$$

with

$$S(u,\phi) = \sum_p \frac{1}{e(\varepsilon)^2} (1-\mathrm{Re}\; u(p)) + \frac{1}{2} \sum_b |(D_u\phi)_b|^2$$

$$+ \sum_x (\lambda(\varepsilon)|\phi(x)|^4 - \frac{1}{4} m^2\varepsilon^2|\phi(x)|^2 - \frac{1}{2}\delta m^2\varepsilon^2|\phi(x)|^2) + E.$$

We are using rescaled coupling constants

$$\lambda(\varepsilon) = \lambda\varepsilon^{4-d} , \qquad e(\varepsilon) = e\varepsilon^{(4-d)/2} ,$$

and D_u is the unit lattice covariant derivative, $D_u\phi = u_b\phi_{b_+} - \phi_{b_-}$.

Note that all nonquadratic pieces of the action have acquired coefficients that are positive powers of ε, for $d < 4$. This becomes more apparent when we expand $u(p)$ in terms of correctly normalized field strength variables. With

$$f_p = \frac{1}{ie(\varepsilon)} \log u(p) ,$$

we have

$$\frac{1}{e(\varepsilon)^2} (1-\mathrm{Re}\; u(p)) = \frac{1}{2} f_p^2 - \frac{1}{4!} e(\varepsilon)^2 f_p^4 + \cdots \quad .$$

The mass counterterm δm^2 diverges like ε^{-1} in three dimensions, logarithmically in ε in two dimensions. Thus $|\phi|^2$ still has a positive power of ε in its coefficient. The fact that the model becomes extremely weakly coupled when viewed on this scale is just a manifestation of its superrenormalizability.

The price we pay for the rescaling, however, is that the model appears almost massless. Mass terms in the basic Gaussian have acquired powers of ε (at any rate they only appear in the Higgs part of the action, and then only with the wrong sign). The masslessness and the associated long-range correlations are circumvented by the method of block spin renormalization transformations. If we fix the block averages of the fields and integrate only over the fluctuations, then the gradient terms in the action will act like a mass. Thus we will obtain an integral which is a small perturbation of a massive Gaussian measure. Such integrals are relatively easy to control--they have well behaved perturbative expansions with remainder terms which can be controlled by means of convergent cluster expansions.

2.3 Average Fields and the Renormalization Transformation

We divide the lattice into blocks of L^d sites each, where L is a small positive integer like 2 or 3. A naive definition of the average of ϕ on a block would be $\Sigma_{x\in \text{Block}} L^{-d} \phi(x)$. However this definition of average is not gauge covariant--each term in the sum transforms according to the gauge transformation at a different point. In order to have an average that transforms according to the gauge function at a single point (say at the corner of the block) we use "parallel transport operators" $u(\Gamma_{y,x})$. Here $\Gamma_{y,x}$ is some contour from y (the corner of the block) to x (an arbitrary point in the block. For example we can take $\Gamma_{y,x}$ as in figure 1.

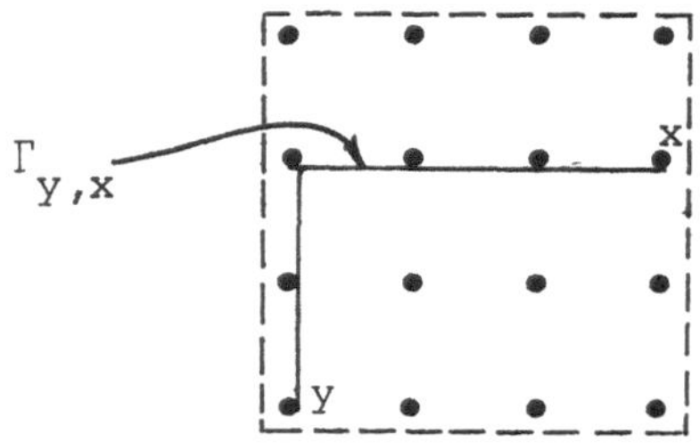

Figure 1.

We denote the gauge covariant average of ϕ by

$$(Q(u)\phi)_y = \sum_{x\in B(y)} L^{-d} u(\Gamma_{y,x})\phi(x) ,$$

where $B(y)$ is the block containing y, a point of the L-lattice. Under the gauge transformation

$$\phi(x) \to \phi(x)e^{i\lambda(x)} , \qquad u_b \to u_b e^{-i(\lambda(b_+)-\lambda(b_-))}$$

we have

$$(Q(u)\phi)_y \to (Q(u)\phi)_y \; e^{i\lambda(y)} .$$

For the gauge field the average should be defined for bonds of the block lattice $L\mathbb{Z}^d$, i.e. for $b' = \langle y,y'\rangle$, y,y' corners of nearest neighbor blocks. For gauge covariance we consider a collection of contours each starting at y and ending at y'. A convenient choice is the following: For each $x \in B(y)$ we define $\Gamma_{y,x,y'} = \Gamma_{y,x} \cup \langle x,x'\rangle \cup \Gamma_{x'y'}$, where $x' = x + y' - y$ and $\Gamma_{x',y}$ is the reverse of $\Gamma_{y,x'}$.

Figure 2.

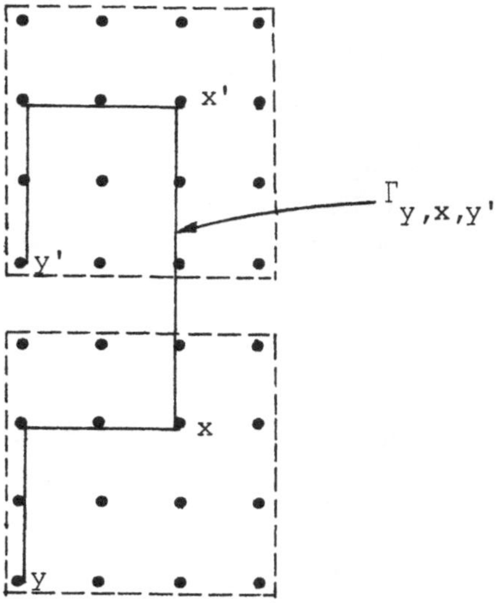

We form the collection of group elements $\{u(\Gamma_{y,x,y'})\}$, as x ranges over $B(y)$. An average of these is defined as follows. We regard the $u(\Gamma_{y,x,y'})$ as points on the unit circle. If all the points lie inside some half-circle, the average is the point in the half-circle such that the sum of the difference angles vanishes. In other words, the argument of the average is the average of the arguments, as long as the jump in the definition of the argument occurs outside the half-circle. If not all points lie inside a half-circle, we add the group elements as complex numbers and divide by the modulus of the result to get back to the unit circle.

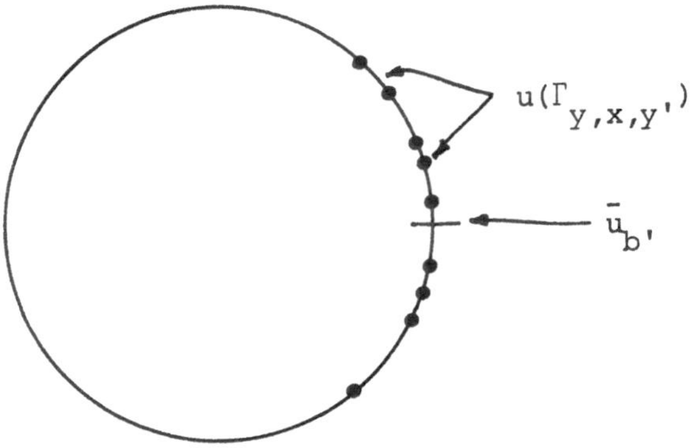

Figure 3.

The second case only occurs when there are large plaquette variables ($u(p)-1$ not small). As these occurrences are suppressed strongly by the action, almost any definition of average would do. The average of $\{u(\Gamma_{y,x,y'})\}$ is denoted $\bar{u}_{b'}$, and it transforms as follows:

$$\bar{u}_{b'} \to \bar{u}_{b'}\, e^{-i(\lambda(b'_+)-\lambda(b'_-))}$$

We now define the renormalization transformation, which is a transformation from densities $\rho(u,\phi)$ to densities $(T\rho)(v,\psi)$. Here $v \in U(1)$, $\psi \in \mathbb{C}$ are block fields defined on bonds, sites of $L\mathbb{Z}^d$. Let us take $\rho(u,\phi) = e^{-S(u,\phi)}$, and define

$$\begin{aligned}\rho_1(v,\psi) &= (T\rho)(v,\psi)\\ &= c\int du\,d\phi\ \delta(v\bar{u}^{-1})\exp[-1/2\langle\psi-Q(u)\phi,\psi-Q(u)\phi\rangle]\ e^{-S(u,\phi)},\end{aligned}$$

where

$$\delta(v\bar{u}^{-1}) = \prod_{b'} \delta((v\bar{u}^{-1})_{b'})$$

and δ is the δ-function at the identity of $U(1)$. The constant c is chosen so that $T1 = 1$. Of course ρ_1 satisfies the basic property

$$\int dv\,d\psi\ \rho_1(v,\psi) = \int du\,d\phi\ e^{-S(u,\phi)}$$

The transformation for the gauge fields consists simply of integrating out all u with fixed averages $\bar{u} = v$. For the scalar field there is no δ-function but the approximate δ-function $\exp[-1/2\langle\psi - Q(u)\phi, \psi - Q(u)\phi\rangle]$ makes it highly probable that $Q(u)\phi$ is close to ψ.

2.4 Gauge Invariance and Gauge Fixing

It is extremely important that we preserve gauge invariance for the block fields v, ψ. We will exploit this in several ways throughout the sequence of renormalization transformations; also at the end when we analyze the effective unit lattice theory, we need gauge invariance to go to the unitary gauge $(\theta = 0)$. Let us verify that $\rho_1(v,\psi)$ is invariant under

$$\psi(y) \to \psi(y)e^{i\lambda(y)} \qquad\qquad v_{b'} \to v_{b'}\, e^{-i(\lambda(b'_+)-\lambda(b'_-))}$$

Extend λ in an arbitrary way to a function on all sites (only the values at corners of blocks are used in the above transformation). We can replace u, ϕ by their gauge transforms because $dud\phi$ is gauge invariant. The action is unchanged, while the gauge transformations of ψ, $Q(u)\phi$ and v, $\bar{u}$ match and drop out of the integrand, leaving $\rho_1(v,\psi)$ in its original form.

From this calculation we see that the integrand is still invariant under gauge transformations λ that vanish at corners of blocks. We wish to calculate the effective action for v, ψ by using a Gaussian approximation for the u,ϕ integral, but the invariance will lead to zero modes and spoil the approximation. Hence we need to fix the gauge for the u,ϕ integral.

We use the simplest possible gauge fix, a kind of axial gauge on blocks. The choice of gauge fix is not very important at this stage, since we are doing an integral that corresponds to one slice of momenta only. We set $u_b = 1$ for each bond in a maximal tree of bonds in each block. The tree (Fig. 4), composed of the contours $\Gamma_{y,x}$, is convenient.

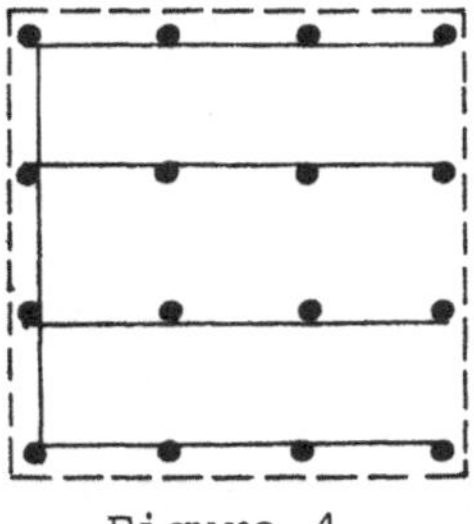

Figure 4.

We thus define

$$\delta_{axial}(u) = \prod_{blocks} \prod_{b\in tree} \delta(u_b)$$

and insert it under the integral. The density $\rho_1(v,\psi)$ is unchanged, as we take $\int du = 1$ by convention.

2.5 Effective Masses

The most important effect of the renormalization transformation is to introduce effective masses into the quadratic forms. In the density

$$\rho_1(v,\psi) = c\int du d\phi \; \delta(v\bar{u}^{-1})\delta_{axial}(u)\exp\Big[-\frac{1}{2}\langle\psi-Q(u)\phi,\psi-Q(u)\phi\rangle$$
$$-\frac{1}{2}\langle D_u\phi,D_u\phi\rangle - \frac{1}{2}\langle f,f\rangle - \sum_x(\lambda(\varepsilon)|\phi(x)|^4 - \frac{1}{4}m^2\varepsilon^2|\phi(x)|^2$$
$$-\frac{1}{2}\delta m^2\varepsilon^2|\phi(x)|^2) - \sum_p\sum_{n=1}^{\infty}\frac{(-1)^n e(\varepsilon)^{2n}}{(2n+2)!} f_p^{2n+2} - E\Big] \tag{1}$$

we have quadratic forms $D_u^* D_u + Q(u)^* Q(u)$ for ϕ, and $\partial^*\partial$ for A (writing $u = e^{ie(\varepsilon)A}$, $f = \partial A$, with $(\partial A)(p) = \Sigma_{b\in p} A_b$). We can prove strictly positive lower bounds on these forms, at least in the small field region, where f is not too large (see the next section). Thus we have, for example

$$\langle\phi,(D_u^* D_u + Q(u)^* Q(u))\phi\rangle \geqslant \sum_x c|\phi(x)|^2$$

for some $c > 0$, and we have a lower bound given by a mass term. The term $Q(u)^* Q(u)$ selects out the "constant" mode which would have been a zero mode for $D_u^* D_u$ alone. The form $\partial^*\partial$ has many zero modes, but the δ-functions $\delta(v\bar{u}^{-1})$ and $\delta_{axial}(u)$ reduce the integration over A to a subspace, on which the desired lower bound holds.

The lower bounds on the quadratic forms allow us to prove exponentially decaying bounds on the corresponding inverse operators, giving the correlations between fields at different sites in the Gaussian approximation. Thus distant fields are almost independent, and we can make localized calculations involving only small portions of the lattice. Of course we have not lost the long range correlations of the original model--they are carried by the external fields v,ψ. The effective masses also allow us to take advantage of the extremely small rescaled coupling constants $\lambda(\varepsilon)$, $e(\varepsilon)$ describing the corrections to the Gaussian. In what follows we show how these simplifying features can be used to compute an effective action for v,ψ. The price to pay for all the simplicity is the need to iterate the renormalization transformation many times ($\log_L \varepsilon^{-1}$ times). However the main features of the calculation are independent of the iteration step.

2.6 Large and Small Fields

As mentioned above, we lack some basic estimates on quadratic forms and propagators when the gauge field is rough. In addition, the fact that A is really a periodic variable makes the Gaussian approximation break down completely wherever there are large fields. Therefore we wish to separate large and small field regions, do perturbative calculations in the small field region, and use only the basic positivity of the action to control the large field integrations.

We define the large field region to be some neighborhood of places where any of the following inequalities hold:

$$\begin{aligned} &|\phi| > \lambda(\varepsilon)^{-1/4}|\log \varepsilon|^p \\ &|D_u\phi| > |\log \varepsilon|^p \\ &|\psi - Q(u)\phi| > |\log \varepsilon|^p \\ &\frac{1}{e(\varepsilon)}|u(p)-1| > |\log \varepsilon|^p \end{aligned} \tag{2}$$

In each case, one of the terms $[\lambda(\varepsilon)|\phi|^4$, $|D_u\phi|^2$, $|\psi-Q(u)\phi|^2$, $\frac{1}{e(\varepsilon)^2}\mathrm{Re}(1-u(p))]$ in the exponential in (1) is larger than $|\log \varepsilon|^{2p}$. Thus we can expect to obtain small factors $\exp(-|\log \varepsilon|^{2p}) \leqslant \varepsilon^\kappa$, κ arbitrary, from integrals in the large field region. Thus large field regions are rare, and contribute very little to the partition function.

We implement these ideas in our integral (1) by inserting a partition of unity under the integral. Thus we write

$$1 = \sum_{\Lambda} \chi_\Lambda \zeta_{\Lambda^c} \quad ,$$

where χ_Λ is an approximate characteristic function restricting all fields to be small in Λ, in the sense of (2), and where ζ_{Λ^c} forces some fields to be large throughout Λ^c.

At this point we do almost nothing in Λ^c except extract the factors $(\varepsilon^\kappa)^{|\Lambda^c|}$. Even in later steps, no calculations are performed there. We always treat Λ^c as a large field region. Thus the flow of information is always along the following lines:

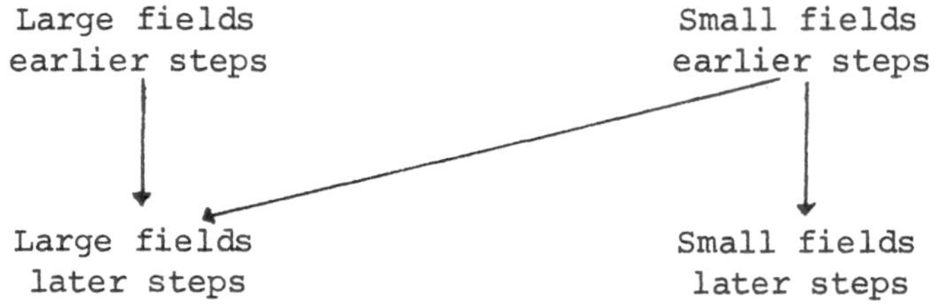

The missing arrow from large to small fields contrasts the procedure described in the lectures of Gawedzki and Kupiainen.

While our treatment seems to be forced by the nonlinearities in our model--the action does not split neatly into a Gaussian piece plus an interaction piece--it is the source of troubles with generalizations to four dimensions. This is seen when we estimate the contribution of large field regions to the partition function. Taking into account the entropy coming from the sum over Λ, we allow for a factor $1+\varepsilon^\kappa$ at each site. At the k-th step, we allow for a factor $1+(L^k\varepsilon)^\kappa$ at each L^k-block. The product over all k and all sites or blocks is bounded by

$$\exp\left[\sum_{k=0}^{\log_L \varepsilon^{-1}} (L^k\varepsilon)^\kappa (L^k\varepsilon)^{-d} |T_\varepsilon|\right] \leqslant \exp(c|T_\varepsilon|) \quad .$$

Here $|T_\varepsilon|$ is the volume of the lattice, which is ε^{-d} times the number of sites in the lattice. Then $|T_\varepsilon|(L^k\varepsilon)^{-d}$ is the number of blocks on the scale k. As long as $\kappa > d$ the sum over k converges, yielding a finite contribution to the vacuum energy. The method breaks down in four dimensional Yang-Mills theory, because the effective coupling constant is expected to be only logarithmically small in ε. This reduces the size of allowable small fields, and so yields less convergence from large fields. In particular, κ is at best a small positive constant, and the sum over k would diverge. The basic problem is that we are not cancelling the vacuum energy contribution from large fields at each step. This works in the superrenormalizable case but apparently not in the borderline case of logarithmically asymptotically free models.

2.7 Translations and the Background Fields

In the small field region the integral in

$$\rho_1(v,\psi) = \sum_{\Lambda} c \int du d\phi \; \chi_\Lambda \zeta_{\Lambda^c} \delta(v\bar{u}^{-1}) \delta_{axial}(u)$$

$$\exp\left[- \frac{1}{2} <\psi-Q(u)\phi,\psi-Q(u)\phi> - \frac{1}{2} <D_u\phi,D_u\phi> - \frac{1}{2} <f,f> - \ldots\right]$$

is a small perturbation of a Gaussian measure. The best way to treat the integral is to translate the fields to the (v,ψ-dependent) minimum of the Gaussian part of the action. The minimum is the background field, and we will do perturbation theory and cluster expansions about this configuration.

More precisely, we do this in two steps--one for u and one for ϕ--since the quadratic form for ϕ depends on the external gauge field. We write $u = u_1 e^{ie(\varepsilon)A'}$, where $u_1 = u_1(v)$ is the unique minimum of $<f,f>$ under the constraints $\delta(v\bar{u}^{-1})\delta_{axial}(u)$. The background field u_1 can be written explicitly in terms of v. The fluctuation field A' is small ($|A'| \leqslant |\log \varepsilon|^p$) in the small field region, and the integral over u is rewritten as an integral over A'.

At this point we expand all the terms in the exponential in powers of A'. This means that the scalar field forms produce new interactions:

$$\frac{1}{2} <\psi-Q(u)\phi,\psi-Q(u)\phi> + \frac{1}{2} <D_u\phi,D_u\phi>$$

$$= \frac{1}{2} <\psi-Q(u_1)\phi,\psi-Q(u_1)\phi> + \frac{1}{2} <D_{u_1}\phi,D_{u_1}\phi> + \text{interaction terms} .$$

The terms coming from the expansion of $<D_u\phi,D_u\phi>$ are the usual scalar field-vector field interaction vertices, and the others are new vertices from the renormalization transformation. All terms are small, as each power of A' comes with a coupling constant $e(\varepsilon)$.

The δ-functions, when written in terms of A', become linear constraints on the A'-integral:

$$\delta(v\bar{u}^{-1})\ \delta_{axial}(u) \quad = \quad \delta(QA')\delta_{axial}(A') \quad .$$

Here Q is a kind of averaging operator for functions on bonds:

$$(QA)_{b'} \quad = \quad \sum_{x\in B(b'_-)} L^{-d}A(<x,x'>) \quad .$$

We put $A(\Gamma) = \Sigma_{b\in\Gamma} A_b$, and as in the definition of $\bar{u}_{b'}$, $x-x' = b'_--b'_+$. In $\delta_{axial}(A')$, all A'_b are set to zero for b in any of the maximal trees in blocks. The averaging procedure for A' is not much different from the simplest procedure for scalar fields (averaging over blocks). It yields the same kind of result: The form $<f,f>$ under the constraints $\delta(QA')\delta_{axial}(A')$ has an effective mass and well-behaved, exponentially decaying propagators. Note that under a gauge transformation $A'\to A'-\delta\lambda$, QA' transforms into $QA'-\bar{\lambda}(b'_+)+\bar{\lambda}(b'_-)$, where $\bar{\lambda}(y) = \Sigma_{x\in B(y)} L^{-d}\lambda(x)$. This contrasts with the transformation law for $\bar{u}$, which involved λ restricted to corners of blocks. Of course the axial gauge conditions broke the invariance under transformations λ that are not constant on blocks, and the two laws are equivalent if λ is constant on blocks.

We next compute the minimum of the quadratic action for ϕ (external field u_1). It is a linear function of ψ, and we denote it $\phi_1 = K_1(u_1)\psi$. Thus we write $\phi = \phi_1 + \phi'$, with ϕ_1 the background field, ϕ' the fluctuation field. The fluctuation field is small: $|\phi'| \leqslant c|\log\varepsilon|^p$.

2.8 Calculation of the Effective Action

To simplify the discussion, let us consider only the term $\Lambda = T_\varepsilon$, i.e., the whole lattice is the small field region. The effect of the translations on the quadratic terms in the action is to split each quadratic form into two forms, one for the block field and one for the fluctuation field. So we write

$$<f,f> = <f^{(1)},\sigma_1 f^{(1)}> + <\partial A',\partial A'>$$

$$<D_{u_1}\phi, D_{u_1}\phi> + <\psi-Q(u_1)\phi,\psi-Q(u_1)\phi>$$

$$= <\psi,\Delta_1(u_1)\psi> + <\phi',(D^*_{u_1}D_{u_1} + Q(u_1)^*Q(u_1))\phi'> \quad ,$$

where

$$f^{(1)}_{p'} \quad = \quad \frac{1}{ie(\varepsilon)} \log v(p') \quad ,$$

and where σ_1, $\Delta_1(u_1)$ are the resulting block field forms.

The block field terms $<f^{(1)},\sigma_1 f^{(1)}>$ and $<\psi,\Delta_1(u_1)\psi>$ are the quadratic terms of the effective action for v,ψ. If we rescale the L-lattice to the 1-lattice, they have properties analogous to the original quadratic terms for u,ϕ: $<f,f>$ and $<\phi,D_u^* D_u\phi>$. The forms are massless, gauge-invariant, and obey lower bounds which have a local form:

$$<f^{(1)},\sigma_1 f^{(1)}> \geqslant c<f^{(1)},f^{(1)}>$$

$$<\psi,\Delta_1(u_1)\psi> \geqslant c\sum_{b'} |u_1(<b'_-,b'_+>)\psi(b'_+) - \psi(b'_-)|^2 \quad .$$

The forms are no longer local, however--they have exponential tails. The terms can be expressed simply as the original quadratic terms evaluated on the configurations u_1 and ϕ_1.

The scalar field self-interaction gives rise to some ϕ'-independent terms when we expand $\phi = \phi_1 + \phi'$. These are just

$$\sum_x (\lambda(\varepsilon)|\phi_1(x)|^4 - \frac{1}{4} m^2\varepsilon^2|\phi_1(x)|^2 - \frac{1}{2}\delta m^2\varepsilon^2|\phi_1(x)|^2) \ ,$$

and the other terms are small, having one power of $\lambda(\varepsilon)$ and three or less powers of ϕ_1, with $|\phi_1| \leqslant \lambda(\varepsilon)^{-1/4}|\log\varepsilon|^p$.

After the translations and expansions, the term $\Lambda = T_\varepsilon$ in our original density (1) becomes

$$c\exp\Big[-\frac{1}{2}<f^{(1)},\sigma_1 f^{(1)}> - \frac{1}{2}<\psi,\Delta_1(u_1)\psi>$$
$$-\sum_x (\lambda(\varepsilon)|\phi_1(x)|^4 - \frac{1}{4} m^2\varepsilon^2|\phi_1(x)|^2 - \frac{1}{2}\delta m^2\varepsilon^2|\phi_1(x)|^2) - E\Big]$$
$$\cdot\int dA'd\phi'\ \delta(QA')\delta_{axial}(A')\chi_\Lambda \exp\Big[-\frac{1}{2}<\partial A',\partial A'>$$
$$-\frac{1}{2}<\phi',(D^*_{u_1}D_{u_1} + Q(u_1)^*Q(u_1))\phi'> - V\Big]$$

Here V contains the ϕ',A'-dependent terms, all of which are bounded by some power of coupling constants $\lambda(\varepsilon)$, $e(\varepsilon)$ for all values of ϕ',A' permitted by χ_Λ. We introduce the Gaussian normalization

$$Z^{(0)}(u_1) = \int dA' d\phi' \, \delta(QA')\delta_{axial}(A')\exp\Big[-\frac{1}{2}\langle\partial A',\partial A'\rangle - \frac{1}{2}\langle\phi',(D^*_{u_1}D_{u_1} + Q(u_1)^*Q(u_1))\phi'\rangle\Big],$$

and the corresponding normalized measure $\langle\cdot\rangle$. Thus we have the remaining terms in the effective action for v,ψ: $\log Z^{(0)}(u_1)$ and $\log\langle\chi_\Lambda e^{-V}\rangle$.

We need to give a very precise expansion for $\log\langle\chi_\Lambda e^{-V}\rangle$ because renormalization cancellations must be exhibited. This is especially important after many steps of the iteration have been performed and the divergences in perturbation theory start to manifest themselves. We have found it most convenient to exhibit cancellations on a purely perturbative level, as the cancellations or finiteness properties due to gauge invariance are fairly subtle. Thus we extract a few orders of perturbation theory using the cumulant expansion,

$$\log\langle\chi_\Lambda e^{-V}\rangle = -\langle V\rangle + \frac{1}{2!}\langle V;V\rangle - \cdots$$

where

$$\langle V;V\rangle = \langle V^2\rangle - \langle V\rangle^2$$

The restrictions in χ_Λ --A', ϕ' smaller than $|\log\varepsilon|^p$ --produce a change which is smaller than any power of ε, and hence is not seen in the perturbation expansion in ε. If we extract n orders of perturbation theory, for some sufficiently large n, then the remainder will be of the order of ε^κ, with $\kappa > d$. As in our large field estimate, this is small enough to be ignored henceforth. However, in order to address the question of whether there is a mass gap, we must exhibit the locality properties of the remainder. This is done with a cluster expansion. The result is

$$\log\langle\chi_\Lambda e^{-V}\rangle = \sum_{j=1}^{n} \frac{1}{j!}\langle(-V;)^j\rangle + \sum_X W(X) \quad ,$$

where $\langle(\cdot;)^j\rangle$ denotes the j-th-truncated correlation. Here $W(X)$ depends only on v,ψ in the connected set X, and $|W(X)| \leqslant \varepsilon^\kappa e^{-c|X|}$ with $|X|$ denoting the number of sites in X.

In the general case, with nonempty large field region, the small field calculations are performed with the values of the fields fixed in a neighborhood of the large field region (conditional integration).

After dropping or expanding out the $W(X)$ terms, we have the following form of the effective action:

$$S^{(1)}(v,\psi) = \frac{1}{2}<f^{(1)},\sigma_1 f^{(1)}> + \frac{1}{2}<\psi,\Delta_1(u_1)\psi> + \mathcal{P}^{(1)}(v,\psi)$$
$$- \log Z^{(0)}(u_1) + E \quad .$$

Here $\mathcal{P}^{(1)}(v,\psi)$ contains the terms $\lambda(\varepsilon)|\phi_1(x)|^4 + \cdots$ as well as the truncated correlations of V. A final step is to rescale all expressions and fields from the L-lattice to the 1-lattice. Relabeling v,ψ by u,ϕ, we are in a position to repeat the process.

2.9 The Effective Action After k Steps

Rather than describe the general step of the procedure, let us simply examine the k-th effective action to get some feel for the behavior of the model under iterated application of renormalization transformations. We stick to the small field region, and use block fields u,ϕ defined on the unit lattice, which is the k-times decimated version of the original lattice rescaled to spacing L^{-k}.

We have background fields $u_k(u)$ and $\phi_k = K_k(u_k)\phi$ defined on the L^{-k}-lattice. These fields represent the fields u,ϕ on the original lattice in a smooth way. (Note that as k increases, the L^{-k}-lattice becomes finer; eventually it is the ε-lattice and the procedure stops.) The background fields are the minimum energy L^{-k}-lattice configurations, given u,ϕ and taking into account only quadratic terms.

As before, the leading terms in the effective action are just the original action evaluated on the configurations u_k,ϕ_k. Thus we have, for example, $\Sigma_x L^{-dk}\lambda(L^k\varepsilon)|\phi_k(x)|^4$. The coupling constants are partially rescaled back to their original values--we have $\lambda(L^k\varepsilon) = \lambda\cdot(L^k\varepsilon)^{4-d}$ with $L^k\varepsilon$ increasing towards 1. The basic picture is that the renormalization transformation is driving the model away from a trivial fixed point, so the coupling constants are growing. It is convenient to introduce a diagrammatic representation for terms in the effective action. The above quadratic term is represented by

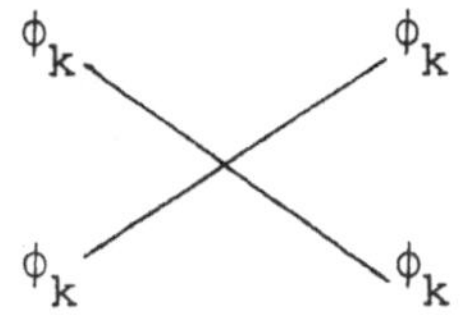

with a standard ϕ^4 vertex summed over the L^{-k}-lattice. The external lines in all diagrams are background fields ϕ_k or $f_k(p) = (ie(L^k\varepsilon)L^{-k})^{-1} \log u_k(p)$.

When ϕ_k and f_k are substituted into the quadratic terms of the original action, we obtain the main quadratic forms for the block fields $\phi(y)$ and $f^{(k)}(p') = (ie(L^k\varepsilon))^{-I} \log u(p')$. We denote these by $\Delta_k(u_k)$ and σ_k, and prove the basic stability bounds for them:

$$<\phi,\Delta_k(u_k)\phi> \geq \sum_{b'} c|u_k(<b'_-,b'_+>)\phi(b'_+) - \phi(b'_-)|^2$$

$$<f^{(k)},\sigma_k f^{(k)}> \geq c <f^{(k)},f^{(k)}> \quad .$$

We also have a sequence of Gaussian normalization factors, $Z^{(0)}(u_k)\cdots Z^{(k-1)}(u_k)$, whose logarithms contribute to the effective action.

Finally we have the higher-order terms in $\lambda(L^k\varepsilon)$ and $e(L^k\varepsilon)$, which are given by diagrams with internal as well as external lines. The vertices in these diagrams are derived from the L^{-k}-lattice action, but there are some new vertices arising from the expansion of $Q(u)$ in the renormalization transformation. The propagators are derived from the original action also, but with the quadratic terms or δ-functions from the renormalization transformations included as well. Thus the scalar field propagator is

$$G_k(u_k) = (D_{u_k}^{L^{-k}*} D_{u_k}^{L^{-k}} + Q_k(u_k)^* Q_k(u_k))^{-1} \quad ,$$

with $Q_k(u_k)$ being the k-th iterate of the averaging operation for scalar fields. The term $Q_k(u_k)^*Q_k(u_k)$ provides an effective mass for this L^{-k}-lattice propagator; thus we can prove its exponential decay. The short-distance behavior of $G_k(u_k)$ is also important for the analysis of ultraviolet divergences.

It is worth noting that our effective action is a purely perturbative one; it is given by the sum of diagrams of order less than some fixed n. The higher order and nonperturbative effects have been expanded out of the action and treated like large field regions. Thus the task of providing uniform bounds on the effective action is reduced to that of controlling the perturbation expansion. Renormalization cancellations are built into the perturbation expansion through the definition of counterterms.

We must consider effective observables as well as effective actions, since the original fields have been integrated out. The situation is quite analogous to that of the action--there are perturbative terms as well as nonperturbative or high-order terms. The latter can be estimated and essentially neglected, while the former generates diagrams like those considered for the action, except for new vertices coming from the observables on the L^{-k}-lattice.

2.10 Changes in Gauge

In the k-th renormalization transformation we perform the same basic steps that were outlined for the first step. However there is one new operation, the change of gauge, that is worth describing. The purpose is to improve the ultraviolet behavior of gauge field propagators in the effective action. This is accomplished by modifying the gauge fix for the fluctuation fields already integrated out.

Since we have been imposing axial gauge conditions each time we integrated out a fluctuation field, the gauge that would naturally occur in the vector field propagators in the k-th effective action is an axial-type gauge, with $A_b = 0$ for b in a maximal tree on each block of L^k sites. For large k these propagators are poorly behaved in the ultraviolet, and we are unable to prove the needed bounds on the perturbation expansion. To obtain better propagators, we need to change the background field u_k by a gauge transformation before expanding in the fluctuation field A'. Note that we have invariance of the effective action under the full group of L^{-k}-lattice gauge transformations of u_k even though that gauge freedom was broken when we integrated out the fluctuation fields. (When we transform u_k, the current fields u, ϕ must also be transformed by the restriction of the gauge function to the unit lattice.)

Unfortunately the required gauge transformation depends nonlocally on the current field u. This introduces nonlocal effects that have to be controlled with additional expansions. After the gauge transformation we have the background field written as

$$u_k = u_{k+1} \exp(ie(L^k \varepsilon) L^{-k} H_k A') \quad ,$$

with H_k a regular, exponentially decaying kernel. Like u_k, the L^{-k}-lattice configuration $H_k A'$ is of minimal energy under certain constraints, but with a Feynman-like gauge fix used to measure energy instead of axial gauge restrictions. With the above form for u_k, we can expand the action with respect to A' as before. The next background field u_{k+1} remains in all expressions.

After integrating out A', we find that the regularity of H_k (and of earlier H_j, $j < k$) yields well-behaved gauge field propagators. In this way we see that it is possible to use one gauge for integrating a field out, and another for representing the trace of that field in the effective action.

3. THE HIGGS MECHANISM

The renormalization transformations are continued until $L^k\varepsilon$ gets close to unity. The estimates begin to break down when $L^k\varepsilon$ approaches the smaller of m^{-1} (the inverse of the classical scalar field mass), and $(8\lambda/e^2)^{1/2}m^{-1}$ (the inverse of the classical vector field mass). At this point the payoff comes--we have a gauge invariant effective action for unit lattice fields with properties expected from perturbation theory. In particular, if we are interested in ultraviolet stability, the bounds on the effective action are independent of ε, and simple estimates for the last integration over u,ϕ suffices to prove ε-independent bounds on the original functional integrals. If we are interested in the mass generation and the infinite volume limit, we must extract mass terms from the effective action in order to integrate over u,ϕ--we can no longer rely on effective masses from renormalization transformations. We can extract mass terms by going to the unitary gauge for u,ϕ. The ultraviolet problem has already been treated; there are no difficulties associated with choosing $\theta = 0$ at each point of the unit lattice.

We conclude with a brief outline of the steps performed in integrating out the last fields and exhibiting the Higgs mechanism. The negative mass-squared term in the scalar potential has become significant, so we can use the "wine bottle" shape of the potential to introduce restrictions on ϕ. The small field region is defined to be where ϕ lies the annulus $\big||\phi| - (m^2/8\lambda)^{1/2}\big| \leqslant |\log e|^p$ and where $|D_{\bar{u}_k}\phi| \leqslant |\log e|^p$. [We define $\bar{u}_k(b') = u_k(\langle b'_-, b'_+\rangle)$.] The large field region has suppression factors coming from the corresponding terms in the action. For simplicity, we put $L^k\varepsilon = 1$.

It turns out that the analogous stability bound for the gauge field is best seen in axial gauge. Thus we again use our freedom to change u_k by a gauge transformation and put it in axial gauge ($u_k(b) = 1$ for b in maximal trees in L^k-blocks). The corresponding form of the action is almost like the standard unit lattice Wilson action because u_k takes the following simple form:

$u_k(b) = u(b')$ for b touching both neighboring blocks corresponding to the endpoints of b'

$u_k(b) = 1$ otherwise.

Actually, $u_k(b)$ differs from $u(b')$ or 1 by a small field; the difference is of the order of $e|\log e|^p$.

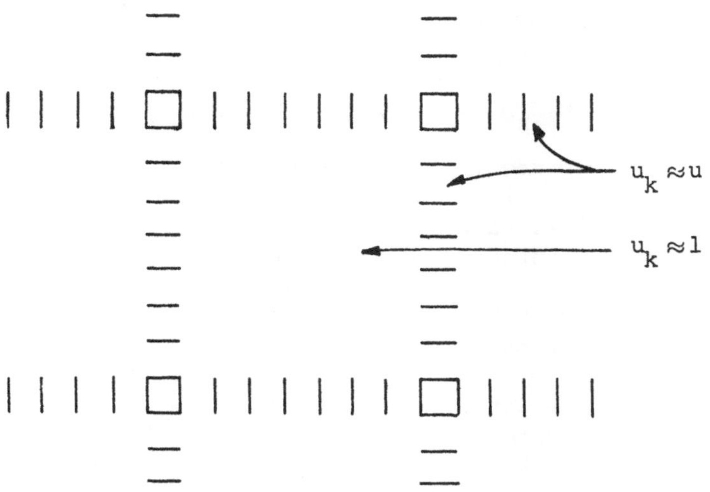

Figure 5.

For $d=3$, these corridors become one higher dimensional. Of course this is a very singular gauge; when written in terms of Lie algebra elements, the concentrations at the corridors between blocks are almost δ-functions.

We recall the basic stability bound for the scalar field quadratic form:

$$\langle\phi, \Delta_k(u_k)\phi\rangle \geq c \sum_{b'} |u_k(\langle b'_-, b'_+\rangle)\phi(b'_+) - \phi(b'_-)|^2 \quad .$$

In axial gauge we have $u_k(\langle b'_-, b'_+\rangle) \approx u(b)$. Since ϕ is limited to a small neighborhood of $|\phi| = (m^2/8\lambda)^{1/2}$, this estimate allows us to restrict u to a small neighborhood of a pure gauge. Thus if we write

$$\phi = r\, e^{i\theta} \qquad\qquad r > 0$$

then we have

$$u = \exp[ie(A - \partial\theta)] \quad , \qquad |A_b| \leq c|\log e|^p$$

in the small field region. In the large field region we have convergence from the term

$$|u_k(\langle b'_-, b'_+\rangle)\phi(b'_-)|^2 \approx \frac{m^2}{8\lambda} |\exp[ieA_b] - 1|^2 \quad .$$

Now that we have the restrictions on A, we can change to a gauge in which the dependence of u_k on A is regular. We have

$$u_k = \exp[ie\varepsilon(H_k A + \partial^{\varepsilon}\theta')], \qquad \theta'(x) = -\theta(y(x)) + (D_k A)_x \quad ,$$

with $D_k(x,b)$ bounded; $y(x)$ is the corner of the block containing x. The term $\partial^{\varepsilon}\theta'$ is removed from u_k by a gauge transformation. The phase of ϕ is canceled--there is no longer any dependence on θ, and it can be integrated out trivially. This leaves us in unitary gauge $\phi = r > 0$. However, there is a small residual phase $\exp[ie(D_k A)_y]$ multiplying each $r(y)$. This cannot be gauged away and is a new interaction.

We expand the action with respect to $H_k A$ and $D_k A$. A shift $r = r_0 + r'$, $r_0 = (m^2/8\lambda)^{1/2}$ to the minimum of the scalar potential allows us to extract an explicit mass term for the scalar field. The gauge field mass term is also extracted at this point. The quadratic form resulting from perturbing $\langle r_0, \Delta_k(u_k) r_0 \rangle$ with respect to $H_k A$ and $D_k A$ contains terms like Figure 6.

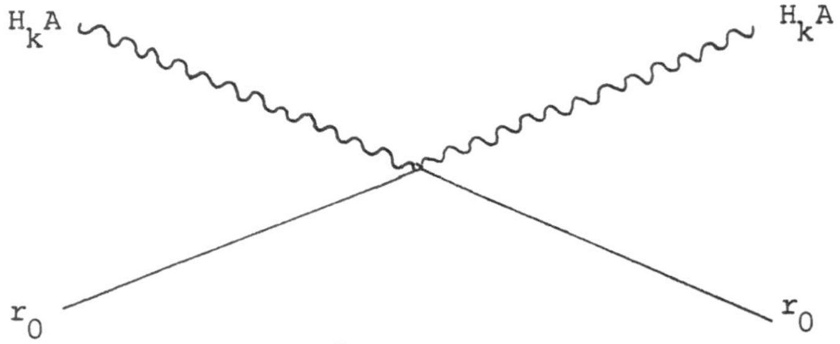

Figure 6.

We combine this form with the kinetic energy term $\langle f^{(k)}, \sigma_k f^{(k)} \rangle = \langle \partial A, \sigma_k \partial A \rangle_k$ to obtain a quadratic form with a strictly positive lower bound $\sim r_0^2 e^2 \langle A, A \rangle$. The coefficient $r_0^2 e^2 = m^2 e^2/8\lambda$ is the square of the semiclassical gauge field mass.

A cluster expansion can now be performed, since the mass terms lead to exponentially decaying propagators for r' and A, and since all interaction terms are small in the small field region. A convergent cluster expansion yields exponentially decaying correlations and the existence of the infinite volume limit.

REFERENCES

1. J. Glimm, A. Jaffe, Positivity of the φ_3^4 Hamiltonian, Fortschritte der Physik 21:327 (1973).
2. G. Benfatto, M. Cassandro, G. Gallavotti, F. Nicoló, E. Olivieri, E. Presutti, E. Scacciatelli, Ultraviolet stability in Euclidean scalar field theories, Commun. Math. Phys. 71:95 (1980).

3. J. Glimm, A. Jaffe, "Quantum Physics: A Functional Integral Point of View," Springer-Verlag, New York-Heidelberg-Berlin (1981).
4. K. Osterwalder, R. Schrader, Axioms for Euclidean Green's functions, I and II, Commun. Math. Phys. 31:83 (1973) and 42:281 (1975).
5. D. Brydges, J. Fröhlich, E. Seiler, On the construction of quantized gauge fields. I. General results, Ann. Phys. 121:227 (1979), II. Convergence of the lattice approximation, Commun. Math. Phys. 71:159 (1980), III. The two-dimensional abelian Higgs model without cutoffs, Commun. Math. Phys. 79:353 (1981).
6. D. Weingarten, J. Challifour, Continuum limit of QED_2 on a lattice, Ann. Phys. 123:61 (1971). D. Weingarten, Continuum limit of QED_2 on a lattice, II, Ann. Phys. 126:154 (1980).
7. K. Ito, Construction of Euclidean $(QED)_2$ via lattice gauge theory. Boundary conditions and volume dependence, Commun. Math. Phys. 83:537 (1982).
8. E. Seiler, "Gauge Theories as a Problem of Constructive Quantum Field Theory and Statistical Mechanics," Lecture Notes in Physics, vol. 159, Springer-Verlag, Berlin-Heidelberg-New York (1982).
9. T. Balaban, $(\mathrm{Higgs})_{2,3}$ quantum fields in a finite volume. I. A lower bound, Commun. Math. Phys. 85:603 (1983). II. An upper bound, Commun. Math. Phys. 86:555 (1982). III. Renormalization, Commun. Math. Phys. 88:411 (1983). Regularity and decay of lattice Green's functions, 89:571 (1983).
10. T. Balaban, Propagators and renormalization transformations for lattice gauge theories, I., Harvard University preprint HUTMP-B139.
11. T. Balaban, Renormalization group approach to nonabelian gauge field theories, in: "Mathematical Problems in Theoretical Physics, Proceedings of the VII-th International Conference on Mathematical Physics, Boulder, 1983," Lecture Notes in Physics, Springer-Verlag, Berlin-Heidelberg-New York, to appear.
12. T. Balaban, D. Brydges, J. Imbrie, A. Jaffe, The mass gap the Higgs models on a unit lattice. Harvard University preprint.

THE SPECTRUM IN LATTICE GAUGE THEORIES

Bernd Berg

II. Institut für Theoretische Physik

Luruper Chaussee 149, D-2000 Hamburg 50

ABSTRACT

A critical status report on spectrum calculations for 4d SU(2) and SU(3) lattice gauge theories (without quarks) is given. Also some conceptual details are discussed.

1. INTRODUCTION *)

So far for QCD the most successful non-perturbative regularization scheme has been the lattice regularization. QCD on the lattice was first introduced by Wilson /2/. (In a rather different context lattice gauge theories were previously defined by Wegner /3/). Wilson also proposed the use of strong coupling (SC) expansions /2,4/ and Monte Carlo (MC) calculations /4/ for the investigation of lattice QCD. (Preliminary results on a MC renormalization group investigation of lattice gauge theory are presented in Ref. /5/).

A MC study of the phase structure of Z^N and U(1) lattice gauge theories was carried out in Ref. /6/ and extended to SU(2) in Ref. /7/. A major break-through was the MC calculation of the string tension by Creutz /8/, who recognized that the scaling region can be reached on rather small sized lattices (typical extent: 8^4).

An outstanding problem is a reliable calculation of the spectrum of non-abelian gauge theories. Three major lines of attack are:

*) These lectures are an updated and extended version of an earlier report /1/.

1.) Strong coupling (SC) expansions /9-16/.

2.) MC calculations with gauge group SU(2) /17-36/ of SU(3) /37-48/.

3.) Lüscher's finite volume approach /49/.

I will give a review of SC and MC results. For the finite volume approach see Münster and Lüscher /43/. Related topics are discussed by Lüscher in these proceedings.

So far MC results for the string tension, the deconfinement temperature and the spectrum are obtained at rather moderate values of the coupling constant β. With our present methods we have very little control about the possible systematic errors involved in connecting these results to the continuum limit $\beta \to \infty$. Large discrepancies may even emerge in rather favourable situations, like the 0^{++} state, where a scaling window *) exists. See Ref. /50/ for an illustration in the 2d O(3) non-linear σ-model.

With nowadays MC methods we only obtain reliable results at β-values in a rather small range $\beta_1 \le \beta \le \beta_2$. This range extends the SC region and indicates a beginning (asymptotic) scaling region. As far as the MC statistics allows, we have at single β-values control over the statistical errors. (An insufficient MC statistics leads to unreliable, fluctuating error bars.) The statistical error bars apply of course to the measured quantitites (for instance correlations at distance $t = 1, 2, \ldots$). The physical quantitites (like glueball masses require extrapolations (for instance $t \to \infty$)). The systematic errors of these extrapolations are only to some extent under control (rather good for the 0^{++} state, bad for other states). Due to the numerical inaccuracy of our results at each single β-value and due to the small β-range over which scaling can be followed, no reliable estimate of the slope of a scaling curve (for instance for the 0^{++} glueball) is possible. The slope determines, however, the numerical accuracy of any continuum extrapolation. Due to "scaling in general" mass ratios may behave better. But as an intriguing fact we get only nice results for the 0^{++} state, and all other glueball states are high in units of $m(0^{++})$. They do therefore not allow very reliable MC results at the considered β-values.

There are two possible viewpoints under which such a review can be written:

a) Phenomenology.

b) Quantum field theory.

I like to emphasize the second point of view. Our final aim

*) For definitions of basic terminology see the next section.

is to calculate the spectrum of continuum non-abelian gauge theories and finally of continuum QCD. This is equivalent to a calculation at very large β on a very large lattice such that the statistical and systematical errors are under control. Only such a calculation would enable us to falsify QCD, and to make predictions, which are predictions of QCD. I think nobody with scientific reputation will claim that present lattice MC calculations do allow this. For understanding the approach of lattice gauge theory to its continuum limit it is, however, natural to begin with the SC limit, where we have everything under control. Then exploratory MC studies give us important informations about the continuation of the SC region to larger β, and consistency checks for the convergence of the SC expansion. In particular the evidence for early asymptotic scaling of the string tension /8/ stimulated very much the interest in these MC calculations. But the connection to very large β and a precise error analysis are still to come.

The phenomenologist is mainly interested in comparing final results with experiments. He is willing to consider results from QCD motivated models. As far as (asymptotic) scaling windows exist, one may make a fit of present MC data to the asymptotic behaviour. In this way one will obtain a "continuum" result with very subjective error bars. In an ugly notation one could speak about "lattice model" results. In that sense one has to understand the $m(0^{++})$ results as reviewed in these lectures and also results about the string tension and the deconfinement temperature as given in the literature. There are heuristic arguments (for instance rather universal behaviour of different actions) for conjecturing at least the order of magnitude to remain correct in the real continuum limit. Therefore one could eventually get from present MC results predictions in the sense of a "good" phenomenology. But even in the sense of phenomenology the relevance of numbers from lattice MC calculations becomes more and more obscure, when there is no (clear) evidence for (asymptotic) scaling. For instance indications of scaling /45/ for the 2^{++} state are so weak, that its acceptance depends entirely on subjective standards of judgement and rigour. For this state there is at least hope for possible improvements by using source methods /27,47,51/ or restoration of Lorentz invariance /36,52/. Other glueball states are more or less a desaster. From this point of view I have severe criticism concerning the representation of MC results (in particular so called "predictions") in Ref. /41,45/.

In these lectures I make an attempt to collect critically all existing MC results on the mass spectrum of lattice gauge theories. "Critically" does of course not mean that I can guarantee the correttness of all reported results. Unfortunately most of the papers in the literature do not present enough "raw" data to allows tracing back possible mistakes. Critically means, I have made attempts to elaborate on inconsistencies and to present all results from an unified point of view.

These lectures are organized as follows:

Lattice gauge theories and basic concepts are defined in section 2. The concept of universality is discussed and several actions are introduced.

In section 3 the MC variational (MCV) method, as proposed by Wilson /53/ and pioneered in Ref. /21,22,23/, is introduced. Most of the MC spectrum results rely on this method.

Other MC methods are sketched in section 4, where a chronological summary of the first MC results is also given.

SU(2) results are presented in section 5 and 6. Mass gap $m(0^{++})$ calculations are given in section 5, and each action is treated in a separate subsection. Practical aspects of the MCV method are discussed in section 5.1. Excited SU(2) glueball states are treated in section 6 and subsections.

SU(3) results are presented in section 7 (mass gap $m(0^{++})$) and section 8 (SU*3) excited states). See also appendix A.

Results for momentum $\vec{p} \neq 0$ states are reported in section 9.

Conclusions are drawn in section 10. For completeness I like to mention other attempts /54/ of calculating the mass-spectrum on non-abelian guage theories. Appendix B contains very recent results.

2. LATTICE GUAGE THEORY AND SCALING

The reader is assumed to be familiar with the Euclidean formulation on non-abelian guage theories on the lattice. The <u>standard action (SA)</u> was introduced by Wilson /1/. Vacuum expectation values are calculated with respect to the partition function

$$Z \int \prod_{p} dU(b) \exp \{- S^{SA}\}$$
$$S^{SA} = - \frac{\beta}{N} \sum_{p} \text{Re Tr } (1-U(\dot{p})). \tag{2.1}$$

The gauge group is assumed to be SU(N), the sum goes over all unoriented plaquettes p, $U(\dot{p})$ is the ordered product of the four link matrices surrounding the plaquette, dU(b) is the Hurwitz measure and the product goes over all links b.

With $U(b) = \exp(-igaA_\mu(x_b))$, $\beta = 2N/g^2$ and x_b the position of link b, we obtain in the classical limit $a \to 0$ the well-known Yang-Mills action. On an infinite lattice the continuum quantum field theory is conjectured to be obtained for $a(\beta) \to 0$. If there is no phase transition at a finite β_c, this happens for $\beta \to \infty$.

The classical limit does not uniquely fix the lattice action. This leads to the concept of universality: All lattice regularizations with the same classical limit are conjectured to lead to identical quantum field theories. MC results can be checked with respect to universality by using different lattice actions. Actions for which MC results are reported in these lectures are listed in the following.

Manton's /55/ action is for SU(2) given by

$$S^M = \frac{\beta}{2} \Sigma\, \Theta_p^2 \,, \tag{2.2}$$

where θ_p is the plaquette angle, related to the plaquette variable U_p through

$$U_p = \cos\theta_p + i\vec{\sigma}\hat{n} \sin\theta_p .$$

With Manton's action the transition from the SC to the weak coupling region was found to be smoother than with the SA /65/.

Mixed actions

$$S^{FS} = - \sum_{\dot{p}} \left[- \frac{\beta_F}{N} \mathrm{Tr}_F \,(U(\dot{p})) - \frac{\beta_N}{N^2-1} \mathrm{Tr}_A \,(U(\dot{p})) \right] \tag{2.3}$$

were investigated by a number of authors /56-58/. $\mathrm{Tr}_F = \mathrm{Tr}$ stands for the trace in the fundamental representation and Tr_A stands for the trace in the adjoint representation. In particular a first order phase transition line was found in the (β_F, β_A) coupling constants plane /56/.

Six-link actions are defined by

$$S = - \frac{\beta}{N} \Big\{ c_o \sum_{\Box} \mathrm{Re}\, \mathrm{Tr}\, \Box + c_1 \sum_{\Box\Box} \mathrm{Re}\, \mathrm{Tr}\, \Box\Box + c_2 \sum \mathrm{Re}\, \mathrm{Tr}\, \Box\!\!\angle + c_3 \sum \mathrm{Re}\, \mathrm{Tr}\, \Box\!\!\angle \Big\}$$

Wilson's /5/ block-spin improved action $S = S^W$ is defined by

$$c_o = 4.376,\ c_1 = -\,0.252,\ c_2 = 0,\ c_3 = -\,0.17 \tag{2.4}$$

Finally the choice

$$c_o = {}^5/_3, \quad c_1 = -{}^1/_{12}, \quad c_2 = c_3 = 0 \tag{2.5}$$

defines Symanzik's /59/ tree improved action (TIA) $S = S^{TI}$, which was for gauge theories calculated in Refs. /60,61/. MC simulations with Symanzik improved action are reviewed in Ref. /62/.

With $\beta = 2N/g^2$ the standard definition of the lattice mass scale is

$$\Lambda_L^{SA} = a^{-1} \, (\beta_o g^2)^{-\frac{\beta_1}{2\beta_2}} \exp\left(-\frac{1}{2\beta_o g^2}\right) \tag{2.6}$$

Here β_o and β_1 are the first coefficients of the perturbative expansion of the β-function /63/

$$\beta_o = \frac{11}{3}\left(\frac{N}{16\Pi^2}\right) \text{and } \beta_1 = \frac{34}{3}\left(\frac{N}{16\Pi^2}\right)^2 . \tag{2.7}$$

The Λ parameters of conventional perturbation theory in the continuum are related to Λ_L^{SA} by means of 1-loop calculations. For instance /64/

$$\Lambda^{MOM}/\Lambda_L^{SA} = 112.5 \exp\left(-\frac{3\Pi^2}{11\,N^2}\right) = \begin{cases} 57.4 & (N = 2) \\ 83.4 & (N = 3) \end{cases} \tag{2.8}$$

Also the relations between the Λ_L parameters for our different lattice actions are known. For Manton's action one obtains /57,65/

$$\Lambda_L^M/\Lambda_L^{SA} = \exp\left[\frac{3\pi^2}{11}\left(\frac{2}{3} - \frac{1}{N^2}\right)\right] = \begin{cases} 3.07 & (N = 2) \\ 4.46 & (N = 2) \end{cases} \tag{2.9}$$

For the gauge group SU(2) the mixed action (2.3) scale is given by /57,58/

$$\Lambda_L^{FA} = a^{-1}\left[\frac{6\pi^2}{11}(\beta_F + 2\beta_A\right]^{\frac{51}{121}} \exp\left[\frac{-3\pi^2}{11}(\beta_F + 2\beta_A) + \frac{15\pi^2}{22}\frac{\beta_A}{\beta_F + 2\beta_A}\right] . \tag{2.10}$$

Various authors have calculated Λ-scales for actions including up to six-link loops. For Wilson's action (2.4) the result

$$\Lambda_L^W/\Lambda_L^{SA} = \begin{cases} 35.3 & (N = 2) \\ 67.8 & (N = 3) \end{cases} \tag{2.11}$$

is reported /66/, and for the TIA (2.5) one obtains /60,66,67/

$$\Lambda_L^{TI}/\Lambda_L^{SA} = \begin{cases} 4.13 & (N = 2) \\ 5.29 & (N = 3) \end{cases} \tag{2.12}$$

As in Ref. /50/ I will distinguish asymptotic scaling and scaling in general. Any physical mass*) calculated on the lattice obeys

$$\{ - a \frac{\partial}{\partial a} + \bar{\beta}(g) \frac{\partial}{\partial g} \} m(g,a) = O(\frac{a^2}{\xi^2} \ell n \frac{a}{\xi}) \tag{2.13}$$

(ξ = correlation length, see next section). The first two terms of the Taylor expansion of the $\bar{\beta}$-function are regularization scheme independent and given by equation (2.7). This implies in the continuum limit that any physical mass becomes proportional to Λ_L $(1 + O(g^2))$. If the order g^2 corrections are negligible within our numerical accuracy, we call the proportionality

$$m = \text{const.}\ \Lambda_L \tag{2.14}$$

asymptotic scaling. Scaling in general means, equation (2.13) holds with some (general) $\bar{\beta}$-function and the r.h.s. ≈ 0 within our numerical accuracy. For mass ratios scaling in general implies

$$\frac{m_1}{m_2} = \text{const.} + O(\frac{a^2}{\xi^2} \ell n \frac{a}{\xi}) \cdot O(g^2) \tag{2.15}$$

Symanzik /59-61/ actions systematically improve the $a^2/\xi^4 \ln a/\xi$ corrections. For the TIA (2.5) $a^4/\xi^4 \ln a/\xi$ holds up to the tree approximation in the coupling constant.

For any physical mass m and reasonable large finite $\beta \geq \beta_m^1$ (asymptotic) scaling will be approximately true, for instance within requirements on the magnitude of allowed numerical deviations. The region $\beta \geq \beta_m^1$ is called scaling region. For different physical masses , (e.g. glueballs) the values of β_m^1 are supposed to be different. It is, however, very likely that the β_m^1 corresponding to the mass gap will be the smallest. The aim of a MC calculation of m on a finite sized lattice is to get reliable results within a finite range $\beta_m^2 \geq \beta \geq \beta_m^1$. If a MC calculation exhibits a range $\beta_m^2 \geq \beta \geq \beta_m^1$, where we find for the considered mass m (asymptotic) scaling, we call the range $\beta_m^2 \geq \beta \geq \beta_m^1$ (asymptotic) scaling window. The actual value of β_m^2 is determined by finite size restrictions, involving the finite size of the lattice and (in the practice of glueball calculations even more important) limitations in the approximation of the considered physical quantity.

---

*) More generally similar arguments hold for n-point Green-functions.

Symanzik's improvement is best tested for mass ratios. Results on glueball mass ratios are reviewed here (and in Ref. /62/). They are not stable, because only the 0^{++} state exhibits asymptotic scaling. Hence it is natural to consider the ratio $\sqrt{K}/m(0^{++})$, K string tension. Recent string tension investigations /68,69,70/ have, however, a strongly decreasing tendency for $\sqrt{K}$ (in units Λ_L). Abstracted numbers become rather meaningless, as asymptotic scaling seems also to be violated /70/. Under these unstable circumstances it seems not useful to me to review estimates of $\sqrt{K}/m(0^{++})$. Of course there is some danger, that $m(0^{++})$ estimates may also become unstable in future. But even then it is certainly useful to have a collection of presently achieved results, which set the standards for possible improvements.

The Euclidean time evolution operator is the transfer matrix T, which connects spacelike planes of the lattice at distance t = a. A strictly positive transfer matrix has been constructed for lattice gauge theories including quarks /71/. In the Euclidean approach the lattice Hamiltonian H is defined by

$$e^{-aH} = T \quad \text{(a lattice spacing).} \tag{3.1}$$

The vacuum $|0\rangle$ is the eigenstate of H with lowest eigenvalue which we normalize to zero implying $H \geq 0$. On a finite lattice the vacuum is always non-degenerate, on an infinite lattice it may be degenerate. The existence of a mass gap m_g above the vacuum means that

$$m_g = -\ln \{\max_{\substack{|\psi\rangle \\ \langle 0|\psi\rangle=0}} \left(\frac{\langle\psi|T|\psi\rangle}{\langle\psi|\psi\rangle} \right) \} > 0 \, . \tag{3.2}$$

This is easily seen by expanding

$$|\psi\rangle = \sum_{n=1}^{\infty} c_n \, |\psi_n\rangle$$

in terms of eigenfunctions $H|\psi_i\rangle = E_i|\psi_i\rangle$ of the Hamiltonian. A complete system of states is spanned by taking products of different spacelike Wilson loops in all possible SU(N) representations

$$O_{i\nu}(t) = \sum_{\vec{x}} O_{i\nu}(\vec{x},t) = \sum_{\vec{x}} \{\chi_\nu(\prod_{c_i} U) - \langle\chi_\nu(\prod_{c_i} U)\rangle\} \tag{3.3}$$

and applying them to the vacuum state $|0\rangle$. The summation over $\vec{x}$ in equation (3.3) implies $\vec{p} = 0$ for the momentum.

In an MCV calculation one truncates the set of operators.

The relevant expectation values

$$\langle O_{i,\nu}(0)T^n O_{j,\mu}(0)\rangle = \langle O_{i,\nu}(0)O_{j,\mu}(na)\rangle \tag{3.4}$$

are calculated by means of MC measurements. Wilson loops as far as considered in the reported MCV calculations are numerated in Figure 3.1. Within a truncated ansatz equation (3.2) yields an upper bound

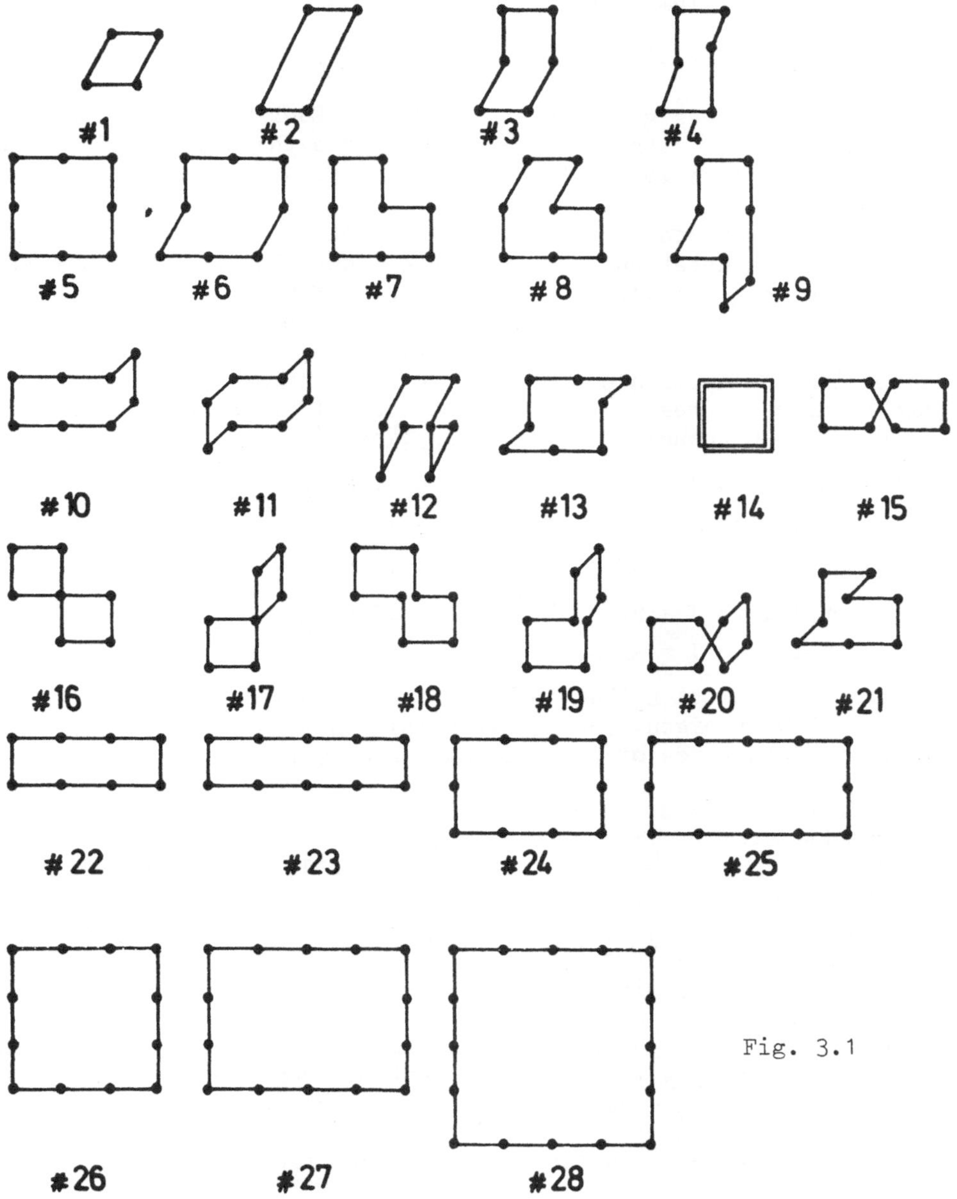

Fig. 3.1

for the mass gap. In a practical MCV calculation this remains only true if the statistical error is smaller than the systematic deviation of the approximation from the exact result.

On the lattice we have an exact cubic symmetry. This leads to important selection rules concerning the possible spin J in the continuum. Bethe /72/ and others constructed the irreducible representations of the 48 element symmetry group O_h of the 3d cube. There are five irreducible representations of the cubic group O:

$$R = A_1(1d),\ A_2(1d),\ E(2d),\ T_1(3d) \text{ and } T_2(3d) \tag{3.5}$$

with dimensions as given in the brackets. Each irreducible representation can occur with parity $P = \pm 1$. A_1 is the scalar and T_1 the vector representation.

Let us now consider states $|\psi>_R$ which transform under the irreducible representation R of the cubic group. In the continuum limit the state $|\psi>_R$ can only couple to spin J if $D_J{}^o \supset R$. Here $D_J{}^o$ is the subduced representation of the irreducible spin J representation D_J in the continuum. Subduced representations are obtained by trivially embedding the cubic group O into the rotation group. Up to J = 12 the subduced representations can be read off from the tables of Altmann and Cracknell /73,43/*). Implications are:

A_1 couples to J = 0, 4, 6, ... → 0 if $E_{J=0} < E_{J=4}$, ...
T_1 couples to J = 1, 3, 4, ... → 1 if $E_{J=1} < E_{J=3}$, ...
E couples to J = 2, 4, 5, ... → 2 if $E_{J=2} < E_{J=4}$, ...
T_2 couples to J = 2, 3, 4, ... → 2 if $E_{J=2} < E_{J=3}$, ...
A_2 couples to J = 3, 6, 7, ... → 3 if $E_{J=3} < E_{J=6}$, ...

Under the assumption that the lowest allowed spin in the sector $|\psi>R$ has the lowest eigenvalue E_J, we obtain in a variational calculation finally eigenstates of spin as indicated by the arrows.

In Ref. /43/ Billoire and I have constructed all irreducible representations of the cubic group on Wilson loops as depicted in Figure 3.1. With the above identifications 16 different continuum spins are obtained: (P = parity, C = C-parity)

$$J^{PC},\ (= 0,\ 1,\ 2,\ 3),\ P = \pm 1,\ C = \pm 1. \tag{3.6}$$

Most glueballs may mix with flavourless $q\bar{q}$ mesons of the quark model. Exceptions are 0^{+-}, 0^{--}, 1^{-+}, 2^{+-} and 3^{-+}, which according to equation (3.6) can be constructed from our Wilson loops. These states are called oddballs. They were for instance considered within potential models of gluon constituents /75/. Our classification does,

*) For non-integer spin subduced representations are calculated in Ref. /74/.

however, not give any dynamical information so far. In particular we do not even know whether all possible states are actually realized in the continuum limit.

In an MCV calculation we will now, by solving an eigenvalue problem /43/, construct an operator

$$O(t) = O^{R^{PC}}(t) = \sum_{i,\nu} c_{i,\nu} \, O^{R^{PC}}_{i,\nu}(t) \tag{3.7}$$

$$(R = A_1,\ A_2,\ E,\ T_1,\ T_2;\ P = \pm 1,\ C = \pm 1)$$

such that the signal $\langle O(t)O(0)\rangle / \langle O(0)O(0)\rangle$ is maximized at fixed distance t, typically t = 1 because of problems with noise (see section 5.1.1).

Finite distance glueball masses are then defined by

$$m(t_1,t_2) = \frac{-1}{t_1-t_2} \ln \frac{\langle O(t_1)O(0)\rangle}{\langle O(t_2)O(0)\rangle} \ , \ (t_1 > t_2). \tag{3.8}$$

Let us introduce the notation

$$m(1) = m(1,0),\ m(2) = m(2,0) \text{ and } \hat{m}(2) = m(2,1) \tag{3.9}$$

for the cases of practical importance. All these definitions are theoretical upper bounds for the real glueball mass. $\hat{m}(2)$ gives of course a better bound than m(2) but has two times larger error bars.

For practical reasons (disk space and computational simplicity) it is often convenient to consider on-diagonal correlations only. Finite distance glueball masse $m_i(t_1,t_2)$, $m_i(1)$, ... from a single operator O_i are defined by replacing $O \rightarrow O_i$ in equation (3.8). Minimization (maximization of the signal) is than simple to take $m = \min\{m_i\}$. In a practical MCV calculation it is suitable to define the best operator to be the one, which gives the lowest value $m_i(1)$ at distance t = 1. There is no major disadvantage as compared with real minimization, if at distance t = 2 statistical noise already overwhelms the effect of minimalization.

Our discussion so far applies to the SA, which has a positive definite trasnfer matrix. In a formal way the MCV method is also applied to the other introduced actions. In case of Manton's action there is no severe problem, as has been discussed in Ref. /30/. This is different for six-link actions, see Ref. /36/. In that case one has to rely on the $t \rightarrow \infty$ limit and the upper bound property is lost for t = 1, 2, ... As all actions are in the same universality class, one may argue the problem to become unimportant in the scaling limit $\beta \to \infty$.

As in Ref. /1/ I like to distinguish the correlation length and the relevant range of interaction. The correlation length ξ is defined to be the inverse mass gap $\xi = 1/m_g$. The thus defined correlation length is often set equal to a length ξ_r, which I like to call "relevant range of interaction", this is /76/ the length to which we can reduce the size of the (lattice) system without qualitatively changing its physical properties. Contrary to the definition of ξ, the definition of ξ_r is quantitatively not precise. For a 1d (spin) system with nearest neighbour interactions ξ_r is certainly of the order of magnitude of ξ, although one can hardly exclude a relation $\xi_r \approx 1\text{-}2\ \xi$. In dimensions $d \geq 1$ one should not identify ξ_r with ξ, but expect the length ξ_r to be of the order of magnitude of the maximum of the function $f(r) = r\ r^{d-1}\ e^{-r/\xi}$. For d=1 we obtain $\xi_r = \xi$, whereas in dimensions $d > 1$ the factor r^{d-1} takes account for the increase of the number of sites (or links) with the distance r from a given point x, and one obtains the order of magnitude

$$\xi_r = d\,\xi\,. \tag{3.10}$$

This equation reconciles some apparently paradoxical features of scaling with a correlation length $\xi < a$, as required by mass gap estimates in 4d lattice gauge theories. The lattice spacing a is not a natural scale, which has to be small compared with the correlation length ξ, in order to achieve scaling. We should, however, have $\xi_r >> a$ for the onset of scaling. As is clear from renormalization group arguments (e.g. /5/), also the detailed form of the action enters decisively in the relation between ξ and ξ_r. Equation (3.10) is a crude estimate for nearest neighbour interactions. For the SA it matches well the order of magnitude between correlation length and the critical length (in tide direction) which causes the finite temperature phase transition in SU(2) /77/ and SU(3) /78/ lattice gauge theories.

In case of Symanzik improved actions the relevant interaction range ξ_r is a natural parameter /34,36/, if one likes to have the first numerical coefficient close to ≈ 1 in the corrections to equation (2.15). Improvement then means: $a^2/\xi^2 \ln a/\xi \to a^4/\xi^4 \ln a/\xi$ etc.

4. VARIOUS MONTE CARLO METHODS AND MASS GAP RESULTS

The first SU(2) mass gap estimates are summarized in Table 4.1 in a chronological order (given years refer to preprints).

Table 4.1

First SU(2) mass gap estimates		m_g/Λ_L^{SA}
1.)	Berg /17/, MC, (p-p), 1980:	290 ± 95
	Bhanot and Rebbi /18/, MC, (p-p), 1980:	270 ± 90
2.)	Münster /10/, SC, u^8, 1981:	140 ± 60
3.)	Engels, Karsch, Satz, Montvay /19/, MC (FT), 1981:	155 ± 54
	Brower, Creutz, Nauenberg /20/, MC (FS), 1981:	120 ± 30
4.)	Berg, Billoire, Rebbi /21/, MCV, ASW, 1981:	200 ± 50
	Falcioni et al., /22/, MCV, Manton's action $\beta = 1.55$, 1981:	150 ± 20 [a)]
	Ishikawa, Schierholz, Teper /23/, MCV, $\beta = 2.3$, 1981:	205 ± 20
5.)	Engels, Karsch, Satz, Montvay /24/, MC (FT), 1982:	200 ± 40
6.)	Mütter, Schilling /25/, MC (BO), ASW, 1982:	200 ± 25
7.)	Berg, Billoire, Rebbi /21, addendum/, MCV, ASW, 1982:	170 ± 30
8.)	K. Seo /12/, SC, u^8, 1982:	$\approx$ 130
9.)	Ishikawa, Schierholz, Teper /26/, MCV, 1982:	195^{+19}_{-38}
⋮		

a) Λ_L^M scale converted by means of equation (2.9)

1.) The early mass gap estimates /17,18/ rely on matching a signal for the onset of scaling coming from plaquette-plaquette (p-p) correlations to the first order SC expansion.

2.) + 8.) Münster /10/ calculated $m(0^+)$ in SC expansion up to order u^8. The expansion variable is the SU(2) character $u = I_2(\beta)/I_1(\beta)$, ($I_n$ modified Bessel function; see Eq. 8.445 of Ref. /79/). Numerically small errors were corrected in the errata and by Seo /12/. The SC expansion is plagued by singularities in the complex coupling constant plane /11/. The estimates 1.), 8.) rely on taking, according to the asymptotic scaling formula (2.6), the tangent to the SC series. This has to be done in a region where the SC expansion breaks down.

3.) + 5.) These investigations rely on MC studies of finite temperature (FT) or finite size (FS) effects, and involve an ideal gas ansatz for low-lying glueball states. Several free parameters may enter, most prominently the multiplicity r of low-lying glueballs.

There are rather large discrepancies between Ref. /20/ and Ref. /24/. Particularly because the authors of Ref. /20/ would get with multiplicity $r = 5$ the result $m_g = 90\ \Lambda_L^{SA}$, whereas the authors of Ref. /24/ obtain the value of the Table by assuming a similar low multiplicity $r = 6$.

All these estimates are not self-consistent in the sense that the asymptotic scaling behaviour is followed over a finite range in the coupling β. The estimates must therefore - and by various intrinsic problems of the used methods - considered to be unreliable. Better are results relying on asymptotic scaling windows (ASW).

4.), 7.) + 8.) These results are MCV $m(0^+)$ estimates and will be discussed in the next sections. In particular an asymptotic scaling window was first found in Ref. /21/.

6.) Mütter and Schilling /25/ study the dependence of the plaquette action on free and fixed boundary conditions (BO). This is a source method which allows signals up to distance $t = 4$. With a similar statistics MCV calculations give signals up to distance $t = 2(3)$. This means source methods enhance the signal considerably, and the BO method leads to a nice ASW. A major disadvantage of this method is that it does not allow projections on momentum $\vec{p} = 0$ and angular momentum $J = 0^+$ states.

Other source methods allow to define momentum and angular momentum, and may eventually become very useful for studying excited glueball states. I like to emphasize the use of the Langevin equation, as pioneered for the SU(2) $m(0^+)$ mass gap in Ref. /27/. The reported improvement factor 10^4 as compared with the MCV method is, however, unrealistic. Finally one may insert a source with well-defined quantum numbers in an ordinary MC calculation /42,51/. For the 0^{++} state Ref. /51/ reports a drastically enhanced signal as compared with a standard MCV calculation.

The results with ASWs /21,24/ are well consistent with a recent high statistics finite size study /36/, yielding

$$m(0^+) = (190 \pm 30)\ \Lambda_L^{SA} \tag{4.1}$$

Finally first SU(3) mass gap estimates are summarized in Table 4.2 in a chronological order.

Table 4.2

First SU(3) mass gap estimates		m_g/Λ_L^{SA}
1.)	Münster /10/, SC, u^8, 1981:	$\approx$ 670
2.)	Hamber and Parisi /37/, MC (FS), 1981:	720 ± 100
3.)	Berg and Billoire /38/, MCV, ASW, 1982:	350 ± 50
4.)	Münster /10, erratum/, SC, u^8, 1982:	$\approx$ 310
	Seo /12/, SC, u^8, 1982:	$\approx$ 310
5.)	Berg and Billoire /39,1/, MCV, ASW, 1982:	280 ± 50
	Ishikawa, Schierholz, Teper /40/, MCV, 1982:	300 ± 17
6.)	Michael and Teasdale /42/, MC, 0^{++} source, ASW,	370 ± 30
7.)	Berg and Billoire /43,44/, MCV, ASW, 1982/3:	280 ± 40
8.)	Ishikawa, Sato, Schierholz, Teper /45/, MCV, ASW, 1983:	309 ± 17
.		
.		
.		

As will be discussed in section 7 results around $m(0^{++}) \approx 280\ \Lambda_L^{SA}$ are rather stable. One may express this in physical units by using an experimental value for Λ_L^{SA} or a physical value for the string tension. At present experimental values /80/ for the Λ-parameters are very inaccurate. Therefore I prefer to rely on the string tension. With $\sqrt{K}$ = 440 MeV the final $m(0^{++})$ estimate depends on the MC calculation of the string tension. Using Creutz /8/ ratios a conventional estimate /81/ is

$$\sqrt{K} = (167 \pm 30)\ \Lambda_L^{SA}, \text{ leading to } m(0^{++}) \approx 740 \text{ MeV}. \tag{4.2}$$

Unfortunately MC estimates of the string tension have a decreasing tendency /68-70/, and Ref. /69/*) would lead to the value

$$m(0^{++}) \approx 1300 \text{ MeV}. \tag{4.3}$$

*) Because of a bias in the used lattice configurations, this MC calculation should be repeated.

5. SU(2) MASS GAP m(0^+)

I will summarize the present status of asymptotic scaling and universality. Each action is treated in a separate subsection. Some practical aspects of the MCV method are discussed in subsection 5.1.

5.1 The standard action

To illustrate the MCV method, I first like to discuss the calculation by Billoire, Rebbi and myself /21/ in some details. In this investigation an asymptotic scaling window (ASW) for the SU(2) mass gap (0^+) was found.

Our MC calculation was carried out on an $4^3 16$ lattice (4^3 is the spacelike box) and SU(2) was approximated by the icosaeder subgroup /82/. Altogether 12 operators were used in the variational calculation, namely Wilson loops up to length L = 6 (i.e. #1 - #4 in Fig. 3.1) in the 1/2, 1 and 3/2 representation. All operators are taken in the A_1^+ representation of the cubic group and glueball mass definitions as introduced in section 3 are used.

In Figure 5.1.1 MC data are compared with results from the SC expansion /10/. Consistency between SC and MC is observed. Different

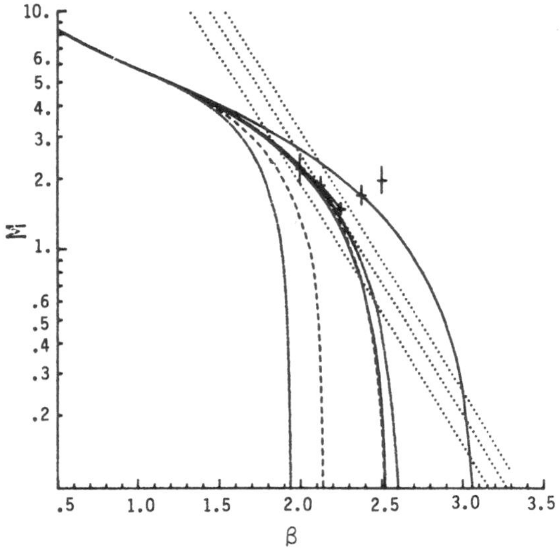

Fig. 5.1.1

SC expansion for the glueball mass. Expansion variables: β(-·-), u(---) and z (——). The lower curve always corresponds to the highest order. The dotted lines correspond to the glueball mass estimate (and error) of Ref. /21/. MC data are from $\hat{m}_1(2)$ on a 4^3x16 lattice

variables are used for plotting the SC expansion. This illustrates the relative freedom of extrapolating the SC series.

On a preciser scale Figure 5.1.2 gives results as obtained from the 1 plaquette operator (in the fundamental representation) alone. The three straight, solid lines represent the continuum estimate and clearly an ASW is seen. The broken line $4\ \ell n(^4/\beta)$ is the lowest order

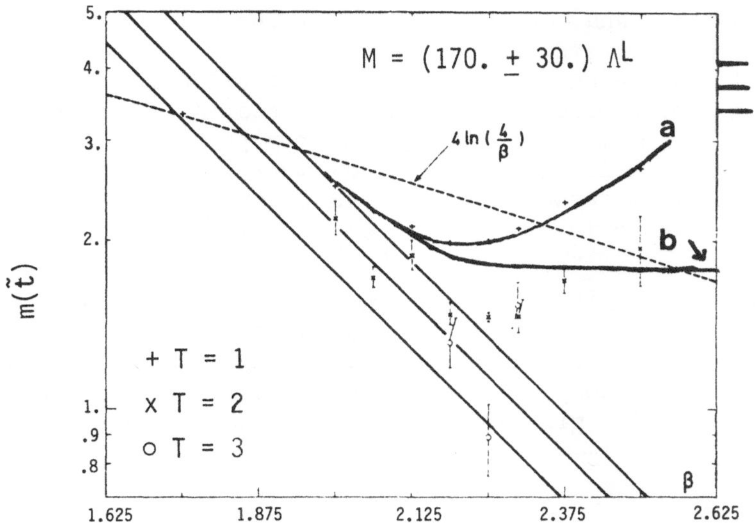

Fig. 5.1.2. Mass gap estimate and data for $\tilde{m}_1(t)(t=1, 2, 3)$.

SC expansion. At the r.h.s. of the Figure the spin wave (SW) behaviour $\beta \to \infty$ is in lowest order indicated for m(1), m(2) and $\hat{m}(2)$. This means MC data for m(1) (etc.) have to follow the shape of line a). This is in contrast to the spin wave behaviour /88/ of the string tension, for which line b) would be typical. Therefore the location of a scaling window for glueballs is in practice more difficult than for the string tension. Even on a large lattice reliable results can only be obtained if one is far enough away from the spin wave region of the used approximation.

Figure 5.1.3 gives the same results as Figure 5.1.2, but after

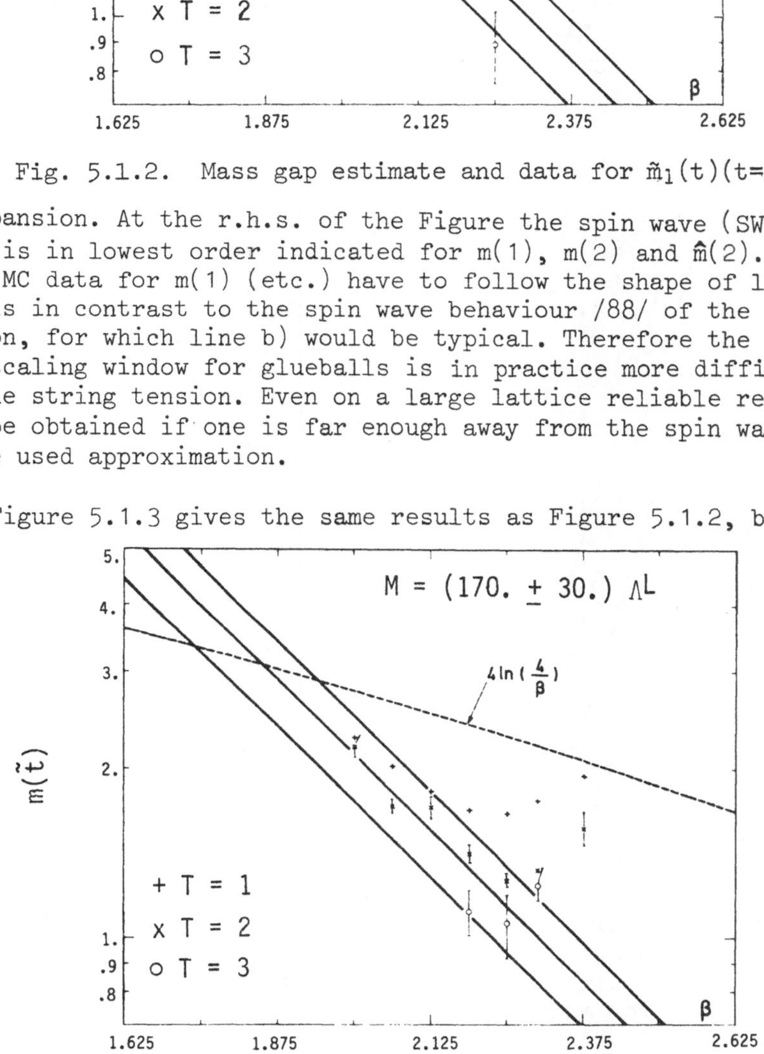

Fig. 5.1.3. Mass gap estimate from $\tilde{m}(t)(t=1, 2, 3)$ after minimization as explained in the text.

maximizing the signal at distance t = 1. The 4 Wilson loop operators up to length L = 6 are used in the fundamental representation. At distance t = 1 the mass gap results are considerably lowered. At that distance we now already get a signal for scaling. This is important, because these results are numerically very precise. At distance t = 2 statistical noice is already more important than minimization. This may indicate that these results are asymptotic in the sense $t \to \infty$. The final estimate is

$$m(0^+) = (170 \pm 30)\ \Lambda_L. \tag{5.1.1}$$

We did not include the higher representations in the final minimizatio because they contribute little to the final wave function, but add noise. Figure 5.1.4 gives the MC data for the 1-plaquette in the $^3/_2$-representation in comparison with the final estimate. Figure 5.1.5 demonstrates finite size effects. A spacelike 2^3 bose is definitely

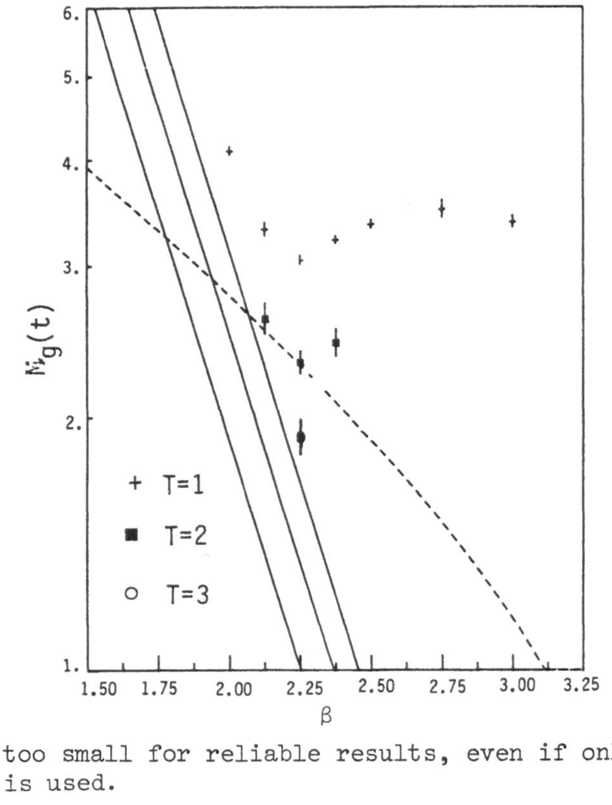

Fig. 5.1.4

too small for reliable results, even if only the 1-plaquette operator is used.

In our MC calculation we have typically carried out 30 000 MC sweeps at each β-value. (A sweep is defined by upgrading each link in the lattice once.) The whole calculation took approximately 40h CDC 7600 computer time. It is convenient to keep the spacelike lattice small, because (for momentum $\vec{p} = 0$ states) the needed computer time is proportional to L_S^3. I like to add some technical remarks concerning the MCV method:

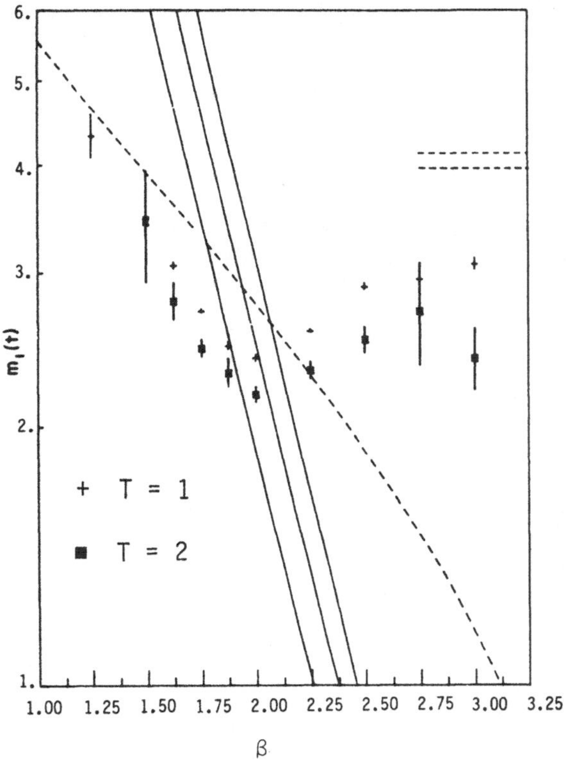

Fig. 5.1.5

1.) Bias: If we have many operators in ou. calculation, some operators may give a large signal due to a statistical fluctuation. An MCV calculation will overestimate these operators and can therefore give a too low result. Within a typical MC statistics minimization works well at distance t = 1 and is sometimes also possible at distance t = 2.

2.) Finite lattice size problems: Let the lattice size be $L_S^3 \cdot L_t$ ($L_t \geq L_S$). An MCV calculation may have finite size difficulties if:

a) Important operators W_{IJ} for the wave function have a size

$$\max\{I,J\} \geq \frac{1}{2} L_S .$$

b) $t > \frac{1}{2}L_t - 0.5$ (Example: t = 3 on a $4^3 \cdot 8$ lattice.)

c) Thermal loops around one of the <u>space</u> direction acquire a non-zero expectation value.

If a,b) or c) happen (as is most of the present exploratory studies) one has to judge whether a serious problem is implied or not. For instance in our calculation /21/ the W_{21} Wilson loop contributes to the wave function, but later studies show that this finite size problem is not important.

Consistent with our results are MCV data of Ishikawa et al. /23/ at $\beta = 2.3$.

Mütter and Schilling /25/ obtain with their BO-method also an ASW and estimate $m(0^+) = (200 \pm 25)\Lambda_L$ in agreement with equation (5.1.1).

Solving numerically Langevin equation, Falcioni et al. /27/ obtained plaquette-plaquette correlation functions up to distance $t = 4$ on an 8^4 lattice. Their estimate, see Figure 5.1.6, is higher

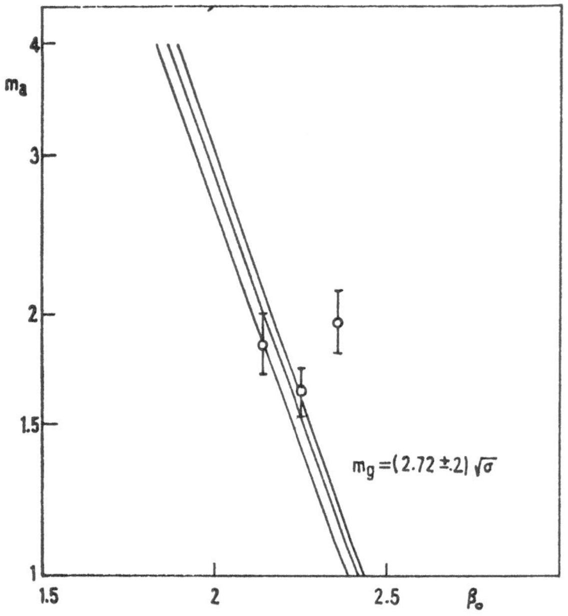

Fig. 5.1.6

but still consistent with MCV results. In view of the claimed drastic improvement Figure 5.1.6 looks rather meager. In my opinion an important test for using the Langevin equation would be a calculation of an excited glueball state. The natural candiate is 2^+.

Brooks et al. /31/ repeated the described MCV calculation on microprocessors. They find agreement.

A finite size study on an 8^4 lattice was attempted by Ishikawa et al. /26,28/. To improve the MC statistics they used "momentum smeared" wave functions. Subtracting the unknown momentum contribution amounts to introducing a new parameter and renders the final results unreliable. Recently a high statistics (40 000 sweeps/β-value) study on an 8^4 lattice was carried out in Ref. /36/. Figure 5.1.7 compares on a physical scale (i.e. in units of Λ_L^{SA}) equivalent

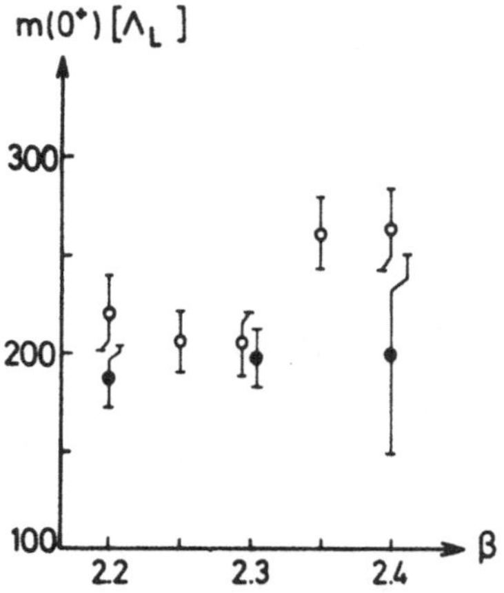

Fig. 5.1.7

results from Ref. /26,28/ (●), (after subtracting the unknown momentum contribution), and from Ref. /36/ (○), (for $\vec{p}$ = 0). Using similar operators as on the 4^3 16 lattice, the scaling region extends now up to β = 2.30 (previously on the 4^3 16 lattice up to β = 2.25) and the final m(0^+) estimate is slightly higher. For reasons see the list of finite lattice size problem, in particular 2.c).

In Ref. /36/ we have calculated on-diagonal correlations between all operators of Fig. III.1, and the statistics allows also t = 3 results. Final results from the best operators are given in Figure 5.1.8 (m(1)× , m(2)●, $\hat{m}$(2)○, m(3,2)□). The scaling region extends

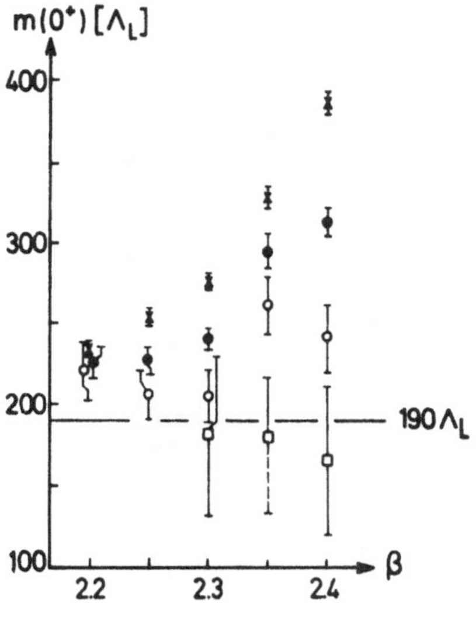

Fig. 5.1.8

now up to β = 2.4, and the ASW has the size

$m_a \approx 2.4$, $m_b \approx 0.9$, $^{m_a}/_{m_b} \approx 2.7$

(m_a, m_b are the values at begin and end of the ASW). The best final estimate is

$$m(0^+) = (190 \pm 30)\ \Lambda_L^{SA} \tag{5.1.2}$$

and well consistent with (5.1.1).

5.2 Manton's action

At β = 1.55 Falcioni et al. /22/ obtained on an 8^4 lattice the MCV result $m(0+) = (49 \pm 7)\ \Lambda_L^M$.

Using their BO-method Mütter and Schilling /29/ found an ASW as depicted in Figure 5.2.1.

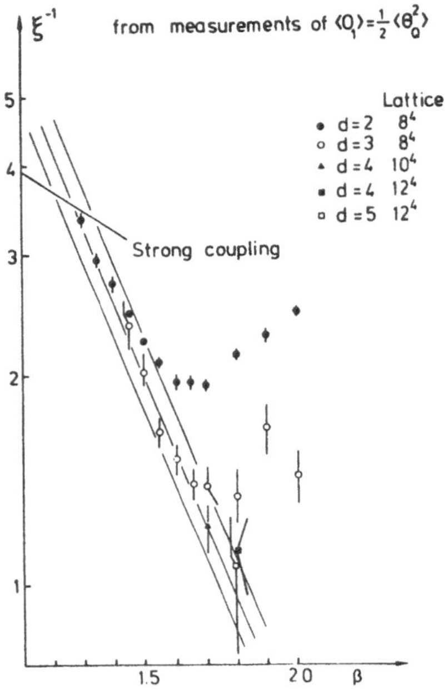

Fig. 5.2.1

Their final estimate is

$$m(0^+) = (46 \pm 6)\ \Lambda_L^M. \tag{5.2.1}$$

This was confirmed by an MCV calculation of Billoire, Koller and myself /30/. Details are similar to the MCV calculation /21/ described in section 5.1.1. Our result

$$m(0^+) = (49 \pm 9)\ \Lambda_L^M \tag{5.2.2}$$

is based on the ASW of Figure 5.2.2.

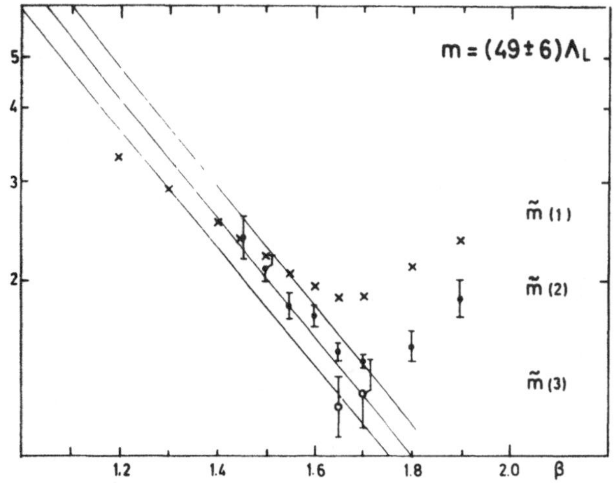

Fig. 5.2.2

This gives MC estimates for the $\Lambda_L^M/\Lambda_L^{SA}$ ratio: $\approx$ 4.0 /25,29/ and $\approx$ 3.5 /21,30/, i.e. within rather large errors consistent with 3.07 (2.9) and hence universality.

5.3 Mixed actions

Freimuth, Mütter and Schilling /32/ used the BO-method. They investigated the value β_A = 1.21 and find no ASW.

Otto and Randeria /33/ did an MCV calculation on the line $\beta_A = -0.24\ \beta_F$. Their motivation is a) to stay far away from the critical line /56/ in the mixed action plane and b) the remormalization flow as computed /83/ in the Migdal-Kadanoff approximation. The line $\beta_A = -0.24\ \beta_F$ is "Migdal-Kadanoff improved".

The results of Ref. /33/ are depicted in Figure 5.3.1.. No ASW is found. The authors claim, however, that scaling in the general sense if true, and that the results are closer to the continuum limit than with the SA. Using (for N = 2) large N resummation /84/ they map their data on the SA axis and find the results of Figure 5.3.2. The mapped mixed action data are plotted with SA data /31/. Reasonable consistency is seen. In particular all data with $\beta > 2.3$ are mixed action data, and consistent with asymptotic scaling.

Similarly the result of Ref. /32/ becomes consistent with SA results if large N resummation techniques are used.

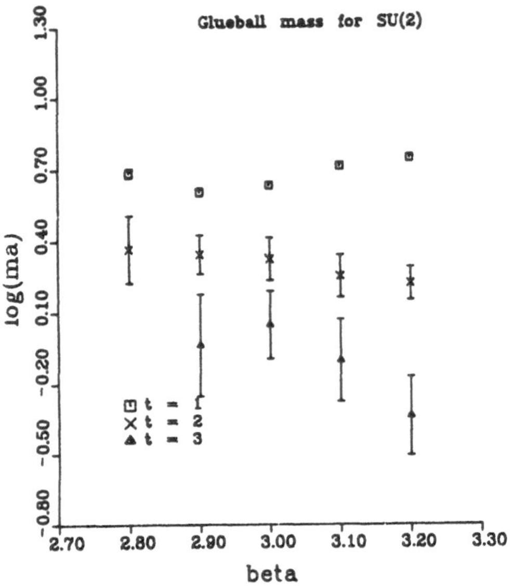

Fig. 5.3.1

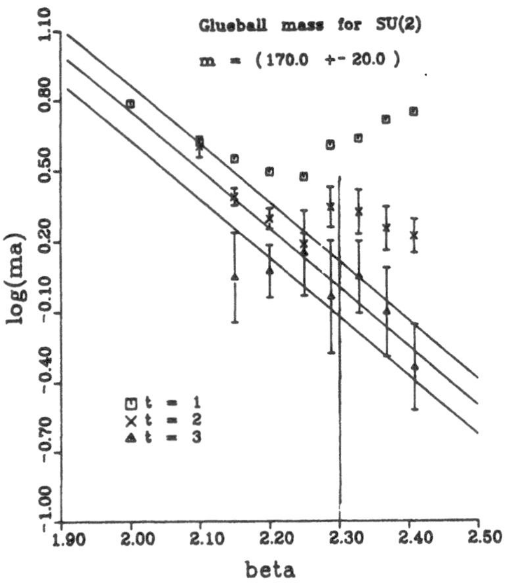

Fig. 5.3.2

5.4 Wilson's action

Fukugita et al. /35/ calculate plaquette-plaquette correlations ($\vec{p}$ = 0) on $4^3 8$ and $5^3 8$ lattices, and obtain the estimate

$$m(0^+) = (10.4 \pm 0.8)\ \Lambda_L^W \qquad (5.4.1)$$

Comparing this with their SA $m(0^+)$ estimate leads to $\Lambda_L^W/\Lambda_L^{SA} \approx 20$. This is a considerably lower value than the perturbative expectation 35.3 (2.11) /66/.

5.5 Symanzik TIA

A high statistics MCV calculation was carried out in Ref. /34,36/. We worked on an $5^3 8$ lattice and measured on-diagonal correlations between the first 21 Wilson loops of Figure 3.1. Our final results /36/ are depicted in Figure 5.5.1 (m(1)x , m(2)Φ, $\hat{m}$(2)Φ).

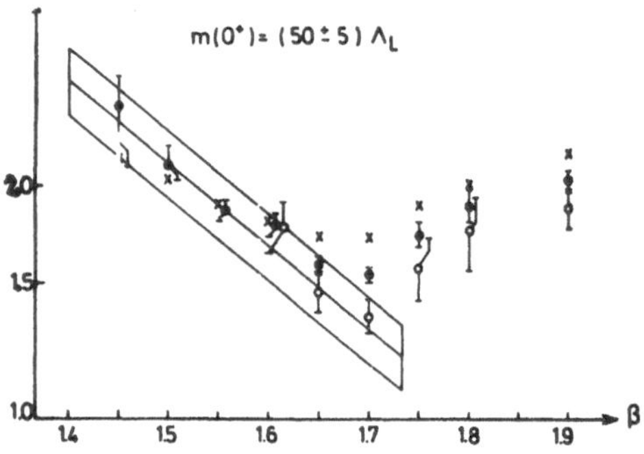

Fig. 5.5.1

This yields the MC estimate $\Lambda_L^{TI}/\Lambda_L^{SA} \approx 3.8$ in good agreement with the perturbative result (2.12). Recently Mütter and Schilling*) obtained a consistent result.

6. SU(2) EXCITED STATES

With the SA one finds restoration of rotation invariance between $\beta = 2.0$ and $\beta = 2.25$ /85/. One may therefore hope to calculate reliably higher spin glueball states. The results are, however, puzzling. All MCV calculations give large masses (in units of $m(0^+)$) at distance $t = 1$. Consequently statistical noise is a severe problem at distance $t = 2$.

6.1 The standard action

A first result is due to Ishikawa, Schierholz and Teper /23/. They did an MCV calculation at $\beta = 2.3$. Using 4^4, $4^3 8$ and 6^4 lattice they find

$$m(2^+) \approx 1.8\ m(0^+). \qquad (6.1.1)$$

(They also report $m(0^-) \approx 1.7\ m(0^+)$, but this number has theoretical problems. See appendix A.)

The same authors did also an MCV calculation on an asymptotic $5^3 \cdot 40$ lattice /28/. Their choice

$$\beta_s = 0.664 \text{ and } \beta_t = 8.5 \qquad (6.1.2)$$

(s = spacelike, t = timelike) is motivated by the Hamiltonian limit $a_t \ll a_s$. In case of the 2^+ state the signal for $m(t, t-1)$ can now be followed up to $t = 4(5)$, as depicted in Figure 6.1.1.

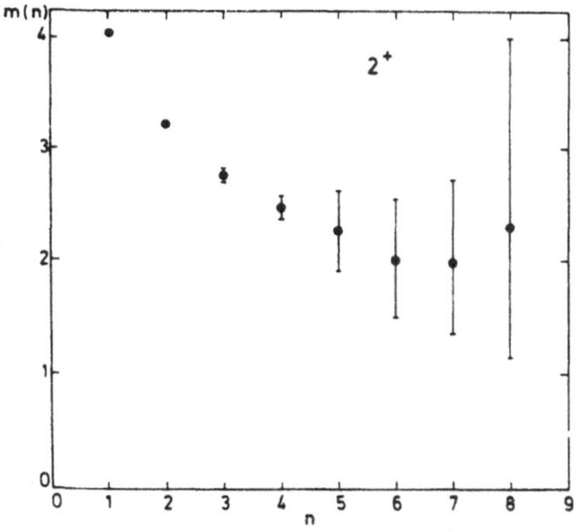

Fig. 6.1.1

*) Private communication. See also appendix B.

This is rather promising, although the a_t-spacing is of course much smaller than a in the Euclidean case. The issue could be finally judged by trying to establish a self-consistent ASW on this aymmetric lattice. So fat the final 2^+ estimate $m(2^+) \gtrsim 1.94\, m(0^+)$ is again high. (Further an $m(0^-)$ result is reported, see appendix A).

In Ref. /36/ Billoire, Meyer, Panagiotakopoulos and I calculated masses for 0^-, 2^+ and 1^+ states. Our lowest state is 2^+. Mass ratios $m(2^+)/m(0^+)$ are rather unstable and there is no clear signal for scaling. The order of magnitude is $m(2^+) \gtrsim 1.8\, m(0^+)$. Data for $m(2^+)$ (in lattice units a^{-1}) are given in Table 6.1.1. Within the large

Table 6.1.1

β	m(1)	m(2)	SC, $O(u^8)$
2.20	3.18 ± 0.04	$3.29^{+\infty}_{-0.45}$	3.17
2.25	3.04 ± 0.04	$2.72^{+0.23}_{-0.16}$	3.09
2.30	3.03 ± 0.04	$3.09^{+0.58}_{-0.26}$	3.01
2.35	2.89 ± 0.04	$2.54^{+0.32}_{-0.20}$	2.93
2.40	2.78 ± 0.03	2.38 ± 0.10	2.86

statistical noise there is no very significant decrease by going from distance t = 1 to t = 2. The m(1) results are remarkably consistent with the highest order SC expansion /10,12/, and I would argue the 2^+ SC expansion to be convergent up to larger values of β than the 0^+ SC expansion. I will come back to this point in case of the SU(3) gauge group.

0^- and 1^+ results are clearly higher than 2^+ results. Consequently correlations at distance t = 2 are only statistical noise. At distance t = 1 the order of magnitude is

$$m(0^-) \gtrsim 2.5\, m(0^+) \qquad (6.1.3)$$

$$m(1^+) \gtrsim 3\quad m(0^+) \qquad (6.1.4)$$

Even more as for the 2^+ state it is obscure, whether these numbers have any meaning for the continuum limit.

6.2 Manton's action

Falcioni et al. /22/ considered 2^+ and 3^+ states at β = 1.55 in their MCV calculation on an 8^4 lattice. Their claim for these masses was of order $m(0^+) \pm \Delta m$, where Δm is roughly 25% of $m(0^+)$. A high statistics MCV analysis (on an $4^3 \cdot 16$ lattice) by Billoire, Koller and myself /30/ excludes now such a result. It is consistent to attribute it to be due to a statistical fluctuation.

In Ref. /30/ we have calculated all correlation functions between operators up to length L = 6 and constructed 1^-, 2^+, 2^- and 3^+ states. Again the 2^+ state is the lowest. Our final results for this state are summarized in Figure 6.2.1. If someone insists he could

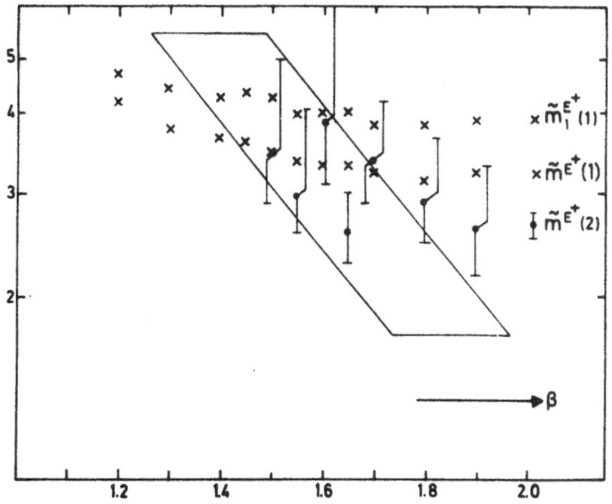

Fig. 6.2.1

find consistency with scaling for $\beta \geq 1.7$, but in the overall picture our data are much more consistent with the non-scaling t = 1 data. Because of statistical errors a final judgement is not possible. All other states are again much higher. Orders of magnitude as reached at distance t = 1 are:

$$m(2^+) \approx 1.7\ m(0^+) \quad (6.2.1)$$

$$m(1^-) \approx 3.4\ m(0^+) \quad (6.2.2)$$

$$m(2^-) \approx 2.6\ m(0^+) \quad (6.2.3)$$

$$m(3^+) \approx 2.7\ m(0^+) \quad (6.2.4)$$

6.3 Wilson's action

Fukugita et al. /35/ report $m(2^+) \approx 1.2\ m(0^+)$ from plaquette-plaquette correlations. In view of a limited MC statistics and a calculation using only the 1-plaquette operator the results are at present rather inconclusive.

6.4 Symanzik TIA

A very high statistics calculation has been done by Billoire, Meyer, Panagiotakopoulos and myself /34,36/. On an $5^3\ 8$ lattice we did up to 160 000 sweeps at selected β-values /36/. From the first

21 operators of Figure 3.1 we constructed 0^-, 2^+ and 1^+ states and did MC measurements of on-diagonal correlations.

In consistency with other actions the lowest state is 2^+. With a more limited statistics /34/ (40 000 sweeps at several β-value) we had an indication of scaling for this state. With our extended statistics the signal disappeared nearly completely as is obvious from Figure 6.4.1 (x m(1), ● m(2), ○ $\hat{m}(2)$). The figure is based on the best

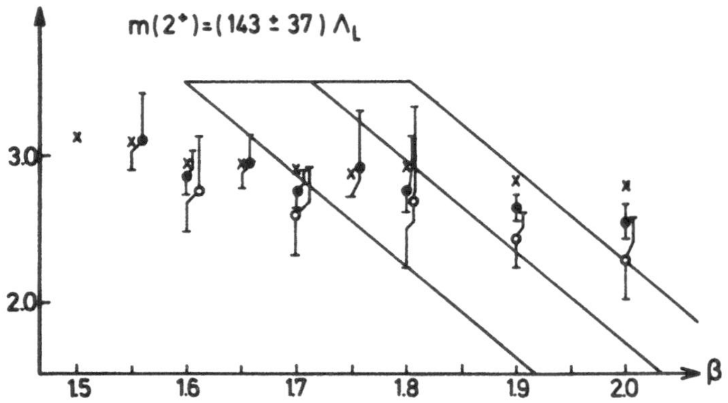

Fig. 6.4.1

operators. This is nearly all β values #7. Our high statistics illustrates clearly the severe problems with statistical noise at distance t = 2 for excited states. For a detailed discussion see Ref. /36/.

The other states are higher and allow only at distance t = 1 correlations out of the statistical noise. At distance t = 1 the obtained orders of magnitude are

$$m(2^+) \approx 1.6\ m(0^+) \tag{6.4.1}$$

$$m(0^-) \approx 2.2\ m(0^+) \tag{6.4.2}$$

$$m(1^+) \approx 2.8\ m(0^+). \tag{6.4.3}$$

This is similar as for the SA and 2^+ for Manton's action.

7. SU(3) MASS GAP $m(0^{++})$

So far MC calculations were done with the SA, for which also the SC expansion /10,12/ exists, and with Symanzik's TIA.

7.1 The standard action

Already correlations at distance t = 1 indicate clearly an ASW, as was first shown by Billoire and myself /38/. Figure 7.1 gives the result of this investigation. It is based on an MCV calculation on an $4^3\ 8$ lattice, using the Wilson loops up to length L = 6. Corre-

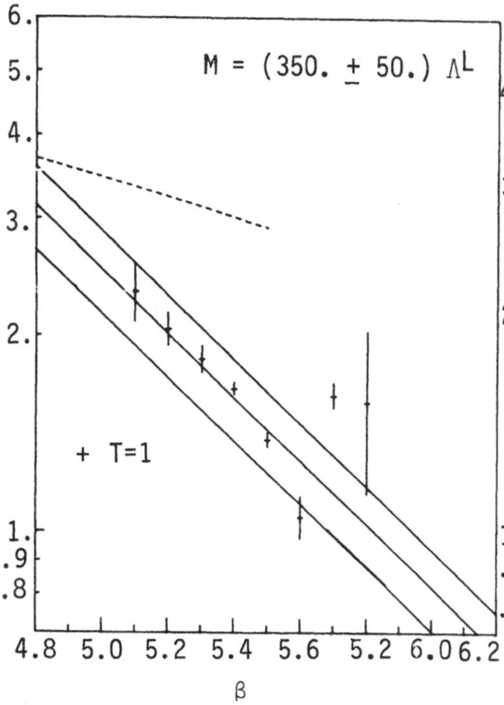

Fig. 7.1.1

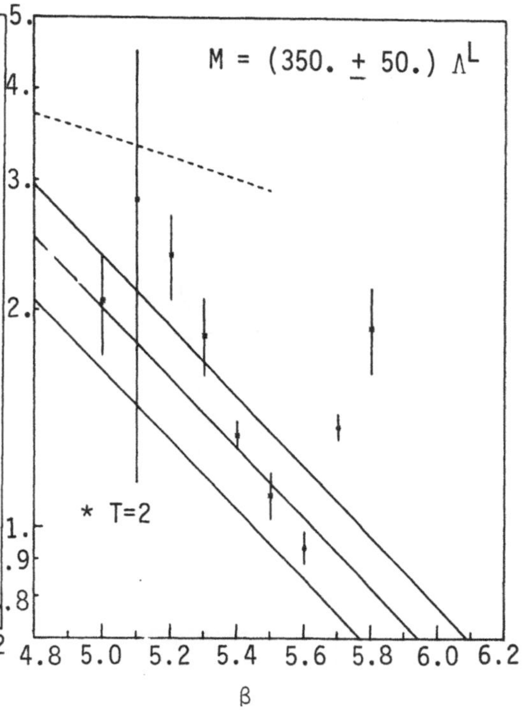

Fig. 7.1.2

lations at distance t = 2 further lower the mass gap estimate as is obvious from Figure 7.1.2 (Ref. /39,1/).

Ishikawa, Schierholz and Teper /40/ studied with high statistics the value $\beta = 5.7$ and obtained a consistent result.

The estimate /39,1/

$$m(0^{++}) = (280 \pm 50)\ \Lambda_L^{SA} \qquad (7.1.1)$$

was further confirmed in Ref. /43/. Figure 7.1.2 exhibits a suspicious overshooting of scaling, which seems to be a finite lattice size effect, as we see from Figure 7.1.3 (Ref. /44/). This figure is based on an MCV calculation on an $5^3\ 8$ lattice using the first 21 operators of Figure 3.1.

Ref. /45/ and to some extent also Ref. /42/ agree well with the estimate (7.1.1). Figure 7.1.4 (Ref. /44/) summarizes on a physical scale (here in units of $m(0^{++})$ at $\beta = 5.7$) results from Ref. /42-45/. The point(■)/45/ is based on correlations at distance t = 3 on a $4^3\ 8$ lattice. The scaling region is nicely extended, but there may be a finite size problem (see section 5.1).

Finally Gupta and Patel /46/ present MCV results at $\beta = 6.0$.

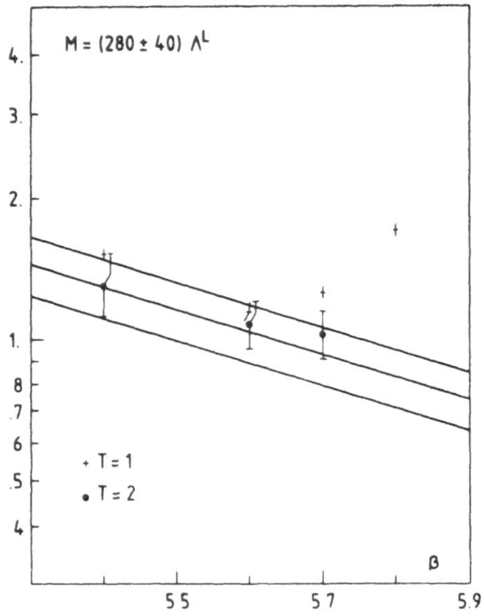

Fig. 7.1.3

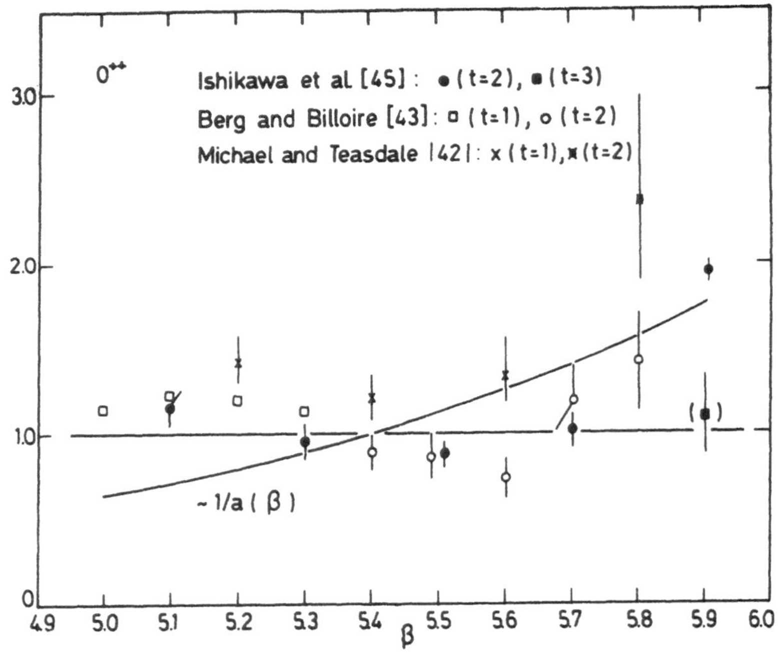

Fig. 7.1.4 *)

The obtained upper bound is rather high. This is not unexpected because of the near spin wave region /21,43/ (see also section 5.1).

*) Slight changes are done to make the Figure clearer.

By the same reason the point ⧫ (t = 2) at $\beta = 5.9$ of Figure 7.1.4 is already high. For the SU(3) SA the Langevin equation is used in Ref. /47/. Similar as in the SU(2) case /27/ results are slightly higher than MCV results. Further exploratory studies would be useful.

7.2 Symanzik TIA

De Forcrand and Roiesnel /48/ used an 6^4 lattice and measured correlations between rectangular Wilson loops W_{IJ} ($1 \le I, J \le 3$). Their result as depicted in Figure 7.2.1 indicates

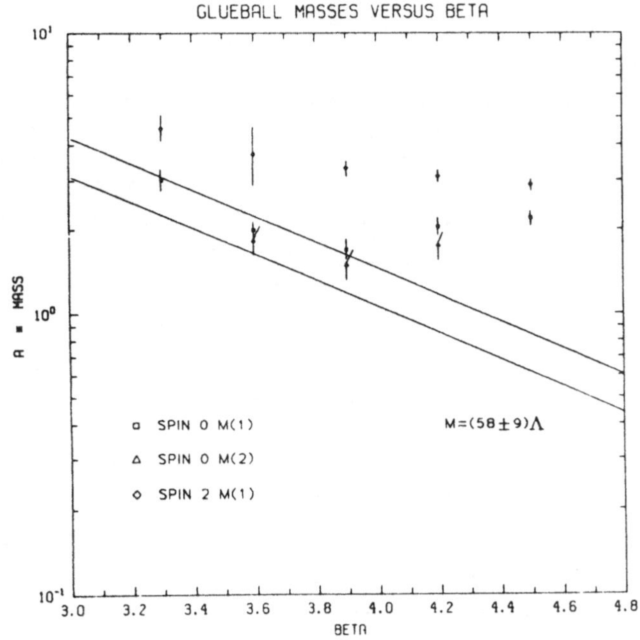

Fig. 7.2.1

$$m(0^{++}) = (58 \pm 9)\ \Lambda_L^{TI}. \qquad (7.2.1)$$

This gives $\Lambda_L^{TI}/\Lambda_L^{SA} \approx 4.8$ in reasonable agreement with the perturbative expectation (2.12).

8. SU(3) EXCITED STATES

We consider the SA.

8.1 SC calculations

SC results in the Hamiltonian formulation were obtained in a pioneering paper by Kogut, Sinclair and Susskind /9/. Their series expansion is up to $O(\beta^8)$ and they extrapolate mass ratios to $\beta \to \infty$ by using diagonal Padé approximants for the ratios. This gives

$$m(1^{+-}) \approx 1.6\ m(0^{++})$$
$$m(2^{++}) \approx 1.0\ m(0^{++})\ . \qquad (8.1.1)$$

More recently the Euclidean SC expansion (based on zero momentum plaquette-plaquette correlations in the possible irreducible representations of the cubic group) was calculated also to $O(\beta^8)$ /10,12/. Padé extrapolations were analysed in Ref. /13/ and suggest

$$m(1^{+-}) \approx 1.8\ m(0^{++})$$
$$m(2^{++}) \approx 1.2\ m(0^{++}). \qquad (8.1.2)$$

A similar result is obtained by Münster /14/. In case of the Hamiltonian formulation he expands the excited states in powers of the correlation length, and in case of the Euclidean formulation an expansion variable closely related to the correlation length is used. The results are depicted in Figure 8.1.1 (Full lines: Euclidean, dashed lines: Hamiltonian). Equations (8.1.1) and (8.1.2) are in good agreement. This indicates universality between Hamiltonian and Euclidean SC extrapolations. Even more remarkable is the consistency with results /49/ in the finite volume weak-coupling range, where Lüschers expansion is reliable.

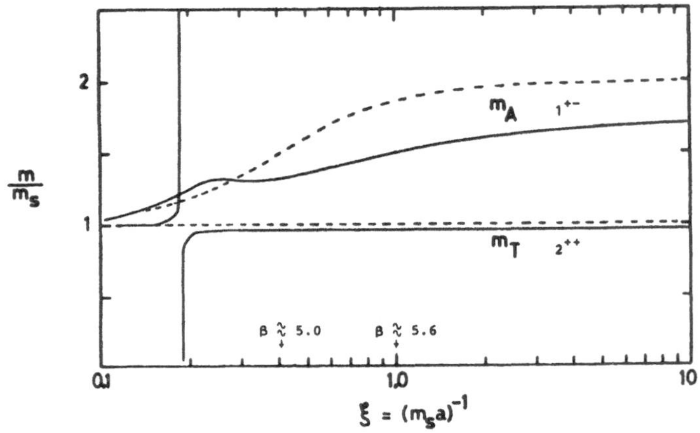

Fig. 8.1.1

There are, however, a number of MCV results, which are inconsistent with equations (8.1.1) and (8.1.2). Therefore a more careful discussion is required. MCV results are presented in the following.

8.2 MCV calculations

First MCV results are due to Billoire and myself /39,1/ and due to Ishikawa, Schierholz and Teper /40/. The calculations were further developed in Ref. /41,43-45/.

Table 8.2.1

State	$\beta = 5.2$	$\beta = 5.6$
0^{--}	$\gtrsim 2.5$	$3.9 \begin{smallmatrix} +0.5 \\ -0.4 \end{smallmatrix}$
1^{+-}	2.4 ± 0.2	4.1 ± 0.3
1^{-+}	$\gtrsim 2.4$	$4.7 \begin{smallmatrix} +0.9 \\ -0.5 \end{smallmatrix}$
2^{++}	1.72 ± 0.05	2.96 ± 0.06
2^{+-}	$2.6 \begin{smallmatrix} +0.6 \\ -0.3 \end{smallmatrix}$	$\gtrsim 4.7$
2^{-+}	2.3 ± 0.2	3.45 ± 0.11
2^{--}	2.4 ± 0.2	$4.6 \begin{smallmatrix} +0.7 \\ -0.4 \end{smallmatrix}$
3^{++}	$\gtrsim 2.3$	3.74 ± 0.24
3^{+-}	$\gtrsim 2.5$	$3.0 \begin{smallmatrix} +0.4 \\ -0.3 \end{smallmatrix}$

Billoire and I /39,1/ considered all states, which can be built from Wilson loops up to length L = 6. On an $4^3\,8$ lattice we did 11 200 sweeps and measurements at $\beta = 5.6$, 5.2. Mass ratios $m(J^{PC})/m(0^{++})$ as obtained at distance t = 1 are given in Table 8.2.1.

Between $\beta = 5.2$ and $\beta = 5.6$ there is clearly no scaling for the mass ratios, whereas according to Figure 7.1.1 $m(0^{++})$ results at distance t = 1 are consistent with asymptotic scaling. At distance $t \geq 2$ statistical noise is a severe problem for excited states. (Modest) results can only be obtained for the 2^{++} state.

Ishikawa et al. /14/ also calculated on an $4^3\,8$ lattice. They considered $\beta = 5.7$ (24 000 sweeps) and $\beta = 5.9$ (7 000 sweeps). Correlations between operators # 1,2 and # 5 of Figure 3.1 are calculated. Their result

$$m(2^+) = (2.25 \pm 0.40)\ m(0^+) \qquad (8.2.1)$$

is estimated from distance t = 2 correlations. For a discussion of their $m(0^{-+}) = (1.18 \pm 0.14)\ m(0^{++})$ result see appendix A. They also give results for several "electric" states. These states involve timelike plaquettes and do not lead to any new quantum numbers in the physical Hilbert space.

Further development:

Billoire and I increased /43/ our MC statistics on the $4^3\,8$

Table 8.2.

β	5.0	5.1	5.2	5.3	5.4	5.5	5.6	5.7	5.8
sweeps	2 800	2 800	11 200	5 600	5 600	5 600	14 000	5 600	2 800

Table 8.2.3

β	5.4	5.6	5.7	5.8
sweeps	3 000	5 250	6 000	1 500

lattice to the one given in Table 8.2.2. We did not find significant changes of the results.

Further we extended our MCV calculation to an 5^3 8 lattice /44/. The statistics are given in Table 8.2.3. In this case we calculated all correlation between the first 21 operators of Figure 3.1. From the irreducible representations of the cubic group on these loops /39,43/ we obtain results for 16 different J^{PC} states. The masses, as obtained in each case from the best operator at distance t = 1, are collected in Table 8.2.4.

Table 8.2.4 The number in brackets identifies the operator in the notation of Fig. 3.1.

J^{PC}/β	5.4	5.6	5.7
0^{++}	1.56 ± 0.06 (8)	1.21 ± 0.04 (21)	1.36 ± 0.06 (13)
0^{+-}	%	4.7 ± 0.7 (19)	6.2 ± 2.2 (19)
0^{-+}	4.22 ± 0.35 (8)	4.39 ± 0.36 (19)	3.92 ± 0.25 (17)
0^{--}	4.6 ± 0.9 (21)	4.9 ± 0.7 (19)	5.04 ± 0.69 (8)
1^{++}	4.79 ± 0.47 (14)	4.67 ± 0.29 (8)	4.56 ± 0.30 (8)
1^{+-}	4.37 ± 0.31 (13)	4.00 ± 0.16 (21)	4.08 ± 0.14 (21)
1^{-+}	4.85 ± 0.52 (19)	4.66 ± 0.30 (9)	4.42 ± 0.23 (19)
1^{--}	4.9 ± 0.6 (19)	5.3 ± 0.6 (21)	5.0 ± 0.30 (10)
$2^{++}(E)$	3.24 ± 0.14 (2)	3.13 ± 0.07 (7)	2.93 ± 0.60 (7)
$2^{+-}(E)$	4.24 ± 0.32 (19)	4.9 ± 0.6 (8)	5.46 ± 0.97 (9)
$2^{-+}(E)$	4.5 ± 0.6 (8)	4.38 ± 0.26 (17)	4.40 ± 0.27 (9)
$2^{--}(E)$	4.6 ± 0.5 (10)	4.69 ± 0.34 (9)	4.83 ± 0.48 (8)
$2^{++}(T_2)$	3.88 ± 0.19 (21)	3.40 ± 0.09 (13)	3.41 ± 0.07 (21)
$2^{+-}(T_2)$	4.13 ± 0.20 (21)	4.9 ± 0.5 (10)	5.0 ± 0.45 (2)
$2^{-+}(T_2)$	4.48 ± 0.39 (3)	4.14 ± 0.19 (8)	4.38 ± 0.21 (8)
$2^{--}(T_2)$	5.14 ± 0.71 (6)	4.76 ± 0.37 (10)	5.11 ± 0.43 (9)
3^{++}	4.8 ± 0.8 (15)	3.90 ± 0.24 (8)	4.13 ± 0.30 (8)
3^{+-}	4.0 ± 0.3 (7)	4.40 ± 0.40 (21)	5.09 ± 0.56 (13)
3^{-+}	%	6.0 ± 2.2 (21)	4.45 ± 0.24 (8)
3^{--}	5.8 ± 1.4 (12)	4.7 ± 0.7 (20)	5.4 ± 0.9 (10)

The lowest state is $2^{++}(E)$. T_2^{++} gives slightly higher results for the same state. This is either because rotation invariance for higher O(3) representations is not yet restored, or because the T_2^{++} representation does not project as well as the E^{++} representation onto the 2^{++} wave function.

In contrary to Ref. /41,45/ (see below) our 1^{-+} result is high.

Of particular phenomenological interest /86/ is the 0^{-+} state. Table 8.2.5 /44/ gives the results

Table 8.2.5 Results for the 0^{-+} state. We give the mass values $\hat{m}_1(1)$. The indice i labels the operators according to the notation of Fig. 3.1

i \ β	5.4	5.6	5.7	5.8
8	4.2 ± 0.4	4.6 ± 0.5	4.0 ± 0.3	4.0 ± 0.4
9	6.5 ± 4.9	4.9 ± 0.6	4.2 ± 0.4	3.7 ± 0.5
17	4.5 ± 0.4	5.2 ± 1.0	3.9 ± 0.3	3.6 ± 0.4
19	5.1 ± 1.1	4.4 ± 0.4	4.1 ± 0.3	3.9 ± 0.5
21	4.8 ± 0.7	5.7 ± 1.5	4.0 ± 0.2	4.1 ± 0.7
minimized	4.1 ± 0.7	4.0 ± 0.5	3.7 ± 0.4	3.4 ± 0.4

after minimization over the 5 contributing operators. In this case minimization is of course already at distance t = 1 biased by the rather large statistical noise and gives a too small value (for an upper bound on the mass), which turns out to be remarkably high. Different 0^{-+} claims are made in Ref. /41,45/ (see appendix A).

Finally results for our lowest state (E^{++}) after minimization are depicted in Figure 8.2.1. Because of large statistical noise at distance t = 2 all results plotted rely on distance t = 1. Up to β = 5.8 there is no clear signal of scaling and even including all 21 operators in the minimization does not lower the results very significantly. The dashed line to the left is the lowest order SC expansion, and the dashed line to the right the leading order spin wave result for the 1-plaquette operator /21,43/. It looks as if the data connect immediately the SC and the spin wave regime.

Table 8.2.6

β	5.1	5.3	5.5	5.7	5.9
sweeps	9 000	19 000	11 000	27 000	25 000

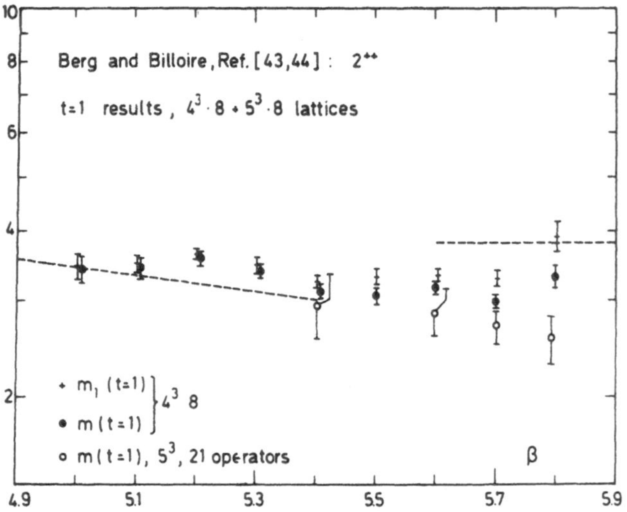

Fig. 8.2.1

Ishikawa, Sato, Schierholz and Teper /45/ extended the statistics of Ref. /40/ to the high statistics of Table 8.2.6. For the 2^{++} state they claim asymptotic scaling at values $\beta \geq 5.5$ (4^3 8 lattice). On a physical scale (units $m(0^{++})$ at $\beta = 5.7$) their results, based on correlations at distance $t = 2$, are given in Figure 8.2.2. True is

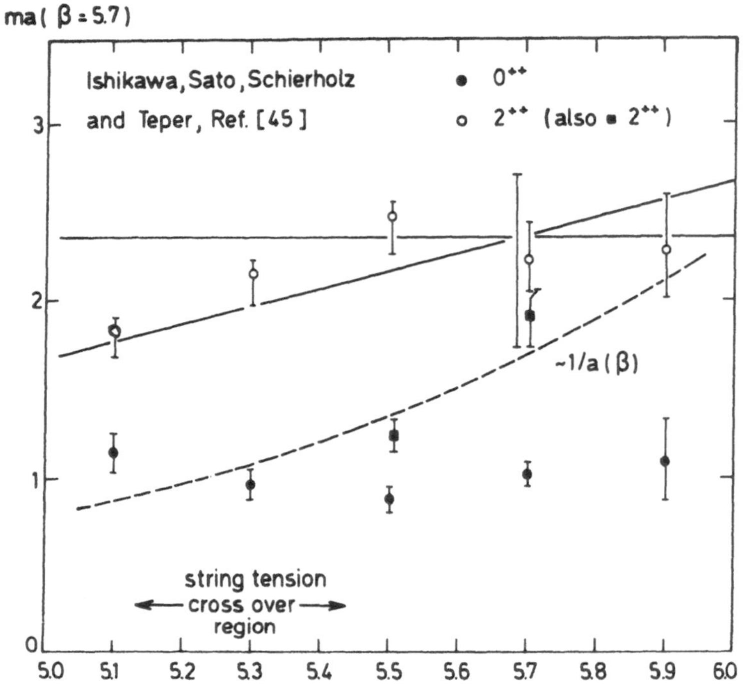

Fig. 8.2.2

that their results do not exclude asymptotic scaling (straight line), but would also be consistent with a rather different behaviour (line with slope).*) The error bars of the noisy 2^{++}, t = 2 correlations may be unreliable. With a much higher statistics such problems are discussed for the SU(2) TIA in Ref. /36/. In the present case results for each of the used three operators are given in the Tables of Ref. /45/. I take in each case the lowest result from a single operator and find the results I, which I have also included in Figure 8.2.3. These results are (much) lower than the minimized results. Realistic error bars could therefore have a magnitude as indicated in Figure 8.2.2 by **I** at β = 5.7.

Ishikawa et al. /41,45/ further "predict" a low lying 1^{-+} oddball from an MCV calculation on an asymmetric 4^3 16 lattice at

$$\beta_s = 3.3 , \qquad \beta_t = 10.68. \tag{8.2.2}$$

4 000 sweeps for equilibrium and 6 000 sweeps for measurements were done. On-diagonal correlations between the operators # 1, # 3, # 5 of Figure 3.1 and the operator of Figure 8.2.3 were measured. Except

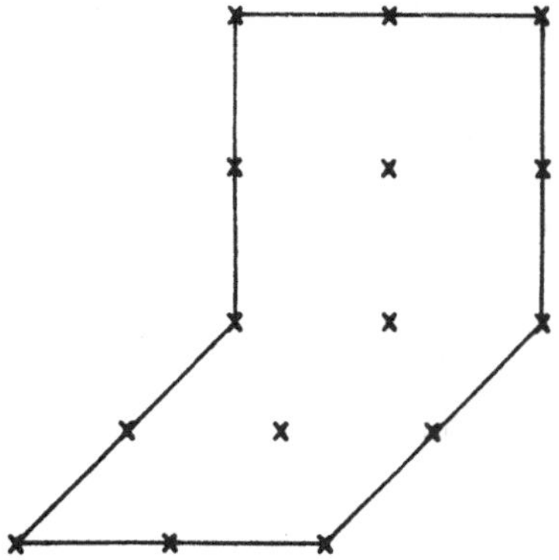

Fig. 8.2.3

for 0^{++} correlations are noise at distance t ≥ 3. In Table 8.2.7 mass ratios $\hat{m}(J^{PC})/\hat{m}(0^{++})$ from the best operators (and for 1^{-+} from all contributing operators) are collected. The operators are labelled by their length. All results are again high in units of $m(0^{++})$ and remain inconclusive as long as no scaling curve on this asymmetric lattice can be obtained. In case of the very much emphasized "prediction" of a low lying 1^{-+} oddball, the authors prefer the distance t = 2 result as obtained from the L = 12 operator. This operator gives a high result at distance t = 1. The distance t = 2 result is therefore argued to be unstable and the error bars are very doubtful. An attempt was made to improve MC statistics by using "momentum smeared" $\vec{p} \neq 0$ wave functions. Such results are, however,

*)In Ref. /45,87/ also a not existing "controversy" about the trivial fact that distance t = 2 masses are expected to be lower than distance t = 1 masses is established and "resolved".

Table 8.2.7

J^{PC}	t = 1	t = 2	L_{OP}
2^{++}	1.78 ± 0.02	1.67 ± 0.11	8
1^{-+}	2.20 ± 0.04	2.59 ± 0.29	6
	2.80 ± 0.04	$1.82\ {}^{+0.37}_{-0.28}$	12
0^{--}	3.36 ± 0.18	$1.9\ {}^{+1.9}_{-0.7}$	6
2^{--}	3.36 ± 0.09	%	12
1^{+-}	2.61 ± 0.004	%	8

unreliable, because the $\vec{p}^2$-contribution cannot be estimated. In fact at distance t = 1 the $\vec{p} \neq 0$ 1^{-+} energy comes out to be lower than the p = 0 1^{-+} energy.

8.3 SC versus MC:

As claimed in Ref. /43/ 1^{+-} and 2^{++} MCV results are in contradiction with Padé (etc.) SC continuum extrapolations, but not with the SC series itself. This remark is illustrated in Figure 8.3.1. MCV data and the SC expansion, taken (without any extrapolation) in the SU(3) character variable u /10/, are plotted together and good agreement is found:

The 1^{+-} state is high. Therefore minimization at distance t = 1 will already introduce a bias. Within the rather large uncertainties the SC expansion fits well the data.

Minimized distance t = 1 2^{++} data /43,44/ are upper bounds which may well lower down to the SC series, and the distance t = 2 2^{++} data /45/ are perfectly described by the SC series up to $\beta = 5.9$.

In contrast the rather reliable 0^{++} MCV data deviate from the 0^{++} SC series and prefer the asymptotic scaling behaviour. At $5.6 < \beta < 5.7$ the 0^{++} SC series becomes negative.

Two possible scenarios are:

1.) The matching of 1^{+-} and 2^{++} data with the SC expansion up to $\beta = 5.9$ is an accident. The 1^{+-} and 2^{++} MCV wave functions are very poor approximations to good wave functions. The final results may lower considerably and eventually become consistent with SC extrapolations like Figure 8.1.1.

2.) 1^{+-} and 2^{++} MCV data give us reasonable good bounds for the masses of these states. The matching with the plain SC expansions is no accident. 1^{+-} and 2^{++} SC expansions have a larger radius of convergence than the 0^{++} expansion.

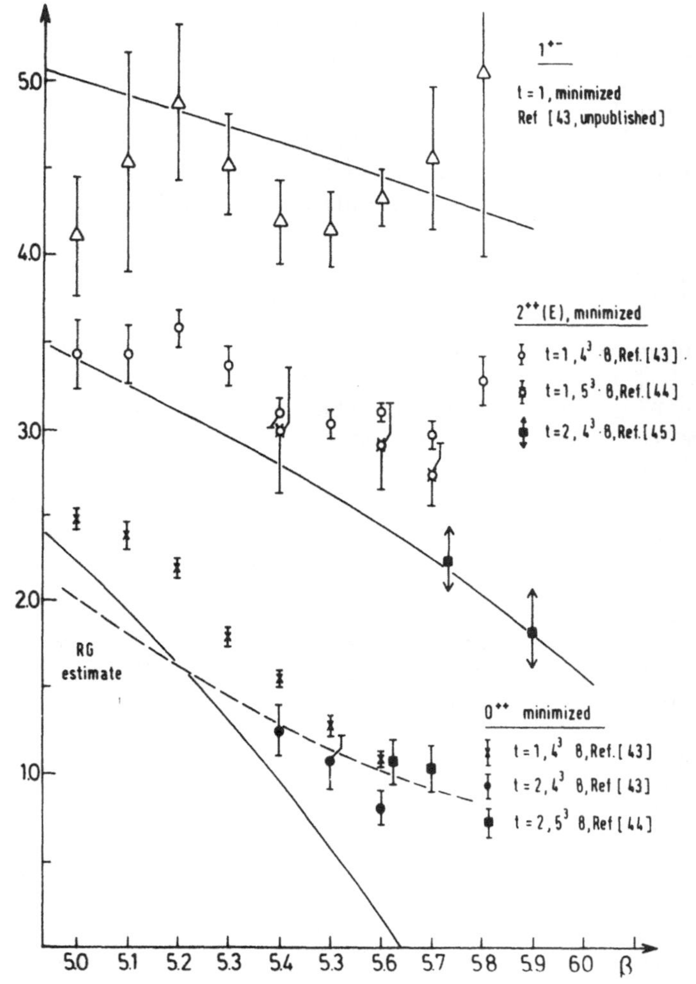

Fig. 8.3.1

Option 2.) is possible due to the fact that the cubic symmetry is an exact symmetry of the lattice. Padè and related extrapolation methods may then be unreliable for mass ratios, because the 0^{++} singularities are artificially mixed with the convergent 1^{+-} and 2^{++} expansions. In fact: The last orders and the diagonal Padès of the 1^{+-} and 2^{++} expansions (no ratios) are rather stable for $5.0 \leq \beta \leq 5.9$, as compared with 0^{++}. SC extrapolations like Figure 8.1.1 deviate already at $\beta = 5.1$ significantly from the plain SC expansion and from MCV results. At such small β-values MCV estimates are, however, argued to be rather good.

For ruling out the Padé extrapolations it is sufficient to establish reliable results at one value $\beta > 5.0$. The issue could be finally settled if source methods /27,51/ would allow us to go to larger distances. If stable results are obtained under going from $t \to t + 1$, it can hardly be doubted that the asymptotic regime $t \to \infty$ is reached.

If we apply for 1^{+-} and 2^{++} states the same SC method as for the 0^{++} state and take (according to the asymptotic scaling formula (2.6)) the tangent to the SC series, we obtain the values

$$m(1^{+-}) \approx 13 \; m(0^{++}) \qquad m(2^{++}) \approx 2.3 \; m(0^{++}). \tag{8.3.1}$$

The 2^{++} value is in reasonable agreement with MCV, because the tangent matches at $\beta \approx 5.8$, where still MCV data exist. The 1^{+-} value is exotically high and may suggest that perhaps not all possible states are realized in the spectrum of physical states $(\beta \to \infty)$. Further discussion is relegated to the conclusions in section 10.

8.4 Symanzik TIA

De Forcrand and Roiesnel /48/ found mass ratios $m(2^{++})/m(0^{++})$ which are slightly lower than those typical for the SA. At $\beta = 3.3$, 3.6 and 3.9 the mass values are respectively $m(2^{++})/m(0^{++}) \approx 1.7$, 1.9 and 2.4. Here the scaling estimate (7.2.1) is used for $m(0^{++})$.

9. MOMENTUM STATES

In the continuum limit the relativistic energy-momentum dispersion

$$E(\vec{p}) \equiv \sqrt{m^2 + \vec{p}^2} \tag{9.1}$$

has to hold. SC and MC investigations of momentum states may study the expected restoration of Lorentz invariance. A solvable example is provided by the 2d Ising model in the scaling limit (for a discussion see e.g. Ref. /89/).

In case of 4d SU(2) lattice gauge theory Kimura and Ukawa /89/ did a SC calculation up to order β^6. Lorentz invariance is not restored within the accessible β-region.

Recently several MCV calculations /90,36,52/*) with momentum $\vec{p} \neq o$ states were carried out.

In Ref. /90/ Panagiotakopoulos and I considered U(1) abelian lattice gauge theory. The 1^{+-} axial vector is closely related to the photon. As depicted in Figure 9.1 (already from correlations at distance $t = 1$) a nice signal is obtained at an energy

*) Only after my lectures I got aware of the results /52/. They are now included for completeness.

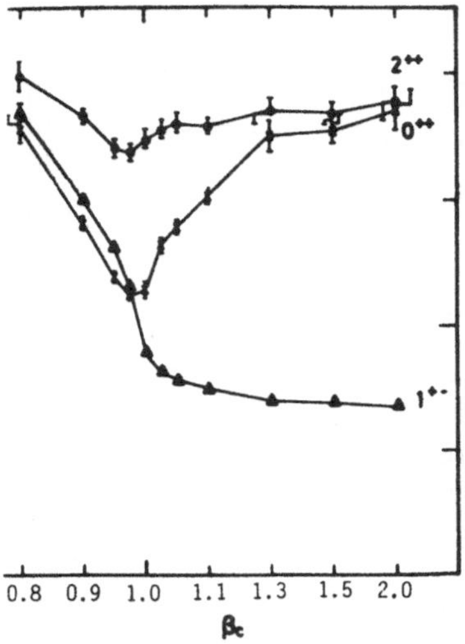

Fig. 9.1. Momentum $|\vec{K}| = \frac{2\pi}{4}$

$$E(\vec{p}) \approx \sqrt{\vec{p}^2}\ , \tag{9.1}$$

if $\beta > \beta_c$ holds. This means MC techniques clearly distinguish the non-confining situation with a massless particle from the confining situation.

In Ref. /36,52/*) restoration of Lorentz invariance was found for non-abelian lattice gauge theories (0^{++} and 2^{++} states). An example for the SU(2) 0^+ state is depicted in Figure 9.2 (Ref. /36/).

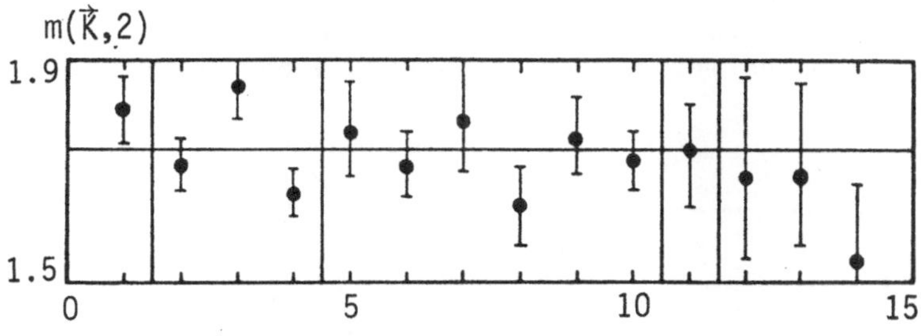

Fig. 9.2. 0^+, $\bar{\beta}$=2.2, bent plaquette

For i = 1, ..., 14 the momenta $\vec{p} = \frac{\vec{n}}{2\pi L}$ (L = 8) are defined in Table 9.1. One finds that energies of different $\vec{p}$, but with identical $|\vec{p}|$

*) There seem to be (minor) discrepancies between Ref. /36/ and Ref. /52/, which cannot be traced back at present.

Table 9.1

i	1	2	3	4	5	6	7	8	9	10	11	12	13	14
n_1	0	1	0	0	1	1	0	1	1	0	1	2	0	0
n_2	0	0	1	0	1	0	1	-1	0	1	1	0	2	0
n_3	0	0	0	1	0	1	1	0	-1	-1	1	0	0	2

behave statistically rather independent. We may therefore assume Lorentz invariance and use states with sufficiently low momenta for increasing the MC statistics. This has already been done by Schierholz and Teper /52/. For SU(2) and SU(3) mass gap 0^{++} estimates they find consistency with results as reported here.

The SU(2) 2^+ results of Ref. /52/ are given in Figure 9.3 (a scale in physical units $\sim \Lambda_L$ is used). There is consistency with the high statistics $\vec{p} = 0$ results of Ref. /36/ and earlier claims of Ref. /26,28/ are ruled out: No asymptotic scaling is found from correlations at distance t = 2. (The 2^+ "scaling" of Ref. /26,28/ relied on using "momentum smeared" wave functions.)

The SU(3) 2^{++} results of Ref. /52/ look differently. Figure 9.4 shows that energies from t = 2 correlations are consistent with asymptotic scaling ($\beta \geqslant 5.5$) and also with the SC observations of section 8.3. A problem, however, remains: The 0^{++} and 2^{++} states mix in higher orders of $\vec{p}^2$ /89/. At large distances the lower 0^{++} state will win, and we do not know how much the presented 2^{++} results are polluted by 0^{++}. Using momentum states for increasing the MC statistics is only save for the lowest state: In U(1) lattice gauge theories the photon and in SU(2), SU(3) lattice gauge theories the mass gap 0^{++}. An independent check of the 2^{++} state, for instance by means of a source method, would therefore be desirable.

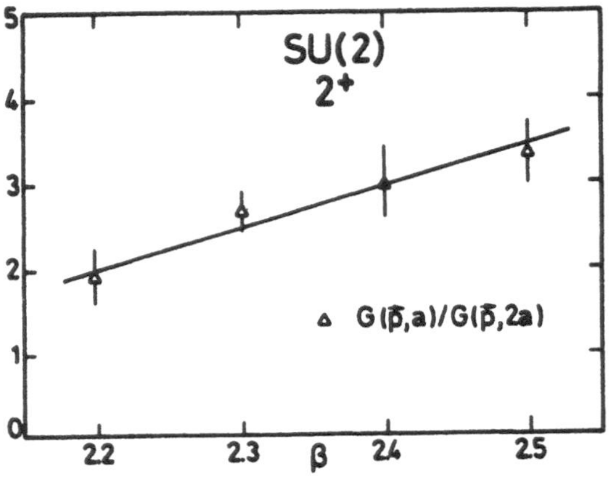

Fig. 9.3

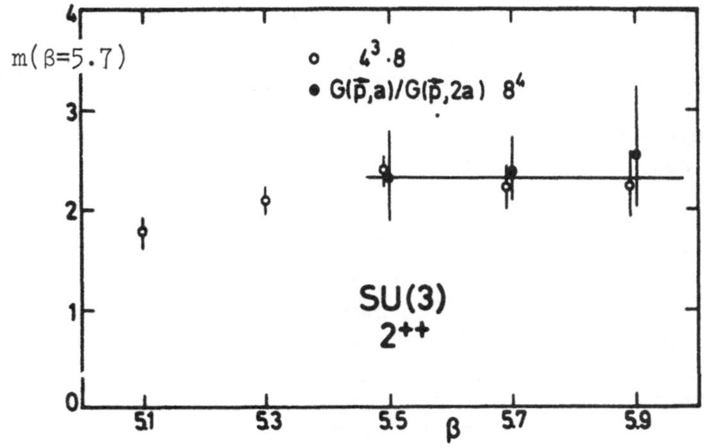

Fig. 9.4

10. SUMMARY AND OUTLOOK

What does it all mean? An optimist will say, SU(3) MCV calculations with the SA action have the region up to $\beta \lesssim 6.0$ under control. For 0^{++}, 1^{+-} and 2^{++} the results are qualitatively depicted in Figure 10.1. For small β the SC expansions (full lines) are valid.

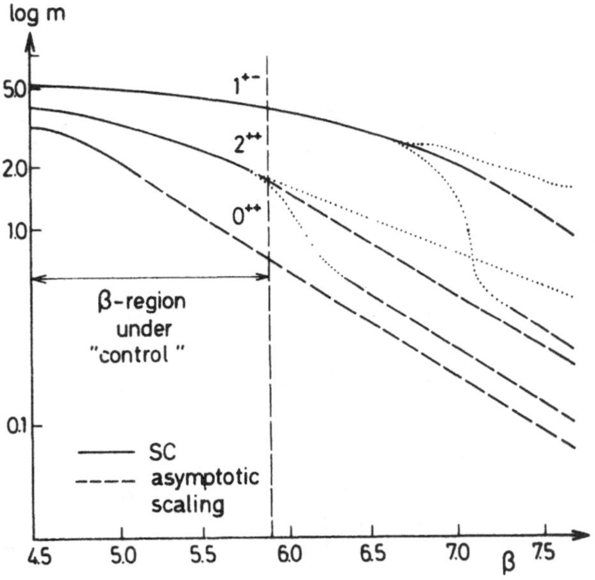

Fig. 10.1

Then several scenarios are possible, a few are sketched now.

10.1 The "standard" scenario: After the SC region a small cross-over region follows. Then asymptotic scaling immediately sets in. The corrections to asymptotic scaling are numerically small and already the tangent to the SC series gives a reasonable continuum estimate. If this receipt is applied to the 1^{+-} state a very high mass (8.3.1) results, whereas the thus obtained 2^{++} estimate would match the order of magnitude of MCV results. We note from Figure 10.1: Only for 0^{++} a notable β-range (scaling window) exists over which MCV calculations can follow asymptotic scaling <u>and</u> distinguish it from the SC expansion.

10.2 Not all possible glueball states are realized in the physical spectrum of the continuum limit. As we have very little dynamical information about non-abelian gauge theories, this possibility can in principle not be excluded. In units of $m(0^{++})$ we obtain with the receipt of 10.1 a very high 1^{+-} and a high 2^{++} state. This might be and indication that these states do not exist. In Figure 10.1 the upper dotted continuations of the 1^{+-} and 2^{++} SC expansions indicate this possibility: 1^{+-} and 2^{++} masses are sent to infinity in the continuum limit.

10.3 1^{+-} and 2^{++} masses come finally down to low values (in units of $m(0^{++})$). In Figure 10.1 the lower dotted continuations of the 1^{+-} and 2^{++} SC expansion indicate this option. This means the SA has to have large a^2/ξ^2-corrections in the sense of Symanzik /59/. At least in the β-region as accessible by MCV calculations we do not observe drastic corrections of glueball mass ratios by using the Symanzik TIA (see sections 6.4 and 8.4).

The pessimist will doubt that MCV calculations have the region up to $\beta \lesssim 6.0$ (and for SU(2) a corresponding region up to $\beta \lesssim 2.5$) under control. He will first of all argue that MCV wave functions for excited states may be very poor so that the results are in fact rather crude upper bounds. SC Padé extrapolations (section 8.1) and Lüschers small volume weak coupling calculations /49/ could then match well together. Next the pessimist will question the asymptotic scaling window of the 0^{++} state. Recent results for the string tension /68-70/ and the deconfinement temperature /92/ violate asymptotic scaling. Mass ratios could, however, still stay constant if we are able to get rid of the $m(0^{++})$ asymptotic scaling.

I am convinced that the disput between our "optimist" and our "pessimist" can be settled with standard MC methods. This requires to go (at present β-values) to a larger distance t and to check present results for stability (see for instance section 9). For the 2^{++} state distance t = 2 results (section 6, 8 and 9) do not indicate a very drastic lowering as compared with distance t = 1 results. Therefore I would consider the evidence for a β-region with $m(2^{++})/m(0^{++}) > 1$ (i.e. in contradiction with equation (8.1.1, 8.1.2)) to be already strong. This would rule out a smooth matching of SC and finite volume calculations. On the other hand a drastic lowering of the 1^{+-} state (and other high states) cannot be excluded. Also one could imagine that precise (large lattice) distance t = 3 results would lower for instance the o^{++} mass gap more at $\beta = 5.7$ than at $\beta = 5.4$. Hence 0^{++} asymptotic scaling could still become violated.

The relevancy of checking these question is obvious. To get a better feeling for important (see Ref. /36/) operators it would be desirable to have spin wave results for a large number of operators (and not only for the 1-plaquette operator /21,43/) at hand.

MC results are reported for a number of equivalent actions. With exception of results /35/ for the Wilson action (2.4) universality is in a rather good shape. In view of the reported low 2^+ mass for this action it seems to be worthwhile to spend some computer time into a high statistics MCV calculation with the Wilson action.

Future efforts will certainly have to concentrate more around developing new methods and less around obtaining quick results. Better methods in combination with new computer generations may open new horizons for numerical computations. Finally serious computer experiments need financial support on the scale of other branches in experimental physics.

ACKNOWLEDGEMENTS

I like to thank A. Billoire, P. Hasenfratz, N. Kimura, M. Lüscher, G. Mack, S. Meyer, I. Montvay, G. Münster, C. Panagiotakopoulos and P. Weisz for many useful discussions. Further I am indebted to G. Münster and G. Schierholz for reading the manuscript. Finally, I would like to thank the organizers for their kind invitation, and I gratefully acknowledge support by the Hamburgische Wissenschaftliche Stiftung and by the Cargèse Institute.

These lectures are dedicated to the memory of Kurt Symanzik, who made many valuable suggestions and was actively involved in numerical calculations in one of his last papers /50/. In particular Symanzik /59/ improved actions have found considerable interest and were investigated by many authors.

APPENDIX A: The 0^{-+} state

Ishikawa et al. report a 0^- lower than 2^+ for SU(2) /23,28/ and similarly for SU(3) a 0^{-+} state lower than 2^{++} /40,41,45/. (The final result is in any case close to an experimental candidat /86/ at 1440 MEV.) This in in contrast to MCV calculations /36,44/ as collected in sections 6 and 8 of this review. There the distance t = 1 0^{-+} mass is always much higher than the 2^{++} mass measured at the same distance.

The low lying 0^{-+} results of Ishikawa et al. rely on correlations of the operator

$$F\tilde{F}(t) = F_{\mu\nu}(t)\varepsilon_{\mu\nu\rho\sigma}F_{\rho\sigma}(t) \tag{A1}$$

For its implementation on the lattice see Ref. /91,28,45/. $F\tilde{F}$ has T-parity −1 and therefore positivity /71/ implies

$$\langle F\tilde{F}(0)\ F\tilde{F}(t)\rangle < 0 \text{ for } t \geq 3. \tag{A2}$$

t = 3 is the smallest distance at which the operators have no overlap. Of course $\langle F\tilde{F}(0)\ F\tilde{F}(0)\rangle \gg 0$ and a possible shape for the correlation function is depicted in Figure A1. Reliable informations about the 0^{-+}

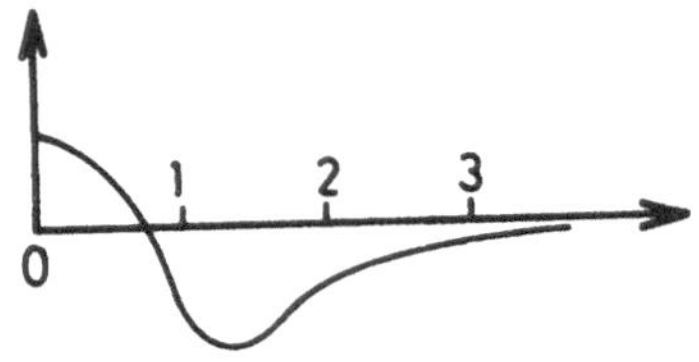

Fig. A1

mass can only obtained by comparing correlations at distance t ≥ 3.

Momentum $\vec{p} = 0$ results are only calculated in Ref. /45/. They rely on the mass definition m(2,1) (3.8), μ = 0 in equation (A1). An overlap problem exists, and we would expect a too low mass. Elsewhere in Ref. /28,45/ MC statistics is increased by using "momentum smeared" wave functions and attempts are made to reduce the overlap problem. But the overlap problem still exists and for 0^+ and 2^+ states results /26,28/ from "momentum smeared" wave functions have turned out to be unreliable (see sections 5 and 9).

In conclusion the results from $F\tilde{F}(t)$ correlations are unreliable. It would be difficult to understand why all five trial operators as used in Ref. /44,36/ should give such a bad signal for 0^{-+}, whereas some of these operators give good signals for 0^{++} and 2^{++}.

APPENDIX B

Very recent results

After writing up this review I received a few new preprints.

König et al. /93/ use the BO-method (section 4, Ref. /25/) to study the SU(3) mass gap. Their result $m_g = (320 \pm 15\%)\Lambda_L^{SA}$ is consistent (slightly higher) with MCV estimates (section 7).

Ukawa and Yang /94/ calculate Λ-parameters for six link actions. Their results are in agreement with equations (2.11, 2.12).

Kimura /95/ calculates Hamiltonian SC expansions for excited states.
Mütter and Schilling published /96/ their SU(2) TIA results.

REFERENCES

1. B. Berg, Invited lecture at the JOHNS HOPKINS WORKSHOP, Florence, 2-4 June 1982, Preprint, CERN-TH.3327 (unpublished).
2. K. Wilson, Phys. Rev. D10 , 2445 (1974).
3. F. J. Wegner, J. Math. Phys. 12, 2259 (1971).
4. K. Wilson, Erice lectures 1975, edited by A. Zichichi, Plenum Press, New York (1977).
5. K. Wilson, Cargèse lectures 1979, edited by G. 't Hooft et al., Plenum Press, New York (1980).
6. M. Creutz, L. Jakobs and C. Rebbi, Phys. Rev. D20, 1915 (1979).
7. M. Creutz, Phys. Rev. Lett. 43, 553 (1979).
8. M. Creutz, Phys. Rev. D21,2308 (1980); M. Creutz, Phys. Rev. Lett. 45, 313 (1980).
9. J. Kogut, D. K. Sinclair and L. Susskind, Nucl. Phys. B114, 199 (1976).
10. G. Münster, Nucl. Phys. B190 [FS3], 439 (1981), errata: B200, [FS4], 536 (1982), B205 [FS5], 648 (1982).
11. M. Falcioni, E. Marinari, M. L. Paciello, G. Parisi and B. Taglienti, Phys. Lett. 102B , 270 (1981); Nucl. Phys. B190 [FS3], 782 (1981).
12. K. Seo, Nucl. Phys. B209, 200 (1982).
13. J. Smit, Nucl. Phys. B206, 309 (1982).
14. G. Münster, Phys. Lett. 121B, 53 (1983).
15. R. S. Schor, Preprint, München, MPI-PAE/PTH 15/83, to appear in Nucl. Phys. B.
16. For a recent review of SC results see: J.-M. Drouffe and J.-B. Zuber, Physics Report, to appear.
17. B. Berg, Phys. Lett. 97B, 401 (1980).
18. G. Bhanot and C. Rebbi, Nucl. Phys. B180 [FS2], 469 (1981).
19. J. Engels, F. Karsch, H. Satz and I. Montvay, Phys. Lett. 102B, 332 (1981).
20. R. C. Brower, M. Creutz and M. Nauenberg, Nucl. Phys. B210 [FS6], 133 (1982).
21. B. Berg, A. Billoire and C. Rebbi, Ann. Phys. (NY) 142, 185 (1982) and addendum 146, 470 (1983).
22. M. Falcioni, E. Marinari, M. L. Paciello, G. Parisi, E. Rapuano, B. Taglienti and Zhang Yi-Cheng, Phys. Lett. 110B, 295 (1982).
23. K. Ishikawa, G. Schierholz and M. Teper, Phys. Lett. 110B, 399 (1982).
24. J. Engels, F. Karsch, H. Satz and I. Montvay, Nucl. Phys. B205, [FS5], 545 (1982).
25. K. H. Mütter and K. Schilling, Phys. Lett. 117B, 75 (1982).
26. K. Ishikawa, G. Schierholz and M. Teper, Z. Phys. C16, 69 (1982).
27. M. Falcioni, E. Marinari, M. L. Paciello, G. Parisi, B. Taglienti and Zhang Yi-Cheng, Nucl. Phys. B215 [FS7], 265 (1983).
28. K. Ishikawa, G. Schierholz and M. Teper, Z. Phys. C19, 327 (1983).
29. K. H. Mütter and K. Schilling, Phys. Lett. 121B, 267 (1983).
30. B. Berg, A. Billoire and K. Koller, Nucl. Phys. B233, 50 (1984).

31. E. Brooks III, G. Fox, S. Otto, M. Randeria, B. Athas, E. De Benedicts, M. Newton and C. Seitz, Nucl. Phys. B220 [FS8], 383 (1983).
32. T. Freimuth, K. H. Mütter and K. Schilling, Phys. Lett. 131B, 415 (1983); 132B, 379 (1983).
33. S. Otto and M. Randeria, Nucl. Phys. B225, 579 (1983).
34. B. Berg, A. Billoire, S. Meyer and C. Panagiotakopoulos, Phys. Lett. 133B, 359 (1983).
35. M. Fukugita, T. Kaneko, T. Niuya and A. Ukawa, Preprint, Kyoto, RIFP-530 (1983).
36. B. Berg, A. Billoire, S. Meyer and C. Panagiotakopoulos, Preprint SACLAY, to appear.
37. H. Hamber and G. Parisi, Phys. Rev. Lett. 47, 1792 (1981).
38. B. Berg and A. Billoire, Phys. Lett. 113B, 65 (1982).
39. B. Berg and A. Billoire, Phys. Lett. 114B, 324 (1982).
40. K. Ishikawa, G. Schierholz and M. Teper, Phys. Lett. 116B, 429 (1982).
41. K. Ishikawa, A. Sato, G. Schierholz and M. Teper, Phys. Lett. 120B, 387 (1983).
42. C. Michael and I. Teasdale, Nucl. Phys. B215 [FS7], 433 (1983).
43. B. Berg and A. Billoire, Nucl. Phys. B221, 109 (1983).
44. B. Berg and A. Billoire, Nucl. Phys. B226, 405 (1983).
45. K. Ishikawa, A. Sato, G. Schierholz and M. Teper, Z. Phys. C21, 167 (1983).
46. R. Gupta and A. Patel, Preprint, Pasadena, CALT-68-973.
47. H. Hamber and U. Heller, Preprint (1983).
48. Ph. de Forcrand and C. Roiesnel, Preprint, e' Ecole, Polytechnique (1983).
49. M. Lüscher, Phys. Lett. 118B, 391 (1982), M. Lüscher, Nucl. Phys. B219, 233 (1983), M. Lüscher and G. Münster, Nucl. Phys. B232, 445 (1984).
50. B. Berg, S. Meyer, I. Montvay and K. Symanzik, Phys. Lett. 126B, 467 (1983), B. Berg, S. Meyer and I. Montvay, Preprint, DESY 83-098, to be published in Nucl. Phys. B.
51. Ph. de Forcrand, to appear (private communication).
52. G. Schierholz and M. Teper, Preprints, DESY 83-106, 83-107.
53. K. Wilson, Closing remarks at the Abingdon Meeting on Lattice Gauge Theories (March 1981).
54. W. A. Bardeen, R. B. Pearson and E. Rabinovici, Phys. Rev. D21, 1037 (1980).
55. N. S. Manton, Phys. Lett. 96B, 328 (1980).
56. G. Bhanot and M. Creutz, Phys. Rev. D24, 3212 (1981).
57. A. Gonzales-Arroyo and C. P. Korthals Altes, Nucl. Phys. B205 [FS5], 46 (1982).
58. G. Bhanot and R. Dashen, Phys. Lett. 113B, 299 (1982).
59. K. Symanzik, in "Mathematical Problems in Theoretical Physics", eds. R. Schrader et al. (Lecture Notes in Physics 153, Springer, Berlin 1982); K. Symanzik, Nucl. Phys. B226, 187 (1983); B226, 205 (1983).

60. P. Weisz, Nucl. Phys. B221, 1 (1983);
P. Weisz and R. Wohlert, Preprint, DESY 83-091, submitted to Nucl. Phys. B.
61. G. Curci, P. Menotti and G. P. Paffuti, Phys. Lett. 130B, 205 (1983).
62. B. Berg, Invited talk given at the "International Symaposium on the Theory of Elementary Particles", Ahrenshoop (DDR), 9.10. - 15.10.1983, Preprint, DESY 83-120, to be published in the proceedings.
63. W. E. Cashwell, Phys. Rev. Lett. 33, 244 (1974);
D. R. T. Jones, Nucl. Phys. B75, 531 (1974).
64. A. Hasenfratz and P. Hasenfratz, Nucl. Phys. B193, 210 (1981), and references given therein.
65. C. B. Lang, C. Rebbi, P. Salomonson and B. S. Skagerstam, Phys. Rev. D26, 2028 (1982).
66. Y. Iwasaki and S. Sakai, Preprint, UTHEP-112 (revised)
67. W. Bernreuther and W. Wetzel, Phys. Lett. 132B, 382 (1983).
68. F. Gutbrod, P. Hasenfratz, Z. Kunszt and I. Montvay, Phys. Lett. 128B, 415 (1983).
69. G. Parisi, R. Petronzio and F. Rapuano, Phys. Lett. 128B, 418 (1983).
70. F. Gutbrod and I. Montvay, Preprint, DESY 83-112.
71. M. Lüscher, Commun. Math. Phys. 54, 283 (1977),
K. Osterwalder and E. Seiler, Ann. Phys. (NY) 110, 440 (1978).
72. H. Bethe, Ann. der Phys. 3, 133 (1929).
73. S. L. Altmann and A. P. Cracknell, Rev. Mod. Phys. 37, 19 (1965).
74. R. C. Johnson, Phys. Lett. 114B, 147 (1982).
75. J. J. Coyne, P. M. Fishbane and S. Meshkow, Phys. Lett. 91B, 259 (1980);
S. Meshkow, Proc. JOHNS HOPKINS WORKSHOP 6 (Florence 1982) p. 185.
MC investigations support an effective gluon mass:
C. Bernard, Phys. Lett. 108B, 431 (1982).
76. K. Wilson and J. Kogut, Physics Report 12, 75 (1974).
77. J. Kuti, J. Polónyi and K. Szláchanyi, Phys. Lett. 98B, 199 (1981);
L. Mc Lerran and B. Svetitsky, Phys. Lett. 98B, 195 (1981); Phys. Rev. D24, 450 (1981);
J. Engels, F. Karsch, H. Satz and I. Montvay, Phys. Lett. 101B, 89 (1981).
78. K. Kajantie, C. Montonen and E. Pietarinen, Z. Phys. C9, 253 (1981),
I. Montvay and E. Pietarian, Phys. Lett. 110B, 148 (1982); 115B, 151
79. I. S. Gradshteyn and I. M. Ryshik, "Tables of Integrals, Series and Products", Academic Press, New York, 1965.
80. For instance: A. J. Buras, Proc. 1981 Int. Symp. on Lepton and photon interactions at high energies, ed. W. Pfeil, Physikalisches Institut, Universität Bonn, p. 636.
81. M. Creutz and K. J. M. Moriaty, Phys. Rev. D26, 2166 (1982).
82. C. Rebbi, Phys. Rev. D21, 3350 (1980).

D. Petcher and D. Weingarten, Phys. Rev. D22, 2465 (1980);
G. Bhanot, C. Lang and C. Rebbi, Computer Phys. Comm. 25, 275 (1982).
83. K. Bitar, S. Gottlieb, C. Zachos, Phys. Rev. D26, 2853 (1982).
84. Yu. Makenko and M. I. Polikarpov, Nucl. Phys. B205 [FS5], 386 (1982);
B. Grossmann and S. Samuel, Phys. Lett. 120B, 383 (1983);
J. Jurkiewicz, C. P. Korthals Altes and J. W. Dash, Preprint, CERN-TH.3621
85. C. Lang and C. Rebbi, Phys. Lett. 115B, 137 (1982).
86. E. D. Bloom, Preprint SLAC-PUB-2976 (1982). Talk presented at the International Conference on Physics in Collision, Stockholm, Sweden 1982;
L. D. Scharre, in Proceedings of the 1981 International Symposium on Lepton and Photon Interactions at High Energies, edited by W. Pfeil (University of Bonn, 1981) p.163 ff.
87. M. Teper, Invited Talk at the International Europhysics Conference on High Energy Physics, Brighton (U.K.), 20 - 27 July 1983, Preprint LAPP-TH-91 (1983).
88. T. Hattori and H. Kawai, Phys. Lett. B105, 43 (1981);
J. P. Kovall and H. Neuberger, Nucl. Phys. B189, 535 (1981).
89. N. Kimura and A. Ukawa, Nucl. Phys. B205 [FS5], 637 (1982).
90. B. Berg and C. Panagiotakopoulos, Phys. Rev. Lett. 52, 94 (1984).
91. P. Di Vecchia, K. Fabricius, G. G. Rossi and G. Veneziano, Nucl. Phys. B192, 392 (1981).
92. T. Celik, J. Engels and H. Satz, Phys. Lett. 129B, 323 (1983).
93. A. König, K. H. Mütter, J. Paech and K. Schilling, Preprint, Wuppertal University, WU B 83-26.
94. A. Ukawa and S.-K. Yang, Preprint, Tokyo University, UT-417 (1983).
95. N. Kimura, Preprint, DESY 84-x, to appear.
96. K. H. Mütter and K. Schilling, Preprint, Wuppertal University, WU B 84-3.

LATTICES, DEMONS AND THE MICROCANONICAL ENSEMBLE†

Gyan Bhanot*

The Institute for Advanced Study
Princeton, N. J.

ABSTRACT

A method proposed recently for computer simulation of dynamical systems is investigated in detail for the Ising model.

I. Preliminaries

Recently, a new method for simulating dynamical systems on the computer was suggested.[1] In this method, one works in the microcanonical rather than the canonical ensemble. I will describe here the method using the 2-d Ising model as an example. The algorithm used is an adaptation of the one discussed by Creutz.[1]

Consider a 2-d Ising system with action

$$E_s = \sum_{<ij>} (1 - S_i S_j) \qquad (1)$$

where the sum runs over nearest neighbor spins S_i on the sites of an L×L Lattice. Following Creutz, imagine a demon with a sack of energy E_d. By decree, E_d is nonnegative and bounded above by E_d^{max}.

†Lecture given at the 1983 Cargèse School, Cargese, France and at the Workshop on Lattice gauge theory, Visegrad, Hungary, Sept., 1983.

*The work described here was done in collaboration with Herbert Neuberger and Michael Creutz.

Given an initial lattice distribution of spins with energy E_s and a demon with energy E_d, one generates a sequence of configurations with constant $E_d + E_s = E_T$. The algorithm that accomplishes this is the following.

The demon hops from site to site trying to flip spins. At each site, it computes ΔE_s, the change in action on flipping the spin. If ΔE_s is positive, the demon must supply energy to flip the spin. He does so if his bag has the necessary energy. Energy supplied to the lattice depletes the demon's own energy. If, on the other hand, ΔE_s is negative, the spin flip lowers the lattice energy. This energy is accepted by the demon (and the spin is flipped) if the demon has space in his bag to hold the extra energy. If ΔE_s is zero, the spin flip is always accepted.

This algorithm, repeated several times over the lattice spins, generates an ensemble of states at fixed E_T. Any correlation function involving the spins is computed as an average over this ensemble of states. The precise connection between correlations computed thus and the same correlations in the canonical ensemble will be discussed later.[2]

Simulating a system at fixed total energy rather than at fixed temperature has some obvious advantages.

First, the whole updating procedure is deterministic, involving no random numbers. This eliminates the usual problems, such as unwanted correlations and finite cycle length, associated with pseudo-random number generators. Further, with multiple spin packing, the updating can be done truly simultaneously on <u>all</u> the spins in a computer word. This is in contrast to the usual procedure,[3] where even if multiple spin packing is used, the spins have to be updated one at a time because an update involves a comparison with a floating point random number. In the microcanonical simulation, the updating algorithm is constructed out of Boolean commands (an example of which will be given later). The computer executes these much faster than an arithmetic command such as an add or a multiply. This makes the programs one writes extremely fast compared to the usual Monte Carlo codes.

Another advantage is that one works directly in terms of correlations (order parameters) and not in terms of unphysical quantities like the bare coupling constants (temperatures). This might be advantageous for block spin renormalization group calculations. Further, metastable states can be studied via microcanonical simulation.[4]

There are also some potential disadvantages. One is that the system may get trapped in a cycle in phase space because of bad choices of the initial state. In general, it is very hard to ensure that such non-ergodic behavior is excluded. One necessary condition for ergodicity is time reversal invariance, i.e. each final state of demon + system must have a unique ancestor. It is easy to convince oneself that the algorithm described above has this property. One should also choose an initial state which is as random as possible. We have made some checks using different initial conditions and compared correlations between them to check for ergodicity. Within statistical error, non-ergodic behavior was not found.

Finally, finite size effects are rather subtle in the microcanonical ensemble. They will be discussed in detail below.

II. The Model and the Computer Code

Consider the 2-d Ising model on a 64 × 64 lattice. Since an Ising spin takes only two values, we maximally utilize storage by using one computer bit per spin. Spin "up" corresponds to the bit being zero and spin "down" to it being one. Our codes were written on a VAX 11/780 which has a 32 bit word length. We store the 64 × 64 lattice in 128 words indexed by an integer I, I=1,2...128. A given row of the lattice is coded into two successive words in the list of 128 words. For example, the word indexed by I = 1 contains in its rightmost bit position the spin corresponding to the lower right corner of the lattice, in the next bit position, the third spin in the bottom row counting from right to left, etc. The word I = 2 contains the even spins of the bottom row and so on. The important thing is that each word contain spins that do not interact with each other via the action (Eqn. (1)) and so may be updated simultaneously.

We use 32 demons, each of which is allowed four energy states labelled 00, 01, 10 and 11 in binary notation. These are stored in two 32 bit words D1 and D2. The ith bit in D1(D2) contains the coefficient of $2^0(2^1)$ in the energy of the ith demon. It is easy to see that the minimum nonzero energy change on the lattice corresponds to a spin flip that has $\Delta E_s = \pm 4$. The demon energy is therefore quantized in units of four. The total energy E_T of the system of spins plus the demons is then given by

$$E_T = E_s + 4E_d \tag{2}$$

Given the demon energies D1 and D2 and the products S•S1, S•S2, S•S3, S•S4 of the spin word S (which is being updated) with its four neighboring spin words S1, S2, S3 and S4, one must construct a Boolean function B which produces the unique final values SN, D1N and D2N for the spin and demon energy words using the algorithm described above. On a bit level, B must reproduce Table I. In this table, the entries under "Initial spin state" are the four corresponding bits of the products S•Si, i=1,2,3,4 (permutations of these bits yield the same result for SN, D1N, D2N). Under "Initial Demon Energy" are given the two bits of the energy of the demon corresponding to the spin being updated. In the table, Y means the spin flip is accepted and N that it is not. The numbers beside Y or N in each box denote the two bits of the final demon energies.

A Fortran code that produces Table I is given below. All variables here are Integer variables and the functions XOR, OR, AND and NOT are the standard Boolean functions.

```
c       Input: S,S1,S2,S3,S4,D1,D2
        I(1) = .not.and(S,S1)
        I(2) = .not.and(S,S2)
        I(3) = .not.and(S,S3)
        I(4) = .not.and(S,S4)
        IR = 0
        REJECT = 0
        D10 = D1
        D20 = D2
        D3 = D2
        D2 = .not.D2
        DO 1 K = 1,4
        IR = or(IR,and(I(K),.not.or(D1,or(D2,D3))))
        D1 = xor(D1,I(K))
        D2 = xor(D2,and(I(K),D1))
        D3 = and(D3,.not.and(D1,D2))
1       continue
        REJECT = or(D3,or(IR,REJECT))
        ACCEPT = .not.REJECT
        SN = xor(S,ACCEPT)
        D1N = or(and(D10,REJECT),and(D1,ACCEPT))
        D2N = or(and(D20,REJECT),and(D2,ACCEPT))
c       Output: SN,D1N,D2N
```

Table I. The Boolean function B that gives new demon energies and spin flips from the products S·Si, (i=1,2,3,4), and the initial demon energies. Y and N stand for site spin flip and no-spin flip respectively.

Initial Spin State	Initial Demon Energy							
	00		01		10		11	
1111	Y	10	Y	11	N	10	N	11
1110	Y	01	Y	10	Y	11	N	11
1100	Y	00	Y	01	Y	10	Y	11
1000	N	00	Y	00	Y	01	Y	10
0000	N	00	N	01	Y	00	Y	01

After each pass through the lattice, the demon energies are scrambled up by shifting the bits of the demon energy words by a random amount.

Using this logic in our program, we achieved a speed of 6×10^5 spin updates/second on a VAX 11/780. Using a more sophisticated code and special tricks, Shapiro has succeeded in gaining an extra factor of 4 in speed. This means that on the VAX 11/780, one can write code which updates about 2.5×10^6 Ising spins per second. This is only a factor of 10 slower than special purpose machines built for the Ising model.[5] On a CDC-7600, a 3-d Ising model program[2] can run at a speed of $\sim 10^7$ spins/sec in the microcanonical ensemble.

III. Computing β

To compare results obtained in the microcanonical ensemble to exact Ising model results, we must relate E_T to the inverse temperature β. This is done via the demons' energy distribution as we will now demonstrate.

The partition function that represents the microcanonical ensemble we generate is given by:*

$$Z = \sum_{Cc} \delta_{E_s(C)+4E_d(c),E_T} \tag{3}$$

where C and c represent lattice and demon configurations respectively. Let $g(E_d)$ and $G(E_s)$ be the number of states of the demons and lattice at energy E_d and E_s respectively. Then

$$Z = \sum_{E_s E_d} G(E_s)g(E_d)\delta_{E_s+4E_d,E_T} \tag{4}$$

Obviously, the probability that the demons have total energy E_d is proportional to $g(E_d)$ times the probability that the lattice has energy $E_T - 4E_d$. Thus,

$$P(E_d) \sim g(E_d)G(E_T - 4E_d) \tag{5}$$

Now comes the crucial step. If the lattice volume V is large, then $G(E_T - 4E_d)$ may be expanded around E_T. Writing,

$$G(E) = e^{VS(E/V)} \tag{6}$$

*Note that Eqn. (3) already assumes several things (eg. ergodity). For a more detailed discussion see Ref. 2.

with S the entropy, we get,

$$G(E_T - 4E_d) = C_{E_T}\, e^{-4E_d\beta_T - O\frac{E_d^2}{V}} \qquad (7)$$

where,

$$\beta_T = \frac{1}{V}\frac{\partial\, \ln G(E)}{\partial E}\Bigg|_{E=E_T} \qquad (8)$$

Thus, to leading order,

$$P(E_d) \sim e^{-4E_d\beta}\, g(E_d) \qquad (9)$$

For our system of 32 demons, we can compute everything explicitly. Thus,

$$Z_{demon} = \sum_{E_d} P(E_d) = \left(1+e^{-4\beta} + e^{-8\beta} + e^{-12\beta}\right)^{32} \qquad (10)$$

From which, we have using Eqn. (9)

$$g(E_d) = \sum_{k=0}^{\min[32,E_d/2]} \binom{32}{E_d - 2k}\binom{32}{k} \qquad (11)$$

and the average demon energy $\overline{\overline{E}}_d$ is,

$$\overline{\overline{E}}_d = 32\left[\frac{1}{1+e^{4\beta}} + \frac{1}{1+e^{8\beta}}\right] \qquad (12)$$

From Eqn. (12), it is clear that a knowledge of $\overline{\overline{E}}_d$ immediately yields β.

Before presenting numerical data, we now discuss the important question of 1/V effects.

IV. 1/V Effects

In the canonical ensemble, the expectation value of any lattice spin operator Θ is given by

$$\langle\Theta\rangle_\beta = \frac{\sum_E G(E)e^{-\beta E}\,\bar{\Theta}(E)}{\sum_E G(E)e^{-\beta E}} \qquad (13)$$

where $\Theta(E)$ is its average value in the sector with energy E. Thus,

$$\bar{\Theta}(E) = \sum_C \delta_{E(C),E}\, \Theta(C) / \sum_C \delta_{E(C),E} \tag{14}$$

In the microcanonical ensemble under discussion, this expectation value (denoted by a double bar) is,

$$\bar{\bar{\Theta}}(E_T) = \frac{\left[\sum_{E_d} g(E_d)\ \bar{\Theta}(E_T-4E_d)\ G(E_T-4E_d)\right]}{\left[\sum_{E_d} g(E_d)\ G(E_T-4E_d)\right]} \tag{15}$$

When $V = \infty$, the averages in Eqns. (13) and (15) are equal if β satisfies (8). This is seen by saddle point expansion of $G(E)e^{-\beta E}$ in Eqn. (13) and Taylor expansion around E_T in Eqn. (15).

The 1/V effects are more subtle to compute. After some algebra,[2] one gets the following compact expression,

$$\langle\Theta\rangle_{\beta^*} = \bar{\bar{\Theta}}(E_T) - \frac{1}{2V}\frac{\partial}{\partial\beta}\left[\frac{(\partial\langle\Theta\rangle/\partial\beta)}{(\partial\langle e\rangle/\partial\beta}\right]_{\beta=\beta^*} \tag{16}$$

where,

$$e = E_s/V \tag{17}$$

and β^* is defined implicitly by

$$\langle e\rangle_{\beta^*} = \bar{\bar{e}}(E_T) \tag{18}$$

V. Numerical Simulation

As a check, we first did a very accurate run at a fixed value of $E_T (E_T = 4096)$. 10,000 initial lattice sweeps were discarded and 40,000 more were done to compute averages. We obtained,

$$\left.\begin{aligned} \bar{\bar{e}} &= 0.50421 \pm 0.00002 \\ \bar{\bar{e}}_{diag} &= 0.37629 \pm 0.00008 \\ \bar{\bar{e}}_{nnn} &= 0.30038 \pm 0.00011 \end{aligned}\right\} \tag{19}$$

Here, $\bar{\bar{e}}_{diag}$ and $\bar{\bar{e}}_{nnn}$ are expectation values of the product of two spins along a unit diagonal and axis-next-nearest neighbor, respectively. The average demon energy gave,

$$\beta = 0.3807 \pm 0.0010 \tag{20}$$

All the errors quoted here are truly statistical. They were measured by blocking our data until the correct $1/\sqrt{N}$ behavior was seen.

These correlation functions can be computed exactly for the Ising model.[6] First, we found (see Eqn. (18)),

$$\beta^* = 0.380383 \pm 0.000009 \qquad (20a)$$

which agrees with (20). However, the exact Ising model result[6] for the other correlations at this temperature are

$$\begin{aligned} \langle e_{diag} \rangle_{\beta^*} &= 0.376460 \pm 0.000025 \\ \langle e_{nnn} \rangle_{\beta^*} &= 0.300722 \pm 0.000025 \end{aligned} \qquad (21)$$

It is clear that Eqn. (19) and Eqn. (21) do not agree within the error bars. It is here that the 1/V effects computed before come to the rescue. Using Eqn. (16)

$$\left.\begin{aligned} \langle e_{diag} \rangle_{\beta^*} &= \bar{\bar{e}}_d + 0.00013 \\ \langle e_{nnn} \rangle_{\beta^*} &= \bar{\bar{e}}_{nnn} + 0.00032 \end{aligned}\right\} \qquad (22)$$

This neatly explains the discrepancy.

Another check we made was to compare the observed demon-energy distribution with expectations from Eqn. (9) and Eqn. (11). This is shown in Figure 1 which is a plot of E_d vs $\ell n[g(E_d)/P(E_d)]$. The slope of this curve yields 4β. Moreover, the coefficient γ of the $O(E_d^2/V)$ term in Eqn. (7) may be estimated from this plot. We find $\gamma \lesssim 3 \times 10^{-5}$. This justifies neglecting γ.

We also did another simulation at various values of E_T discarding 200 sweeps and averaging over 1000. Figures (2) and (3) show $\bar{\bar{e}}$ and $\bar{\bar{e}}_{diag}$ as functions of β (which is computed from the average demon energy). The open circles correspond to an initial start near $\beta \sim 0$ and the dark points to one near $\beta = \infty$. The solid curve is the exact (canonical) result. At first sight, the results of Figures (2) and (3) do not seem terribly impressive. However, notice that the deviations in both curves from the exact result are highly correlated. This makes one suspect that the source of the discrepancy is in the estimate of β. Indeed, if we plot $\bar{\bar{e}}_{diag}$ vs β^* (see Figure (4)), the agreement with the exact result (solid line) is much better. The important lesson to learn is that most of the discrepancy between the exact result and the data in Figures (2) and (3) comes from the difference between β and β^*. Notice that this is not an intrinsic issue of the microcanonical ensemble. It is only important when comparing this ensemble to the canonical

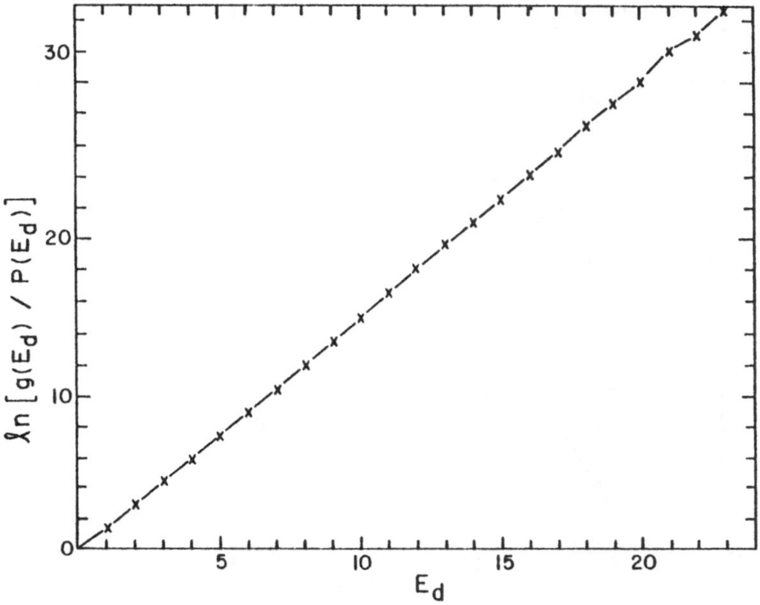

Figure 1. Plot of $\ell n[g(E_d)/P(E_d)]$ (see Eqn. (9)) vs E_d for a run at $E_T = 4096$.

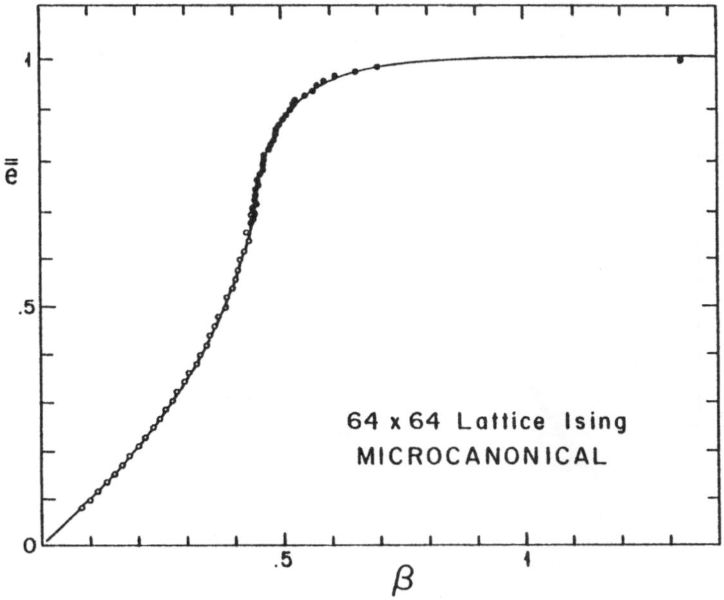

Figure 2. Nearest neighbor spin-correlation function $\bar{\bar{e}}$ versus β determined from the demon energy. The solid line is the exact result.

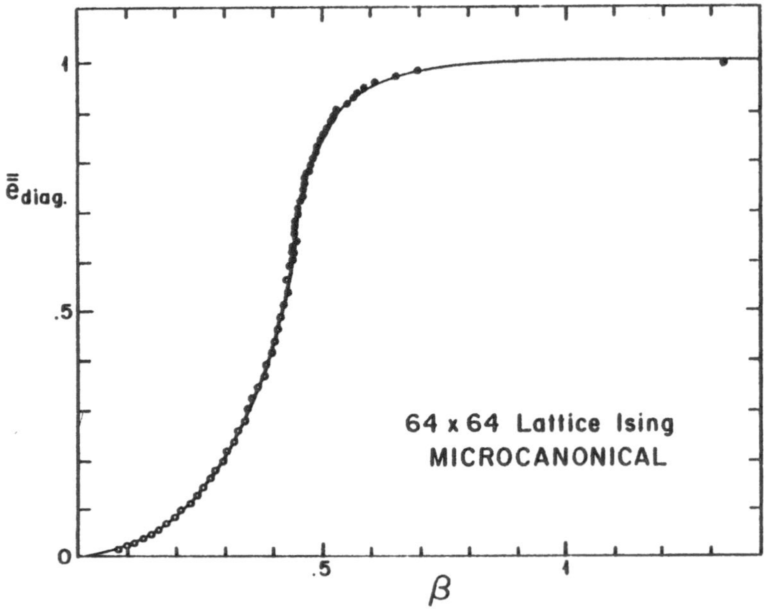

Figure 3. Nearest-neighbor-along-unit-diagonal spin-correlation $\bar{\bar{e}}_{diag}$ vs β.

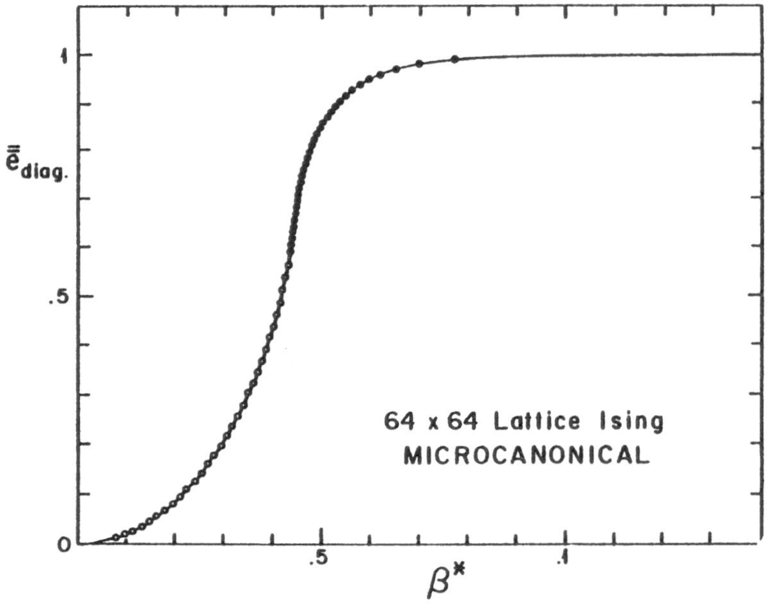

Figure 4. The data of Figure 3 plotted against β^* (see Eqn. (18)).

(i.e., when it is necessary to compute β). In Figure (4) there are of course the other 1/V corrections we have discussed, see Eqn. (16). But these are too small to be seen on this plot.

In Figure (5) we plot the quantity $\langle e_{diag} \rangle_\beta^{exact} - \bar{\bar{e}}_{diag}$ - (1/V corrections). This represents the intrinsic statistical error of the microcanonical ensemble. In Figure (6) we plot the corresponding quantity (viz. $\langle e_{diag} \rangle_\beta^{exact} - \langle e_{diag} \rangle_\beta$) for the canonical ensemble. This data was taken by running a Metropolis Monte Carlo at the same β values as the data in Figure (2) with hot and cold lattice starts and using the same amount of computer time as for the microcanonical runs (discarding 70 sweeps and averaging over 350 per β). A comparison of the scales on the ordinates in Figures (5) and (6) says that the statistical error in the microcanonical runs, for the same amount of computer time, is much smaller than for the canonical. This is mainly, though not entirely, from the difference in speed of the two programs.

Finally, in Figure (7), we plot $\bar{\bar{e}}_{diag}$ versus $\bar{\bar{e}}$ along with the exact curve in the canonical ensemble. The agreement is very good. This reflects the point already made that the major source of error

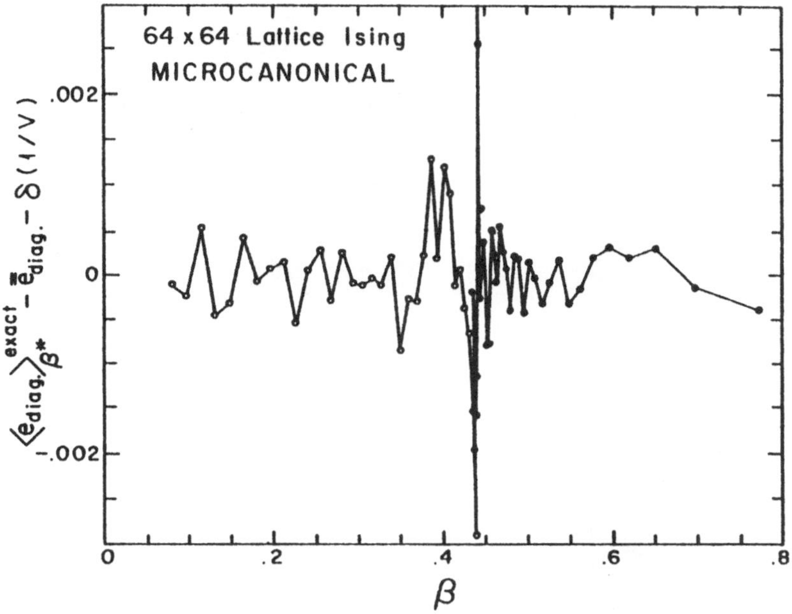

Figure 5. Statistical errors in the microcanonical ensemble for $\bar{\bar{e}}_{diag}$.

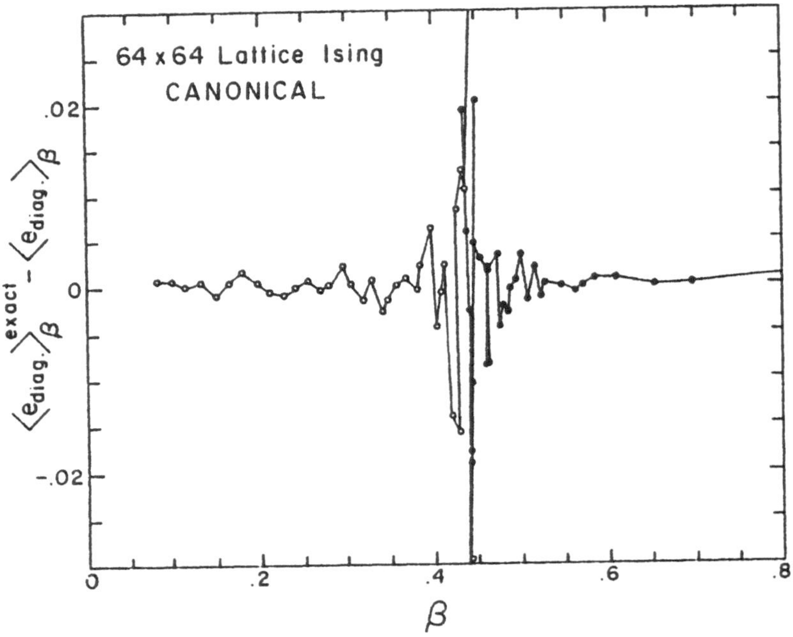

Figure 6. Statistical errors in the canonical ensemble for $<e_{diag}>$ using the same amount of CPU time as in Figure 5.

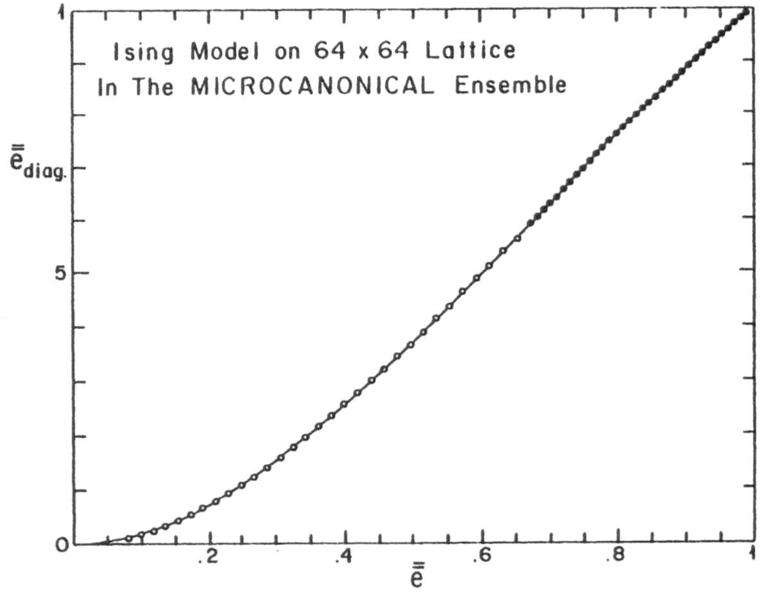

Figure 7. $\bar{\bar{e}}_{diag}$ versus $\bar{\bar{e}}$. The solid line is the exact (canonical result. This graph should be compared with Figures 2, 3 and 4.

in the microcanonical simulation is in estimating β from the demon energy and that this is not an issue if we measure correlations as a function of the natural variable $\overline{\overline{e}}$ instead of β.

There are many subtle issues that have been left unaddressed in this paper. These are discussed in detail in Ref. 2.

I am grateful to the Institute for Advanced Study and the organizers of the Cargese School for travel grants and the organizers of the Visegrad School for their hospitality. This work is supported in part by the DOE under grant # DE-AC02-76ER02220.

REFERENCES

1. M. Creutz, Phys. Rev. Lett. 50 (1983) 1411. D. Callaway and A. Rahman, Phys. Rev. Lett. 49 (1982) 613, and ibid., Argonne preprint, ANL-HEP-PR-83, 04, Jan., 1983.
2. For more detail see "Microcanonical simulation of Ising systems," G. Bhanot, M. Creutz and H. Neuberger, IAS preprint, Dec., 1983, to appear in Nucl. Phys. B[FS].
3. L. Jacobs and C. Rebbi, "Multi-spin Coding: A very efficient technique for Monte Carlo simulations of Spin Systems," J. of Comp. Phys. 1 (1981) 203.
4. U. Heller and N. Seiberg, Phys. Rev. D27 (1982) 2980.
5. R. B. Pearson, J. L. Richardson and D. Toussaint, "A special purpose machine for Monte Carlo simulation," Santa Barbara preprint, NSF-ITP-81-139, Jan., 1982; ibid., "A fast processor for Monte Carlo simulation," Santa Barbara preprint NSF-ITP-82-98, Oct., 1982.
6. B. M. McCoy and T. T. Wu, "The two dimensional Ising model," Harvard Univ. Press, Cambridge, Mass., 1973.

QUANTUM FIELD THEORY IN TERMS OF RANDOM WALKS AND RANDOM SURFACES

Jürg Fröhlich

Theoretical Physics
ETH-Hönggerberg
CH-8093 Zürich, Switzerland

In these notes I wish to summarize some recent developments in relativistic quantum field theory (RQFT) which have led to the idea that some class of RQFT in the Euclidean description, i.e. at imaginary time, should be viewed as classical statistical mechanics of interacting random walks and random surfaces. Put differently, the construction of many models of RQFT is a problem in stochastic geometry. We shall see that random walks describe fluctuations of matter degrees of freedom - quarks, leptons, Higgs particles - while random surfaces describe fluctuations of the "field of force", i.e. of the degrees of freedom of gauge fields. More precisely, random surfaces describe the fluctuations of the (chromo-electric) flux sheets of gauge theories in the confinement phase.

In terms more familiar to experts in statistical physics, the construction and analysis of the Euclidean Green's functions of a large class of RQFT models involving matter- and gauge fields turns out to be equivalent to a problem in the theory of critical phenomena in systems of "polymer chains" and random surfaces, "membranes".

The results reported in the following have been worked out in most enjoyable and, I believe, rather successful collaborations with D. Brydges, B. Durhuus, T. Jonsson, A. Sokal, T. Spencer, and others, and may be found in refs.[1-11]. These colleagues ought to be regarded as co-authors of these notes, but they bear no responsibility for possible errors. For further related results one should consult

* based on the author's lectures at Cargèse

[12-15]. The developments in [1-11] which I review in the following originated in [16-18,12]. Tom Spencer deserves much credit for having gotten many of us, including myself, interested in Symanzik's work [16]. I am also indebted to Giorgio Parisi for some illuminating comments and for reminding me of his interesting paper with Drouffe and Sourlas [19]. I hope that these remarks explain the genesis of most of the ideas sketched in these notes.

After completion of part of these lecture notes the tragic news reached me that our colleague and friend, Prof. K. Symanzik, has suddenly died. He was a veteran of the Cargèse schools. The material discussed in my notes is partly directly motivated, partly much inspired by Symanzik's deep work [16]. I should like to dedicate these notes to his memory, although I realize that he might not have approved of their rather soft style and that I cannot meet his standards of quality and understanding.

As remarked above, the construction of many RQFT models involving gauge fields can be reduced, essentially, to studying critical points in statistical systems of interacting random surfaces. The theory of critical behaviour in random surface models is not well developed, yet. Some of the results described in these notes represent a contribution to that theory which, we feel, might be somewhat interesting. As an example, I shall discuss a mean field theory of random surfaces [19] and prove that it yields unconventional values, $\gamma = 1/2$ and $\nu = 1/4$, for conventional critical exponents; ($\eta = 0$). I shall then show that mean field theory supplies rigorous bounds on correlations and exponents in a model of planar random surfaces [7,8]. The study of this planar random surface model is motivated by attempts towards constructing a theory of relativistic strings and by the well known qualitative features of pure SU(N) lattice gauge theory in the large-N-limit [20-25]. But the model might be of interest in a statistical mechanical context, too. And, in any event, it is fun to study it.

There is one problem with the main results summarized below: One could say that the (generalized) free field has struck again. Now, this sounds rather discouraging - and it is. However, some of the generalized free fields that we shall meet are "interesting" generalized free fields, (glue balls), whose spectral properties depend on the dimension of space-time. Moreover, some of our results, or rather conjectures, may be used to estimate the leading contribution in $1/N$ to glue ball - glue ball scattering, at least in a certain approximation to gauge theory. (Incidentally, the ideas alluded to here can be extended to include mesons, as well!)

In the first part of these notes we shall see that ultraviolet problems are one of the ways in which the free field may strike, (triviality of $\lambda\phi^4_{4+\varepsilon}$); another one is "large N". I shall indicate some consequences of the possible (or certain?) triviality of $\lambda\phi^4_4$ for unified theories of electro-weak interactions (comments on [26]), namely an upper bound on the Higgs mass. Moreover, liking science fiction, I shall propose yet another speculation about how quantum gravity may cure ultraviolet diseases which involves Anderson localization, caused by the fluctuating metric of space-time. (This is basically just a comment on [27].) But, apart from such digressions, these notes are intended to transmit serious information.

In the first part of these notes, I briefly review the random walk representation of scalar RQFT models, in particular of $\lambda\phi^4$, and I sketch how random walks can be used to solve problems in RQFT. I also make a comment on how quantum field theory can be used to solve problems concerning critical properties of random walks, [28-30]. I discuss problems arising in trying to construct $\lambda\phi^4_{4+\varepsilon}$ in the broken symmetry phase and describe some heuristic implications for gauge theories with Higgs fields such as the standard model. Fermions are not included in this review of pure matter QFT, (but see other contributions in this volume).

In the second part, I first recall some general qualitative features of non-abelian gauge theories, such as QCD, and formulate corresponding theoretical challenges. I then discuss some basic difficulties encountered in the application of perturbation theory, strong coupling-, 1/N- and ε-expansions to non-abelian gauge theories (Sect. 2.1). Subsequently I derive a flux-sheet (random-surface) representation of SU(2) lattice gauge theory which is quite useful in a qualitative analysis of confinement and which avoids some of the problems besetting the strong coupling expansion (Sect. 2.2). I review some results on what Mack has called dielectric gauge theories [31], i.e. "linear" versions of lattice guage theory, which were first proposed in [32]; see also [33,21]. Finally, I sketch extensions of those techniques to U(N)- and O(N)-models, for general values of N, and introduce, as an approximation to large-N-gauge theory, the planar random surface (PRS)model [7,8,21], (Sect. 2.3).

In the third part, I sketch a general approach to the "statistical mechanics of random surfaces" of which the models studied in part 2 are special examples. I formulate theoretical problems associated with the phenomena of "roughening", "topological complexity", "crumpling" and "collapse into trees" of random surfaces, (Sect.3.1). In Sect. 3.2, I discuss the PRS model in some detail and argue that it appears to have a continuum limit which is a good, albeit non-interacting RQFT, non-canonical in dimension $<d_c$, with $d_c \geqq 6$. Some speculations may be found in Sect. 3.3.

While, in the first part, I review theories of matter fields, the second and third part are devoted to the analysis of pure gauge theories. In a fourth part, matter- and gauge fields should be combined. But this is postponed to another occasion, (hoping there will be one). For some results in the spirit of this article see, however, [6,34].

ACKNOWLEDGMENTS

I thank the organizers of the Cargèse School for inviting me to present the material described below in the form of three lectures, and the friends and colleagues mentioned above for discussions and collaboration.

§ 1. Random Walk Representation of Scalar RQFT Models, Triviality of $\lambda\phi^4_{4+\varepsilon}$, and Consequences for the Standard Model

Some of the important topics in the theoretical physics of the eighties appear to be

1). Consistency questions in local RQFT.
   - Low-energy properties of QCD.
   - Beyond $SU(3)_c \times SU(2)_W \times U(1)$; the rôles of supersymmetry and gravity in particle physics at very high energies.
2) Dynamics of macroscopic systems, turbulence, "chaos".
3) Physics of disordered systems, incommensurate structures.
4) The rôle of computers in theoretical physics.

The main themes of these notes concern the first two circles of problems of topic 1):

- Stochastic-geometric analysis of RQFT in the Euclidean description, [35].
- Consistency of scalar RQFT models, in particular triviality of the continuum $\lambda\phi^4_{4+\varepsilon}$ model.
- Consequences for the standard model of electro-weak interactions.
- Confinement and other low-energy properties of non-abelian gauge theories.

In this first part, I discuss the first three themes. I now wish to summarize the main results concerning $\lambda\phi^4$ theory in four dimensions:

I only speak of those $\lambda\phi_4^4$ theories which can be constructed as continuum (scaling) limits of lattice $\lambda\phi^4$ theory with ferromagnetic nearest-neighbor couplings, (i.e. the field strength is kept non-negative, and Osterwalder-Schrader positivity is maintained). Moreover, $\lambda \geqq 0$. (There has been some discussion of antiferromagnetic couplings and negative λ, but that has not yet led to success, except in the large-N limit, where λ can be chosen to be negative [36].) I am aware of the fact that my premises are somewhat restrictive, but I have reasons to think that the conclusions are generally valid. (See, however, [37] for some speculations.)

Results

A) No non-Gaussian renormalization group fixed point can be reached starting from lattice $\lambda\phi_4^4$ theory; i.e. there is no non-trivial, scale-invariant $\lambda\phi_4^4$ theory. This means, heuristically, that the β-function has no non-trivial zeros.

B) If there are logarithmic violations of any one of the mean-field scaling laws, as predicted by the perturbative renormalization group [38] and numerical data for the self-avoiding walk model [3], then

- any continuum limit of lattice $\lambda\phi_4^4$ theory is a (generalized) free field;

- there is no broken-symmetry phase, i.e. $\langle\phi_{\text{ren.}}\rangle = 0$, or $= \infty$.

In dimension $d > 4$ these conclusions hold whenever the two-point function has a well-defined continuum limit. See [12,13,2,3,6].

The theoretical framework underlying our results and the subsequent discussion combines the following elements:

(i) Every RQFT admits an equivalent Euclidean (imaginary-time) description [39], and, in the case of scalar- and gauge fields, the Euclidean description is usually given by a Euclidean field theory (EFT) [40,41].

(ii) EFT, formulated in terms of functional integrals, is a special case of classical statistical mechanics. All EFT models considered in these notes can be constructed as continuum (scaling) limits of classical lattice systems. In this sense, the construction of the continuum limit of an EFT model is a problem in the theory of critical phenomena in classical lattice systems. See [42,43] for general summaries of rigorous results.

(iii) A large class of classical lattice systems, in particular classical spin systems, can be represented as gases of interacting random walks (Brownian paths). Lattice gauge theory in the confinement phase can be represented as a gas of interacting random surfaces; [16,1,2,18,6].

Among the theoretical physicists who have invented and developed these ideas (or ideas underlying them) one should mention A. Einstein; K. Wilson, K. Symanzik, L. Kadanoff, F. Wegner, and others.

It is one of the goals of these notes to make the synthesis between RQFT, EFT, critical phenomena in classical lattice systems and Brownian motion (and random surface theories) more precise and describe some important consequences thereof.

The connection between RQFT and EFT - Osterwalder-Schrader reconstruction [39,41] and further developments in that direction - are well known nowadays and are taken for granted in what follows.

We now briefly describe some of the other structures mentioned in (i) - (iii).

1.1 Random Walks on a Simple, (Hyper) Cubic Lattice

Consider a particle diffusing on the lattice $\mathbb{Z}^d$ by jumping from site to site. For simplicity, we suppose that it makes jumps of length one, with equal probability in all 2d directions. The trajectory of such a particle is a (nearest-neighbor) walk, ω . All walks, ω, of equal length, $|\omega|$ (= # jumps), have equal a priori probability. Let $n_j(\omega)$ be the total number of visits of a walk ω at some site $j \in \mathbb{Z}^d$. Clearly

$$|\omega| + 1 = \sum_{j \in \mathbb{Z}^d} n_j(\omega) \tag{1.1}$$

It is convenient to introduce a variable β, with $\ell n\beta$ conjugate to $|\omega|$, and to assign a statistical weight $\beta^{|\omega|+1}$ to ω. The critical value of β is

$$\beta_{\text{crit.}} = 1/2d , \tag{1.2}$$

because 2d is the number of directions in which the particle may jump. We introduce a generating function, $G_\beta(x,y)$, of those walks,

ω, which start at x and end at y, indicated by $\omega: x \to y$, as follows:

$$G_\beta(x,y) = \sum_{\omega:x \to y} \beta^{|\omega|+1} = \sum_{\omega:x \to y} \prod_{j \in \mathbb{Z}^d} \beta^{n_j(\omega)} \tag{1.3}$$

This is clearly the Green's function of

$$-\Delta + m^2, \text{ with } m^2 = \beta^{-1} - 2d , \tag{1.4}$$

where Δ is the finite difference Laplacian.

I.e.

$$G_\beta(x,y) = [(-\Delta + m^2)^{-1}]_{xy} . \tag{1.5}$$

Typical problems concerning random walks on $\mathbb{Z}^d$ are:

(a) Given a subset $\Omega \subset \mathbb{Z}^d$ and a walk ω with starting point $x \notin \Omega$. What is the probability that ω hits Ω ?

(b) What is the probability that n walks, $\omega_1, \omega_2, \ldots, \omega_n$, such that $\omega_k : \theta x_k \to \theta y_k$ and ω_k has weight $\beta(\theta)^{|\omega_k|+1}$, where $\beta(\theta)^{-1} - 2d = m(\theta)^2 = m_*^2 \theta^{-2}$, and $k = 1, \ldots, n$, intersect in at least one common point, $z \in \mathbb{Z}^d$, asymptotically, as $\theta \to \infty$? How does this probability depend on d and n ?

(c) If two walks, ω and ω', start at a common point, e.g. at 0, and have length n, how many times do they intersect each other?

etc.

(d) Are there analogous problems for random surfaces? Is there a standard, or "free" theory of random surfaces? If so, what kind of "potential theory" goes with it?

Answers to questions (a)-(c) are sketched in Sect. 1.4, answers to (d) in Sect. 3.2.

1.2 Classical Lattice Spin Systems

Consider now a system of classical spins, $\phi_x = (\phi_x^1, \ldots, \phi_x^N) \in \mathbb{R}^N$,

$x \in \mathbb{Z}^d$. Each spin has an a priori distribution

$$d\,\lambda(\phi_x) = e^{h\phi_x^1}\, g(\phi_x^2)\, d^N\phi_x \tag{1.6}$$

where

$$g(\phi^2) \overset{\text{e.g.}}{=} \begin{cases} \delta(|\phi|^2 - 1), \text{ or} \\ \exp[-\frac{\lambda}{4}|\phi|^4 + \frac{\mu^2}{2}|\phi|^2], \end{cases} \tag{1.7}$$

and h and μ^2 are real numbers, $\lambda \geq 0$.

The classical Hamilton function, H, is defined, for example, as

$$H_\Lambda = - \sum_{\langle j,\ell\rangle}^{\Lambda} \phi_j \cdot \phi_\ell, \tag{1.8}$$

where the sum ranges over all nearest-neighbor pairs $\langle j,\ell\rangle$ in $\Lambda \subseteq \mathbb{Z}^d$. The equilibrium state is given by the probability measure

$$d\,\mu_\Gamma(\phi) = \lim_{\Lambda \nearrow \mathbb{Z}^d} Z_{\Gamma,\Lambda}^{-1}\, e^{-\beta H_\Lambda(\phi)}\, B_{\partial\Lambda}(\phi) \prod_{j\in\Lambda} d\lambda(\phi_j), \tag{1.9}$$

where Γ denotes the parameters (β,λ,μ^2), with $\beta = (kT)^{-1}$ the inverse temperature, $\{B_{\partial\Lambda}\}$ is a family of functions, depending only on $\{\phi_j\}_{j\in\partial\Lambda}$, which prescribe the boundary conditions, and $Z_{\Gamma,\Lambda}$ is the usual partition function.

The properties of such systems can be derived from the correlation functions

$$\langle\phi_{x_1} \cdots \phi_{x_n}\rangle_\Gamma \equiv \int \prod_{j=1}^{n} \phi_{x_j}\, d\mu_\Gamma(\phi). \tag{1.10}$$

We also define

(1) the inverse correlation length (mass)

$$m(\Gamma) \equiv \lim_{|x|\to\infty} -\frac{1}{|x|} \ln \langle\phi_o;\phi_x\rangle_\Gamma, \tag{1.11}$$

where $\langle A;B\rangle_\Gamma \equiv \langle A\cdot B\rangle_\Gamma - \langle A\rangle_\Gamma \cdot \langle B\rangle_\Gamma$;

(2) the susceptibility

$$\chi\,(\Gamma) \equiv \sum_x \langle \phi_o ; \phi_x \rangle_\Gamma \; ; \tag{1.12}$$

(3) the dimensionless (four-point) coupling:

If $\langle \phi_x \rangle_\Gamma = 0$ define

$$u_{4,\Gamma}(x_1,x_2,x_3,x_4) = \langle \phi_{x_1} \cdots \phi_{x_4} \rangle_\Gamma - \sum_{\text{pairings},p} \langle \phi_{x_{p(1)}} \phi_{x_{p(2)}} \rangle_\Gamma \langle \phi_{x_{p(3)}} \phi_{x_{p(4)}} \rangle_\Gamma , \tag{1.13}$$

and

$\bar{u}_{4,\Gamma} \equiv \sum_{x_2,x_3,x_4} u_{4,\Gamma}\,(x_1, x_2, x_3, x_4)$. Then the dimensionless four-point coupling is defined by

$$\lambda_r\,(\Gamma) \equiv - \bar{u}_{4,\Gamma}\chi(\Gamma)^{-2}\, m(\Gamma)^d \; . \tag{1.14}$$

Among the important questions which one wants to ask about such systems are

(a') Is there a phase transition? Is it accompanied by symmetry breaking? For example: Consider $\langle \phi_x \rangle_\Gamma$. When β is small, $\lim_{h \searrow 0} \langle \phi_x \rangle_\Gamma = 0$. Now, increase the value of β . Is there some β_o such that, for $\beta > \beta_o$, $\lim_{h \searrow 0} \langle \phi_x \rangle_\Gamma \equiv M(\Gamma) \neq 0$?

(b') Is there a critical point $\beta_c \leq \beta_o$ such that, for $h = 0$, $m(\Gamma) \searrow 0$, as $\beta \nearrow \beta_c$; (and $m\,(\Gamma_c) \searrow 0$, as $h \searrow 0$).

If a critical point exists one may ask

(c') How does the approach to the critical point look like? Is the scaling hypothesis valid at $\beta = \beta_c$? (For simplicity, we assume here that only β is varied.)

If so one may introduce <u>critical exponents</u>

$$\langle \phi_x \phi_y \rangle_{\Gamma_c} \underset{|x-y| \to \infty}{\sim} |x-y|^{-(d-2+\eta)} \tag{1.15}$$

$$m(\Gamma) \sim (\beta_c - \beta)^{\nu} \tag{1.16}$$

$$\chi(\Gamma) \sim (\beta_c - \beta)^{\gamma} \tag{1.17}$$

One may then ask

(d') What are the values of the critical exponents? Why are $\gamma \geqq 1$, $\nu \geqq 1/2$, $\eta \geqq 0$? Why are $\gamma = 1$, $\nu = 1/2$, $\eta = 0$ in more than four dimensions?

I shall sketch answers to these questions in Sect. 1.5.

1.3 Euclidean Field Theory

If in all the formulas of Sect. 1.2 $\mathbb{Z}^d$ is replaced by $\mathbb{R}^d$, $\sum_{x \in \mathbb{Z}^d} (\cdot)$ by $\int_{\mathbb{R}^d} d^d x\, (\cdot)$, etc. we obtain the basic heuristic formulas defining Euclidean field theory. Instead of a spin we now speak of a Euclidean field, ϕ_x, and the Hamiltonian function, H, is now called (Euclidean) action A. It is, formally, given by

$$A(\phi) = \int d^d x \left\{\frac{\beta}{2} |(\nabla\phi)_x|^2 + \frac{\lambda}{4} |\phi_x|^4 - \frac{\mu^2}{2} |\phi_x|^2\right\} \tag{1.18}$$

The Euclidean vacuum functional is given by the formal probability measure

$$d\mu_\Gamma(\phi) = Z_\Gamma^{-1}\, e^{-\ (\phi)} \prod_{x \in \mathbb{R}^d} \mathcal{D}\phi_x, \quad \Gamma \equiv (\beta, \lambda, \mu^2) , \tag{1.19}$$

and the properties of the theory are coded into the so-called Euclidean Green's- or Schwinger functions

$$S_\Gamma(x_1, \ldots, x_n) \equiv \int \prod_{i=1}^{n} \phi_{x_i}\, d\mu_\Gamma(\phi) \tag{1.20}$$

This is the Euclidean Gell-Mann-Low formula. The reconstruction theorem of Osterwalder and Schrader [39] says that if the S_Γ have certain properties they uniquely determine the Wightman distributions of an RQFT by analytic continuation in the time variables.

As in 1.2, we define

$$m = m(\Gamma) \equiv \lim_{|x|\to\infty} - \frac{1}{|x|} \ln S_\Gamma(0,x) \quad \text{(mass)} \tag{1.21}$$

$$\chi = \chi(\Gamma) \equiv \int d^d x \; S_\Gamma(0,x) \quad \text{(susceptibility)} \tag{1.22}$$

and

$$u_{4,\Gamma}(x_1,\ldots,x_4) \equiv S_\Gamma(x_1,\ldots,x_4) - \sum_p S_\Gamma(x_{p(1)},x_{p(2)}) S_\Gamma(x_{p(3)},x_{p(4)}),$$

$$\bar{u}_{4,\Gamma} \equiv \int d^d x_2 \, d^d x_3 \, d^d x_4 \; u_{4,\Gamma}(x_1,x_2,x_3,x_4) \;,$$

$$\lambda_r \equiv \bar{u}_4 \, \chi^{-2} \, m^d \tag{1.23}$$

(physical coupling constant)

Some of the main questions which one likes to ask about EFT are:

(a'') Do formulas (1.18) - (1.20) have any rigorous, mathematical meaning? Are there any models of EFT which are non-Gaussian, i.e. $\lambda_r \neq 0$, (more precisely which are not equivalent to a free or generalized free field)? How can we construct such models?

(b'') Supposing that a non-Gaussian model exists what are its main physical properties.

To provide some partial answers to these two questions is the main purpose of these notes.

As our strategy towards settling (a'') - one among several possible strategies - we shall adopt the principle that EFT models are regularized at short distances by putting them onto a lattice. In this way classical lattice spin systems of the sort described in Sect. 1.2 are the lattice approximations to models of EFT. The question then arises how one should go about constructing the continuum limit of lattice spin systems? We consider the example introduced in eqns. (1.6) - (1.8). By re-scaling ϕ_x we may always set $\beta = 1$. We know from [44] that, in three or more dimensions, there is a line $\mathcal{L}_c = \{(\lambda(t), \mu^2(t)) : -\infty < t < \infty\}$ such that, for all $\Gamma \equiv (1,\lambda,\mu^2) \in \mathcal{L}_c$, and for N = 1, 2,

$$m^2(\Gamma) = \chi^{-1}(\Gamma) = 0 \tag{1.24}$$

Moreover, $m^2(\Gamma)$ is continuous in Γ , [45,5].

We may thus choose a function $\Gamma(\theta)$ of a scale parameter θ ranging over $[1,\infty)$ such that

$$\theta\, m(\Gamma(\theta)) \xrightarrow[\theta \to \infty]{} m^* \geqq 0 . \tag{1.25}$$

Furthermore we choose a function $\alpha(\theta)$ such that, for $x \neq y$, $|x-y| < \infty$ and θ large enough, the re-scaled two-point function, $\alpha(\theta)^2 \langle \phi_{\theta x}\, \phi_{\theta y} \rangle_{\Gamma(\theta)}$, is bounded away from 0 and ∞ . Note that θx and θy must lie in the lattice $\mathbb{Z}^d$, hence x and y belong to a lattice of lattice spacing θ^{-1} . If x and y are in $a \cdot \mathbb{Z}^d$ then $\langle \phi_{\theta x}\, \phi_{\theta y} \rangle_{\Gamma(\theta)}$ is defined for all $\theta = n\, a^{-1}$, n = 1,2,3,4, ... Hence the <u>scaling limit</u>, $\theta \to \infty$, <u>is the same as the continuum limit</u>, $a \to 0$, <u>and, since</u> $m(\Gamma(\theta)) \sim \theta^{-1} \to 0$, <u>the parameters</u>, $\Gamma(\theta)$, <u>of the lattice model are required to approach a critical point, as</u> $\theta \to \infty$.

One has the following

<u>Theorem</u>

Consider one of the classical lattice spin systems introduced in (1.6)-(1.8), Sect. 1.2, with N=1 or 2. If, for a suitably chosen sequence $\{\theta_k\}_{k=1}^{\infty}$ diverging to $+\infty$, $\langle\theta\rangle_{\Gamma(\theta_k)} = 0$, $m(\Gamma(\theta_k)) > 0$, for all k, and

$$\lim_{k\to\infty} \alpha(\theta_k)^2 \langle \phi_{\theta_k x}\, \phi_{\theta_k y} \rangle_{\Gamma(\theta_k)} \equiv S(x,y) \tag{1.26}$$

exists as a non-zero measure with support strictly larger than $\{x = y\}$ then there is a subsequence $\{\tilde{\theta}_k\}_{k=1}^{\infty}$ of $\{\theta_k\}_{k=1}^{\infty}$ such that, in the sense of distributions,

$$S(x_1,\ldots, x_n) = \lim_{k\to\infty} \alpha(\tilde{\theta}_k)^n \langle \phi_{\tilde{\theta}_k x_1} \ldots \phi_{\tilde{\theta}_k x_n} \rangle_{\Gamma(\tilde{\theta}_k)}$$

exists, for all n. Moreover, for n odd, $S(x_1,\ldots,x_n) = 0$, while for $n = 2\ell$ even

$$\frac{1}{(2\ell-1)!!} G(x_1,\ldots,x_n) \leqq S(x_1,\ldots,x_n) \leqq G(x_1,\ldots,x_n), \tag{1.27}$$

where $G(x_1,\dots,x_n) \equiv \sum_{\text{pairings } p} S(x_{p(1)},x_{p(2)}) \dots S(x_{p(2\ell-1)},x_{p(2\ell)})$

(1.28)

Remarks. (1) The critical exponent η can also be defined by the relations

$$\alpha(\theta)^2 \sim \theta^{d-2+\eta}, \text{ and } S(x,y) \underset{x \approx y}{\sim} |x-y|^{-(d-2+\eta)} \tag{1.29}$$

From this and (1.26) one obtains immediately

$$\gamma = \nu\ (2-\eta) \tag{1.30}$$

[See definitions (1.15)-(1.17). We may again suppose, for simplicity, that only β is varied.]

(2) From the above discussion we see that the construction of EFT models in the continuum limit and the analysis of critical phenomena in classical lattice systems are closely related problems: If we know enough about the approach to the critical point, including the values of various critical exponents, in a lattice spin system with symmetric transfer matrix we can use that information to construct an EFT model as the scaling limit of that lattice system.

Conversely, if Euclidean field theories which arise as scaling limits of lattice systems can be constructed directly, at least approximatively, we can use them to calculate critical exponents of lattice systems. For example, the short-distance behaviour of the connected two-point functions of the fields ϕ_x and ϕ_x^2 of such an EFT determine the exponents η, ν and γ introduced above. This is behind the field-theoretic methods of e.g. [38].

1.4 Random Walks as EFT

In sect. 1.1 we have defined simple random walks, ω, with fixed end-points and have assigned a statistical weight

$$\mathfrak{z}_\beta(\omega) = \beta^{|\omega|}, \quad 0 < \beta \leqq \frac{1}{2d} \tag{1.31}$$

to each such walk. We may obtain this weight from an integral over "local times". Let $m^2 = \beta^{-1} - 2d$. Let $n_j(\omega)$ be the total number of visits of ω at site j, and let $t_j = t_j(\omega)$ be the total time ω has spent at site j . This is a random variable with distribution

$$e^{-\beta^{-1}t_j}\, d\nu_{n_j(\omega)}(t_j) \quad , \tag{1.32}$$

where

$$d\nu_n(t) = \begin{cases} \delta(t)\, dt, & \text{if } n = 0 \\ \dfrac{t^{n-1}}{(n-1)!}\,\theta(t)\, dt, & \text{if } n = 1,2,3, \ldots . \end{cases} \tag{1.33}$$

Note that (1.32) is the distribution of the sum $\sum_{k=1}^{n_j(\omega)} t_j^{(k)}$, of exponential waiting times, $t_j^{(k)}$, (with distribution $e^{-\beta^{-1}t_j^{(k)}}\, dt_j^{(k)}$).

We define

$$\left.\begin{aligned} d\nu_\omega(t) &\equiv \prod_{j\in\mathbb{Z}^d} d\nu_{n_j(\omega)}(t_j) \;, \qquad \text{and} \\ d\rho_\omega(t) &\equiv \Big(\prod_{j\in\mathbb{Z}^d} e^{-\beta^{-1}t_j}\Big)\, d\nu_\omega(t) \end{aligned}\right\} \tag{1.34}$$

Then

$$\mathfrak{z}_\beta(\omega) = \int \ldots \int d\,\rho_\omega(t)$$

Let Δ be the finite difference Laplacian.

The heat kernel is given by

$$[e^{T(\Delta-m^2)}]_{xy} = \sum_{\omega:x\to y} \int \ldots \int d\rho_\omega(t) \quad \delta\big(\sum_j t_j - T\big) \;,$$

and the propagator of the free lattice field by

$$G_\beta(x,y) = [(-\Delta+m^2)^{-1}]_{xy} = \sum_{\omega:x\to y} \int \ldots \int d\rho_\omega(t) \;. \tag{1.35}$$

Next, we introduce a "connected four-point function"

$$G^{c}_{\beta,\lambda}(x_1,x_2;\, y_1,y_2) = \sum_{\substack{\omega_1: x_1 \to x_2 \\ \omega_2: y_1 \to y_2}} \int \dots \int d\rho_{\omega_1}(t^1)\, d\rho_{\omega_2}(t^2)\, I_\lambda(t^1,t^2), \tag{1.36}$$

where

$$I_\lambda(t^1,t^2) = \exp\Big(-\lambda \sum_j t^1_j \cdot t^2_j\Big) - 1$$

Note that $-1 \leqq I_\lambda \leqq 0$, with

$$I_\lambda(t^1,t^2) = 0 \quad \text{if } \omega_1 \cap \omega_2 = \emptyset, \quad \forall \lambda \in [0,\infty), \tag{1.37}$$

and

$$\lim_{\lambda\to\infty} I_\lambda(t^1,t^2) = -1 \text{ if } \omega_1 \cap \omega_2 \neq \emptyset .$$

Furthermore

$$I^{(2n+1)}_\lambda(t^1,t^2) \leqq I_\lambda(t^1,t^2) \leqq I^{(2n)}_\lambda(t^1,t^2), \tag{1.38}$$

where $I^{(m)}_\lambda(t^1,t^2)$ arises by replacing the exponential by its Taylor series expansion up to m^{th} order in λ.

Now

$$p_m(x_1,y_1) \equiv -m^4 \lim_{\lambda\to\infty} \sum_{x_2,y_2} G^{c}_{\beta,\lambda}(x_1,x_2;\, y_1,y_2), \tag{1.39}$$

with, we recall,

$$m^{-2} \equiv (\beta^{-1} - 2d)^{-1} = \sum_y G_\beta(x,y)$$

is the probability that two walks, ω_1 and ω_2, starting at points x_1 and y_1, intersect somewhere.

For this theory, i.e. simple random walk, it is easy to construct the scaling limit and to identify it with Brownian motion. In the last section we have learnt that passing to the scaling limit is equivalent to approaching a critical point of the lattice theory in such a way that the mass, in physical units (e.g. cm), stays fixed. Thus quantities, like the intersection probability of Brownian paths, referring to the properties of Brownian motion can be calculated from the theory of simple random walk on the lattice by letting the "bare mass" m tend to 0.

We want to illustrate this principle by analyzing the m-dependence of $p_m(x_1,y_1)$. Since $p_m(x_1,y_1)$ is a probability, it is a pure number. Moreover, m has dimensions of $(\text{length})^{-1}$. Thus, the average intersection probability

$$p_m \equiv \sum_{y_1 \in \mathbb{Z}^d} m^d \, p_m(x_1,y_1) \tag{1.40}$$

is a dimensionless quantity. It corresponds to the physical coupling constant, λ_r, introduced in (1.23).

The main results on p_m are summarized in the following

Theorem

1) In four dimensions,

$$p_m \sim |\log m|^{-1}$$

2) In two or three dimensions, p_m is strictly positive, as $m \searrow 0$.

This theorem says that, in the continuum limit in four dimensions, two Brownian paths never intersect, while they do intersect in less than four dimensions. As a corollary one can deduce from this result that, in $d \leqq 4$ dimensions, two Brownian paths starting at the same point intersect each other infinitely often with probability 1.

A correct proof of part 1) has first been given by Lawler [46]. (The lower bound on p_m, $\sim |\log m|^{-1}$, appears in [47,3,13]. A streamlined version of Lawler's proof has been worked out by Felder [30].)

Part 2) of the theorem is due to Erdös and Taylor [47], with an improvement due to Sokal. In the following we give a heuristic "proof" of this theorem using a renormalization group (flow-) equation. For this purpose, we expand the r.h.s. of (1.39) in powers of λ, using the following "splitting lemma".

<u>Lemma</u>

$$\sum_{\omega:x\to y} \int d\rho_\omega (t) \; t_j \; F(t)$$

$$= \sum_{\omega_1:x\to j} \sum_{\omega_2:j\to y} \int d\rho_{\omega_1} (t^1) d\rho_{\omega_2} (t^2) \; F(t^1+t^2),$$

<u>with</u> $t^i = \{t_j^i\}_{j\in\mathbb{Z}^d}$, F <u>some function of</u> $\{t_j\}_{j\in\mathbb{Z}^d}$.

The proof follows by inspection; but see [4]. This lemma can now be used to evaluate the $d\rho_{\omega_1}(t^1) \cdot d\rho_{\omega_2}(t^2)$-integrals of

$$I_\lambda(t^1,t^2) = \sum_{n=1}^\infty \frac{(-\lambda)^n}{n!} \sum_{j_1,\dots,j_n} t_{j_1}^1 \; t_{j_1}^2 \; \cdots \; t_{j_n}^1 \; t_{j_n}^2$$

After resumming each term over all resulting random walks the final outcome can conveniently be described in terms of Feynman diagrams:

$\longleftrightarrow$ $p_m \equiv \lambda_r$,   o $\longleftrightarrow$ a bare vertex,

$\underline{\qquad}$ $\longleftrightarrow$ a bare propagator, $G_\beta(x,y)$. All vertices, but one,

are to be summed over all possible positions. The final result then is

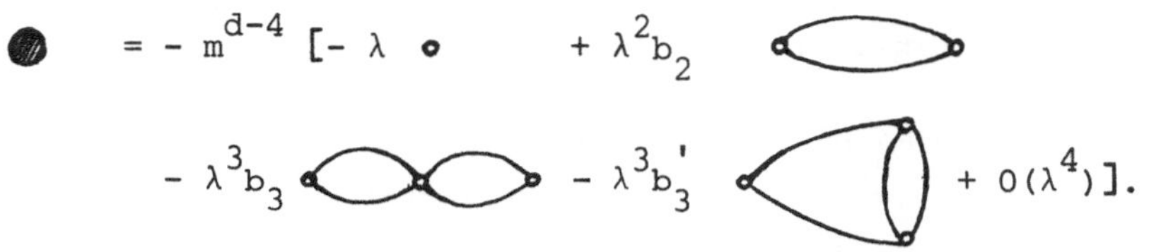

$$\text{⬤} = -m^{d-4} [-\lambda \; \text{o} \qquad + \lambda^2 b_2 \; \text{⬯}$$

$$\qquad\qquad - \lambda^3 b_3 \; \text{⧖} - \lambda^3 b_3' \; \text{⬱} + O(\lambda^4)]. \tag{1.41}$$

This expansion resembles Feynman perturbation theory for λ_r in (lattice) $\lambda\phi^4$-theory, with the difference that <u>no</u> self-energy diagrams occur. Note that, by (1.38), this expansion yields an <u>upper bound</u> on p_m if broken off after the $(2n+1)^{st}$ order, while it yields a <u>lower bound</u> when broken off after the $(2n)^{th}$ order. A similar result holds for $\lambda\phi^4$ theory, [4].

It is natural to ask whether expansion (1.41) comes from some (in d=4 dimensions) renormalizable quantum field theory related to $\lambda\phi^4$? The answer is <u>yes</u>! Indeed, consider fields $\vec{\phi}_1,\dots,\vec{\phi}_p$, with $\vec{\phi}_j = (\phi_j^{(1)}, \dots, \phi_j^{(n)})$, j=1,...,p, which have a self-interaction given by

$$A_{\text{int.}} = \frac{1}{4} (\lambda_1 \sum_{j=1}^{p} |\vec{\phi}_j|^4 + \lambda_2 \sum_{i \neq j} |\vec{\phi}_i|^2 |\vec{\phi}_j|^2) \tag{1.42}$$

(This action keeps its form under renormalization in four dimensions.) Then

$$G_\beta(x,y) = \lim_{\substack{\lambda_1 \to 0 \\ n \to 0}} \langle \phi_1^{(1)}(x)\, \phi_1^{(1)}(y) \rangle_\Gamma \,, \tag{1.43}$$

where $\Gamma = (\lambda_1, \lambda_2, m)$, and $\langle(\cdot)\rangle_\Gamma$ is the expectation in the measure $d\mu_\Gamma$ introduced in (1.9) and (1.19) ; see also (1.10).

Furthermore,

$$G^c_{\beta,\lambda}(x_1,x_2 \,;\, y_1,y_2)$$

$$= \lim_{\lambda_2 \to \infty} \lim_{\substack{\lambda_1 \to 0 \\ n \to 0}} \langle \phi_1^{(1)}(x_1)\, \phi_1^{(1)}(x_2) \,;\, \phi_2^{(1)}(y_1)\, \phi_2^{(1)}(y_2) \rangle_\Gamma \tag{1.44}$$

A proof of this and more general results is contained in [29]. Next, we introduce two dimensionless, physical coupling constants, $\tilde{\lambda}_r$ and λ_r: $\tilde{\lambda}_r$ is associated with $\langle \phi_1^{(1)}(x_1);\phi_1^{(1)}(x_2);\phi_1^{(1)}(x_3);\phi_1^{(1)}(x_4)\rangle_\Gamma$, and $\lambda_r \equiv p_m$. We propose to calculate the β-functions for the flow of $\tilde{\lambda}_r$ and λ_r under variations of $|\log m|$; (they are denoted W, $\tilde{W}$, respectively). From (1.41) we get, with $\varepsilon \equiv 4-d$ <u>and in the limit</u> $n \to 0$,

$$W(\lambda_r) \equiv \frac{d\lambda_r}{d|\log m|} = -\varepsilon\, \lambda_r + a_2 \lambda_r^2 + O(\lambda_r^3\,). \tag{1.45}$$

All coefficients in this expansion are finite, as $m \searrow 0$, (renormalizability of the four-dimensional theory!), and $a_2 > 0$. The function $\tilde{W}(\tilde{\lambda}_r,\lambda_r)$ can also be calculated by standard Feynman perturbation theory and has a finite power series expansion in ε, $\tilde{\lambda}_r$ and λ_r. The n- and p-dependence of W and $\tilde{W}$ in lowest orders in ε is easily determined [48]. From this one infers that the <u>fixed point of the renormalization flow</u>, when $n \gtrsim 0$, $p \gtrsim 2$, ε small, is given by

$$\tilde{\lambda}_r^* = \frac{n(p-1)\varepsilon}{pn(n+8)-16(n-1)}\,, \quad \lambda_r^* = \frac{(4-n)\varepsilon}{pn(n+8)-16(n-1)}\,, \tag{1.46}$$

to lowest order in ε . See Brézin et al. [48]. Thus, when $n \to 0$ $\tilde{\lambda}_r^* \to 0$, hence $\tilde{\lambda}_r \to 0$, for arbitrary $\varepsilon \geqq 0$, and

$$\lambda_r \equiv p_m \sim \begin{cases} |\log m|^{-1} & , \; d = 4\,, \\ \varepsilon & , \; d < 4\,, \end{cases} \tag{1.47}$$

as $m \searrow 0$.

This agrees with mathematically rigorous results of Lawler (d=4) and Erdös and Taylor (d=3); [46,47].

Very similarly, we can also calculate intersection probabilities for $N > 2$ Brownian paths. For this purpose we consider a lattice field theory with self-interaction

$$A_{\text{int.}} = \lambda\, |\vec{\phi}_1|^2 \cdots |\vec{\phi}_N|^2 + \mathcal{E}_n\,, \tag{1.48}$$

$$\vec{\phi}_j = (\phi_j^{(1)}, \ldots, \phi_j^{(n)}), \quad j = 1,\ldots,N.$$

The term $\mathcal{E}_n$ becomes irrelevant as $n \to 0$.

On the lattice,

$$G_\beta(x,y) = \lim_{n\to 0} \langle \phi_j^{(1)}(x)\, \phi_j^{(1)}(y) \rangle_{\lambda,m}\,,$$

with $m^2 = \beta^{-1} - 2d$, and the <u>average intersection probability</u>, $p_m^{(N)}$, is given by

$$p_m^{(N)} = \sum_{x_2,\ldots,x_N} (m^d)^{N-1} \sum_{x_1'\ldots,x_N'} \chi(\beta)^{-N}\, G^c_{\beta,\lambda}(x_1,x_1';\ldots;x_N,x_N')\,, \tag{1.49}$$

where $\chi(\beta) = m^{-2} = \sum_y G_\beta(x,y)$, and

$$G^c_{\beta,\lambda}(x_1,\ldots;x_N,x'_N) = \lim_{\lambda\to\infty}\ \lim_{n\to 0} \langle \phi^{(1)}_1(x_1)\phi^{(1)}_1(x'_1);\ \ldots\ ;\ \phi^{(1)}_N(x_N)\phi^{(1)}_N(x'_N)\rangle_{\lambda,m} \tag{1.50}$$

This theory is renormalizable when

$$(m^d)^{N-1} = \chi(\beta)^{-N} = m^{2N}\ ,\ \text{i.e. in}$$

$$d = {}^{2N}\!/_{N-1} > 2 \text{ dimensions.} \tag{1.51}$$

Using a renormalization group (flow-)equation, as above - see (1.45) - we obtain

$$\left.\begin{array}{l} 0 < p_m^{(N)} \lesssim |\log m|^{-1}\ ,\ \text{when } d = {}^{2N}\!/_{N-1}, \\ \text{and} \quad p_m^{(N)} \geqq \text{const., for all } m, \text{ when } d = 2. \end{array}\right\} \tag{1.52}$$

The first part of (1.52) can be made rigorous, at least when N = 3 [46]. It casts some doubts on the possibility of constructing a non-trivial $\lambda\phi^6$ theory in three dimensions.

It is tempting to conjecture that for the theory of simple random walk, the ε-expansion and a 1/N-expansion are asymptotic. In the direction of such results, Felder has derived a non-perturbative expression and some estimates, in particular strict positivity for all values of λ_r, for the β-function, $W(\lambda_r)$; [30].

We believe, we have indicated here how useful and powerful field-theoretic methods can be in the analysis of random walk problems. (This is also true in the physics of polymer chains, branched polymers, etc., as is well known. See e.g. [28,14].) In the next subsection we show how powerful random-walk methods can be in the analysis of quantum field theory.

1.5 EFT as a Gas of Random Walks with Hard Core Interactions

In this section we review a random walk representation of the Ising model and lattice $\lambda\phi^4$ theory originally proposed by Symanzik [16]. It consists of a somewhat cleverly partially resummed high-temperature expansion. It has first been analyzed mathematically in [1] and was then used to derive many results, old and new, on $\lambda\phi^4$-theory in dimension $d \geqq 2$; see [2-6]. Some part of these as well as some other results were first proven by Aizenman [12] by different, though related methods and extended in [13].

We consider a classical spin system (= lattice EFT) as defined in (1.6)-(1.10). For simplicity we consider one-component spins, but our methods work also for 0- and 2-component spins, some of them even in the general case.

The starting point of our random walk representation is the following sequence of trivial identities:

$$\phi \int d\nu_{n-1}(t) g(\phi^2+2t) = -\frac{\partial}{\partial\phi} \int d\nu_n(t)\, g(\phi^2 + 2t), \quad n = 1,2,3, \ldots, \text{ with } d\nu_n(t) \text{ as in (1.33).} \tag{1.53}$$

Let $\underline{n} = \{n_j\}_{j \in \mathbb{Z}^d}$, $n_j = 0,1,2,\ldots$, for all j.

We define

$$\left.\begin{aligned} [(\cdot)]_\Gamma(\underline{n}) &= \int (\cdot)\, e^{-\beta H(\phi)} \prod_{j \in \mathbb{Z}^d} g_{n_j}(\phi_j^2)\, d\phi_j \,, \\ Z_\Gamma(\underline{n}) &= [1]_\Gamma(\underline{n}) \,, \end{aligned}\right\} \tag{1.54}$$

where

$$\left.\begin{aligned} g_n(\phi^2) &\equiv \int d\nu_n(t) g(\phi^2+2t) \,, \\ H(\phi) &= -\sum_{\langle i,j \rangle} \phi_i \cdot \phi_j - \sum_j h_j \phi_j \,, \end{aligned}\right\} \tag{1.55}$$

and $\Gamma = (\beta, \{h_j\})$. As an example, we now consider the unnormalized two-point function:

$$[\phi_x \phi_y]_\Gamma(\underline{n}) = \int \ldots \int \Big(\prod_{j \neq x} g_{n_j}(\phi_j^2)\, d\phi_j\Big)\, \phi_y \cdot$$

$$\cdot \int \phi_x\, g_{n_x}(\phi_x^2)\, e^{-\beta H(\phi_x, \ldots)}\, d\phi_x$$

$$= \beta \int \ldots \int \left(\prod_{j \neq x} g_{n_j}(\phi_j^2)\, d\phi_j \right) \phi_y \,\cdot$$

$$\cdot \int g_{n_x+1}(\phi_x^2) \left[-\frac{\partial H}{\partial \phi_x} \right](\phi)\, e^{-\beta H(\phi)}$$

and we have used (1.53) with $\phi = \phi_x$ and (1.54).

By (1.55),

$$-\frac{\partial H}{\partial \phi_x}(\phi) = h_x + \sum_{x_1 : |x_1 - x| = 1} \phi_{x_1}$$

Thus

$$\begin{aligned} [\phi_x \phi_y]_\Gamma(\underline{n}) &= \beta \{ h_x [\phi_y]_\Gamma (\underline{n} + \delta_x) + \\ &+ \sum_{x_1 : |x_1 - x| = 1} [\phi_{x_1} \phi_x]_\Gamma (\underline{n} + \delta_x) , \end{aligned} \tag{1.56}$$

with $\delta_x(j) \equiv \delta_{xj} = \begin{cases} 1, & x = j \\ 0, & \text{otherwise} . \end{cases}$

Similarly,

$$\begin{aligned} [\phi_y]_\Gamma(\underline{n}) &= \beta \{ h_y [1]_\Gamma (\underline{n} + \delta_y) + \\ &+ \sum_{y_1 : |y_1 - y| = 1} [\phi_{y_1}]_\Gamma (\underline{n} + \delta_y) \} \end{aligned} \tag{1.57}$$

By iterating (1.56) and (1.57) until only terms proportional to $[1]_\Gamma(\underline{n} + \cdot)$ are left we finally get

$$\begin{aligned} [\phi_x \phi_y]_\Gamma(\underline{n}) &= \sum_{x',y'} h_{x'} h_{y'} \sum_{\substack{\omega_1 : x \to x' \\ \omega_2 : y \to y'}} \beta^{|\omega_1| + |\omega_2|} [1]_\Gamma (\underline{n} + \underline{n}(\omega_1) + \underline{n}(\omega_2)) \\ &+ \sum_{\omega : x \to y} \beta^{|\omega|} [1]_\Gamma (\underline{n} + \underline{n}(\omega)) , \end{aligned} \tag{1.58}$$

where $\underline{n}(\omega) = \{n_j(\omega)\}_{j\in\mathbb{Z}^d}$ and $n_j(\omega)$ is the total number of visits of the walk ω at site j, and $|\omega| = -1 + \sum_j n_j(\omega)$ is the number of jumps made by ω. We define

$$\mathcal{Z}_\Gamma(\omega_1,\ldots,\omega_n) = \beta^{|\omega_1|+\ldots+|\omega_n|} Z_\Gamma^{-1} Z_\Gamma\left(\sum_{\alpha=1}^{n} \underline{n}(\omega_\alpha)\right) \tag{1.59}$$

Identity (1.58) then yields

$$\langle\phi_x\phi_y\rangle_\Gamma = \sum_{x',y'} h_{x'} h_{y'} \sum_{\substack{\omega_1: x\to x'\\ \omega_2: y\to y'}} \mathcal{Z}_\Gamma(\omega_1,\omega_2) + \sum_{\omega: x\to y} \mathcal{Z}_\Gamma(\omega)$$

This is the random walk representation promised at the beginning of this section.* It generalizes to arbitrary n-point functions in a straightforward manner. When $h_j = 0$, for all j, this generalization is particularly elegant and simple:

$$\langle\phi_{x_1}\cdots\phi_{x_{2n}}\rangle_\Gamma = \sum_{\substack{\text{pairings}\\ p}} \sum_{\substack{\omega_\alpha: x_{p(2\alpha-1)}\to x_{p(2\alpha)}\\ \alpha=1,\ldots,n}} \mathcal{Z}_\Gamma(\omega_1,\ldots,\omega_n) \tag{1.61}$$

For the connected four-point function, $u_{4,\Gamma}$, introduced in (1.13), this yields

$$u_{4,\Gamma}(x_1,x_2,x_3,x_4) = \sum_p \sum_{\substack{\omega_1: x_{p(1)}\to x_{p(2)}\\ \omega_2: x_{p(3)}\to x_{p(4)}}} [\mathcal{Z}_\Gamma(\omega_1,\omega_2) - \mathcal{Z}_\Gamma(\omega_1)\mathcal{Z}_\Gamma(\omega_2)] \tag{1.62}$$

Finally, we wish to derive a <u>local-time representation</u> for the "correlation functions" $\mathcal{Z}_\Gamma(\omega_1,\ldots,\omega_n)$:

---

*This derivation [6] of the random walk representation was inspired by a discussion with E.H. Lieb which we gratefully acknowledge.

Using the simple identity

$$d\nu_{n+m}(t) = \int \left(\frac{d\nu_n}{dt}\right)(t-s)\, d\nu_m(s)$$

it is straightforward to verify that

$$\mathcal{Z}_\Gamma(\omega_1,\ldots,\omega_n) = \beta^{|\omega_1|+\ldots+|\omega_n|} \int\ldots\int \prod_{\alpha=1}^{n} d\nu_{\omega_\alpha}(t^\alpha)\, \mathcal{Z}_\Gamma\left(\sum_{\alpha=1}^{n} t^\alpha\right), \quad (1.63)$$

where

$$d\nu_\omega(t) = \prod_j d\nu_{n_j(\omega)}(t_j)\,,$$

and

$$\mathcal{Z}_\Gamma(t) = Z_\Gamma^{-1} \int \left(\prod_j g(\phi_j^2 + 2t_j)\, d\phi_j\right) e^{-\beta H(\phi)} \qquad (1.64)$$

The variables t_j^α have the interpretation of <u>waiting</u> or <u>local times</u> for the walks ω_α .

<u>Examples</u>

1) For $g(\phi^2) = \exp[-(2d+m^2)\,\phi^2]$, $h_j = 0$, for all j , formulas (1.60) and (1.63) yield the usual <u>local- (or "proper"-) time representation</u> of the bare propagator.

2) Setting $g(\phi^2) = \exp[-(2d+m^2)\phi^2]$ and $H(\phi) = -\sum_{\langle i,j\rangle} \phi_i \cdot \phi_j + \sum_j \left(\frac{\lambda}{4\beta}\,\phi_j^4 - h\phi_j\right)$, we may use formulas (1.59)-(1.61) to generate ordinary Feynman perturbation theory in λ and h for lattice $\lambda\phi^4$ theory, by resumming over all walks ω_α that have been formed to reconstitute bare propagators.

3) Setting $H(\phi) = -\sum_{\langle i,j\rangle} \phi_i \cdot \phi_j$ and expanding $\exp[-\beta H(\phi)]$ in its Taylor series, we may use formulas (1.53), ..., for arbitrary $g(\phi^2)$, to generate the full-fledged high-temperature expansion which rewrites lattice $\lambda\phi^4$ theory, for example, as a gas of closed random walks interacting via soft-core exclusion [16,1]. This representation shows that one may interpret $\mathcal{Z}_\Gamma(\omega_1,\ldots,\omega_n)$ as a correlation func-

tion of n polymer chains, $\omega_1, \ldots, \omega_n$, in a gas of polymer loops, interacting via soft-core exclusion.

4) In [4,15] the random walk representation has been used to develop a "skeleton expansion" and prove "skeleton bounds" for lattice $\lambda\phi^4$ theory. These results have enabled Brydges, Sokal and the author to give a simple construction of continuum $\lambda\phi_2^4$ - and $\lambda\phi_3^4$-theories, exhibit a mass gap, at weak coupling, and establish the asymptotic nature of perturbation theory [5]; see also [15] for some additional results.

5) A modified random-walk representation has turned out to be rather useful to prove Borel summability of the 1/N-expansion for non-critical O(N) non-linear σ-models on the lattice, [49].

We now come to our first major application of the random-walk representation: a proof of triviality of $\lambda\phi_d^4$ and the non-existence of a broken symmetry phase in $d \overset{>}{(=)} 4$ dimensions; (for the applications alluded to in 4) and 5), above, see [4,5,14,15,49]). Our proof of these results [2,3,6] is based on the following simple facts, (see [12]for an alternative, prior approach) : We set

$$H(\phi) = - \sum_{\langle i,j \rangle} \phi_i \cdot \phi_j - \sum_j h_j \phi_j , \quad h_j \geq 0 ,$$

for all $j \in \mathbb{Z}^d$, and

$$g(\phi^2) = \exp \left[- \frac{\lambda}{4} \phi^4 + \frac{\mu^2}{2} \phi^2\right] ,$$

with $\lambda > 0$ and μ^2 real.

This lattice field theory satisfies Griffiths inequalities [50], in particular

$$\langle \phi_{x_1} \ldots \phi_{x_k} \cdot \exp[-\sum_j t_j \phi_j^2] \rangle_\Gamma \leq \langle \phi_{x_1} \ldots \phi_{x_k} \rangle_\Gamma \langle \exp[-\sum_j t_j \phi_j^2] \rangle_\Gamma$$

if $t_j \geq 0$, for all j . These inequalities and the local-time representation of $[(\cdot)]_\Gamma(\underline{n})$ yields inequalities (I) and (II), below:

$$\text{(I)} \quad Z_\Gamma(\underline{n})^{-1} [\phi_{x_1} \ldots \phi_{x_k}]_\Gamma(\underline{n}) \leq \langle \phi_{x_1} \ldots \phi_{x_k} \rangle_\Gamma , \tag{1.65}$$

for all $\underline{n}$. When $h_j = 0$, for all j, we can also formulate this inequality as follows.

$$\sum_{p} \sum_{\substack{\omega_\alpha : x_{p(2\alpha-1)} \to x_{p(2\alpha)} \\ \alpha=1,\ldots,k}} \mathcal{Z}_\Gamma(\omega_1,\ldots,\omega_k,\omega'_1,\ldots,\omega'_\ell)$$

$$\leqq \Big(\sum_{p} \sum_{\omega_\alpha:\ldots} \mathcal{Z}_\Gamma(\omega_1,\ldots,\omega_k)\Big)\, \mathcal{Z}_\Gamma(\omega'_1,\ldots,\omega'_\ell) \ . \qquad (1.66)$$

See [2] , and [6] for the case where $h_j \gneqq 0$.

(II) If $\{\omega_1\ldots\omega_k\} \cap \{\omega'_1\ldots\omega'_\ell\}$ is empty then

$$\mathcal{Z}_\Gamma(\omega_1,\ldots,\omega_k,\omega'_1,\ldots,\omega'_\ell) \geqq \mathcal{Z}_\Gamma(\omega_1,\ldots,\omega_k)\, \mathcal{Z}_\Gamma(\omega'_1,\ldots,\omega'_\ell) \qquad (1.67)$$

See [2] . A clever use of the transfer-matrix formalism [44], along with correlation inequalities [17], yields

(III) $0 \leqq \langle\phi_x \, ; \, \phi_y\rangle_\Gamma \leqq c\beta^{-1} |x-y|^{2-d}$, for some $c < \infty$. (1.68)

For the time being we set $h_j \equiv 0$. Recalling (1.62) we see that (I) yields the Lebowitz inequality

$$0 \geqq u_{4,\Gamma}(x_1,x_2,x_3,x_4) \ ,$$

and, by (1.62) and (II),

$$\geqq \sum_{p} \sum_{\omega_1 \cap \omega_2 \neq \phi} [\mathcal{Z}_\Gamma(\omega_1,\omega_2) - \mathcal{Z}_\Gamma(\omega_1)\, \mathcal{Z}_\Gamma(\omega_2)]$$

$$\geqq - \sum_{p} \sum_{z} \Big\{ \sum_{\omega_1 \cap \omega_2 \ni z} \mathcal{Z}_\Gamma(\omega_1)\, \mathcal{Z}_\Gamma(\omega_2) \Big\} \ , \qquad (1.69)$$

since $\mathcal{Z}_\Gamma(\omega_1,\omega_2) \geqq 0$,

$$\geqq - \beta^2 \sum_{p} \sum_{z;z',z''} \sum_{\substack{\omega'_1 : x_{p(1)} \to z \\ \omega''_1 : z' \to x_{p(2)}}} \sum_{\substack{\omega'_2 : x_{p(3)} \to z \\ \omega''_2 : z'' \to x_{p(4)}}} \mathcal{Z}_\Gamma(\omega'_1,\omega''_2) \cdot$$

$$\cdot\, \mathcal{Z}_\Gamma(\omega'_2,\omega''_2)$$

by (I)

$$\geq - \beta^2 \sum_p \sum_{z;z',z''} \langle\phi_{x_{p(1)}} \phi_z\rangle_\Gamma \langle\phi_{z'} \phi_{x_{p(2)}}\rangle_\Gamma \cdot \langle\phi_{x_{p(3)}} \phi_z\rangle_\Gamma \langle\phi_{z''} \phi_{x_{p(4)}}\rangle_\Gamma + \mathcal{E} ,$$

where z' and z'' are arbitrary nearest neighbors of z , and $\mathcal{E}$ is a correction arising from terms with $z = x_j$, for some $j = 1,2,3,4$. Diagrammatically,

$$0 \geq u_{4,\Gamma}(x_1,x_2,x_3,x_4) \geq - \beta^2 \sum_p \sum_{z;z',z''}$$ 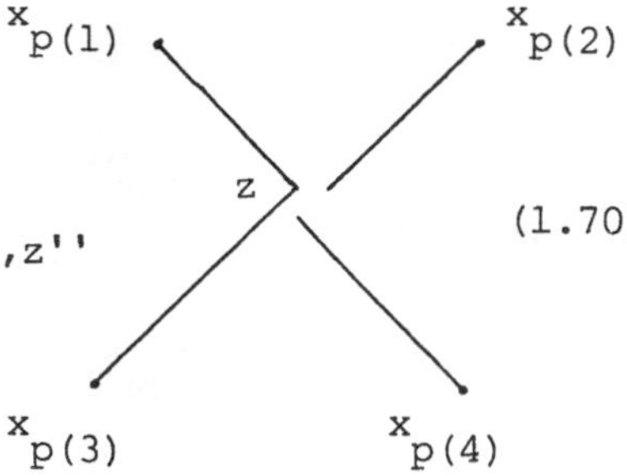

(1.70)

$$+ \mathcal{E}$$

with

$$\mathcal{E} =$$ 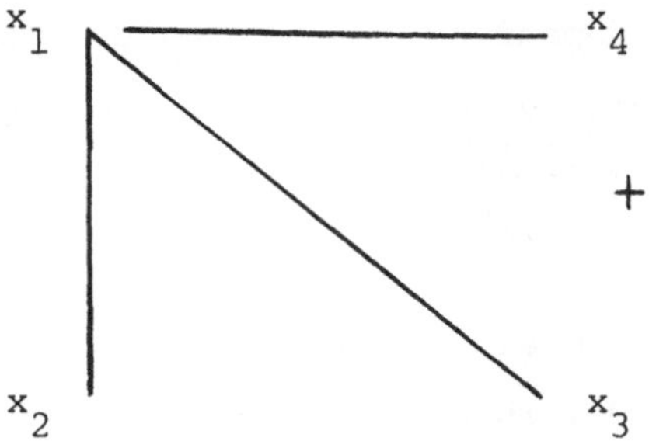

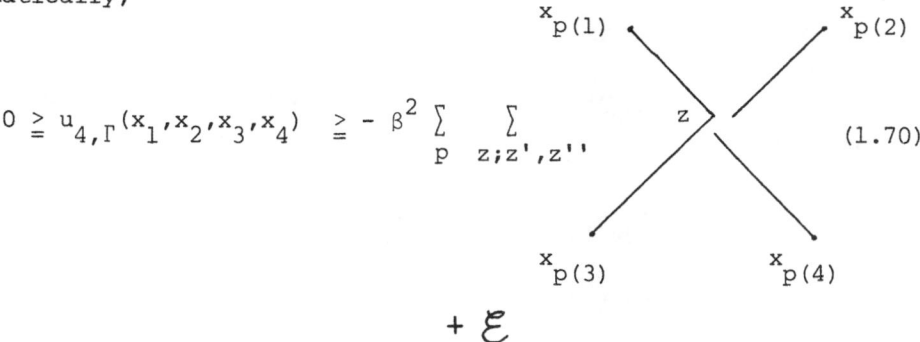

$+ \cdots .$ (1.71)

Introducing the rescaled four-point Ursell function,

$$u_{4,\theta}(x_1,\ldots,x_4) = \alpha(\theta)^4 \, u_{4,\Gamma(\theta)}(\theta x_1,\ldots,\theta x_4) ,$$

with $\Gamma(\theta)$ approaching the critical surface, as $\theta \to \infty$, see (1.25) - (1.28), we derive from (1.70), (1.71) and inequality (III) that

$$0 \geq u_{4,\theta} (x_1,x_2,x_3,x_4) \geq - \text{const } \theta^{4-d} , \tag{1.72}$$

whenever $|x_i - x_j| \geqq \delta$, for $i \neq j$, for some arbitrarily small, but positive δ . This proves that, in dimension $d > 4$,

$$S^c(x_1,\ldots,x_4) \equiv \lim_{\theta\to\infty} u_{4,\theta}(x_1,\ldots,x_4) = 0 \; ! \tag{1.73}$$

It is easy to extend this result to show that

$$\lim_{\theta\to\infty} \alpha(\theta)^n \; u_{n,\theta}(x_1,\ldots,x_n) = 0 \; , \tag{1.74}$$

for all $n > 2$.

In four dimensions, assuming that, for $|x-y| \geqq \delta > 0$ and uniformly in θ ,

$$\alpha(\theta)^2 \langle\phi_{\theta x} \; \phi_{\theta y}\rangle_{\Gamma(\theta)} \leqq k(\varepsilon,\delta) \; |x-y|^{-\varepsilon} \; , \tag{1.75}$$

for some constants $\varepsilon > 0$ and $k(\varepsilon,\delta) > \infty$, we obtain the bounds

$$0 \geqq u_{4,\theta} \; (x_1,\ldots,x_4) \geqq - \text{const.} \; (\alpha(\theta)^{-1}\theta)^p \; , \tag{1.76}$$

for some $p > 0$ depending on ε , provided $|x_i - x_j| \geqq \delta > 0$, for $i \neq j$. From this inequality we deduce that the continuum limit ($\theta\to\infty$) in four dimensions is trivial, unless

$$\alpha(\theta) \sim \theta^{(d-2)/2} = \theta \; , \text{ for } d = 4 \; ,$$

i.e. the (ultraviolet) dimension of the field ϕ_{ren} is canonical, in particular $\eta = 0$, where η is the exponent of $\langle\phi_x\phi_y\rangle_\Gamma$ defined in (1.15). If ϕ_{ren} has canonical (free-field) dimension and the Green's functions, $\{S(x_1,\ldots,x_{2n})\}_{n=0}^{\infty}$, in the continuum limit are scale-invariant then the theory is a free-field theory. This is a general theorem due to Pohlmeyer [51].

In [3,13] inequality (1.70) has been improved as follows:

$$0 \leqq \lambda_r \leqq \text{const.} \; \beta^2 \; m(\Gamma)^d \; \frac{\partial\chi(\Gamma)}{\partial\beta} \tag{1.77}$$

When $d = 4$ and $\eta = 0$ this inequality implies that in the continuum limit, i.e. when Γ approaches the critical surface, λ_r tends to 0 , unless mean-field theory gives the exact behaviour of $m(\Gamma)$ and $\chi(\Gamma)$, as Γ approaches the critical surface, in particular $\nu = 1/2$ and $\gamma = 1$. Moreover, it can be shown that the violations of the mean-field scaling laws in four dimensions are at most logarithmic [13]; see also [14].

In [3] a numerical analysis of the four-dimensional Edwards model (0-component $\lambda|\vec{\phi}|^4$ theory) has been initiated, using a novel Monte Carlo procedure, related to one in [52] and originally proposed in [53]. Although the accuracy of the numerical calculations is rather difficult to estimate, they provide strong support for logarithmic violations of the mean-field law for the mass, (much in accordance with the perturbative renormalization group). See also [54].

We now wish to summarize these results: While triviality of $\lambda\phi_4^4$ is not a rigorous mathematical theorem (yet), there is strong theoretical and numerical evidence for that contention. Assuming that $\lambda\phi_4^4$ can be constructed as a scaling limit of ferromagnetic lattice theories with arbitrary charge-, mass- and field strength renormalization compatible with (1.75) we can prove that there is no non-trivial, scale-invariant continuum theory (which, heuristically, implies that $W(\lambda_r)$ has no non-trivial roots!). In particular, a non-trivial $\lambda\phi_4^4$ theory would have to be asymptotically free, (in conflict with predictions of the perturbative renormalization group). Moreover, any $\lambda\phi_4^4$ continuum theory is trivial, unless mean-field theory is exact for lattice $\lambda\phi_4^4$ theory; see (1.77).

Before we review some results about broken-symmetry $\lambda\phi_d^4$ theory ($d \geqq 4$) we wish to sketch an intuitive interpretation of the above results: By (1.69)

$$\lambda_r = \chi(\Gamma)^{-2} \sum_{x_2,x_3,x_4} m(\Gamma)^d \, u_{4,\Gamma}(x_1,x_2,x_3,x_4)$$

$$\leqq \quad 3\, p_\Gamma ,$$

where $\chi(\Gamma) = \sum_y \sum_{\omega: x\to y} \mathfrak{z}_\Gamma(\omega)$, and p_Γ is the average intersection probability for two "field-theoretic" walks, ω_1 and ω_2, with weights $\mathfrak{z}_\Gamma(\omega_1), \mathfrak{z}_\Gamma(\omega_2)$. Moreover, inequality (III),

$$0 \leqq \langle \phi_x \phi_y \rangle_\Gamma = \sum_{\omega: x\to y} \mathfrak{z}_\Gamma(\omega) \leqq c\, \beta^{-1} |x-y|^{2-d} ,$$

(see (1.68)) says, roughly speaking, that the Hausdorff dimension of a field-theoretic walk, ω, is at most 2 . Hence one expects that $p_\Gamma \searrow 0$, as $m(\Gamma) \searrow 0$, in dimension $d > 4$. This is the contents of inequalities (1.70)-(1.72).

Next, we review some results concerning the broken-symmetry phase of $\lambda\phi^4$-theory. We thus suppose that $h_j \equiv h \geqq 0$, for all j. The random walk representations for the one-point and the two-point function are

$$\langle\phi_x\rangle_\Gamma = \sum_y h\, G_\Gamma(x,y) , \tag{1.78}$$

where

$$G_\Gamma(x,y) \equiv \sum_{\omega : x\to y} \mathfrak{z}_\Gamma(\omega), \tag{1.79}$$

and

$$\langle\phi_x\phi_y\rangle_\Gamma = \sum_{x',y'} h^2 \sum_{\substack{\omega_1 : x\to x' \\ \omega_2 : y\to y'}} \mathfrak{z}_\Gamma(\omega_1,\omega_2) + G_\Gamma(x,y) \tag{1.80}$$

Inequality (I) - see (1.66) - yields

$$\sum_{x'} h \sum_{\omega_1 : x\to x'} \mathfrak{z}_\Gamma(\omega_1,\omega_2) = Z_\Gamma(\underline{n}(\omega_2))^{-1} [\phi_x]_\Gamma (\underline{n}(\omega_2))\, \mathfrak{z}_\Gamma(\omega_2)$$

$$\leqq \langle\phi_x\rangle_\Gamma \cdot \mathfrak{z}_\Gamma(\omega_2)$$

Hence, by (1.78) and (1.79)

$$\langle\phi_x\phi_y\rangle_\Gamma \leqq \langle\phi_x\rangle_\Gamma \langle\phi_y\rangle_\Gamma + G_\Gamma(x,y) \tag{1.81}$$

Multiplying by h and summing over y we obtain, using (1.78) ,

$$h\,\chi(\Gamma) \equiv \sum_y h \langle\phi_x;\phi_y\rangle_\Gamma \leqq \langle\phi_x\rangle_\Gamma . \tag{1.82}$$

One simple consequence of this inequality is that, for $h > 0$, the expectation value $\langle(\cdot)\rangle_\Gamma$ clusters integrably fast.

Using inequality (II) - see (1.67) - we can also prove that

$$<\phi_x;\phi_y>_\Gamma \geqq G_\Gamma(x,y) - \beta^2<\phi>_\Gamma^2 \sum_z G_\Gamma(x,z)\ G_\Gamma(y,z) + \mathcal{E}'.^{*} \tag{1.83}$$

Hence, after multiplying by h and summing over y, we get using (1.78)

$$h\chi(\Gamma) \geqq <\phi>_\Gamma - \frac{\beta^2}{h} <\phi>_\Gamma^4 \tag{1.84}$$

Now, if a continuum $(\lambda\phi^4 + h\phi)$-theory exists as a scaling limit $(\theta\to\infty)$ of a lattice $(\lambda\phi^4+h\phi)$-theory then

$$\left.\begin{aligned} <\phi>_{\Gamma(\theta)} &\sim \alpha(\theta)^{-1}, \text{ and} \\ h = h(\theta) &\sim \theta^{-d}\alpha(\theta) . \end{aligned}\right\} \tag{1.85}$$

Thus, the second term on the r.h.s. of (1.83) and (1.84) is <u>irrelevant</u>, as $\theta\to\infty$, if $d > 4$, or in $d = 4$ dimensions provided $\alpha(\theta)^{-1}\ \theta \underset{\theta\to\infty}{\to} 0$. In these cases we conclude from (1.82) and (1.84) that

$$<\phi_{ren}> = h_{ren}\ \chi_{ren}, \tag{1.86}$$

where the subscript "ren" indicates that the scaling limit, $\theta\to\infty$, has been taken. Hence if the continuum theory has a positive physical mass so that $\chi_{ren.} < \infty$ then

$$<\phi_{ren}> \to 0, \text{ as } h_{ren} \to 0 . \tag{1.87}$$

This shows that, in a continuum $\lambda\phi^4$ theory with positive physical mass in $d \underset{(=)}{>} 4$ dimensions, either $<\phi_{ren}>$ is divergent, or the $\phi_{ren}\to-\phi_{ren}$ symmetry remains unbroken.

A somewhat lengthy, but straightforward analysis based on inequalities (I) - (III) proves that a one- or two-component $\lambda|\vec{\phi}|^4$ - theory in > 4 dimensions, or in four dimensions provided $\alpha(\theta)^{-1}\theta \to 0$, as $\theta\to\infty$, with

$$\alpha(\theta)\ <\phi>_{\Gamma(\theta)} \underset{\theta\to\infty}{\to} <\phi_{ren}> < \infty ,$$

---

* The error term, $\mathcal{E}'$, is irrelevant for critical behaviour and is henceforth neglected.

converges to a Gaussian (i.e. free- or generalized-free-field) theory in the continuum limit, (i.e. as $\theta \to \infty$).

For detailed proofs of these and further results concerning $h \to 0$ see [6]. (Some results for $h > 0$ were announced without proof in [2]).

Finally, we wish to point out some heuristic consequences of our triviality results for the standard model of electroweak interactions: Heuristically, the triviality results sketched above suggest that the β-function for the renormalized $|\phi|^4$ vertex, λ_r, in the Glashow-Weinberg-Salam model in four dimensions converges to a strictly positive function, as $g_r \searrow 0$ and $g'_r \searrow 0$. (Logically, this does, of course, not follow from our results, since we can only analyze one- and two-component fields, with $g_r = g'_r = 0$, and because in four dimensions our analysis is somewhat incomplete.) If the renormalized gauge coupling constant, g_r, is so small that perturbation theory in g_r for the coefficient functions in the Callan-Symanzik equation would appear to be trustworthy then the perturbative expressions of the β-functions and the strict positivity of the β-function for λ_r, for $g_r = g'_r = 0$, imply

$$\lambda_r \leqq a\, g_r^2 \tag{1.88}$$

for some calculable constant, a. The standard mass formula in the GWS model is

$$\left(\frac{m_H}{m_W}\right)^2 = b\, \frac{\lambda_r}{g_r^2}\ , \tag{1.89}$$

for some constant b, where m_H is the Higgs mass and m_W the mass of the W boson. Combining (1.88) and (1.89) we obtain

$$m_H^2 \leqq a \cdot b \cdot m_W^2\ . \tag{1.90}$$

This yields $m_H \lesssim 300$ GeV, for $m_W \approx 75$ GeV.

If experiments indicated that this bound is violated we would have reasons to expect "new physics" in an energy range that was sometimes expected to be a "desert".

The calculations on which the above conclusions are based and physical implications are contained in [26]. We think that it would be interesting to redo such calculations with added precision and compare them with rigorous bounds, like (1.86).

There are at least three further important applications of the random-walk representation which we wish to mention in passing:

(1) One may use the random-walk representation and Griffiths inequalities, much like (1.65), (1.66), (1.70), to give an extremely simple construction of the $\lambda\phi^4$ quantum field model in two and three space-time dimensions, in the single-phase region [4,5].

(2) Combining random walk representations of a large class of lattice theories, in particular $\lambda\phi^4$, lattice gauge theories with matter fields, including fermions, etc. with simple expansion methods one may carry out the particle structure analysis in such theories [34]. These methods enable us to prove existence of stable particles, including composite particles such as mesons and baryons in lattice gauge theories. Unfortunately, the expansions are not powerful enough, yet, to ensure convergence of estimates in the vicinity of critical points.

(3) Brydges and Spencer [55] have recently succeeded in using random-walk representations as a concrete and powerful analytical tool in connection with a renormalization group analysis.

§ 2. A Random-Surface Representation of Lattice-Gauge Theory

It is widely believed, nowadays, that QCD is the correct theory of strong interactions at not too high energies. This theory has the appealing feature that its high-energy behaviour is perturbatively close to free field theory. This is what is known as asymptotic freedom. (Although this property is quite crucial in attempts to prove mathematical consistency of non-abelian gauge theories, there are still problems with so-called "large-field regions".) In renormalization group language, asymptotic freedom is the statement that the ultraviolet stable fixed point of a non-abelian pure gauge theory is a Gaussian "glue-wave" state. If the vacuum angle θ vanishes this fixed point appears to admit a single relevant local operator compatible with gauge-invariance, i.e. there appears to be only one "unstable direction", parametrized by the Λ parameter, or by the glue ball mass. One thus expects a unique theory.

At low energies, however, pure gauge theories, or QCD, are unfortunately very complicated theories:

• The low-energy degrees of freedom - glue balls, hadrons - must be described by complicated composite operators of the basic field variables, $(A_\mu^a\,, q,\bar{q})$. For an economical description the concept of a point-like localized field may not be sufficient. The interactions between quarks at low energies are very strong.

· The lower- and upper critical dimensions, d_ℓ and d_u, of non-abelian gauge theories are not known rigorously, although one believes that $d_\ell = 4$, and this is proven for the U(1) theory [56]. From experience with the XY_2-, the Heisenberg_2- and more complicated, two-dimensional non-linear σ-models, for which $d_\ell = 2$, it is difficult to work out the infrared properties of quantum field theories when $d = d_\ell$.

· All the known expansion methods (perturbation theory, strong coupling -, 1/N- and ε-expansions) appear to have diseases when applied to gauge theories at low energies, as briefly discussed below.

To establish the qualitative and quantitative properties of non-abelian gauge theories at low energies such as

- chiral symmetry breaking
- quark confinement
- mass spectrum (Regge trajectories, charmonium spectroscopy, ...)
- low-energy scattering of composite particles; saturation of

forces is therefore a very difficult task.

In this section we propose a novel stochastic-geometrical tool for a qualitative analysis of quark confinement and aspects of the mass spectrum in lattice gauge theory which works well at strong coupling and appears to avoid the diseases of the strong-coupling expansion: a <u>random-surface</u> (chromo-electric flux-sheet) <u>representation</u> of pure gauge theory in the confinement phase. Gauge theories with matter fields are not included in our review, but see [34]. Our random-surface representation, (see [6], and [18] for some earlier results), is analogous to and inspired by the random-walk representation discussed in the last section.

Before we develop our random surface representation [6] we want to comment on the problems with standard expansion methods.

2.1. "Diseases" of Some Standard Expansions in Non-Abelian Gauge Theories

(a) <u>Renormalized perturbation theory</u>.* Since QCD and non-abelian pure gauge theories are asymptotically free, perturbation theory is

*I wish to thank M. Lüscher and B. Berg for very helpful discussions on the problems discussed here.

expected to be reliable at short distances. It has been proposed, therefore (see [57] and refs. given there), to replace physical space, $\mathbb{R}^3$, by a compact manifold, e.g. $\mathbb{T}^3_L$, of diameter L so small that perturbative calculations become accurate and to then extrapolate the results by finite size scaling and related techniques [57]. This idea has been very successful in two-dimensional non-linear σ-models [58]. Here we wish to describe some possible difficulties with this idea in four-dimensional, non-abelian gauge theories.

Space-time is chosen to be

$$S^1_L \times \mathbb{T}^2_{L'} \times \mathbb{R} , \qquad L \leqq L' , \tag{2.1}$$

with space given by $S^1_L \times \mathbb{T}^2_{L'}$ and a non-compact time direction. For perturbation theory to be reliable one must require that, at least, $\Lambda \cdot L$ is small, where Λ is the Λ-parameter; (i.e. some perturbative (high-energy) mass scale). In practical calculations it will be advantageous to set $L' = L$. However, it is convenient to first consider the case where $L' \to \infty$, before L becomes large. In this limit, after analytic continuation to imaginary time, the theory is isomorphic, mathematically, to a gauge theory at <u>positive physical temperature</u> $T = 1/kL$, where k is Boltzmann's constant. Let

$$z = \beta \Lambda , \quad \beta = \frac{1}{kT} = L \tag{2.2}$$

There are theoretical arguments indicating that gauge theory has a "deconfining" transition to a <u>perturbative phase</u> when z drops below some critical value $z_c > 0$. This has been proven rigorously for SU(N) lattice gauge theories, for arbitrary N (and with bounds which show the correct large-N behaviour) by Borgs and Seiler [59]. That transition corresponds to singularities in various physical quantities, such as most probably particle masses, on the <u>positive real z-axis, at</u> $z = z_c$.

When $L' < \infty$ one has good reasons to expect that there are no singularities on the positive z-axis, because physical space is compact. However, the singularity previously encountered at $z = z_c > 0$, for $L' = \infty$, will not simply disappear. Rather, we expect it to split into a pair of complex singularities, $z_o(L')$ and $\overline{z_o(L')}$, which move off into the complex plane, as L' is lowered, continuously in L'. For large L we expect that $z_o(L') \to z_o$, as $L' \searrow L$, where Re z_o is close to z_c and Im z_o is small.

Now, perturbation theory permits one to calculate certain quantities in gauge theory at small values of $|z|$, but particle physics (at zero temperature and density) requires understanding gauge theory in the limit $z \to +\infty$ (z real)! This forces one to analytically continue the theory, as calculated perturbatively, from small values of $|z|$ to large, positive values of z, passing through the window between z_o and $\bar{z}_o$. One can argue that this might be almost as hard as calculating the low-temperature behaviour in the three-dimensional Ising model from the high-temperature series in a finite volume: While this is possible, in principle, it is difficult in practice.

Thus, the usefulness of perturbation theory in the analysis of non-abelian gauge theories at low energies might be limited to large physical temperatures.

(b) The strong-coupling expansion. The strong-coupling expansion in lattice gauge theory is an expansion in g^{-2}, where g is the gauge coupling constant. Hence it is reliable when g is large, where it predicts a positive mass gap and confinement. Asymptotic freedom predicts that, in four dimensions at zero physical temperature, the continuum limit is reached when $g^2 \to 0$, ($g^{-2} \sim |\log am|$, where m is a typical physical mass scale and $a \sim \theta^{-1}$ is the lattice spacing). The phenomenon which obstructs the continuation of the strong coupling expansion to large values of g^{-2} of surface-(flux sheet-) related quantities, such as the string tension (or the Regge slope) is surface roughening [60]. Roughening appears to introduce singularities in the string tension on the positive g^{-2} axis. Suitably chosen mass ratios (e.g. a baryon-meson mass ratio) might, however, be free from such problems.

There is a simple, albeit not terribly intuitive way of characterizing the roughening transition which seems to have escaped the attention of many workers in this field: Let $V(g;L)$ be the static $q\bar{q}$ potential in a non-abelian, pure gauge theory felt by two colour sources separated by a distance L. One can prove quite easily that, for large g, i.e. in the phase where the flux sheet is rigid,

$$V(g;L) = \alpha(g)\, L + \gamma(g) + \delta(g)\, e^{-L/\xi(g)}, \qquad (2.3)$$

where α, δ and ξ are finite, positive constants.

On the basis of experience with simple models (e.g. the Gaussian sheet) and approximate calculations in string theory [61] one can argue that, for small g, where capillary waves are abundant

$$V(g;L) = \alpha(g)\, L + \gamma(g) + \frac{\kappa(g)}{L}, \qquad (2.4)$$

where $\alpha(g) > 0$ and $\kappa(g)$ is a dimensionless number, ($\kappa(g) \simeq (d-2)\pi/24$, [61]). The correction term $\frac{\kappa(g)}{L}$ comes from the one-dimensional Casimir effect. The regimes (2.3) and (2.4) are separated by the roughening transition, occurring at some positive value, g_R, of g. The form (2.4) is expected to be valid for all $g < g_R$. (The power L^{-1} is related to the behaviour of the propagator of two-dimensional free, massless fermions.) A more intuitive characterization of the roughening transition is sketched later.

(c) The $\frac{1}{N}$ expansion.* Consider a family of non-abelian gauge theories with gauge group SU(N), N = 2,3,4,..., e.g. pure gauge theories. As originally proposed by 't Hooft [62], it is attractive to try to expand low-energy quantities in these theories, like masses, the string tension, or glue-ball scattering amplitudes, in powers of 1/N. For the lattice theories, one may try to expand each term in a high-temperature expansion in powers of 1/N. However, when one tries to do that one encounters problems: Terms in the high-temperature expansion of SU(N) - lattice gauge theories are indexed by two-dimensional "surfaces" immersed in the lattice. These "surfaces" typically have rather complicated topologies, in particular, non-trivial $(d-3)^{rd}$ homology groups with coefficients in $\mathbb{Z}_N$, for all N, appear, provided the dimension d of the lattice is at least four. As the reader easily convinces himself by considering simple examples, the numerical weights of these surfaces, as computed for different SU(N)-theories, tend to depend discontinuously on N; see [11]. It is tempting to conjecture, therefore, that in four or more dimensions there is no systematic 1/N expansion to arbitrary order in 1/N, because the topological complexities of virtual flux sheets may introduce discontinuities in 1/N and may, in fact, make it impossible to extend 1/N to arbitrary real or complex values. A detailed study of such difficulties in the simpler q-states Potts gauge theory and some comments on SU(N) lattice gauge theory may be found in [11]. (Of course one still hopes that there is continuity in 1/N at N = ∞ and that the N = ∞ theory has some intrinsic simplicity.)

(d) The ε-expansion. Mean-field theory for lattice gauge theory [63] predicts a first-order (discontinuous) deconfining transition at a finite, positive value of g^2. Numerical experiments suggest that non-abelian lattice theories with Wilson action may have a first order deconfining transition in five and more dimensions. It is conceivable that complicated lattice actions, probably destroying reflection (Osterwalder-Schrader) positivity, may lead to continuous, deconfining transitions, e.g. in five dimensions; (see also Sect. 2.2, (1c)). However, it is fair to say that it appears difficult to invent lattice gauge theories with continuous deconfining transitions

*These remarks have been worked out with M. Aizenman whom I thank for inspiration.

in large dimension, d, which,as d approaches some upper critical dimension, d_u, are described by some simple universal theory. This would be a prerequisite in attempting to set up an ε-expansion, i.e. an expansion in powers of d_u-d, $(d < d_u)$. We shall argue that, in a specific lattice string theory, such an ε-expansion presumably exists, in contrast to lattice gauge theory.

But even if it were possible to construct a well-defined ε-expansion for (lattice) gauge theory it would presumably not be of much use in the analysis of four-dimensional theories, since, most probably, $4 = d_\ell$, and one cannot expect that an ε-expansion in powers of d_u-d yields any reasonable predictions about the theory in $d = d_\ell$.

There are several lines out of the rather frustrating situation described above:

(i) Give up.

(ii) Use super-computers to analyze lattice gauge theories numerically.

(iii) Try to guess an approximate low-energy effective dynamics of non-abelian gauge theories in the confinement phase. Investigate specific, but simplified models, such as string theory, in some detail, with the hope of learning some general qualitative features common to confining theories.

(iv) Use hard and powerful tools, such as real-space renormalization group methods without approximations, in order to rigorously control the theory.

In the remainder of these notes we describe some ideas and results in line (iii). Presumably, line (iv) will eventually have to be followed, supplemented by (ii) to get quantitative information, but this is still too hard, at present.

Rather many theoreticians seem to have a certain preference for (i) and move to other topics.

2.2. The Flux-Sheet Representation of Pure SU(2) Lattice Gauge Theory, [6]

In this section we describe a method to approximately calculate the effective low-energy dynamics of pure lattice gauge theories in the confinement phase which have as their gauge group, G, a sphere. Hence G = $\mathbb{Z}_2$, or U(1) , or SU(2). The intuition behind our method (see [6] for a detailed description) is that, in the confinement phase, the chromoelectric flux between two static sources is focused, i.e. is essentially a thin (but wiggly) tube. Its space-(imaginary) time history is therefore a thin sheet, the flux sheet. The flux-sheet has basically no internal degrees of freedom, but exhibits surface fluctuations (capillary waves) much like the interface in e.g. the three-dimensional Ising model below the critical temperature.

We thus propose to rewrite the expectation of a Wilson loop (= space-(imaginary) time portrait of two static sources) in a pure gauge theory, in the confinement phase, as a statistical sum over random surfaces, the flux-sheets. It turns out that if G is a sphere that representation becomes particularly simple, and, moreover, the statistical weight of every surface contributing to that sum is <u>positive.</u> It is this positivity property which appears to permit us to get round the problems posed by surface roughening and to develop a "classical statistical mechanics of random surfaces." In principle, flux-sheet representations can be derived for all theories with $G = U(N)$, or $G = O(N)$, (see Sect. 2.3 and refs. given there), but they are rather cumbersome, and the statistical weights need <u>not</u> be positive, <u>unless</u> G is a sphere.

Clearly, such representations for $G = U(\infty)$ theories would be of interest. In the $N \to \infty$ limit, only planar, orientable surfaces are expected to contribute. However, we shall analyze, instead, a simplified model which shares some qualitative features with $U(\infty)$ gauge theory but is easier; (see Sect. 3, and [21,7,8]).

In the following we set $G = SU(2)$; (theories with $G = U(1)$, or $G = \mathbb{Z}_2$ are simpler). With each bound $b = \langle x,y \rangle$ of $\mathbb{Z}^d$ we associate a group element $g_b \in SU(2)$ with $g_{\bar{b}} = g_b^{-1}$, for $\bar{b} = \langle y,x \rangle$; [64]. We set

$$g_{\mathcal{L}} = \prod_{b \in \mathcal{L}}^{\circlearrowleft} g_b \ ,$$

where $\mathcal{L}$ is a loop in $\mathbb{Z}^d$, and $\prod^{\circlearrowleft}$ indicates a path-ordered product. The Wegner-Wilson loop observable is defined by

$$W(\mathcal{L}) = \operatorname{tr}\,(g_{\mathcal{L}}) \ .$$

The Wilson action of a pure SU(2) theory is given by

$$A(g) = -\sum_p \operatorname{tr}\,(g_{\partial p}) \ ,$$

where p is an arbitrary plaquette (unit square) in $\mathbb{Z}^d$.

The Euclidean vacuum functional of the theory is given by the measure

$$d\mu_{\bar\beta}(g) = Z_{\bar\beta}^{-1}\, e^{-\bar\beta A(g)} \prod_{b\in\mathbb{Z}^d} dg_b \quad , \tag{2.5}$$

where dg denotes Haar measure.

It is convenient to use the following standard parametrization of SU(2):

$$g_b = \sum_{\alpha=0}^{3} \phi_b^\alpha \, \sigma_\alpha^E \quad , \tag{2.6}$$

with $\sigma_0^E = \begin{pmatrix}1&0\\0&1\end{pmatrix}$, $\sigma_j^E = i\sigma_j$, $j = 1,2,3$, where σ_j is the jth Pauli matrix, and

$\vec\phi_b \equiv (\phi_b^0, \phi_b^1, \phi_b^2, \phi_b^3)$ is a unit vector, i.e.

$$|\vec\phi_b| = \sqrt{\sum_{\alpha=0}^{3} (\phi_b^\alpha)^2} = 1 \; . \tag{2.7}$$

We define

$$\vec\phi_{\bar b} = (\phi_b^0 , - \phi_b^1 , - \phi_b^2 , - \phi_b^3) \tag{2.8}$$

Then

$$g_{\bar b} = g_b^{-1} = \sum_{\alpha=0}^{3} \phi_{\bar b}^\alpha \, \sigma_\alpha^E \; .$$

Next, we define

$$\left.\begin{aligned} \phi(\alpha_{\mathcal{L}}) &\equiv \prod_{b\in\mathcal{L}} \phi_b^{\alpha(b)} \quad , \\ \sigma(\alpha_{\mathcal{L}}) &\equiv \prod_{b\in\mathcal{L}}^{\circlearrowleft} \sigma_{\alpha(b)} \quad . \end{aligned}\right\} \tag{2.9}$$

In the $\vec\phi$ -variables, the measure $d\mu_{\bar\beta}(g)$ is given by

$$d\mu_{\bar\beta}(\vec\phi) = Z_{\bar\beta}^{-1} \exp\,[\bar\beta \sum_p \sum_\alpha \mathrm{tr}(\sigma(\alpha_{\partial p}))\,\phi(\alpha_{\partial p}))] \cdot$$

$$\cdot \prod_b g\,(|\vec\phi_b|^2)\,d^4\phi_b \ , \tag{2.10}$$

where

$$g(|\vec\phi|^2) = \delta(|\vec\phi|^2 - 1).$$

The properties of this theory are coded into expectation values of products of Wilson loops, such as

$$\langle W(\mathcal{L})\rangle_{\bar\beta} \equiv \int \sum_\alpha \mathrm{tr}(\sigma(\alpha_{\mathcal{L}}))\ \phi(\alpha_{\mathcal{L}})\,d\mu_{\bar\beta}(\vec\phi), \tag{2.11}$$

$$\langle W(\mathcal{L}_1)\ W(\mathcal{L}_2)\rangle_{\bar\beta}\ , \ \ldots\ .$$

In the following, the function $g(|\vec\phi|^2)$ on the r.h.s. of (2.10) may be any even function on $\mathbb{R}^4$ for which

$$\int \exp(c|\vec\phi|^4)\ g(|\vec\phi|^2)\ d^4\phi < \infty\ , \tag{2.12}$$

for any $c > 0$. (This freedom may be useful in block-field renormalization transformations and other studies of the continuum limit. It has been proposed in [32] and studied, subsequently, quite systematically in [31].) We now define

$$g_n(|\vec\phi|^2) = \int d\nu_n(t)\ g(|\vec\phi|^2 + 2t)\ , \tag{2.13}$$

with

$$d\nu_n(t) = \begin{cases} \delta(t)\ dt & , \quad \text{if } n = 0 \\ \dfrac{t^{n-1}}{(n-1)!}\ \theta(t)dt\ , & \quad n = 1,2,3,\ \ldots\ . \end{cases}$$

Then

$$\phi^\alpha\, g_n(|\vec\phi|^2) = -\frac{\partial}{\partial\phi^\alpha}\, g_{n+1}\,(|\vec\phi|^2)\ ; \tag{2.14}$$

see (1.33), (1.53).

We set

$$A(\vec{\phi}) = - \sum_p \sum_\alpha \operatorname{tr}(\sigma(\alpha_{\partial p}))\phi(\alpha_{\partial p}) \tag{2.15}$$

and define

$$[(\cdot)]_{\overline{\beta}}(\underline{n}) = \int (.)\, e^{-\overline{\beta}A(\vec{\phi})} \prod_{b \in \mathbb{Z}^d} g_{n_b}(|\vec{\phi}_b|^2)\, d^4\phi_b\,, \tag{2.16}$$

where $\underline{n} = \{n_b\}_{b\in\mathbb{Z}^d}$.

Let b_o be some bond of $\mathcal{L}$. Using (2.9) and (2.14)-(2.16) we obtain the identidy

$$[W(\mathcal{L})]_{\overline{\beta}}(\underline{n}) = \sum_\alpha \operatorname{tr}(\sigma(\alpha_{\mathcal{L}})) \int\ldots\int \Big(\prod_{b\neq b_o} g_{n_b}(|\vec{\phi}_b|^2)\, d^4\phi_b \Big) \cdot$$

$$\cdot \prod_{\substack{b\in\mathcal{L}\\ b\neq b_o}} \phi_b^{\alpha(b)} \int \phi_{b_o}^{\alpha(b_o)}\, g_{n_{b_o}}(|\vec{\phi}_{b_o}|^2)\, e^{-\beta A(\vec{\phi})}\, d^4\phi_{b_o}$$

$$= \overline{\beta} \sum_{p:\partial p \ni b_o} \sum_\alpha \operatorname{tr}(\sigma(\alpha_{\mathcal{L}}))\, \operatorname{tr}(\sigma(\alpha_{\partial p})) \cdot$$

$$\cdot \int e^{-\beta A(\vec{\phi})}\, \phi(\alpha_{\mathcal{L}\circ\partial p}) \prod_b g_{n_b + \delta_{bb_o}}(|\vec{\phi}_b|^2)\, d^4\phi_b\,. \tag{2.17}$$

Next, we use the fact that $\{2^{-1/2}\sigma_\alpha^E\}_{\alpha=0}^3$ is an orthonormal basis in $\mathbb{M}_2(\mathbb{C})$, hence

$$\sum_\alpha \operatorname{tr}(A(\sigma_\alpha^E)^*)\, \operatorname{tr}(\sigma_\alpha^E B) = 2\, \operatorname{tr}(AB) \tag{2.18}$$

which yields

$$\sum_{\alpha(b_o)} \operatorname{tr}(\sigma(\alpha_{\mathcal{L}}))\, \operatorname{tr}(\sigma(\alpha_{\partial p})) = 2\, \operatorname{tr}(\sigma(\alpha_{\mathcal{L}\circ\partial p}))\,.$$

From this and (2.17) we finally get

$$[W(\mathcal{L})]_{\bar\beta}(\underline{n}) = 2\bar\beta \sum_{p:\partial p \ni b_o} [W(\mathcal{L} \circ \partial p)]_{\bar\beta}(\underline{n}+\delta_{bb_o}) . \tag{2.19}$$

This identity can now be iterated until only terms are left in which $\mathcal{L}$ has been deformed to the empty loop, i.e. after deformations by plaquettes $p_1, \ldots, p_m$, with

$$\mathcal{L} \circ \partial p_1 \circ \ldots \circ \partial p_m = \emptyset .$$

Here we use the fact that in applying identity (2.14) a derivative, $-(\partial/\partial\phi_b^{\alpha(b)})$, may be carried over to differentiate either $e^{-\bar\beta A(\vec\phi)}$ or any power of $\phi_b^{\alpha(b)}$ still present in the integrand. It may then happen that all monomials $(\phi_b^{\alpha(b)})^k$ in the integrand are differentiated away ($k \to 0$) after many applications of (2.19). Also, note that, when $n_b \neq 0$,

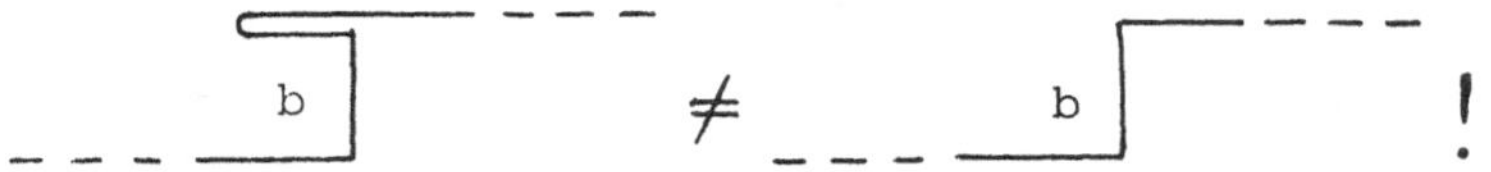

A set of plaquettes $\{p_k\}_{k=1}^m$ with the property that $\mathcal{L} \circ \partial p_1 \circ \ldots \circ \partial p_m = \emptyset$ forms a connected two-dimensional "surface", S, immersed in the lattice $\mathbb{Z}^d$ with the property that $\partial S = \mathcal{L}$. Note that different (ordered) sequences, $\{p_k\}_{k=1,2,3,\ldots}$ may lead to the same surface S. Let $n_b(S)$ denote the total number of times S passes through a given bond, b, and $\underline{n}(S) = \{n_b(S)\}_{b \in \mathbb{Z}^d}$. Then, by iteration of (2.19) and the above observations, we obtain

$$[W(\mathcal{L})]_{\bar\beta}(\underline{n}) = \sum_{S:\partial S = \mathcal{L}} (2\bar\beta)^{|S|} [1]_{\bar\beta}(\underline{n} + \underline{n}(S)) , \tag{2.20}$$

where $|S|$ is the total number of plaquettes contained in S, counted with multiplicity. Note that in (2.20) the precise way in which $\mathcal{L}$ has been deformed to the empty loop sweeping over the surface S does not enter. (However, if some surface $S' \subsetneq S$ has been swept out already then S shall determine which plaquettes are glued to each bond, b, of $\partial S' \setminus \mathcal{L}$ next. For more precise definitions see [6].) We now define

$$\mathcal{Z}_{\bar\beta}(S_1,\ldots,S_m) = (2\bar\beta)^{\sum_{j=1}^{m}|S_j|} \frac{[1]_{\beta}\,(\sum_{j=1}^{m}\underline{n}(S_j))}{[1]_{\bar\beta}\,(\underline{0})} \tag{2.21}$$

As in (1.63), it follows easily from the definition that

$$\mathcal{Z}_{\bar\beta}(S_1,\ldots,S_m) = (2\bar\beta)^{\sum|S_j|} \int \prod_{j=1}^{m} d\nu_{S_j}(t^j)\,\mathcal{Z}_{\bar\beta}(\sum t^j), \tag{2.22}$$

where

$$d\nu_S(t) = \prod_b d\nu_{n_b(S)}(t_b),$$

and

$$\mathcal{Z}_{\bar\beta}(t) = Z_{\bar\beta}^{-1} \int e^{-\bar\beta A(\vec\phi)} \prod_b g(|\vec\phi_b|^2+2t_b)\,d^4\phi_b . \tag{2.23}$$

The variables $n_b(S)$ give the total number of times S passes through b and form a $\mathbb{Z}_+$-valued vector field on $\mathbb{Z}^d$. The t_b are local-, or waiting times for the surface S. They form an $\mathbb{R}_+$-valued vector field on $\mathbb{Z}^d$. Given some path γ in $\mathbb{Z}^d$, $t_\gamma \equiv \sum_{b\in\gamma} t_b$ measures the total "time" during which $\gamma \cap S$ is non-empty. Alternatively, let D be a $(d-1)$-dimensional surface in the dual lattice, $(\mathbb{Z}^d)^*$. Let $t_D \equiv \sum_{b:b^*\in D} t_b$. Then t_D measures the total "time" during which $D \cap S \neq \emptyset$.

From (2.20) and (2.21) we now get

$$\langle W(\mathcal{L})\rangle_{\bar\beta} = \sum_{S:\partial S=} \mathcal{Z}_{\bar\beta}(S), \tag{2.24}$$

$$\langle W(\mathcal{L}_1)\,W(\mathcal{L}_2)\rangle_{\bar\beta} = \sum_{\substack{S_1:\partial S_1=\mathcal{L}_1\\ S_2:\partial S_2=\mathcal{L}_2}} \mathcal{Z}_{\bar\beta}(S_1,S_2) + G_{\bar\beta}(\mathcal{L}_1,\mathcal{L}_2), \tag{2.25}$$

where

$$G_{\bar{\beta}}(\mathcal{L}_1,\mathcal{L}_2) \equiv \sum_{S:\partial S=\mathcal{L}_1\cup\mathcal{L}_2} \mathcal{Z}_{\bar{\beta}}(S) , \tag{2.26}$$

etc. Since the weights $\mathcal{Z}_{\bar{\beta}}(S)$ are <u>positive</u>, the representation (2.24)-(2.26) ought to be <u>valid for all</u> $\bar{\beta}$ <u>below</u> $\bar{\beta}_c$, (the deconfinement transition). It has rather interesting consequences:

(1) <u>Confinement (area decay of</u> $\langle W(\mathcal{L})\rangle_{\bar{\beta}}$) <u>and exponential decay properties:</u>

(1a) If

$$g(|\vec{\phi}|^2) = \delta(|\vec{\phi}|^2 - 1)$$

then in (2.22) all t-integrations range only over the domain determined by $0 \leqq t_b^j$, $\sum_j t_b^j \leqq 1/2$, for all b . From this fact it is easy to derive some bounds on $\mathcal{Z}_{\bar{\beta}}(S_1,\ldots,S_m)$ and to prove that, for $\bar{\beta} \leqq \text{const}\,(d-2)^{-1}$,

$$\left.\begin{array}{l} \langle W(\mathcal{L})\rangle_{\bar{\beta}} \text{ exhibits area decay, and} \\ G_{\bar{\beta}}(\mathcal{L}_1,\mathcal{L}_2) \text{ has exponential decay in } \mathrm{dist}(\mathcal{L}_1,\mathcal{L}_2) \end{array}\right\} \tag{2.27}$$

(1b) If $g_{\bar{\beta}}$ is <u>log-concave</u>, and

$$\left.\begin{array}{l} 0 \leqq g_{\bar{\beta}}(|\vec{\phi}|^2) \leqq e^{-c\bar{\beta}|\vec{\phi}|^4} , \\ \text{for some constant } c \geqq \text{const.}\,(d-1) \end{array}\right\} \tag{2.28}$$

then one can derive from (2.22)-(2.25), the log-concavity inequality

$$g(|\vec{\phi}|^2 + 2t) \leqq g(|\vec{\phi}|^2)\, g(2t)$$

and standard combinatorial arguments (concerning the entropy of lattice surfaces [7,8])that, <u>for all values of</u> $\bar{\beta}$,

$$\left.\begin{array}{l}\langle W(\mathcal{L})\rangle_{\bar{\beta}} \quad \text{has area decay, and} \\ G_{\bar{\beta}}(\mathcal{L}_1,\mathcal{L}_2) \text{ has exponential decay in dist } (\mathcal{L}_1,\mathcal{L}_2).\end{array}\right\} \tag{2.29}$$

This is thus a lattice gauge model with permanent confinement. It was first proposed in [32] and has recently been analyzed independently in [31], by methods based on [1,18].For a related proposal see [33].*

(1c) Let $$g_{\bar{\beta},\sigma}(|\vec{\phi}|^2) = e^{\sigma|\vec{\phi}|^2}\, \overset{\circ}{g}_{\bar{\beta}}(|\vec{\phi}|^2), \tag{2.29}$$

where $\overset{\circ}{g}_{\bar{\beta}}$ is log-concave and satisfies (2.28). In this model permanent confinement and exponential decay of $G_{\bar{\beta}}(\mathcal{L}_1,\mathcal{L}_2)$ hold for all $\sigma \leqq 0$. It is an interesting speculation that, for fixed $\bar{\beta}$, and a suitable choice of $\overset{\circ}{g}_{\bar{\beta}}$, we can drive this theory to a critical point by increasing σ to positive values. This might be an alternative route towards constructing a continuum SU(2)- (or U(1)-) gauge theory whith permanent confinement.

With some experience the proofs of the claims in (1a)-(1c) follow by inspection, the only somewhat novel ingredient being entropy estimates for lattice random surfaces [6,7,8].(These considerations have greatly profited from discussions, correspondence and collaboration with D. Brydges and B. Durhuus.)

The reader may now wonder whether the replacement of Haar measure, $\delta(|\vec{\phi}|^2-1)d^4\phi$, by a general distribution $g(|\vec{\phi}|^2)d^4\phi$, as in (1b) and (1c), is compatible with basic priciples of relativistic quantum field theory, such as Osterwalder-Schrader positivity [39]. Indeed, it is straightforward to verify that this is the case. All models in (1a)-(1c) satisfy Osterwalder-Schrader positivity. Proofs of these and related results may be found in [31,65,6].

(2) Correlation inequalities for $G = \mathbb{Z}_2$, U(1).

If $G = \mathbb{Z}_2$ or = U(1) then one can prove correlation inequalities of the Ginibre type; [50].

Among novel consequences of these inequalities are:

(I) $$\sum_{S:\partial S=\mathcal{L}} \mathcal{Z}_{\bar{\beta}}(S,S_1,\ldots S_k) \leqq \langle W(\mathcal{L})\rangle_{\bar{\beta}} \cdot \mathcal{Z}_{\bar{\beta}}(S_1,\ldots,S_k)$$

(II) If $(S_1\ldots S_k) \cap (S_1',\ldots S_\ell') = \emptyset$ then

---

* The problem with these "dielectric gauge theories" is that their choice is not canonical, and that they might be regularization-dependent.

$$\mathcal{Z}_{\bar{\beta}}(S_1,\ldots,S_k,S_1',\ldots,S_\ell') \geqq \mathcal{Z}_{\bar{\beta}}(S_1,\ldots,S_k)\,\mathcal{Z}_{\bar{\beta}}(S_1',\ldots,S_\ell')$$

As a consequence of (I) we get

$$0 \leqq \langle W(\mathcal{L}_1);\, W(\mathcal{L}_2)\rangle_{\bar{\beta}} \leqq G_{\bar{\beta}}(\mathcal{L}_1,\mathcal{L}_2)\ .$$

This inequality is useful to prove exponential cluster properties; (II) may serve to estimate intersection probabilities for random surfaces. While all these inequalities may have quite valuable consequences they are certainly much less striking than the ones discussed in Sect. 1.5. For more details see [6].

(3) The main virtue of the random-surface representation is perhaps that it uncovers qualitative mechanisms underlying the roughening and the deconfining transition. See Sect. 3 and [6,8].

2.3. Extensions to U(N)- and O(N)-Models, and the Planar-Surface (Weingarten) Model

We choose G = U(N), the analysis for O(N) being similar. We propose to sketch a random-surface representation for the Wilson loop in compact U(N) lattice gauge theory*. As proposed in [65] it is sometimes advantageous to write a gauge field, $\{g_b \in U(N)\}_{b\in\mathbb{Z}^d}$, in the form

$$g_b = W_+(b)W_-(b)^* \ , \tag{2.30}$$

with $W_\pm(b) \in U(N)$. The action is then given by

$$A(W_+,W_-) = -\sum_p \mathrm{tr}\,\Big(\prod_{b\in\partial p}^{\circlearrowleft} W_+(b)\,W_-(b)^*\Big) \tag{2.31}$$

The Haar measure, dg, can be written as

$$dg = \delta(W_+W_+^* - \mathbb{1})\,\delta(W_-W_-^* - \mathbb{1}) \prod_{i,j=1}^{N} dW_+^{ij}\,\overline{dW_+^{ij}}\,dW_-^{ij}\,\overline{dW_-^{ij}}\ . \tag{2.32}$$

The vacuum measure is thus given by

$$d\mu_{\bar{\beta},N}(W_+,W_-) = Z_{\bar{\beta},N}^{-1}\, e^{-\bar{\beta}N\,A(W_+,W_-)} \prod_b dg_b(W_+,W_-)\ . \tag{2.33}$$

---

*There is no unique such representation, and the following discussion is somewhat superficial. More work is needed.

One may now introduce Lagrange multiplier fields [16,1,18,25], λ_+ and λ_-, to rewrite the δ-functions on the r.h.s. of (2.32):

$$\delta(W_+W_+^* - \mathbb{1}) = \int \mathcal{D}\lambda_+ \, e^{N\operatorname{tr}[\lambda_+(\mathbb{1} - W_+^* W_+)]}, \qquad \delta(W_-W_-^* - \mathbb{1}) = \int \mathcal{D}\lambda_- \, e^{N\operatorname{tr}[\lambda_-(\mathbb{1} - W_- W_-^*)]} \tag{2.34}$$

where $\lambda_\pm$ is the uniform measure on the linear space of antisymmetric matrices. In these identities W_+ and W_- are, a priori, general complex $N \times N$ matrices. In calculating expectations w.r. to $d\mu_{\overline{\beta},N}$ we insert the r.h.s. of (2.34 and then temporarily interchange integrations over $\underline{W}_+$, $\underline{W}_-$ and $\underline{\lambda}_+, \underline{\lambda}_-$. We denote the resulting, formal $(\underline{\lambda}_+, \underline{\lambda}_-)$-dependent expectation by $\langle(\cdot)\rangle_{\overline{\beta},N}(\underline{\lambda}_+, \underline{\lambda}_-)$ and define

$$\mathcal{Z}_{\overline{\beta},N}(\underline{\lambda}_+,\underline{\lambda}_-;\underline{W}_+,\underline{W}_-) = \prod_b \{ e^{N\operatorname{tr}[\lambda_+(b)(\mathbb{1} - W_+^*(b)W_+(b))]} \cdot e^{N\operatorname{tr}[\lambda_-(b)(\mathbb{1} - W_-(b)W_-(b)^*)]} \}$$

Then, formally,

$$\langle W(\mathcal{L})\rangle_{\overline{\beta},N} = Z^{-1}_{\overline{\beta},N} \int \mathcal{D}\underline{W}_+ \mathcal{D}\underline{W}_- \int \mathcal{D}\underline{\lambda}_+ \mathcal{D}\underline{\lambda}_- \, \mathcal{Z}_{\overline{\beta},N}(\underline{\lambda}_+,\underline{\lambda}_-;\underline{W}_+,\underline{W}_-) \cdot \langle W(\mathcal{L})\rangle_{\overline{\beta},N}(\underline{\lambda}_+,\underline{\lambda}_-) \tag{2.35}$$

By simple integrations by part with respect to a Gaussian measure we derive the identity

$$\langle W(\mathcal{L})\rangle_{\overline{\beta},N}(\underline{\lambda}_+,\underline{\lambda}_-) = \sum_{S:\partial S = \mathcal{L}} \overline{\beta}^{|S|} N^{2-2H(S)} \cdot \prod_b \rho_S(\lambda_+(b), \lambda_-(b)), \tag{2.36}$$

where $\sum_S$ ranges over all connected, orientable surfaces with plaquettes glued together along half-bonds, (each half-bond belonging to precisely two plaquettes), see [25], $H(S)$ denotes the number of handles, and $\rho_S(\lambda_+(b), \lambda_-(b))$ is a product of factors $N^{-1}\operatorname{tr}([\lambda_+(b)^{-1} \lambda_-(b)^{-1}]^{n_\alpha})$, with $\sum n_\alpha = n_b(S)$. See [25] for some explicit calculations.

Inserting the identity

$$\lambda_{\pm}^{-n} = \lim_{\varepsilon \downarrow 0} \int_0^\infty \frac{t^{n-1}}{(n-1)!} e^{t(\lambda_{\pm}-\varepsilon)} dt , \tag{2.37}$$

we may convert (2.36) into a local-time representation.

It is easy to see that, before integrating over λ_+ and λ_-, the random-surface representation obtained from (2.36) and (2.37) does not converge absolutely. For finite N, it is quite a job to analyze that representation after integration over λ_+ and λ_-. Apart from the observation that a formal representation of $\langle W(\mathcal{L})\rangle_{\bar{\beta},N}$ as a certain sum over random surfaces bounded by $\mathcal{L}$ exists, no useful results have been obtained. However, when N tends to ∞, formally, only planar surfaces ($H(S) = 0$) contribute to the r.h.s. of (2.36), and convergence properties greatly improve, [7,8]. By analogy with the N-vector model Kostov argues [25], rather interestingly, that, in integrating the r.h.s. of (2.36) over λ_+ and λ_- in the limit $N = \infty$, we may use the factorization property, i.e.

$$\lim_{N\to\infty} \langle W(\mathcal{L})\rangle_{\bar{\beta},N} = \sum_{S:\partial S=\mathcal{L}} \mathcal{Z}_{\bar{\beta},\infty}(S) , \tag{2.38}$$

where $\mathcal{Z}_{\bar{\beta},\infty}(S)$ vanishes, unless S is connected, orientable and planar, in which case

$$\mathcal{Z}_{\bar{\beta},\infty}(S) = \bar{\beta}^{|S|} \prod_b \overline{\rho_S(\lambda_+(b), \lambda_-(b))} , \tag{2.39}$$

where $\overline{(\cdot)}$ indicates averaging over λ_+ and λ_-. However, it seems difficult to justify this proposal and to actually evaluate, or estimate $\overline{\rho_S(\lambda_+(b),\lambda_-(b))}$, even in the limit $N = \infty$. Moreover, since the weights $\mathcal{Z}_{\bar{\beta},\infty}(S)$ are not in general non-negative, the representation (2.38) will stop converging absolutely when $\bar{\beta}$ becomes large, (thus blocking access to the continuum limit), in particular it may break down at the U(1)-or a Gross-Witten type transition [66]. See [25] for more details. (Convergence for small β follows from [7], assuming (2.39).) For these reasons, it is useful to consider a simplified version of the large - N-gauge theory: We return to the g_b-variables. However, g_b may now be an arbitrary complex N×N matrix. Haar measure, dg_b, on U(N) is replaced by a Gaussian measure

$$\exp[-N\mathrm{tr}\,(g_b^* g_b)] \prod_{i,j=1}^{N} dg_b^{ij}\, d\overline{g_b^{ij}}\,, \tag{2.40}$$

for all b. All else remains as in U(N) lattice gauge theory. This yields a model originally proposed by Weingarten [21] as a lattice approximation to the Nambu-Goto string. It is perfectly meaningless. However, its formal $N \to \infty$ limit is an interesting theory, the planar random-surface theory:

$$\lim_{N\to\infty} N^{-1}\langle W(\mathcal{L})\rangle_{\bar{\beta},N} = \sum_{S:\partial S=\mathcal{L}} \mathcal{Z}_{\bar{\beta}}(S) \tag{2.41}$$

$$\lim_{N\to\infty} \langle W(\mathcal{L}_1);W(\mathcal{L}_2)\rangle_{\bar{\beta},N} = \sum_{S:\partial S=\mathcal{L}_1\cup\mathcal{L}_2} \mathcal{Z}_{\bar{\beta}}(S) \tag{2.42}$$

$$\langle W(\mathcal{L}_1);\ldots;W(\mathcal{L}_n)\rangle_{\bar{\beta},N} \sim N^{2-n}\,, \tag{2.43}$$

$$\text{with}\quad \mathcal{Z}_{\bar{\beta}}(S) = \bar{\beta}^{|S|} = e^{-\beta|S|}\,, \tag{2.44}$$

$(\beta \equiv \log \bar{\beta}^{-1})$, and each S is a connected, orientable, planar random surface immersed in $\mathbb{Z}^d$; see [7,8]. The r.h.s. in (2.41) and (2.42) converge for all $\beta > \beta_o$, for some finite, positive β_o independent of $\mathcal{L}$, $\mathcal{L}_1$ and $\mathcal{L}_2$, resp., [7], but there is no way by which one may give mathematical sense to the ℓ.h.s., as pointed out by G. Felder.

A possibility to nevertheless make sense out of the finite-N Weingarten model might be to define it via "stochastic quantization" : Consider the stochastic process, $g(\cdot)$, defined by the equation

$$(\partial g_b/\partial\tau)(\tau) = -(\partial A/\partial g_b)(\tau) + \eta_b(\tau)\,, \tag{2.45}$$

where η is a white-noise process. If $\exp[-A(g)]$ is integrable, then

$$\lim_{T\to\infty} T^{-1}\int_o^T d\tau\, F(g(\tau)) = Z^{-1}\int F(g)\,\exp[-A(g)]dg\,, \tag{2.46}$$

for any bounded, measurable function F. But equ. (2.45) may make sense even if the action functional, A, is not bounded below, as in the Weingarten model. It is an interesting speculation that, for the Weingarten model on a finite lattice,

$$\lim_{T\to\infty}\lim_{N\to\infty} T^{-1}\int_o^T d\tau\, F(g(\tau)) = \langle F\rangle_{PRS}\,, \tag{2.47}$$

where $\langle(\cdot)\rangle_{PRS}$ is an expectation directly defined for the planar-random-surface theory. This speculation is at the basis of the numerical work in [67].

The PRS model defined by the r.h.s. of (2.41) and (2.42) has very appealing properties which we study in the next section. One ought to regard it as the right generalization of Brownian motion, from random paths to random surfaces.

There is a basic issue concerning the random surface representations derived in Sects. 2.2 and 2.3 which we have not discussed, yet: They are not unique! For the $\mathbb{Z}_2$-gauge model, for example, there are at least two different representations; one has $\bar{\beta}$ as an "expansion parameter", the other one tanh $\bar{\beta}$, [6]. Both seem quite useful. Furthermore, before resummation, these representations might be regularization-dependent. (Superficially, they certainly depend on the choice of the underlying lattice, and, in examples (1b) and (1c) of Sect. 2.2, of the choice of $g(|\vec{\phi}|^2)$. In fact, the dynamics of the "radial" degrees of freedom, $|\vec{\phi}_b|^2$, might be regularization-dependent if it survives the scaling limit.) Hence, before a direct physical significance can be attached to the random surface representations of Sects. 2.2 and 2.3, one ought to establish some universality properties of these representations in the scaling limit. (See also Sects. 3.1. and 3.2.)

§ 3. The Statistical Mechanics of Random Surfaces

We shall exclusively consider discrete random surfaces (RS) immersed in $\mathbb{Z}^d$. (For proposals of continuum theories see, however, [68,61], and footnote 1.)

An RS model is specified by the following data:

(1) A countable ensemble, $\mathcal{E}$, of connected RS immersed in $\mathbb{Z}^d$.

Examples.

(a) All RS contributing to the RS representation of SU(2) lattice gauge theory derived in (2.24), (2.25).

(b) An ensemble of "self-avoiding" RS arising in the RS representation of $\mathbb{Z}_2$-lattice gauge theory [6].

(c) All connected, orientable planar RS.

(d) The clusters of Bernoulli plaquette percolation, [10].

(e) Graphs of an integer-valued function on $(\mathbb{Z}^2)^*$ which describe the interfaces of the solid-on-solid model [60c].

(2) A statistical weight

$$\mathcal{Z}_\beta : \mathcal{E} \ni S \longmapsto \mathcal{Z}_\beta(S) > 0, \tag{3.1}$$

e.g. $\mathcal{Z}_\beta(S) = e^{-\beta|S|}$, or $\mathcal{Z}_\beta(S)$ as in (2.24), (2.25), (with $\beta = |\log \bar{\beta}|$).

The basic quantities in such a model are the following:

(i) $$n_{\mathcal{L}}(A) = \#\{S \in \mathcal{E} : |S| = A, \partial S = \mathcal{L}\}, \tag{3.2}$$

where $\mathcal{L} = \mathcal{L}_1 \cup \ldots \cup \mathcal{L}_n$ is a union of lattice loops. For the ensembles defined in (b)-(d) above,

$$n_{\mathcal{L}}(A) = \exp[\beta_o A + o(A)], \tag{3.3}$$

for some finite positive constant β_o independent of $\mathcal{L}$. See [7]. If $\mathcal{E}$ is the ensemble of all connected, orientable RS with precisely H handles then (3.3) holds, for some constant β_o independent of $\mathcal{L}$ and H; [8].

When $\mathcal{L} = \partial p$ then numerical results [67] suggest that

$$n_{\partial\beta}(A) \sim A^{\varepsilon} e^{\beta_o A}, \tag{3.4}$$

with $\varepsilon \simeq -1.5$, within 12%, for d = 3. Moreover, (3.4) holds with $\varepsilon = -1.5$ in mean-field theory [19 ,8], (i.e. presumably for large enough d.)

(ii) Let B_R be a ball of radius R in $\mathbb{Z}^d$ centered at some plaquette p_o. We set $S_R \equiv S \cap B_R$, $S \in \mathcal{E}$. We define the Hausdorff dimension, δ_H, of infinitely extended RS as follows:

$$\delta_H = \lim_{R\to\infty} \lim_{A\to\infty} \frac{1}{n_{\partial p_o}(A)} \sum_{\substack{S \in \mathcal{E} \\ \partial S = \partial p_o \\ |S| = A}} \left(\frac{\log|S_R|}{\log R} \right) \tag{3.5}$$

(In [8] we have introduced a "true" Hausdorff dimension, δ, and an exponent α measuring recurrence, such that $\delta_H = \delta\,\alpha$. Presumably $\alpha = 1$ for d large enough.) There is a second notion of Hausdorff dimension, $\delta(\beta)$, to be distinguished from δ_H: Let $\mathcal{S}_o(\mathcal{L}_{L\times T})$ be the class of all RS in $\mathcal{E}$ with the properties that $\partial S = \mathcal{L}_{L\times T}$ = a rectangle in a coordinate plane, centered at 0, with sides of length L and T, and $S \ni p_o \ni 0$. Let $Z_\beta(L,T) = \sum_{S \in \mathcal{S}_o(\mathcal{L}_{L\times T})} \mathcal{Z}_\beta(S)$.

We define

$$\delta(\beta) = \lim_{R\to\infty} \lim_{L,T\to\infty} Z_\beta(L,T)^{-1} \sum_{S \in \mathcal{S}_o(\mathcal{L}_{L\times T})} \left(\frac{\log|S_R|}{\log R}\right) \mathcal{Z}_\beta(S) \tag{3.6}$$

(iii) we define loop correlations

$$G_\beta(\mathcal{L}_1,\ldots,\mathcal{L}_n) = \sum_{\substack{S\in\mathcal{E} \\ \partial S = \mathcal{L}_1 \cup \ldots \cup \mathcal{L}_n}} \mathcal{Z}_\beta(S). \tag{3.7}$$

The "$q\bar{q}$ potential" is defined by

$$V(\beta;\, L) = \lim_{T\to\infty} -\frac{1}{T} \log G_\beta(\mathcal{L}_{L\times T}), \tag{3.8}$$

and the string tension (or surface tension) by

$$\alpha(\beta) = \lim_{L\to\infty} \frac{1}{L} V(\beta;\, L). \tag{3.9}$$

The inverse correlation length (glue-ball mass) is defined by

$$m(\beta) = \lim_{a\to\infty} -\frac{1}{a} \log G_\beta(\partial p,\, \partial p_a) \tag{3.10}$$

where ∂p_a is a copy of ∂p, translated in a coordinate direction by a units. These limits can be shown to exist [6,7,8,10]. A susceptibility χ is defined by

$$\chi(\beta) = \sum_{p'} G_\beta(\partial p,\, \partial p') \tag{3.11}$$

For $\mathcal{E}$ as in (c), $G_\beta(\mathcal{L}_1,\ldots,\mathcal{L}_n)$ and $\chi(\beta)$ are finite for $\beta > \beta_o$, while they diverge for all $\beta < \beta_o$. For $\mathcal{E}$ as in (a), (b) and (d), β_o is defined to be the deconfinement transition point of the corresponding lattice theory. An important question is now whether β_o is a critical point in the sense that

$$m(\beta) \searrow 0, \text{ as } \beta \searrow \beta_o? \tag{3.12}$$

If this is the case we may introduce critical exponents:

$$\left.\begin{aligned} &m(\beta) \sim (\beta-\beta_o)^{\nu},\ \chi(\beta) \sim (\beta-\beta_o)^{-\gamma} \\ &G_\beta(\partial p,\, \partial p_a) \sim a^{-(d-2+\eta)},\ \text{for } 1 \ll a \ll m(\beta)^{-1}, \end{aligned}\right\} \tag{3.13}$$

and if

$$\alpha(\beta) \searrow 0, \text{ as } \beta \searrow \beta_o,$$

$$\alpha(\beta) \sim (\beta - \beta_o)^{\mu} \tag{3.14}$$

Assuming a weak form of scaling we have

$$\gamma = \nu(2-\eta) \tag{3.15}$$

3.1 Typical Phenomena in the Statistical Mechanics of Random Surfaces

In this section we briefly describe some typical phenomena encountered in the statistical mechanics of RS most of which have no analogue in the theory of interacting random walks and -paths.

(1) Surface roughening, [60]. Let $P_\beta(h|\mathcal{L}_{LxT})$ be the probability that a surface $S \in \mathcal{E}$ with $\partial S = \mathcal{L}_{LxT}$ intersects a (d-2)-dimensional plane perpendicular to the plane containing $\mathcal{L}_{LxT}$ and passing through 0 at a distance h from 0. For models like the ones in (b) - (e) one can prove [60a] that for $\beta >> \beta_o$

$$P_\beta(h) = \lim_{L,T\to\infty} P_\beta(h|\mathcal{L}_{LxT}) \tag{3.16}$$

exists and decays exponentially in h. For the solid-on-solid model, see (e), one can prove [60c] that there exists $\beta_R > 0$ such that, for $\beta < \beta_R$, $P_\beta(h) = 0$, for all $h < \infty$, and

$$\sum_{h=0}^{\infty} h^2 \, P_\beta(h|\mathcal{L}_{LxL}) \sim \log L. \tag{3.17}$$

Heuristically one expects that in a large class of models, including (a)-(e), there exists some $\beta_R > \beta_o$ such that

(i) for $\beta > \beta_R$, the behaviour is described by (2.3) and (3.16);

(ii) for $\beta_R > \beta > \beta_o$, the behaviour is described by (2.4) and (3.17). See [60]. The roughening transition is expected to be of ∞ order. The capillary surface waves present when $\beta < \beta_R$ restore continuous translational symmetry broken by the lattice. The roughening transition is thus an artefact of the lattice. Random walks (for comparison) are always in the rough phase.

(2) Topological complexity. Random "surfaces" have intrinsic topology: They may be non-orientable, may have "pockets" (e.g. in

$\mathbb{Z}_N$- and SU(N)-theories, with $N \geq 3$; pockets arise when N plaquettes are glued together along a common link) and non-trivial $(d-3)^{rd}$ homology with coefficients in $\mathbb{Z}_N$, when $d \geq 4$, [11]. But even "true", orientable surfaces (rather than 2-cell-, resp. (d-2)-cell complexes) may have a complicated topology characterized by the Euler characteristic, (the number of handles). Topological complexity has at least two consequences: The number of RS of area A with arbitrary topology grows very fast (powers of A!). There are "topological anomalies" in the $\frac{1}{q}$ -expansion of the q-states gauge Potts model and possibly in the $\frac{1}{N}$-expansion of SU(N) lattice gauge theory, [11].

(3) Crumpling at β_o. Experience with the solid-on-solid model [60c,8] suggests that, for all $\beta > \beta_o$, the Haussdorff dimension of typical RS is

$$\delta(\beta) = 2. \tag{3.18}$$

In particular, surface roughening does not increase the Hausdorff dimension, and the concept of a flux sheet makes sense for all $\beta > \beta_o$. However, at β_o, $\delta(\beta)$ jumps to a value $\delta_H > 2$, a phenomenon which we call "crumpling". In the planar random surface (PRS) model, δ_H appears to approach 4, as $d \to \infty$; (presumably $\delta_H = 4$, if $d \geq 8$). See [19,8]. When $\beta < \beta_o$ (deconfined phase) experience with plaquette percolation [10] and the three-dimensional Ising model suggests that a typical RS is space-filling with probability 1, δ_H diverges, and the concept of a flux sheet becomes meaningless. (This the perturbative regime. In continuum gauge theories the flux sheet is always crumpled on very short distance scales, with $\delta_H = \infty$. See also footnote 2.)

(4) Collapse to tree-like surfaces, [19,8]. Mean-field theory for the PRS model - see (c) above - predicts that, as the dimension $d \to \infty$,

$$\alpha(\beta) \searrow \alpha^* > 0, \text{ while } m(\beta) \searrow 0, \text{ as } \beta \searrow \beta_o; \tag{3.19}$$

see [19]. One might expect, therefore, that, in this and related RS models, there exists a critical dimension, $d_c < \infty$, such that (3.19) holds, for all $d > d_c$, while for $d < d_c$

$$\alpha(\beta) \searrow 0 \text{ and } m(\beta) \searrow 0, \text{ as } \beta \searrow \beta_o, \text{ [8].} \tag{3.20}$$

Thus when $d > d_c$, typical random surfaces collapse to tree-like structures when watched from far distances, (i.e. in the scaling limit), and, in the mean-field limit, all surfaces degenerate to trees.

An alternative to this collapse mechanism apparently realized in certain lattice gauge theories in $d \geq 5$ dimensions, is that the deconfining transition is first order [63].

3.2 The Planar Randon-Surface (PRS) Model

In this section we summarize some of the results concerning a simplified model reminiscent of large-N lattice gauge theory whose formal continuum limit determines a good relativistic quantum field theory and which avoids the problems connected with the strong coupling expansion, (Sect. 2.1, (b)), and topological complexity (Sect. 2.1, (c)): The PRS model introduced in Sect. 2.3. It ought to be a good toy model to study confinement and Regge behaviour of the glueball spectrum, as well as mechanisms (1), (3) and (4) of Sect. 3.1.

The PRS model has the following appealing features:

(i) $\alpha(\beta) \geqq c_1 (\beta-\beta_o) > 0$, and

$m(\beta) \geqq c_2 (\beta-\beta_o) > 0$,

for all $\beta > \beta_o$; i.e. this model has permanent confinement and mass generation, [7].

(ii) Using the fact that $\mathfrak{z}_\beta(S) = e^{-\beta|S|}$ we see that $\chi(\beta)$, defined in (3.11), behaves like

$$\chi(\beta) \sim (\beta-\beta_o)^{-(2+\varepsilon)},$$

where ε is the exponent of $n_{\partial p}(A)$ introduced in (3.4). Thus

$$\gamma = 2 + \varepsilon. \tag{3.21}$$

The numerical calculation in [67] and mean-field theory [19,8] suggest that

$$\varepsilon \geqq -2, \text{ for } d \geqq 3. \tag{3.22}$$

Assuming (3.22), we conclude that

$$\chi(\beta) \nearrow \infty, \quad \text{as } \beta \searrow \beta_o. \tag{3.23}$$

It is proven in [8] that

$$\chi(\beta) \nearrow \infty, \quad \text{as } \beta \searrow \beta_o \Leftrightarrow m(\beta) \searrow 0, \quad \text{as } \beta \searrow \beta_o; \tag{3.24}$$

i.e. (3.22) implies that β_o is a critical point. This is a necessary prerequisite for the construction of a continuum ($\equiv$ scaling) limit.

(iii) In mean-field theory,

$$G_\beta(\partial p_1, \partial p_2, \partial p_3) = \text{const.} \sum_p \prod_{j=1}^{3} G_\beta(\partial p_i, \partial p). \tag{3.24}$$

More generally $G_\beta(\partial p_1,\ldots,\partial p_n)$ is given by a sum over all tree diagrams with vertices of order 3, propagators given by $G_\beta(\partial p, \partial p')$, and external lines ending in $p_1,\ldots, p_n$.

By (3.11) and (3.24),

$$\frac{d\chi(\beta)}{d\beta} = - \text{const. } \chi(\beta)^3 , \tag{3.25}$$

hence $\gamma = 1/2$ and $\varepsilon = -\frac{3}{2}$. Moreover,

$$G_\beta(\partial p, \partial p_a) \sim a^{-(d-2)}, \tag{3.26}$$

i.e. $\eta = 0$; (see (3.13)). By (3.15)

$$\nu = 1/4. \tag{3.27}$$

Finally,

$$\alpha(\beta) \searrow \alpha^* > 0, \quad \text{as } \beta \searrow \beta_o. \tag{3.28}$$

(iv) It is quite remarkable that mean-field theory provides rigorous bounds: $\beta_o \geqq \beta_o$ (mean-field theory),

$$G_\beta(\partial p_1, \partial p_2, \partial p_3) \geqq \text{const. } \sum_p \prod_{j=1}^{3} G_\beta(\partial p_i, \partial p) \tag{3.29}$$

$$G_\beta(\partial p_1,\ldots,\partial p_n) \geqq \text{const. } \sum \text{tree diagrams.} \tag{3.30}$$

By (3.29) and (3.11),

$$-\frac{d\chi(\beta)}{d\beta} \geqq \text{const. } \chi(\beta)^3, \text{ i.e. } \gamma \leqq \frac{1}{2}. \tag{3.31}$$

Furthermore, a somewhat subtle proof shows

<u>Theorem</u>. $G_\beta(\mathcal{L}, \mathcal{L}')$ <u>satisfies Osterwalder-Schrader (reflection) positivity</u>.

In field theoretic jargon this means that the scaling limit satisfies unitarity and the spectrum condition, in statistical mechanics jargon that the model possesses a positive transfer matrix.

As a standard corollary we deduce from this theorem that if the scaling limit of $G_\beta(\partial p, \partial p')$ exists then

$$\eta \geqq 0. \tag{3.32}$$

Now, from (3.29) and the assumption that the three-loop correlation

$G_\beta(\partial p_1, \partial p_2, \partial p_3)$ has a finite scaling limit we derive that

$$\eta \geqq 2 - \frac{d}{3}, \text{ hence}$$

$$\eta \geqq \max\left(2 - \frac{d}{3}, 0\right). \tag{3.33}$$

In particular, in four dimensions, there is a large anomalous dimension $\eta \geqq \frac{2}{3}$.

Comparison with the Weingarten model, see (2.41) - (2.43), shows that the scaling limits of

$$N^{2-n} G_\beta(\mathcal{L}_1, \ldots, \mathcal{L}_n), \tag{3.34}$$

represent the leading contribution in $\frac{1}{N}$ to connected, physical n-loop functions.

(v) Property (3.28) suggests the following Conjecture: There exists a critical dimension $d_c < \infty$ such that, for $d < d_c$, $\alpha(\beta) \searrow 0$, as $\beta \searrow \beta_o$, $(\alpha(\beta) \sim (\beta-\beta_o)^\mu)$, while, for $d > d_c$, $\alpha(\beta) \searrow \alpha^* \gtrless 0$, as $\beta \searrow \beta_o$.

Now, if this is true fairly convincing, but non-rigorous arguments yield

Quasi-Theorem. In dimension $d > d_c$, the exponents γ, ν and η take their mean-field values, and mean-field theory is asymptotically exact.

Comparison with (3.33) would then give

$$d_c \geqq 6. \tag{3.35}$$

By (3.33) and (3.15)

$$\gamma = \nu(2-\eta) \leqq \nu \frac{d}{3}, \text{ for } d \leqq 6. \tag{3.36}$$

(vi) If $\chi(\beta) \nearrow \infty$, as $\beta \searrow \beta_o$, then under some suitable assumptions of scaling, $(\beta-\beta_o)^{-1} \sim -\frac{d}{d\beta} \log \chi(\beta)$

$$= \chi(\beta)^{-1} \left(\sum_p \sum_{S:\partial S = \partial p_o \cup \partial p} |S| e^{-\beta|S|}\right)$$

$$\sim \chi(\beta)^{-1} \left(\sum_p \operatorname{dist}(p_o, p)^{\delta_H} G_\beta(\partial p_o, \partial p)\right)$$

$$\sim m(\beta)^{-\delta_H} \sim (\beta-\beta_o)^{-\nu\delta_H},$$

as $\beta \searrow \beta_o$. Hence

$$\nu = {}^1/\delta_H \tag{3.37}$$

Similarly, when $d < d_c$,

$$\mu = {}^2/\delta_H = 2\,\nu \tag{3.38}$$

In mean-field theory $\nu = {}^1/4$, i.e. $\delta_H = 4$. Since dim S = 2, $\delta_H \geqq 2$, hence

$$\nu \leqq \frac{1}{2}. \tag{3.39}$$

By (3.36) and (3.39),

$$\gamma \leqq {}^1/3, \quad \text{in } d = 2$$

which casts some doubts on the numerical results in [67]. On the basis of these results and mean-field theory it is reasonable to expect that

$$\gamma \geqq \frac{1}{3}, \quad \text{for } d \geqq 3. \tag{3.40}$$

This and (3.39) yield

$$\frac{2}{3} \leqq 2 - \eta, \quad \text{i.e.} \quad \eta \leqq \frac{4}{3} \tag{3.41}$$

which is compatible with (3.33). And, using (3.33) and (3.37),

$$\delta_H = \frac{2-\eta}{\gamma} \leqq \frac{d}{3\gamma}. \tag{3.42}$$

In four dimensions, presumably $\gamma \simeq \frac{1}{2}$, hence

$$\delta_H < 4, \text{ in four dimensions,} \tag{3.43}$$

(in particular $d_c > 4$, in agreement with (3.35)!). (See also footnote 2.)

All the results for the PRS model reported here have been worked out with B. Durhuus and T. Jónsson in [7,8]. These results suggest the conjecture that the scaling limit of this model exists, and is non-trivial for $d < d_c$. Since the two-loop Green's function $G_\beta(\mathcal{L}, \mathcal{L}')$ is Osterwalder-Schrader (reflection) positive and by (2.43), that scaling limit defines a relativistic quantum field theory with all the required properties [39,41]; in particular, there are no tachyons in its particle spectrum. However, the theory is non-interacting (Gaussian), in particular the scattering matrix is $\mathbb{1}$; (see (3.34) and (2.43)). For $d < d_c$, we have heuristic reasons to expect that the mass spectrum has Regge behaviour, but we have no precise results to report on this, yet.

Finally, we wish to remark that there are close analogies between the PRS model and the planar - $\lambda\ \phi^3$ theory, $\lambda > 0$. These matters are presently investigated by G. Felder and the author. (see Footnote 1.)

3.3 Science Fiction on the Role of Gravity in Curing the Short Distance Problems of Quantum Field Theory

If gravity is added to our considerations then the random paths and random surfaces which we have studied abundantly in Sects. 1 - 3 are really immersed in a fluctuating space-time manifold with its own dynamical degrees of freedom which, at short distances, may be highly curved and have a complicated topology. These properties of space-time may lead to unexpected behaviour of

$$G_\Gamma(x,y) = \int_{\omega: x\to y} \mathfrak{z}_\Gamma(\omega), \tag{3.44}$$

and

$$G_{\bar\beta}(\mathcal{L}) = \int_{S:\partial S=\mathcal{L}} \mathfrak{z}_{\bar\beta}(S), \tag{3.45}$$

when $|x-y|$ and diam ($\mathcal{L}$) are very small, because the statistics and geometrical properties of short paths, ω, and small random surfaces, S, may be very different from what they are in flat space. (E.g. when $\mathcal{L} = \emptyset$, a considerable fraction of surfaces, S, contributing to (3.45) may be non-contractible.) A phenomenon related to Anderson localization may then occur on very short distance scales. As an example, the spectrum of the Laplace-Beltrami operator on a fluctuating space-(imaginary) time manifold, the Green's function of which has a random path representation like (3.44), may be pure-point, at high "energies", with probability 1. (Here "energy" denotes the value of the spectral parameter.) This would mean that scalar quanta of very high "energy" do not propagate through space-time foam, and that the density of states at high "energies" is smaller than in flat space-time. All these effects might improve the short-distance behaviour of quantized fields.

REFERENCES

1. D. Brydges, J. Fröhlich and T. Spencer, Commun.Math.Phys. 83, 123 (1982); (see also:
D. Brydges and P. Federbush, Commun.Math.Phys. 62, 79 (1978); and
A.J. Kupiainen, Commun.Math.Phys. 73, 273 (1980)).

2. J. Fröhlich, Nucl.Phys. B 200 [FS4], 281 (1982).

3. C. Aragão de Carvalho, S. Caracciolo and J. Fröhlich, Nucl.Phys. B 215 [FS7], 209 (1983).

4. D. Brydges, J. Fröhlich and A. Sokal, Commun.Math.Phys. 91, 117 (1983).

5. D. Brydges, J. Fröhlich and A. Sokol, Commun.Math.Phys. 91, 41 (1983.) (For the original results by Glimm and Jaffe see ref. 43.)
For summaries of some of the results in refs.1-5 see also:
J. Fröhlich and T. Spencer, exposé n°586, Séminaire Bourbaki, Astérisque 92-93, 159 (1982).
D. Brydges, in "Gauge Theories: Fundamental Interactions and Rigorous Results", P. Dita et al. (eds.), Boston: Birkhäuser 1982.

6. D. Brydges, J. Fröhlich and A. Sokal, papers in preparation.

7. B. Durhuus, J. Fröhlich and T. Jonsson, Nucl.Phys. B 255 [FS9], 185 (1983).

8. B. Durhuus, J. Fröhlich and T. Jonsson, "Critical Properties of a Model of Planar Random Surfaces", Phys.Letts. B, in press.; and papers in preparation.

9. J. Fröhlich, C.-E. Pfister and T. Spencer, in "Stochastic Processes in Quantum Field Theory and Statistical Physics", Lecture Notes in Physics 173, Berlin-Heidelberg-New York: Springer-Verlag 1982.
J. Frohlich, Physics Reports 67, 137 (1980).

10. M. Aizenman, J.T. Chayes, L. Chayes, J. Fröhlich and L. Russo, Commun.Math.Phys. 92, 19 (1983).

11. M. Aizenman and J. Fröhlich, "Topological Anomalies in the n-Dependence of the n-States Potts Lattice Gauge Theory", Nucl. Phys.B [FS], in press.

12. M. Aizenman, Commun.Math.Phys. 86, 1-48 (1982); and paper in preparation.

13. M. Aizenman and R. Graham, Nucl.Phys. B225 [FS9], 261(1983).

14. A. Bovier, G. Felder and J. Fröhlich, Nucl.Phys. B 230 [FS10], 119 (1984).

15. A. Bovier and G. Felder, "Skeleton Inequalities and the Asymptotic Nature of Perturbation Theory for ϕ^4-Theories in Two and Three Dimensions", Commun.Math.Phys., in press.

16. K. Symanzik, "Euclidean Quantum Field Theory", in "Local Quantum Theory", R. Jost (ed.), New York and London: Academic Press, 1969.

17. A. Sokal, Ann.Inst. Henri Poincaré A37, 317(1982).

18. B. Durhuus and J. Fröhlich, Commun.Math.Phys.75, 103 (1980).

19. J.M. Drouffe, G. Parisi and N. Sourlas, Nucl.Phys. B 161, 397 (1980).

20. M. Jacob (ed.), "Dual Theory", Amsterdam: North Holland, 1974. (Among the original papers are: Y. Nambu, in "Symmetries and Quark Models", R. Chaud (ed.), New York: Gordon and Breach, 1970; and T. Goto, Progr.Theor.Phys. 46, 1560 (1971).)

21. D. Weingarten, Phys.Lett. 90B, 280 (1980).

22. T. Eguchi and H. Kawai, Phys.Lett. 110B, 143 (1982).

23. B. De Wit and G. 't Hooft, Phys.Lett.69B, 61(1977).
I. Bars and F. Green, Phys.Rev. D20, 3311 (1979).
D. Weingarten, Phys.Lett.90B, 285 (1980).

24. S. Coleman, "1/N", Erice Lectures 1979, to be published.

25. I.K. Kostov, "Multicolor QCD in Terms of Random Surfaces", Preprint.

26. T.P. Cheng, E. Eichten and L.-F. Li, Phys.Rev. D9, 2259 (1974).
L. Maiani, G. Parisi and R. Petronzio, Nucl.Phys. B136, 115 (1978).
D.J.E. Callaway, Preprint CERN, 1983.

27. C. Itzykson, these proceedings, and refs. to be found there.

28. P.-G. de Gennes,"Scaling Concepts in Polymer Physics",Ithaca and London: Cornell University Press, 1979; and refs. given there.

29. D. Brydges, J. Fröhlich, A. Sokal, unpublished.

30. G. Felder, unpublished.

31. G. Mack, these proceedings; and "Dielectric Gauge Theory", DESY Preprint 1983.

32. J. Fröhlich, talks at the Trieste meeting (Dec. '82) and at CERN (spring '83); see ref. [6].

33. J.-M. Drouffe, Nucl.Phys. B218, 89 (1983).

34. J. Bricmont and J. Fröhlich, Phys.Lett 122 B, 73 (1983), and papers in preparation.

35. The expression "stochastic geometry", is borrowed from D.G. Kendall and E.F. Harding (eds.), "Stochastic Geometry", New York: Wiley, 1977. The use of stochastic geometry in physics has a rather long standing tradition. Recently, it has been advertized explicitly in:
J. Fröhlich, in "Recent Developments in Gauge Theories", G. 't Hooft et al.(eds.), p. 53, New York and London: Plenum Press, 1980.
M. Aizenman, Commun.Math.Phys. 73,83 (1980). (See also refs. 2,10, 12.)

36. G. 't Hooft, these proceedings, and refs. given there.

37. G. Gallavotti and V. Rivasseau, Phys.Lett. 122B, 268 (1983); preprint 1983.
V. Rivasseau, in preparation.

38. C.G. Callan, Phys.Rev. D2, 1541 (1970).
K. Symanzik, Commun.Math.Phys. 18, 227 (1970).
A.I. Larkin and D.E. Khmel'nitskii, J.E.T.P. (Soviet Phys.) 29, 1123 (1969).
E. Brézin, J. Le Guillou and J. Zinn-Justin, Phys.Rev. D8, 2418 (1973).

39. K. Osterwalder and R. Schrader, Commun.Math.Phys. 31, 83 (1973), Commun.Math.Phys. 42,281 (1975).
For related results see also:
V. Glaser, Commun.Math.Phys. 37,257 (1974).

40. K. Symanzik, "A Modified Model of Euclidean Quantum Field Theory", NYU preprint, IMM-NYU (1964); J.Math.Phys. 7,510 (1966).

41. E. Nelson, J.Funct.Anal.12, 97 (1973); J. Funct.Anal. 12, 211 (1973); in: "Partial Differential Equations", D.C. Spencer (ed.), Providence, R.I.: Publ. A.M.S., 1973.

42. B. Simon, "The $P(\phi)_2$ Euclidean (Quantum) Field Theory", Princeton, N.J.: Princeton University Press, 1974.

43. J. Glimm and A. Jaffe, "Quantum Physics", Berlin-Heidelberg-New York: Springer-Verlag, 1981.

44. J. Fröhlich, B. Simon and T. Spencer, Commun.Math.Phys. 50,79 (1976).

45. J. Rosen, Adv.Appl.Math. 1, 37(1980).
J. Glimm and A. Jaffe, Commun.Math.Phys. 52, 203(1977); ref. 43.

46. G.F. Lawler, Commun.Math.Phys. 86,539 (1982); preprint, Duke University, 1983.

47. P. Erdös and S.J. Taylor, Acta Math. Acad. Sci. Hung. 11, 137 (1960), 11, 231 (1960).

48. E. Brézin, J. Le Guillou and J. Zinn-Justin, Phys.Rev. B10, 892 (1974).

49. A.J. Kupiainen, see ref. 1;
J. Fröhlich, A. Mardin and V. Rivasseau, Commun.Math.Phys. 86, 87 (1982).

50. R.B. Griffiths, J.Math.Phys. 8, 478, 484 (1967); Commun.Math. Phys.6, 121 (1967).
J. Ginibre, Commun.Math.Phys.16, 310 (1970).

51. K. Pohlmeyer, Commun.Math.Phys.12, 204 (1969).

52. B. Berg and D. Förster, Phys.Lett. 106B,323 (1981).

53. J. Fröhlich, see ref. 35.

54. A. Berretti and A. Sokal, in preparation.

55. D. Brydges and T. Spencer, in preparation.

56. A. Guth, Phys.Rev. D21, 2291, (1980).
J. Fröhlich and T. Spencer, Commun.Math.Phys. 81,527 (1981).
M. Göpfert and G. Mack, Commun.Math.Phys. 82, 545 (1982).

57. M. Lüscher, Nucl.Phys. B219, 233 (1983).

58. M. Lüscher, Phys.Lett. 118B, 391 (1982); Ann.Phys.(NY) 142,359 (1982).

59. C. Borgs and E. Seiler, Nucl.Phys. B215 [FS7], 125 (1983); Commun.Math.Phys. 91, 329 (1983).

60. a. R.L. Dobrushin, Theor.Prob.and Appl. 17,582, (1972), 18, 253 (1973).
b. H. van Beijeren, Phys.Rev.Letts. 38,993 (1977).
c. J. Fröhlich and T. Spencer, Commun.Math.Phys. 81, 527 (1981).
d. C.Itzykson, M.E. Peskin and J.-B. Zuber, Phys.Lett. 95B, 259 (1980).
e. A. Hasenfratz, E. Hasenfratz and P. Hasenfratz, Nucl.Phys. B180 [FS2] 353 (1981).
f. M. Lüscher, G. Münster and P. Weisz, Nucl.Phys. B180 [FS2], 1 (1981).
See also refs. 18 and 9.

61. M. Lüscher, K. Symanzik and P. Weisz, Nucl.Phys. B173, 365 (1980).

62. G. 't Hooft, Nucl.Phys. B72, 461 (1974); Nucl.Phys. B75, 461 (1974).

63. J.-M. Drouffe, Nucl.Phys. B170 [FS1], 211 (1980).
E. Brézin and J.-M. Drouffe, Nucl. Phys. B200 [FS4], 93 (1982).
H. Flyvbjerg, B. Lautrup, J.-B. Zuber, Phys.Lett. 110B, 279 (1982).

64. K. Wilson, Phys.Rev. D10, 2445 (1974); see also: F. Wegner, J.Math.Phys. 12, 2259 (1971).
R. Balian, J.-M. Drouffe and C. Itzykson, Phys.Rev. D10, 3376 (1974), D11, 2098 (1975), D11, 2104 (1975).

65. D. Brydges, J. Fröhlich and E. Seiler, Ann.Phys. (NY) 121, 227 (1979).

66. Ref.56;
D. Gross and E. Witten, Phys.Rev. D21, 446(1981).

67. H. Kawai and Y. Okamoto, Phys.Lett. 130B, 415(1983).

68. A.M. Polyakoff, Phys.Lett. 103B, 207 (1981).
B. Durhuus, P. Olesen and J.L. Petersen, Nucl.Phys. B198, 157 (1982).
O. Alvarez, Nucl.Phys. B216, 125 (1983).

FOOTNOTES

1. We have recently noticed a close analogy between the PRS model and the $N\to\infty$ limit of $-\lambda tr(\phi^3)$-theory, where ϕ is a hermitian NxN matrix field. In fact, the two models have the same mean-field theory and they may have the same infrared stable fixed points in any dimension. This suggests that $d_c=6$. The planar $-\lambda tr(\phi^3)$-theory is superrenormalizable in the ultraviolet and has a convergent perturbation expansion, for $d<6$. Its particle spectrum is expected to exhibit Regge behaviour.

2. After completion of these notes a preprint by Billoire, Gross and Marinari has appeared. They propose to study simplicial, planar random surfaces immersed in $\mathbb{R}^d$ with a fixed number of vertices and weights prescribed by the Nambu-Goto-,or the Schild-Eguchi action. They argue that $\delta_H = \infty$. This indicates that their model is always in the phase where the random surfaces are completely crumpled; (see Sect. 3.1, (3)). For d=3, their model is equivalent to a two-dimensional, free massless field for which one can prove that the Hausdorff dimension of its graph diverges, as the continuum limit is taken. (This has been checked by G. Felder.) However, the notion of Hausdorff dimension used here is a measure for the short-distance properties of the surfaces and is not directly related to important critical exponents such as η.

RIGOROUS RENORMALIZATION GROUP AND LARGE N

Krzysztof Gawędzki[+1] and Antti Kupiainen[2]

[1]C.N.R.S., I.H.E.S.
F-91440 Bures-sur-Yvette, France

[2]Department of Technical Physics
Helsinki University of Technology
SF-02150 Espoo 15, Finland

In this lecture we discuss an attempt to obtain a rigorous control of the models of equilibrium statistical mechanics and Euclidean quantum field theory with the spins (fields) transforming under the fundamental representation of O(N) for N large. Our approach is based on a renormalization group (RG) analysis, see [16]. The idea is to use the leading orders of the 1/N expansion to deduce a qualitative behaviour of the RG effective interactions, borrowing heavily on the existing heuristic arguments [17, 18], and to perform a non-perturbative analysis of the corrections with the use of analyticity techniques developed by us [16]. The most interesting models for which our approach might be ultimately applicable include

- critical point of $(\vec{\varphi}^2)^2$ in three dimensions with its non-canonical long distance behaviour,
- O(N) σ-model in two dimensions with conjectured mass generation and asymptotically free renormalizable continuum limit,
- tricritical point of $(\vec{\varphi}^2)^3$ in three dimensions with a presumed non-Gaussian short distance behaviour.

[+]On leave from the Department of Mathematical Methods of Physics, Warsaw University, Poland

Up to now, our efforts have concentrated on the first model. We still have difficulties in coping with non-localities of the effective interactions and have been able to treat fully only the approximate hierarchical version of the model where these problems do not arise.

In order to perform a rigorous $1/N$ analysis, we needed a convenient approach to large N models. This, in our opinion, is provided by the (improved) collective field method of [13]. To be specific, consider a theory in finite (say, periodic) volume $\Lambda \subset \mathbb{Z}^d$ with the $O(N)$ invariant Hamiltonian

$$H(\vec{\phi}) = \frac{N}{2} \sum_{x\in\Lambda} v(\tfrac{1}{N}\vec{\phi}_x^2) + \frac{1}{2} \sum_{<xy>\subset\Lambda} (\vec{\phi}_y - \vec{\phi}_x)^2 \tag{1}$$

where the last term is the kinetic term given by the sum over pairs of lattice nearest neighbors and may be also written as $\frac{1}{2}<\vec{\phi}|-\Delta|\vec{\phi}>$. H generates the Gibbs state

$$\frac{1}{\mathcal{N}} e^{-H(\vec{\phi})} \prod_{x\in\Lambda} d\vec{\phi}_x \ . \tag{2}$$

For simplicity we have set the inverse temperature $\beta = 1$. Let us introduce the $O(N)$ invariant bilinears (collective fields)

$$K_{xy} = \frac{1}{N} \vec{\phi}_x \cdot \vec{\phi}_y \ . \tag{3}$$

The state (2) (restricted to the $O(N)$-invariant sector) may be rewritten in terms of K's as the Gibbs state generated by an effective Hamiltonian

$$H_{eff}(K) = NH^0_{eff}(K) + H^1_{eff}(K) \tag{4}$$

where the second term contains lower order terms in N, see below. The obvious gain from such a transformation is that the N-dependence becomes explicit and the $N \to \infty$ limit semiclassical. $H^0_{eff}(K)$ has been worked out in [13]. Our improvement consists of finding $H^1_{eff}(K)$, which in fact is N-independent. Thanks to this, we can use the collective fields not only to study the $N \to \infty$ limit but also to generate the $1/N$ expansion and to estimate non-perturbative corrections.

In order to obtain $H_{eff}(K)$ let us start with the expectation of an $O(N)$ invariant functional $\vec{\phi} \to F(\frac{1}{N}\vec{\phi}_\cdot \cdot \vec{\phi}_\cdot)$ with respect to

(2):

$$\langle F\rangle \equiv \frac{1}{N}\int F(\tfrac{1}{N}\vec{\phi}\cdot\vec{\phi})\, e^{-\frac{N}{2}\sum_x v(\frac{1}{N}\vec{\phi}_x^2) - \frac{1}{2}\langle\vec{\phi}|-\Delta|\vec{\phi}\rangle}\, D\vec{\phi}$$

$$= \frac{1}{N}\int F(K)\, e^{-\frac{N}{2}\sum_x v(K_{xx})}\, \delta(K - \tfrac{1}{N}\vec{\phi}\cdot\vec{\phi})\, e^{-\frac{1}{2}\langle\vec{\phi}|-\Delta|\vec{\phi}\rangle}\, DK\, D\vec{\phi} \qquad (5)$$

where the DK integration goes over real symmetric matrices $K = (K_{xy})$, $x,y \quad \Lambda$. Representing the δ-function by an oscillatory integral over another set of real symmetric matrices $M = (M_{xy})$, we obtain

$$\langle F\rangle = \frac{1}{N}\int F(K)\, e^{-\frac{N}{2}\sum_x v(K_{xx}) + \frac{Ni}{2}\mathrm{Tr}(K-\frac{1}{N}\vec{\phi}\cdot\vec{\phi})M - \frac{1}{2}\langle\phi|-\Delta|\vec{\phi}\rangle}\, DM\, DK\, D\vec{\phi}$$

$$= \frac{1}{N}\int F(K)\, e^{-\frac{N}{2}\sum_x v(K_{xx}) + \frac{Ni}{2}\mathrm{Tr}KM}\, \det(-\Delta+iM)^{-\frac{N}{2}}\, DM\, DK. \qquad (6)$$

where in the last step, the Gaussian $D\vec{\phi}$ integral has been performed. Exponentiating the determinant on the right of (6), we obtain

$$\langle F\rangle = \frac{1}{N}\int F(K)\, e^{-\frac{N}{2}[\Sigma_x v(K_{xx}) - i\,\mathrm{Tr}KM + \mathrm{Tr}\log(-\Delta+iM)]}\, DK\, DM. \qquad (7)$$

This seems already convenient for studying the $N \to \infty$ limit whose semiclassical character is clearly exhibited by (7). The problem may, however, arise, if the expression in brackets has many stationary points, as to which one gives the leading contribution when $N \to \infty$. This is not easily decided since this expression is complex. In the past, wrong choices led to appearance of tachyons in the $(\vec{\varphi}^2)^2_4$ solution [6, 23] and to the wrong ultraviolet stable RG fixed point in the tricritical $(\vec{\varphi}^2)^3_3$ theory [20].

This problem is solved, if we perform explicitly the M integration in (7). This reduces to the computation of the Fourier transform of $\det(-\Delta+iM)^{-N/2}$ which is related to the integrals appearing in the mathematical literature in the context of generalizations of the Riemann ζ function [22, 25] as well as in the physical literature [24]. The result is [12]

$$\langle F\rangle = \frac{1}{N}\frac{1}{\Gamma(\frac{N}{2})\Gamma(\frac{N-1}{2})\ldots\Gamma(\frac{N-|\Lambda|+1}{2})}\int_{K\geq 0} F(K) \qquad (8)$$

$$\cdot\, e^{-\frac{N}{2}\sum_x v(K_{xx}) - \frac{N}{2}\mathrm{Tr}K(-\Delta) + \frac{N-|\Lambda|-1}{2}\mathrm{Tr}\log K}\, DK$$

from which we may read off (see (4))

$$H^{0}_{eff}(K) = \frac{1}{2}\sum_{x} v(K_{xx}) + \mathrm{Tr}\, K(-\Delta) - \mathrm{Tr}\log K , \qquad (9)$$

$$H^{1}_{eff}(K) = (|\Lambda| + 1)\,\mathrm{Tr}\log K , \qquad (10)$$

forgetting for the moment the Γ-functions in (9) which may be absorbed into the normalizing factor $\mathcal{N}$. The clear gain from (8) with respect to (7) is the reality of the functional integral. Now, the $N \to \infty$ limit is determined by the absolute minimum of (real) $H^{0}_{eff}(K)$. In fact, for a large class of v's it is enough to consider $K = (-\Delta + m^2)^{-1}$ for $m^2 > 0$. This gives us a clearcut way to choose between different stationary points of $H^{0}_{eff}(K)$, at least in finite volume. In fact, our approach may be used, see [12], to substantiate the variational ansatz of [1, 2], where it has been first pointed out that the ultraviolet stable fixed point in the $O(\frac{1}{N})$ β-function for the tricritical $(\vec{\varphi}^2)^3_3$ found in [20] disappears if the minimal stationary point is choosen for the $N \to \infty$ solution.

Let us notice that for $N < |\Lambda|$, $(\det K)^{\frac{N-|\Lambda|-1}{2}}$ appearing in (8) has a non-integrable singularity on the boundary of $\{K \geq 0\}$. This singularity is, however, regularized, see [12], due to the poles of the Γ-functions in front of the integral, so that the right hand side of (8) converges in distributional sense for sufficiently regular $F(K)$ and gives the result analytic in N. In fact, the distribution in question is for integral $N > 0$ supported by positive matrices K of rang $\leq N < |\Lambda|$ which is easily understandable in virtue of (3).

The relation (8) casts some doubt on the interchangeability of the large N and the thermodynamical limits: the semiclassical regime seems to settle only for $N >> |\Lambda|$. This might, however, be a superficial effect. For example, in the high temperature phase it has been proven in [14, 15] that the limits may be interchanged. Also our results for the hierarchical model described below suggest that this still holds at the critical point.

Let us now suppose that after n steps of the RG transformation of the block spin type, see [16], we obtain an effective $O(N)$ invariant Gibbs state

$$\frac{1}{\mathcal{N}}\, e^{-\frac{N}{2} V_n(\frac{1}{N} :\vec{\phi}\cdot\vec{\phi}:)}\, d\mu_{G_n}(\vec{\phi}) \qquad (11)$$

where $d\mu_{G_n}(\vec{\phi})$ is the Gaussian measure with mean zero and covariance G_n giving a distribution of the n-th block spin field in the free $v = 0$ case. We may decompose each field ϕ into the block spin part and fluctuation field

$$\phi = A\,\phi' + Z \tag{12}$$

where

$$\phi'_y = L^{(d-2+\eta)/2}\, L^{-d} \sum_{x:|x^\mu - Ly^\mu| < \frac{1}{2}L} \phi_x \tag{13}$$

is the block spin field at the step $n+1$. η is called the anomalous dimension. It is convenient to choose the kernel A so that in the free case $A\phi'$ is the most probable field with prescribed block spin values (13). In this case $A\phi'$ and Z become statistically independent:

$$d\mu_{G_n}(\phi) = d\mu_\Gamma(Z)\, d\mu_{L^n G_{n+1}}(\phi') . \tag{14}$$

Hence, the distribution of the block spin field ϕ' on step $n+1$ is given by the fluctuation field integral as

$$\frac{1}{N}\left(\int e^{-\frac{N}{2}V_n\left(\frac{1}{N}A(\vec{\phi}'_\cdot\cdot\vec{\phi}'_\cdot)A^+ + \frac{1}{N}A(\vec{\phi}'_\cdot\cdot\vec{Z}_\cdot) + \frac{1}{N}(\vec{Z}_\cdot\cdot\vec{\phi}'_\cdot)A^+ + \frac{1}{N}\vec{Z}_\cdot\cdot\vec{Z}_\cdot\right)} d\mu_\Gamma(\vec{Z})\right) d\mu_{L^n G_{n+1}}(\vec{\phi}') . \tag{15}$$

Denoting the term in paranthesis by $e^{-\frac{N}{2}W_{n+1}\left(\frac{1}{N}\vec{\phi}'_\cdot\cdot\vec{\phi}'_\cdot\right)}$ we set

$$e^{-\frac{N}{2}V_{n+1}\left(\frac{1}{N}\vec{\phi}'_\cdot\cdot\vec{\phi}'_\cdot\right)} = e^{-\frac{N}{2}W_{n+1}\left(\frac{1}{N}\vec{\phi}'_\cdot\cdot\vec{\phi}'_\cdot\right) + \frac{1}{2}(1-L^{-\eta})\langle\vec{\phi}'|G^{-1}_{n+1}|\vec{\phi}'\rangle} \tag{16}$$

so that the distribution of $\vec{\phi}'$ may be written again as

$$\frac{1}{N}\, e^{-\frac{N}{2}V_{n+1}\left(\frac{1}{N}\vec{\phi}'_\cdot\cdot\vec{\phi}'_\cdot\right)} d\mu_{G_{n+1}}(\vec{\phi}') . \tag{17}$$

The subtraction of (16) takes care of the marginal contributions of the kinetic energy type to W_{n+1}. η should be chosen so that such contributions stay finite when $n \to \infty$. Now, we may rewrite the integral of (15) in terms of the collective fields

$$\frac{1}{N}\vec{\phi}'_{y_1}\cdot\vec{\phi}'_{y_2} = K'_{y_1y_2} , \tag{18}$$

$$\frac{1}{N}\vec{\phi}'_y\cdot\vec{Z}_x = (K'^{1/2}\,K_2\,\Gamma^{1/2})_{yx} , \tag{19}$$

$$\frac{1}{N}\vec{Z}_{x_1}\cdot\vec{Z}_{x_2} = (\Gamma^{1/2}(K_1 + K_2^+K_2)\Gamma^{1/2})_{x_1x_2} . \tag{20}$$

A chain of transformations similar to (5) to (8), which we do not perform here, gives

$$e^{-\frac{N}{2}W_{n+1}(\frac{1}{N}\vec{\phi}'_\cdot\cdot\vec{\phi}'_\cdot)} = \text{const}\ \frac{1}{\Gamma(\frac{N-L^{-d}|\Lambda|}{2})\dots\Gamma(\frac{N-(1+L^{-d})|\Lambda|+1}{2})}$$

$$\int\limits_{K_1\geq 0}DK_1\int DK_2\ e^{-\frac{N}{2}V_n\,(AK'A^++AK'^{\frac{1}{2}}K_2\Gamma^{\frac{1}{2}}+\Gamma^{\frac{1}{2}}K_2^+K'^{\frac{1}{2}}A^++\Gamma^{\frac{1}{2}}(K_1+K_2^+K_2)\Gamma^{\frac{1}{2}})} \tag{21}$$

$$\cdot\ e^{-\frac{N}{2}\mathrm{Tr}(K_1+K_2^+K_2)\ +\ \frac{N-(1+L^{-d})|\Lambda|-1}{2}\ \mathrm{Tr}\ \log K_1} .$$

This complicated expression again exhibits the semiclassical character of the large N limit. Its full rigorous anlaysis is still missing due to the difficulties in dealing with its nonlocality caused by exponentially decaying tails of the kernels A and $\Gamma^{1/2}$. What can be analyzed, however, is a simple approximation to (21) in which A and Γ are taken local. For A we take

$$A_{xy} = L^{\frac{2-d}{2}}\ \delta_{[x/L]y}$$

where $[x/L]$ denotes the point in $\mathbb{Z}^d$ with coordinates closest to those of x/L. For Γ we shall choose a matrix factorizing over $L\times\ldots\times L$ blocks of the lattice. To be more specific, take Γ within each block equal to L^d times the projection onto a fixed configuration with values ±1 which sums up to zero (L is taken even). This gives simplest formulae. The great virtue of this approximation is that once V_n is local, $V_n(K) = \sum_x v_n(K_{xx})$, V_{n+1} will also be: $V_{n+1}(K') = \sum_y v_{n+1}(K'_{yy})$. v_{n+1} is given by a simple recursion which is a local version of (21):

$$e^{-\frac{N}{2} v_{n+1}(k')} = \text{const} \int_0^\infty dk_1 \int dk_2 \, e^{-\frac{1}{4}L^d N \sum_\pm v_n(L^{2-d}k' \pm 2 L^{\frac{2-d}{2}} k'^{\frac{1}{2}} k_2 + k_1 + k_2^2)} \cdot e^{-\frac{N}{2}(k_1 + k_2^2) + \frac{N-3}{2} \log k_1} . \tag{22}$$

(22) may be easily obtained directly from the recursion

$$e^{-\frac{N}{2} v_{n+1}(\frac{1}{N}\vec{\varphi}'^2)} = \text{const} \int e^{-\frac{1}{4}L^d N \sum_\pm v_n(\frac{1}{N}(L^{2-d}\vec{\varphi}' \pm \vec{z})\cdot(L^{2-d}\vec{\varphi}' \pm \vec{z})) - \frac{1}{2}\vec{z}^2} d\vec{z} \tag{23}$$

which is the expression in the original variables, by changing the variables to

$$\frac{1}{N}\vec{\varphi}^2 = k' , \ \frac{1}{N}\vec{\varphi}'\cdot\vec{z} = k'^{1/2}k_2 , \ \frac{1}{N}\vec{z}^2 = k_1 + k_2^2 . \tag{24}$$

It may be considered as a local approximation to (21). This is close in spirit to the original Wilson's argument [26, 27] used to obtain a very similar recursion analyzed in [17, 18] at $N = \infty$. (22) and (23) may also be viewed as the exact block spin RG transformation for a model being a version of Dyson's hierarchical one [7] where the kinetic energy term in (1) is replaced by a hierarchical expression mimicking the original one, see [9].

The first crucial observation made by Ma [17, 18] is that (22) is exactly soluble in the $N \to \infty$ limit. Let us summarize the essential lessons to be learnt by studying Ma's solution which we find very instructive. Take e.g. $v_o(\frac{1}{N}\vec{\varphi}^2) = \lambda(\frac{1}{N}\vec{\varphi}^2 - \mu^2)^2$. Then at $N = \infty$, $v_n(\frac{1}{N}\varphi^2) = \tilde{v}_n(\frac{1}{N}\varphi^2 - \mu_n^2)$ where $\tilde{v}_n$ has the minimum at zero. For $2 \le d < 4$, $\tilde{v}_n$ converge quickly (geometricly in n) to a fixed point shape $\tilde{v}_\infty$ being a convex function, having the minimum at zero and behaving like $-L^{2-d}x + \text{const}$ for $x \to -\infty$ and like $\text{const}\, x^{\frac{d}{d-2}}$ $(2 < d < 4)$ or $e^{\text{const}\, x}$ $(d=2)$ for $x \to +\infty$, see Fig. 1.

Thus the dynamics of the $N = \infty$ recursion is essentially determined by the one of μ_n^2. The latter is given by

$$\mu_n^2 = \begin{cases} L^{(d-2)n}(\mu_o^2 - (L^{2-d}-1)^{-1}) + (L^{2-d}-1)^{-1} , & 2 < d < 4 , \\ \mu_o^2 - n , & d = 2 . \end{cases} \tag{25}$$

As a result, for $2 < d < 4$, if $\mu_o^2 < \mu_{o\infty}^2 \equiv (L^{2-d}-1)^{-1}$, μ_n^2 becomes quickly negative, which means that v_n as a function of $\vec{\varphi}$ becomes a single well potential, even if it started from a typical symmetry breaking shape with an $O(N)$ symmetric valley. This is the high temperature phase. If $\mu_o^2 > \mu_{o\infty}^2$ then the valley becomes more and more pronounced. This corresponds to the low temperature Goldstone phase. Finally, if $\mu^2 = \mu_{o\infty}^2$ then v_n converge to the fixed point. This is the critical point. In $d = 2$, however, for any μ_o^2 the potential becomes slowly single well: we are always in the high temperature phase [21, 4], see [19].

Let us write

$$v_n(\tfrac{1}{N}\vec{\varphi}^2) = a_n(\tfrac{1}{N}\vec{\varphi}^2-\mu_{o\infty}^2) + \tfrac{1}{2}\lambda_n(\tfrac{1}{N}\vec{\varphi}^2-\mu_{o\infty}^2)^2 + \tilde{v}_n(\tfrac{1}{N}\vec{\varphi}^2-\mu_{o\infty}^2) \tag{26}$$

where $\tilde{v}_n(0) = \tilde{v}_n'(0) = \tilde{v}_n''(0) = 0$. Take for definiteness $d = 3$. For the $N = \infty$ fixed point, $a = 0$, $\lambda = \lambda_\infty = L^{-3}(L-1)$. The flow of the $N = \infty$ recursion around the fixed point is sketched in Fig. 2 (the dotted lines). The a direction is locally the only relevant one. When N is large but finite, we may hope that the $N = \infty$ picture gets only slightly distorted (the solid lines on Fig. 2). This in fact has been proven in [10] with the use of analyticity techniques to estimate the corrections of higher order in $1/N$ from small field contributions to (22) and non-perturbative corrections from large field contributions, compare [16]. The first exact statement is the following:

Assume that

(A). $e^{-\frac{N}{2}\tilde{v}_n(\tilde{k})}$ is analytic for $|\mathrm{Im}\, k^{1/2}| < \frac{1}{2}\mu_{o\infty}^{-1}\varepsilon$ where

$k \equiv \tilde{k} + \mu_{o\infty}^2$ and for $|\mathrm{Im}\, k^{1/2}| < \frac{1}{2} L^{-\frac{1}{2}} \mu_{o\infty}^{-1}\varepsilon$,

$|\tilde{k}| \geq \varepsilon$ (see Fig. 3)

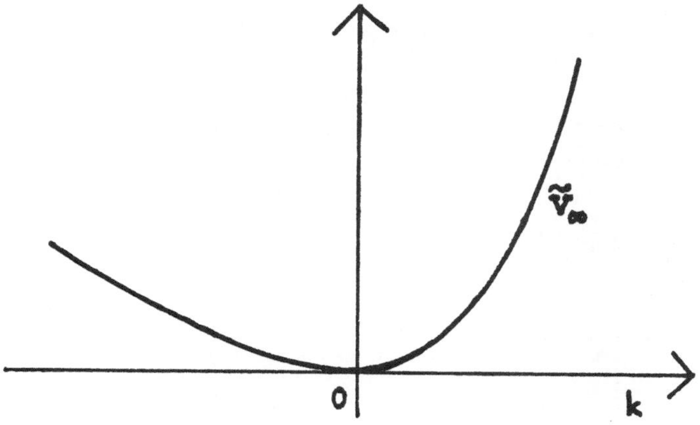

Fig. 1

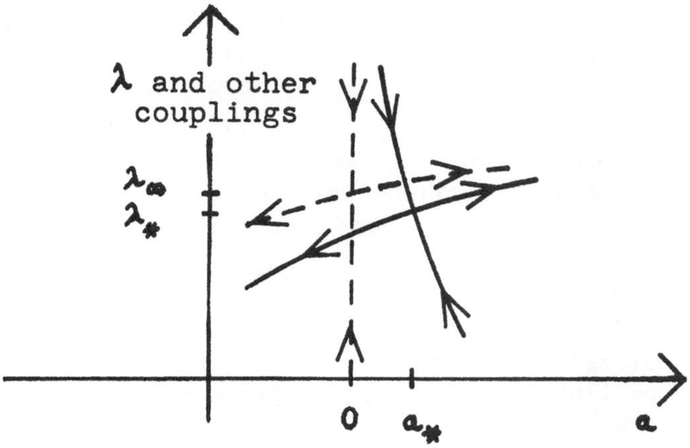

Fig. 2

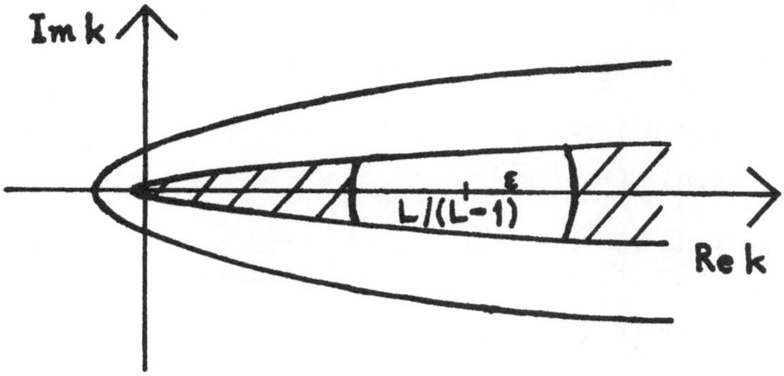

Fig. 3

$$\left| e^{-\frac{N}{2}\tilde{v}_n(\tilde{k})} \right| \leq e^{\frac{4N}{25}\lambda_\infty(\mathrm{Re}\,\tilde{k})^2 + \frac{N}{4}\lambda_\infty(\mathrm{Im}\,\tilde{k})^2} \tag{27}$$

(B). For $|\tilde{k}| < \varepsilon$, $\tilde{v}$ is analytic and

$$|\tilde{v}_n''| \leq \varepsilon^{3/4} \tag{28}$$

(C). $|\lambda_n - \lambda_\infty| \leq \varepsilon^{3/2}$. (29)

Suppose that L is big and ε and $\frac{1}{N}$ are small. Then, if $|a_n| \leqslant \varepsilon^2$, v_{n+1} as given by (22) also satisfies (A) - (C) and $a_{n+1} = La_n + O(\varepsilon^{7/3})$.

Loosely speaking, the above states that if v_n is close to the $N = \infty$ fixed point, so is v_{n+1} at large N except for the departure in the relevant mass direction. It also gives a flavor of the analyticity properties of v_n we use. The question arises how to control the unstable direction. Only for a specific, critical value μ^2_{oc} depending on λ_o we may expect the convergence to the fixed point of v_n's. μ^2_{oc} may be expanded into powers of $1/N$, the leading term being $\mu^2_{o\infty}$. The choice of $\mu^2_{o\infty}$ is a problem of infrared renormalization analogous to the ultraviolet problem of choosing bare couplings so as to obtain given values of the physical ones. Instead of attempting a resummation of the perturbation expansion for μ^2_{oc} , we use the following non-perturbative trick borrowed from [3] to locate μ^2_{oc} (it may also prove very convenient in the ultraviolet case):

1. Admit all μ^2_o such that a_o is in $[-\varepsilon^2, \varepsilon^2] \equiv I_o$,
2. Choose a closed interval $I_1 \subset I_o$ such that if a_o runs through I_1, a_1 sweeps $[-\varepsilon^2, \varepsilon^2]$,
3. Go on applying the previous result choosing at each step $I_{n+1} \subset I_n$ such that for a_o running through I_{n+1} , a_{n+1} sweeps $[-\varepsilon^2, \varepsilon^2]$.
4. For $a_o \in \bigcap_{n=o} I_n$, we stay close to the $N = \infty$ fixed point

forever. μ^2_{oc} corresponds to this value of a_o.

5. Once μ^2_{oc} is located, show that for this value a distance between subsequent iterations of (22) contracts and converges to the finite N fixed point results.

Once the convergence to the fixed point is obtained, one can easily study the hierarchical model correlation functions and demonstrate their non-Gaussian long distance behaviour [11].

Coming back from the hierarchical approximation to the full model, we have to control also the non-localities of the effective interactions. This seems more complicated than in the case where a conventional perturbative argument is used instead of the 1/N one, see [16]. Nevertheless, we hope that it can be done by combining the ideas presented above with a random walk expansion for (21) à la [5], see also [8]. Since a complete argument is still missing, it remains to be seen whether our hopes are realistic.

REFERENCES

1. Bander, M., Bardeen, W.A., Moshe, M.: Spontaneous breaking of scale invariance and the ultraviolet fixed point in O(N) symmetric (ϕ^6_3) theory. Fermilab preprint.
2. Bardeen, W.A., Moshe, M.: Phase structure of the O(N) vector model. Phys.Rev. D28, 1372-1385 (1983).
3. Bleher, P.M., Sinai, Ja.G.: Investigation of the critical point in models of the type of Dyson's hierarchical models. Comm.Math. Phys. 33, 23-42 (1973).
4. Brézin, E., Zinn-Justin, J.: Renormalization of the nonlinear σ model in 2+ε dimensions - application to the Heisenberg ferromagnets. Phys.Rev.Lett. 36, 691-694 (1976).
5. Brydges, D., Fröhlich, J., Spencer, T.: The random walk representation of classical spin systems and correlation inequalities. Commun.Math.Phys. 83, 123-150 (1982).
6. Coleman, S., Jackiw, R., Politzer, H.D.: Spontaneous symmetry breaking in the O(N) model for large N. Phys.Rev. D10, 2491-2499 (1974).
7. Dyson, F.J.: Existence of a phase-transition in a one-dimensional Ising ferromagnet. Commun.Math.Phys. 12, 91-107 (1969).
8. Fröhlich, J., Mardin, A., Rivasseau, V.: Borel summability of the 1/N expansion for the N-vector [O(N) non-linear σ] models. Commun.Math.Phys. 86, 87-110 (1982).
9. Gawędzki, K., Kupiainen, A.: Rigorous renormalization group and asymptotic freedom. IHES preprint.
10. Gawędzki, K., Kupiainen, A.: Non-Gaussian fixed points of the block spin transformation. Hierarchical model approximation. Commun.Math.Phys. 89, 191-220 (1983).

11. Gawędzki, K., Kupiainen, A.: Non-Gaussian Scaling limits. Hierarchical model approximation. IHES preprint.
12. Gawędzki, K., Kupiainen, A.: in preparation.
13. Jevicki, A., Sakita, B.: Collective field approach to the large N limit: Euclidean field theories. Nucl.Phys. B185, 89-100 (1981).
14. Kupiainen, A.: On the 1/n expansion. Commun.Math.Phys. 73, 273-294 (1980).
15. Kupiainen, A.: 1/n expansion for a quantum field model. Commun. Math.Phys. 74, 199-222 (1980).
16. Kupiainen, A.: Contribution to this volume.
17. Ma, S.: Renormalization group in the large N limit. Phys.Lett. A43, 475-476 (1973).
18. Ma, S.: Modern theory of critical phenomena. London, Amsterdam, DonMills-Ontario, Sydney, Tokyo: Benjamin 1976.
19. Mack, G.: Contribution to this volume
20. Pisarski, R.D.: Fixed-point structure of $(\varphi^6)_3$ at large N. Phys.Rev.Lett. 48, 574-576 (1982).
21. Polyakov, A.M.: Interaction of Goldstone particles in two dimensions. Applications to ferromagnets and massive Yang-Mills fields. Phys.Lett. 59B, 79-81 (1975).
22. Siegel., C.L.: The volume of the fundamental domain for some infinite groups. Trans.Am.Math.Soc. 39, 209-218 (1936).
23. Schnitzer, H.J.: Nonperturbative effective potential for $\lambda\phi^4$ theory in the many-field limit. Phys.Rev. D10, 1800-1822 (1974).
24. Speer, E.R.: Dimensional and Analytical Renormalization. In: Renormalization Theory. Eds. Velo, G., Wightman, A.S., Dordrecht, Boston: D. Reidel 1976, pp. 25-93.
25. Weil, A.: Basic number theory. Berlin, Heidelberg, New York: Springer 1973, p. 200.
26. Wilson, K.G.: Renormalization group and critical phenomena. II. Phase-space cell analysis of critical behavior. Phys.Rev. B4, 3184-3205 (1971).
27. Wilson, K.G.: Renormalization of a scalar field theory in strong coupling. Phys.Rev. D6, 419-426 (1972).

ON KÄHLER'S GEOMETRIC DESCRIPTION OF DIRAC FIELDS

M. Göckeler and H. Joos

Deutsches Elektronen-Synchrotron DESY

Hamburg

1. THE LATTICE APPROXIMATION OF FREE DIRAC-KÄHLER FIELDS

The lattice approximation of gauge fields is based on their geometric interpretation. In a similar way, a differential geometric generalization of the Dirac equation due to E. Kähler [1] seems to be an appropriate starting point for the lattice approximation of matter fields [2,3]. It is the purpose of this lecture to illustrate several aspects of this approach.

The Dirac-Kähler Equation (DKE) is given by

$$(d - \delta + m)\Phi = 0. \qquad (1.1)$$

Φ is an inhomogeneous complex differential form in D dimensions (D even for simplicity)

$$\Phi = \sum \frac{1}{n!} \varphi_{\mu_1,\dots,\mu_n}(x)\, dx^{\mu_1} \wedge \dots \wedge dx^{\mu_n} \equiv \sum_H \varphi(x,H)\, dx^H, \qquad (1.2)$$

d the exterior differentiation: $d = dx^\mu \wedge \partial_\mu$, δ the generalized divergence: $\delta = -e^\mu \lrcorner\, \partial_\mu$ with the contraction operator $e^\mu \lrcorner$: $e^\mu \lrcorner\, dx^\nu = g^{\mu\nu}$ etc. The sum over H runs over all ordered index sets $H = \{\mu_1, \mu_2, \cdots, \mu_h\}$ with $\mu_1 < \mu_2 < \cdots < \mu_h$, $h \leq D$. The DKE can be considered on general manifolds [4,5]. Since we are mainly interested in lattice approximations, we restrict ourselves to flat spaces. From $(d - \delta)^2 = -(d\delta + \delta d) = \Box$ it follows that $d - \delta$ is a square root of the Laplacian $\Box$ like the Dirac operator $\gamma^\mu \partial_\mu$.

The relation [2,4,6] between DKE and Dirac equation is clarified by the definition of an associative Clifford product for differential forms

$$dx^\mu \vee dx^\nu = dx^\mu \wedge dx^\nu + g^{\mu\nu} \tag{1.3}$$

or $$dx^\mu \vee \phi = dx^\mu \wedge \phi + e^\mu \lrcorner \phi \; . \tag{1.4}$$

The left regular representation of this Clifford algebra can be decomposed into several copies of the irreducible representation given by the γ-matrices: $\gamma^H = \gamma^{\mu_1} \gamma^{\mu_2} \ldots \gamma^{\mu_h}$. For this, we introduce the basis [2] $Z_a^{(b)}$ with the property of the matrix $Z = (Z_a^{(b)})$

$$dx^\mu \vee Z = \gamma^{\mu T} Z \; , \quad Z \vee dx^\mu = Z \gamma^{\mu T} \; . \tag{1.5}$$

The explicit transformation formulas are given by

$$(Z_a^{(b)}) = \sum_H (-1)^{\binom{h}{2}} (\gamma_H)^T dx^H \; , \quad \phi = \sum_H \varphi(x,H)\, dx^H = \sum_{a,b} \psi_a^{(b)}(x) Z_a^{(b)}$$
$$\varphi(x,H) = \mathrm{Tr}\left(\psi(x) \gamma_H^+\right) \; , \quad \psi_a^{(b)}(x) = 2^{-D/2} \sum_H \varphi(x,H) (\gamma^H)_a{}^b \; . \tag{1.6}$$

It follows immediately from the form of the DK operator $d - \delta = dx^\mu \vee \partial_\mu$ and these formulas that the Dirac components $\psi_a^{(b)}(x)$ of a solution of the DKE satisfy the Dirac equation

$$(\gamma^\mu \partial_\mu + m)\psi^{(b)}(x) = 0\ , \quad b=1,2,\dots,2^{D/2}\ . \tag{1.7}$$

In this sense, the DKE in four dimensions is equivalent to a fourfold degenerate Dirac equation.

It is standard mathematical knowledge [7] that differential forms get mapped on linear functions on the space of lattice elements ("cochains"), when a manifold is approximated by a lattice. The formula

$$^r\phi(C_r) = \int_{C_r} {}^r\phi \tag{1.8}$$

defines the mapping of an r-form on an r-cochain (C_r = r-dimensional lattice cell). Stokes' theorem

$$\int_{C_{r+1}} d\,{}^r\phi = \int_{\Delta C_{r+1}} {}^r\phi \tag{1.9}$$

implies the mapping $d \rightarrow \check{\Delta}$ (dual boundary operator). Similarly, δ corresponds to the dual coboundary operator $\check{\nabla}$. A list of lattice continuum correspondences is the following:

continuum	lattice
weighted sums of points, curves, surfaces,...	chains
boundary	boundary
differential forms ϕ	cochains ϕ
exterior differentiation d, $d^2=0$	dual boundary operator $\check{\Delta}$, $\check{\Delta}^2=0$
codifferential δ, $\delta^2=0$	dual coboundary operator $\check{\nabla}$, $\check{\nabla}^2=0$
Laplace operator $-(d\delta+\delta d)$	Laplace operator $-(\check{\Delta}\check{\nabla}+\check{\nabla}\check{\Delta})$
wedge product $\wedge$	cup product $\wedge$

contraction operator $e^{\mu}\lrcorner$ | contraction operator $e^{\mu}\lrcorner$

Clifford product $\vee$ | Clifford product $\vee$

The meaning of these correspondences is considered to be known or to be intuitively clear. May be, the cup product, the Clifford product and the contraction operator need some comments. Their definitions will be given for hypercubic lattices with lattice spacing a. They require the introduction of coordinates for lattice elements: $(x,\emptyset) \equiv$ lattice point x, $(x,\mu) \equiv$ link connecting x and $x+e_\mu$, $(x,\mu\nu) \equiv$ plaquette spanned by x, $x+e_\mu$, $x+e_\nu$, etc., where e_μ denotes the lattice unit vectors. A basis of elementary cochains is defined by

$$d^{x,H}((x',H')) = a^{h}\, \delta^{x}_{x'}\, \delta^{H}_{H'} \, . \tag{1.10}$$

Similarly to the wedge product of elementary differential forms

$$dx^{H} \wedge dx^{K} = \hat{\rho}_{H,K}\, dx^{H\cup K} \qquad (\hat{\rho}_{H,K} = 0, \pm 1) \tag{1.11}$$

a cup product of elementary cochains is defined by

$$d^{x,H} \wedge d^{y,K} = \hat{\rho}_{H,K}\, \delta^{x+e_H, y}\, d^{x,H\cup K} \qquad \Big(e_H = \sum_{\mu\in H} e_\mu\Big) \, . \tag{1.12}$$

The δ-function in the lattice points describes a matching condition (for examples see Fig. 1). The cup product for general cochains is defined by the distributive law. The matching condition is requested for the product formula of the dual boundary operator:

$$\check{\Delta}(\Phi \wedge \Omega) = (\check{\Delta}\Phi) \wedge \Omega + (A\Phi) \wedge \check{\Delta}\Omega \, , \tag{1.13}$$

where $A\, {}^{r}\Phi = (-1)^{r}\, {}^{r}\Phi$.

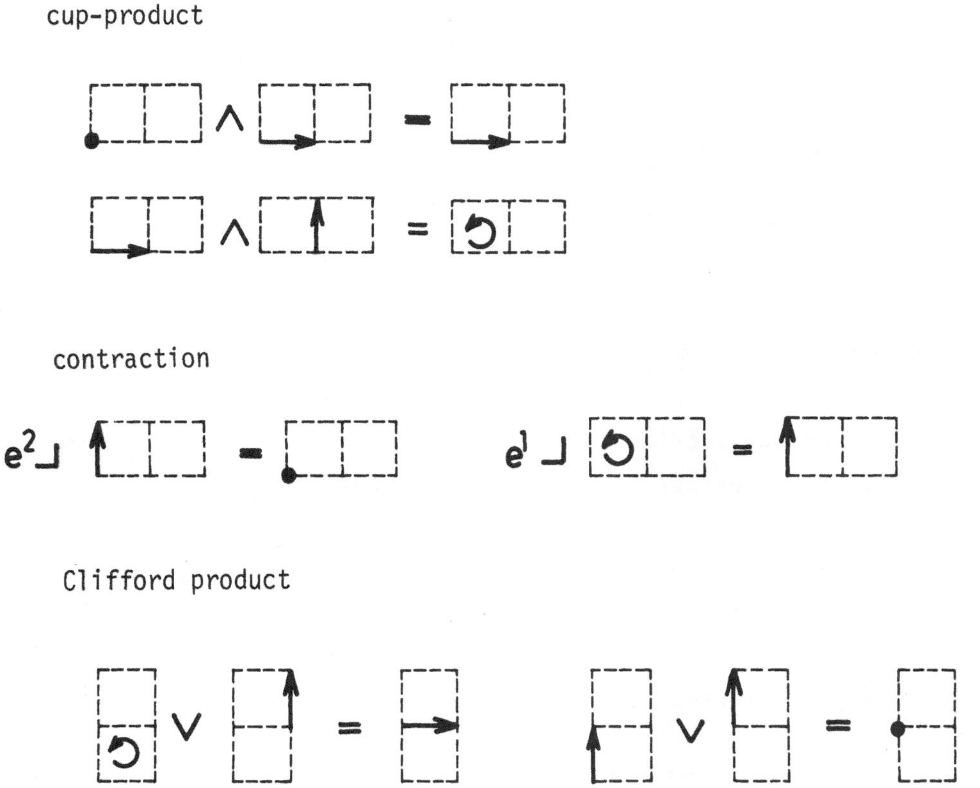

Fig. 1. Examples of the cup product, contraction operator and Clifford product. The Figure illustrates the matching condition for the elementary chains on which the respective cochains are different from zero.

We shall see later that the non-locality introduced by the matching conditions is responsible for many difficulties of the description of Dirac particles on the lattice.

The correspondences of the contraction operator and the Clifford product are expressed by the following formulas:

$$e^{\mu} \lrcorner dx^{H} = \vec{\rho}_{\mu,H} dx^{H-\mu} \rightarrow e^{\mu} \lrcorner d^{x,H} = \vec{\rho}_{\mu,H} d^{x,H-\mu} , \tag{1.14}$$

$$dx^{H} \vee dx^{K} = \check{\rho}_{H,K} dx^{H\Delta K} \rightarrow d^{x,H} \vee d^{y,K} = \check{\rho}_{H,K} \delta^{x+e_H,y} d^{x+e_\Lambda, H\Delta K} \tag{1.15}$$

$$\Lambda = H \cap K , \quad H \Delta K = H \cup K - \Lambda .$$

The coefficients $\vec{\rho}_{\mu,H}$, $\check{\rho}_{H,K}$ take the same values 0, $\pm$ 1 in the continuum and on the lattice. The action of the contraction operator on the lattice is illustrated in Fig. 1. The Clifford product of elementary cochains is different from zero only if a matching condition is satisfied. The resulting elementary cochain is shifted by e_Λ. The geometric meaning of matching condition and result shift is shown by examples in Fig. 1. Contrary to the continuum case, this Clifford product is not associative. Instead we have the relation

$$(d^{x-e_{K\cap L},H} \vee d^{y,K}) \vee d^{z-e_{H\cap K},L} = d^{x,H} \vee (d^{y,K} \vee d^{z,L}) . \tag{1.16}$$

Further properties are given in Ref. 2.

The lattice continuum correspondences allow for an immediate transcription of the DKE on the lattice:

$$(\check{\Delta} - \check{\nabla} + m)\phi = 0 . \tag{1.17}$$

Since $(\check{\Delta} - \check{\nabla})^2 = -(\check{\nabla}\check{\Delta} + \check{\Delta}\check{\nabla})$ is the correct Laplacian, the lattice approximation of the DKE does not introduce spectrum doubling. However, as shown above, the DKE is equivalent to a $2^{D/2}$-fold degenerate Dirac equation. This multiplicity agrees with that of the Susskind reduction of the naive lattice approximation of the Dirac equation [8,9]. Indeed, there exists the following equivalence between the DKE and the Susskind formulation. Let us superimpose on the lattice $\boxplus_x^a$ with lattice spacing a, unit vectors e_μ, a lattice $\boxplus_y^{a/2}$ with half the lattice constant. The lattice sites of $\boxplus_y^{a/2}$ can be written as $y = x + \frac{1}{2} e_H$. Then the Dirac spinor $\psi(y)$ defined by

$$\psi_a(y) \equiv \psi_a(x+\tfrac{1}{2}e_H) \equiv \psi_a(x,H) = \sum_i \gamma^H_{ai} \; \varphi_i(x,H) \tag{1.18}$$

satisfies the naive Dirac equation iff $\sum_{x,H} \varphi_i(x,H)\, d^{x,H}$ satisfies the DKE. The restriction to a fixed i is called the Susskind reduction of the naive Dirac equation [8,10]. Thus the continuum limit of the Susskind fermions are Dirac-Kähler fields.

The lattice continuum correspondence allows the straightforward transcription of the DKE on a general triangulation of space-time (see also the lecture by C. Itzykson). Although in this case the number of degrees of freedom per point on the lattice is larger than the corresponding quantity in the continuum, only the correct number of degrees of freedom survives in the continuum limit, as was shown explicitly for the triangular lattice in two dimensions in Ref. 11. This fact permits a physical interpretation of the flavour degrees of freedom of the DKE, in contrast to the situation for the naive approximation of the Dirac equation on non-cubic lattices [12,13].

In the following paragraphs we shall elaborate on the symmetries of the Euclidean DKE in the continuum and on the lattice for D = 4. We begin with the remark that, due to the associativity of $\vee$ in the continuum, Clifford multiplication with constant forms C from the right transforms solutions of the DKE into solutions:

$$0 = ((d-\delta+m)\phi)\vee C = (dx^\mu \vee \partial_\mu \phi + m\phi)\vee C = dx^\mu \vee \partial_\mu(\phi \vee C) + m\phi \vee C \,. \tag{1.19}$$

As Eq. (1.5) shows, this corresponds to a transformation of the Dirac fields $\psi^{(b)}$, which acts on the index b:

$$\phi \vee C = \sum_{a,b,d} \hat{C}_{bd}\, \psi_a^{(d)}(x)\, Z_a^{(b)} \quad , \tag{1.20}$$

where $\hat{C}_{ab} = \sum_H C(H)\,(\gamma_H^T)_{ab}$ if $C = \sum_H C(H)\, dx^H$.
For $\hat{C} \in \mathcal{U}(4)$ we call this symmetry operation S-flavour transformation.

This symmetry can be used to distinguish the different Dirac fields $\psi^{(b)}$. We may define a "reduction group" with $\vee$ as group multiplication and acting on ϕ by $\vee$-multiplication from the right. If we use the Weyl representation of γ-matrices, this group is given by

$$\mathcal{R} = \{1, \tau, \varepsilon, \varepsilon \vee \tau\} \equiv \{1, i\,dx^{12}, dx^{1234}, -i\,dx^{34}\} \simeq \mathbb{Z}_2 \times \mathbb{Z}_2 \; . \tag{1.21}$$

Then the ϕ's corresponding to the different S-flavour components $\psi^{(b)}$ transform according to the four irreducible representations of $\mathcal{R}$:

$$\hat{\tau} \cdot \phi \equiv \phi \vee \tau = \sigma_\tau^{(b)} \phi \; , \quad \hat{\varepsilon} \cdot \phi \equiv \phi \vee \varepsilon = \sigma_\varepsilon^{(b)} \phi \; , \quad \sigma_\tau^{(b)}, \sigma_\varepsilon^{(b)} = \pm 1 \; . \tag{1.22}$$

Alternatively, we can consider the corresponding projection operators:

$$P^{(b)} = \frac{1}{4}(1 + i\,\sigma_\tau^{(b)} dx^{12}) \vee (1 + \sigma_\varepsilon^{(b)} dx^{1234}) \; . \tag{1.23}$$

The DKE is invariant under the rotation group (Lorentz group), which transforms the components of ϕ as antisymmetric tensor fields. The infinitesimal rotation in the μ-ν-plane may be written as

$$L^{\mu\nu}\phi = (x^{\mu}\partial^{\nu} - x^{\nu}\partial^{\mu})\phi + \frac{1}{2}(S^{\mu\nu}\vee\phi - \phi\vee S^{\mu\nu}) \tag{1.24}$$

with $S^{\mu\nu} = dx^{\mu}\wedge dx^{\nu}$. How is this tensorial transformation law compatible with the spinorial transformations of Dirac fields? According to Eq. (1.5), $\frac{1}{2}S^{\mu\nu}\vee\phi$ represents an infinitesimal spinor transformation of the Dirac components: $\frac{1}{2}S^{\mu\nu}\vee\phi \sim \frac{1}{2}\gamma^{\mu}\gamma^{\nu}\psi^{(b)}$, whereas $\frac{1}{2}\phi\vee S^{\mu\nu}$ corresponds to an S-flavour transformation. Therefore a tensor transformation can be considered as a product of a spinor transformation and a corresponding SU(2) x SU(2) S-flavour transformation.

S-flavour transformations and Euclidean transformations generate the relevant symmetry group $\mathcal{G} = \{(C, S, a)\}$ of the DKE. Its action on the differential forms ϕ is given by:

$$((C,S,a)\phi)(x) = S\vee\phi(\Lambda^{-1}(S)(x-a))\vee C^{-1} . \tag{1.25}$$

$C \in U(4)$ is represented by a constant differential form according to Eq. (1.20). S is a finite spinor transformation acting by Clifford multiplication from the left. It is generated by the infinitesimal transformation $\frac{1}{2}S^{\mu\nu} = \frac{1}{2}dx^{\mu}\wedge dx^{\nu}$ and a reflection $R = dx^4$. To exhibit the action of the Euclidean group $\{\Lambda(S)\in O(4)$, a translation$\}$ on space-time we have given the argument of ϕ explicitly. Eq. (1.25) implies the group multiplication

$$(C,S,a)\circ(C',S',a') = (CC', SS', \Lambda(S)a' + a) . \tag{1.26}$$

For later comparison with the symmetries of the lattice DKE we represent the elements of $\mathcal{G}$ also as products of tensor transformations (S,S,a) and S-flavour transformations (F,1,0):

$$[F,\Lambda(S),a] \equiv (F,1,0)\circ(S,S,a) = (F,1,a)\circ(S,S,0) = (FS,S,a) . \tag{1.27}$$

In this notation the composition law reads

$$[F,\Lambda(S),a]\circ[F',\Lambda(S'),a'] = [FSF'S^{-1},\Lambda(S)\Lambda(S'),\Lambda(S)a'+a]. \quad (1.28)$$

In the case of vanishing mass m, the symmetry group can be extended by chiral transformations, which are infinitesimally generated by

$$\Phi \rightarrow \Phi + \varepsilon \vee \Phi \vee C, \quad (1.29)$$

where $\varepsilon = dx^{1234}$ and C an infinitesimal S-flavour transformation. Therefore the relevant symmetry group in this case is composed of the Euclidean group and a U(4) x U(4).

We begin the discussion of the symmetries of the DKE on the lattice by asking: What remains of the S-flavour transformations on the lattice [14]? With help of the definition of the lattice Clifford product, Eq. (1.15), one can show that

$$\check{\Delta} - \check{\nabla} = d^{\mu} \vee \partial^{-}_{\mu} \quad (1.30)$$

with $d^{H} = \sum_{x} d^{x,H}$, $\partial^{-}_{\mu}\Phi = \sum_{x,H} (\varphi(x,H) - \varphi(x-e_{\mu},H))\, d^{x,H}$.
Although the Clifford product on the lattice is not associative, we get from Eq. (1.16) by summation over x and z:

$$d^{H} \vee (d^{y,K} \vee d^{L}) = (d^{H} \vee d^{y,K}) \vee d^{L} \quad (1.31)$$

and hence with (1.30)

$$((\check{\Delta} - \check{\nabla})\Phi) \vee d^{L} = (\check{\Delta} - \check{\nabla})(\Phi \vee d^{L}) . \quad (1.32)$$

This relation shows the invariance of the lattice DKE under right Clifford multiplication with constant cochains d^{L}. For the product of such lattice flavour transformations we calculate:

$$(\phi \vee d^K) \vee d^L = \check{\rho}_{K,L}\, T^{-1}_{e_{L \cap K}} (\phi \vee d^{K \Delta L}) \tag{1.33}$$

with $T_{e_H} d^{x,K} = d^{x - e_H, K}$.

Therefore the symmetry group generated by these transformations d^H can be identified with the subgroup

$$\{[\pm\gamma^H, 1, \tfrac{1}{2} e_H + \sum_\mu n_\mu e_\mu] \mid n_\mu \in \mathbb{Z},\ H = \text{ordered index set}\} \tag{1.34}$$

of the continuum symmetry group $\mathcal{G}$ by the correspondence

$$d^H \sim [(\gamma^H)^+, 1, \tfrac{1}{2} e_H] . \tag{1.35}$$

Thus lattice flavour transformations are always combined with translations in a manner which is called nonsymmorphous in the language of crystallography [15]. This is a consequence of the matching condition in the definition of the Clifford product and is illustrated geometrically in Fig. 1. Of course, the lattice DKE is also invariant under multiplication of Φ with a constant phase $e^{i\alpha}$. This U(1) symmetry is the only continuous remnant of the U(4) S-flavour symmetry of the continuum DKE. The restriction of the Euclidean symmetry realized by tensor transformations to the hypercubic symmetry is straightforward. Therefore, the description (1.27) is particularly suited to represent the relevant symmetry group $\mathcal{G}_L$ of the lattice DKE as a subgroup of $\mathcal{G}$:

$$\mathcal{G}_L = \{[e^{i\alpha}\gamma^H, \Lambda, \tfrac{1}{2} e_H + a] \mid \Lambda \in W_4,\ a \in \mathcal{T},\ e^{i\alpha} \in U(1)\} , \tag{1.36}$$

W_4 = hypercubic group, $\mathcal{T}$ = group of lattice translations.
We shall show later that this symmetry can be preserved in an interacting theory. The significance of this result is not yet fully explored. One would hope that the $\mathcal{G}_L$ -symmetry leads to the restoration of the symmetry group $\mathcal{G}$ in the continuum limit. Furthermore, the representations of $\mathcal{G}_L$ should be used to classify

the wave functions of hadrons. These expectations are based on the analogy with pure gluon dynamics and the description of low-lying glueball states by representations of the cubic group[16]. The symmetries of Susskind fermions have been studied in a completely different approach by Parisi and Zhang Yi-Cheng [17].

The lattice equivalent of the reduction group generated by id^{12}, d^{1234} is contained in $\mathcal{G}_L$. Because of the intertwining of the reduction group with the translations the decomposition of Φ in Dirac fields is possible only in momentum space:

$$\varphi(x,H) = \int_{-\pi/a}^{\pi/a} d^4p \; e^{-\frac{i}{2} p \cdot e_H} \, \mathrm{Tr}\left(\gamma_H^+ \psi(p)\right) e^{-ip\cdot x} , \tag{1.37}$$

$$\left[-\frac{2i}{a} \sum_\mu \sin\left(\tfrac{1}{2} a p_\mu\right) \gamma_\mu + m\right] \psi^{(b)}(p) = 0 . \tag{1.38}$$

In the case m = 0, we have an additional U(1) symmetry:

$$\mathcal{U}'(\alpha)\Phi = e^{i\alpha \mathcal{A}} \Phi , \text{ i.e. } \mathcal{U}'(\alpha)^\tau \Phi = e^{i\alpha(-1)^\tau} {}^\tau\Phi . \tag{1.39}$$

From the equation

$$\mathcal{A}\Phi = (\varepsilon \vee \Phi) \vee \varepsilon = \varepsilon \vee (\Phi \vee \varepsilon) , \quad \varepsilon = d^{1234} , \tag{1.40}$$

and Eq. (1.5) we see that $\mathcal{U}'(\alpha)$ is a flavour-dependent chiral transformation. A flavour-independent chiral invariance exists only in a discrete form: $\Phi \to \varepsilon \vee \Phi$. It is intertwined with translations: $\varepsilon \vee (\varepsilon \vee \Phi) = T_{e_{1234}} \Phi$.

Symmetries are related to conserved currents. The geometric lattice continuum correspondence is particularly useful for the construction of conserved currents on the lattice. In order to demonstrate this we need some definitions and formulas due to Kähler [1]. The scalar product of two differential forms is the D-form

$$(\Omega, \Phi)_0 = ((B\Omega)\vee\Phi)\wedge\varepsilon = \sum_p (B\,{}^p\Omega)\wedge \star\,{}^p\Phi$$
$$= \sum_p \frac{1}{p!}\,\omega_{\mu_1,\dots,\mu_p}(x)\,\varphi^{\mu_1,\dots,\mu_p}(x)\,dx^{12\dots D}\,. \tag{1.41}$$

The analogue on the Euclidean lattice is the D-cochain

$$(\Omega, \Phi)_0 = \sum_p (B\,{}^p\Omega)\wedge\star\,{}^p\Phi = \sum_{x,H} \omega(x,H)\,\varphi(x,H)\,d^{x,12\dots D}\,. \tag{1.42}$$

The first derived scalar product is the (D-1)-form

$$(\Omega, \Phi)_1 = e_\mu \lrcorner (dx^\mu \vee \Omega, \Phi)_0 = \sum_p ({}^p\Omega\wedge\star B\,{}^{p+1}\Phi + {}^p\Phi\wedge\star B\,{}^{p+1}\Omega) \tag{1.43}$$

with the lattice correspondence

$$(\Omega, \Phi)_1 = \sum_p ({}^p\Omega\wedge\star B\,{}^{p+1}\Phi + {}^p\Phi\wedge\star B\,{}^{p+1}\Omega)\,. \tag{1.44}$$

Here we used the additional notations on the lattice and in the continuum:

$$B\,{}^p\Phi = (-1)^{\binom{p}{2}}\,{}^p\Phi\,, \qquad \star\,\Phi = \Phi\vee\varepsilon\,. \tag{1.45}$$

The following identities are consequences of the product rule for exterior differentiation and Eq. (1.13), respectively:

$$((d-\delta)\Omega, \Phi)_0 + (\Omega, (d-\delta)\Phi)_0 = d(\Omega, \Phi)_1 \quad \text{in the continuum,} \tag{1.46}$$

$$((\check{\Delta}-\check{\nabla})\Omega,\Phi)_0+(\Omega,(\check{\Delta}-\check{\nabla})\Phi)_0=\check{\Delta}(\Omega,\Phi)_1 \quad \text{on the lattice.} \tag{1.47}$$

The conservation law for a current j_μ, $\partial^\mu j_\mu = 0$, is expressed in terms of the current form $j = j_\mu(x)dx^\mu$ as $\delta j = 0$ or $d \star j = 0$. Therefore one sees that

$$j = \star^{-1}(\Omega,\Phi)_1 \tag{1.48}$$

is conserved if

$$(d-\delta-m)\,\Omega = 0, \tag{1.49}$$

$$(d-\delta+m)\,\Phi = 0 . \tag{1.50}$$

This result is valid on the lattice, too, if one replaces all quantities by their lattice analogues. Examples are (D = 4):

$$\tfrac{1}{4}\star^{-1}(\bar{\Phi},\Phi)_1 = \sum_b \bar{\psi}^{(b)}(x)\,\gamma_\mu\,\psi^{(b)}(x)\,dx^\mu , \tag{1.51}$$

flavour currents $\quad \frac{1}{4}\star^{-1}(\bar{\Phi},\Phi \vee C)_1$, (1.52)

axial current $\quad \frac{1}{4}\star^{-1}(\bar{\Phi},\varepsilon\vee\Phi)_1$, (1.53)

where $\bar{\Phi} = \star\,\Phi^*$ satisfies Eq. (1.49) and has the Dirac components [2]

$$\bar{\psi}_a^{(b)}(x) = \frac{1}{4}\sum_H \bar{\varphi}(x,H)\left(\gamma^{H*}\right)_a^{\ b} . \tag{1.54}$$

The two interpretations of differential forms either as Dirac fields, or conventionally as tensor fields, provoke an attempt to formulate fermion boson symmetry, i.e. supersymmetry, in this mathematical language [18]. The formula (1.48) allows the construction of conserved fermion boson currents $j_\alpha^{(i)}$ from a bosonic form Φ and a fermionic form Ψ, which are solutions of the DKE, and a constant form $W_\alpha^{(i)}$:

$$j_\alpha^{(i)} = \frac{1}{4} \mathcal{A}^{-1} (\Psi \vee W_\alpha^{(i)}, \bar{\Phi})_1 \quad . \tag{1.55}$$

Because of the powerful lattice continuum correspondences, the formulation of supersymmetry with help of differential forms may serve as a starting point of its lattice approximation [19].

2. INTERACTING DIRAC-KÄHLER FIELDS

The coupling of a Dirac-Kähler field to a gauge field is straightforward in the continuum. To be specific, we consider the gauge group SU(3) of QCD in four dimensions and the coupling of the gluon field A_μ to a coloured Dirac-Kähler field Φ . The operators d and δ get their covariant definition

$$d_A \Phi = dx^\mu \wedge (\partial_\mu - A_\mu)\Phi \, , \tag{2.1}$$

$$\delta_A \Phi = - e^\mu \lrcorner (\partial_\mu - A_\mu)\Phi \, , \tag{2.2}$$

$$A_\mu(x) = i g \frac{\lambda^a}{2} A_\mu^a(x) \; . \tag{2.3}$$

This leads to the DKE and its Lagrangian with interaction:

$$(d_A - \delta_A + m)\Phi = 0 \, , \tag{2.4}$$

$$S = \int \left\{ \frac{1}{4} (\bar{\Phi}, (d_A - \delta_A + m)\Phi)_0 - \frac{1}{g^2} \mathrm{Tr}(\mathcal{F}, \mathcal{F})_0 \right\} \; . \tag{2.5}$$

The term $-g^{-2}\,\mathrm{Tr}(\mathcal{F},\mathcal{F})_0$ is the usual gluonic Lagrangian expressed in terms of the field strength 2-form

$$\mathcal{F} = \frac{1}{2}(\partial_\mu A_\nu - \partial_\nu A_\mu + [A_\mu, A_\nu])\, dx^\mu \wedge dx^\nu \,. \tag{2.6}$$

Since right multiplication by a constant form $\Phi \to \Phi \vee C$ leaves the DKE invariant also in the case with interaction, the decomposition of Φ in Dirac components proceeds in the same way as given in Eq. (1.6). Thus the action S describes four degenerate Dirac fields coupled to the gauge field A_μ.

There are different schemes for the lattice approximation of the DKE with guage interaction. The definition of the "coarse" coupling is based on a straightforward transcription of formulas (2.1), (2.2) and (2.4), (2.5) to the lattice $\boxplus_x^a$ with volume 4-chain $\mathcal{V} = \sum_x (x,1234)$:

$$(\check{\Delta}_A \Phi)(x,H) \tag{2.7}$$
$$= \frac{1}{a}\sum_\mu \hat{\varsigma}_{\mu,H-\mu}\left(\mathcal{U}^{-1}(x,\mu)\,\varphi(x+e_\mu, H-\mu) - \varphi(x,H-\mu)\right),$$

$$(\check{\nabla}_A \Phi)(x,H) \tag{2.8}$$
$$= -\frac{1}{a}\sum_\mu \vec{\varsigma}_{\mu,H\cup\mu}\left(\varphi(x,H\cup\mu) - \mathcal{U}(x-e_\mu,\mu)\,\varphi(x-e_\mu, H\cup\mu)\right)$$

and

$$(\check{\Delta}_A - \check{\nabla}_A + m)\Phi = 0 \,, \tag{2.9}$$

$$S = \frac{1}{4}(\bar{\Phi}, (\check{\Delta}_A - \check{\nabla}_A + m)\Phi)_0(\mathcal{V}) + S_G(\mathcal{U}) \,. \tag{2.10}$$

In this formulation we have covariance with respect to local gauge transformations g(x):

$$\varphi(x,H) \to g(x)\,\varphi(x,H), \quad \mathcal{U}(x,\mu) \to g(x+e_\mu)\,\mathcal{U}(x,\mu)\,g^{-1}(x) \,. \tag{2.11}$$

Another coupling scheme is based on the equivalence between the lattice DKE and the Susskind reduction of the naive lattice Dirac equation (Eq. (1.18)). For this we consider the gauge fields and gauge transformations on the finer lattice $\boxplus_Y^{a/2}$. The lattice approximations of the gauge transformations act at the center of the lattice cells:

$$\varphi(x,H) \to g(x+\tfrac{1}{2}e_H)\,\varphi(x,H) \equiv g(x,H)\,\varphi(x,H) \equiv g(y)\,\varphi(y)\,. \tag{2.12}$$

Now the covariant dual boundary and coboundary operator must be defined as

$$\begin{aligned}(\check{\Delta}'_A\,\phi)(x,H) \\ = \frac{1}{a}\sum_{\mu}\hat{\rho}_{\mu,H-\mu}\big(U^{-1}(x+\tfrac{1}{2}e_H,\mu)\,\varphi(x+e_\mu,H-\mu) - U(x+\tfrac{1}{2}e_{H-\mu},\mu)\,\varphi(x,H-\mu)\big),\end{aligned} \tag{2.13}$$

$$\begin{aligned}(\check{\nabla}'_A\,\phi)(x,H) \\ = -\frac{1}{a}\sum_{\mu}\vec{\rho}_{\mu,H\cup\mu}\big(U^{-1}(x+\tfrac{1}{2}e_H,\mu)\,\varphi(x,H\cup\mu) - U(x-e_\mu+\tfrac{1}{2}e_{H\cup\mu},\mu)\,\varphi(x-e_\mu,H\cup\mu)\big).\end{aligned} \tag{2.14}$$

The DKE and S have the general form (2.9), (2.10). In the following we shall treat several problems using the coarse and the Susskind coupling. This will allow a comparison of advantage and disadvantage of the two coupling schemes.

From the expression of the action the perturbation series is derived in the standard way. We shall not give the details of this procedure. Only one special feature of the geometric formulation should be mentioned: The n-point Green's functions $\langle\phi\cdots\phi\bar{\phi}\cdots\bar{\phi}\rangle$ become now multi-differential forms in the continuum and multi-cochains on the lattice [20,21]. For example, the free propagator in the Euclidean formulation reads:

$$\langle\phi(x)\,\bar{\phi}(y)\rangle = 4\,(d-\delta+m)_y\,\Delta_F(x-y)\sum_H dx^H \circ dy^H \tag{2.15}$$

or on the lattice:

$$\langle \Phi \bar{\Phi} \rangle = \sum_{\substack{x,H \\ y,K}} \langle \varphi(x,H)\, \bar{\varphi}(y,K) \rangle\, d^{x,H} \circ d^{y,K}$$
$$= \sum_{x,y} 4(\breve{\Delta} - \breve{\nabla} + m)_y\, \tilde{\Delta}(x-y) \sum_H d^{x,H} \circ d^{y,H} \; . \tag{2.16}$$

Here Δ_F is the well-known free 2-point function for a scalar field, whereas $\tilde{\Delta}$ denotes the corresponding lattice expression:

$$\tilde{\Delta}(x) = (2\pi a)^{-4} \int_{-\pi}^{\pi} d^4\beta\, \frac{e^{-i\beta\cdot x/a}}{q_0(\beta) + m^2} \; , \tag{2.17}$$

$$q_0(\beta) = \sum_\mu \left(\frac{2}{a} \sin(\tfrac{1}{2}\beta_\mu)\right)^2 \; .$$

As a first problem, where we find a difference between the two coupling schemes, we consider flavour symmetry in interacting theories. For the Susskind coupling one can prove the gauge covariant generalization of Eq. (1.32)

$$((\breve{\Delta}' - \breve{\nabla}')_A \Phi) \vee d^H = (\breve{\Delta}' - \breve{\nabla}')_{T_H A} (\Phi \vee a^H), \tag{2.18}$$

where $(\breve{\Delta}' - \breve{\nabla}')_{T_H A}$ denotes the Dirac-Kähler operator with the shifted gauge field: $U(y,\mu) \to U(y - \tfrac{1}{2} e_h, \mu)$. Therefore the simultaneous transformation

$$\Phi \to \Phi \vee d^H, \; \bar{\Phi} \to \bar{\Phi} \vee d^H, \; U(y,\mu) \to U(y - \tfrac{1}{2} e_H, \mu) \tag{2.19}$$

is a symmetry of the total action. It is the subdivision of $⊞_x^a$ into $⊞_y^{a/2}$ which allows this extension of the nonsymmorphous lattice flavour transformations, Eq. (1.35), to gauge fields. In the case of the coarse coupling it is not possible to define such an extension. A similar consideration may be applied to the chiral symmetry transformation

$$\Phi \to \varepsilon \vee \Phi, \quad \bar{\Phi} \to -\varepsilon \vee \bar{\Phi} \; , \tag{2.20}$$

under which the Susskind action with m = 0 is invariant if the gauge field is transformed as

$$, \quad \mathcal{U}(y,\mu) \to \mathcal{U}(y+\tfrac{1}{2}e_{1234},\mu) \ . \tag{2.21}$$

It is a central question in the Dirac-Kähler-Susskind approach how one should interpret the S-flavours. Is their geometric origin a hint at a deeper meaning [5,22]? At the moment, this is an open problem. Here we consider the appearance of S-flavours as an unavoidable nuisance connected to the lattice approximation of fermions. Following a suggestion by Susskind [23] we use them for the construction of a model of QCD with the four flavours u,d,s,c. To make the model realistic one has to give the flavours different masses. We start from the continuum action [24]

$$\int \Big\{ \tfrac{1}{4}(\bar{\Phi},(d_A-\delta_A)\Phi)_0 + \tfrac{1}{4}\sum_{b=1}^{4} m_b(\bar{\Phi},\Phi\vee P^{(b)})_0 - \tfrac{1}{g^2}\mathrm{Tr}(\mathcal{F},\mathcal{F})_0 \Big\} \ , \tag{2.22}$$

which becomes the standard QCD action

$$\int d^4x \left\{ \sum_{b=1}^{4} \bar{\psi}^{(b)}(x)(\gamma^\mu D_\mu + m_b)\psi^{(b)}(x) + \tfrac{1}{4} F^a_{\mu\nu} F_a^{\mu\nu} \right\} \tag{2.23}$$

by the transition to Dirac components according to Eqs. (1.6), (1.54). The flavour projections $P^{(b)}$ are defined in Eq. (1.23). The lattice approximation of this action within the coarse coupling scheme is

$$S = \tfrac{1}{4}(\bar{\Phi},(\check{\Delta}_A - \check{\nabla}_A)\Phi)_0(\mathcal{V}) + M(\bar{\Phi},\Phi) + S_G(\mathcal{U}) \ . \tag{2.24}$$

The mass term (see also Ref. 25)

$$M(\bar{\Phi},\Phi) = a^4 \sum_x \sum_{H,H'} \bar{\varphi}(x,H)\, M_{H,H'}\, \varphi(x,H') \tag{2.25}$$

is constructed with help of the same matrix $M_{H,H'}$ as in the continuum:

$$\int \frac{1}{4} \sum_{b=1}^{4} m_b (\bar{\Phi}, \Phi \vee P^{(b)})_0 = \int \sum_{H,H'} \bar{\varphi}(x,H) M_{H,H'} \varphi(x,H') d^4x . \tag{2.26}$$

Because $M(\bar{\Phi}, \Phi)$ is built up by products of fields at the same point, it is obviously invariant under the gauge transformations (2.11).

If we base the realistic model on the Susskind coupling, $\check{\Delta}_A$ and $\check{\nabla}_A$ in (2.24) have to be replaced by $\check{\Delta}'_A$ and $\check{\nabla}'_A$, Eqs. (2.13), (2.14). A suitable mass term has the form (compare Ref. 26)

$$M'(\bar{\Phi}, \Phi, U) = \frac{1}{16} \sum_{\substack{K=\emptyset,\{12\},\\ \{34\},\{1234\}}} \varepsilon_K M_K 2^{-k} \sum_{L \subset K} (\bar{\Phi}, (T_{e_L} \Phi) \vee d^K)_0 (\mathcal{U})_{\text{gauge inv.}} \tag{2.27}$$

analogous to the continuum mass term, which, by means of Eq. (1.23), can be written as

$$\int \frac{1}{4} \sum_{b=1}^{4} m_b (\bar{\Phi}, \Phi \vee P^{(b)})_0 = \int \frac{1}{16} \sum_{\substack{K=\emptyset,\{12\},\\ \{34\},\{1234\}}} \varepsilon_K M_K (\bar{\Phi}, \Phi \vee dx^K)_0 \tag{2.28}$$

$(\varepsilon_\emptyset = \varepsilon_{1234} = 1 , \ \varepsilon_{12} = -\varepsilon_{34} = i)$.

Since the gauge transformations (2.12) of the factors $\Phi, \bar{\Phi}$ in M' refer to different points, this expression is made gauge invariant with help of the appropriate parallel transporters U. M' is invariant under the flavour transformations (1.35) restricted to the reduction group (1.21). The introduction of the translation operators T_{e_L} in M' is necessary in order to enlarge the symmetry group of M' (see Eq. (2.30)).

These lattice approximations of QCD only make sense, if in the continuum limit the full symmetry group $\mathcal{G}$ is restored. Whether this happens, might be questionable, because the mass terms do not only restrict the flavour symmetry but also affect the hypercubic symmetry. Of course, the full symmetry is restored in the free case. In order to see what happens in the interacting theory we consider one-loop calculations of the fermion self-energy. Such a calculation was performed by Mitra and Weisz [27] in the massless model with coarse coupling to an SU(N) gauge field. They found a self-mass of the structure

$$\sum_{\mu} (\bar{\Phi}, (A\Phi) \vee d^{\mu})_0(\mathcal{V}) \cdot \frac{3}{4} \frac{g^2}{16} \frac{N^2-1}{2N} \frac{1}{a} \int_{-\pi}^{\pi} \frac{d^4q}{(2\pi)^4} \frac{1}{\sum_{\lambda} \sin^2(\frac{1}{2} q_{\lambda})} \quad . \tag{2.29}$$

In order to get a continuum limit with vanishing physical mass we would need in the action (2.24) such a mass counterterm, which is not of the form (2.26). An analogous term was found by O. Napoly in his proof of the absence of Goldstone bosons for a U(N) gauge theory with coarse coupling in the large N and strong coupling limit [28]. The invariance of the massless Susskind action under the chiral transformation (2.20), (2.21) forbids mass counterterms, as was already noted by Sharatchandra, Thun and Weisz [9]. One-loop calculations in the realistic model with Susskind coupling and mass term M', Eq. (2.27), indicate that the self-masses have the same structure as M'. This is essentially due to the symmetry properties of M'. Its symmetry group contains the lattice reduction group and in addition the transformations

$[(\gamma^{C\{\mu\}})^+, \Lambda_\mu, \frac{1}{2} e_{C\{\mu\}}]$, $\quad \Lambda_\mu$ = reflection with respect to the hyperplane $x_\mu = 0$, $C\{\mu\} = \{1234\} - \mu$,

(2.30)

$[(\gamma^{\mu\nu})^+, \Lambda_{\mu\nu}, \frac{1}{2} e_{\mu\nu}]$, $\quad \Lambda_{\mu\nu}$ = rotation about π in the μ-ν-plane.

The invariance of M' under the transformations (2.30) is achieved by means of the translations T_{e_L}, $L \subset K$. Thus the higher symmetry of the Susskind coupling as compared to the coarse coupling is advantageous in the formulation of a realistic model of lattice QCD.

We add some remarks on the axial anomaly. The geometric lattice continuum correspondence leads to a straightforward definition of the axial current on the lattice:

$$j_A = \frac{i}{8}\left(\star^{-1}(\bar{\Phi}, \varepsilon \vee \Phi)_1 - \star^{-1}(\varepsilon \vee \bar{\Phi}, \Phi)_1\right) \tag{2.31}$$

which is conserved in the free case. In the case with gauge interaction, j_A is not gauge invariant because of the point splitting introduced by the Clifford product. It was shown that the gauge invariant completion with the coarse coupling produces the correct anomaly for four flavours [29]. For the Susskind coupling there is a similar result by Sharatchandra, Thun and Weisz [9]. However, they use a current which is different from the geometrically motivated expression (2.31).

It was the purpose of this lecture to show that Kähler's geometric description of Dirac fields is a suitable starting point for the treatment of fermions on the lattice. In particular, the lattice continuum correspondence is useful for constructing the lattice analogues of continuum quantities. We have studied the reduction of the continuum symmetry to the lattice symmetry of the free DKE. In the interacting theory, this symmetry can be preserved, if the Susskind coupling is used. This is a definite advantage of this particular type of coupling. Among the questions which should be studied from the geometrical point of view we mention the hopping parameter expansion, the calculation of the hadron spectrum by strong coupling approximation and numerical methods etc. These problems are partially treated in the equivalent scheme of the

Susskind reduction of the naive lattice Dirac equation. However, the Dirac-Kähler approach with its clear geometrical meaning might lead to improved methods and a deeper understanding.

REFERENCES

1. E. Kähler, Rend. Math. Ser. V, 21, 425 (1962).
2. P. Becher and H. Joos, Z. Physik C - Particles and Fields 15, 343 (1982).
3. J.M. Rabin, Nucl. Phys. B201, 315 (1982).
4. W. Graf, Ann. Inst. H. Poincaré, Sect. A 29, 85 (1978).
5. T. Banks, Y. Dothan, and D. Horn, Phys. Lett. 117B, 413 (1982).
6. I.M. Benn and R.W. Tucker, Comm. Math. Phys. 89, 341 (1983).
7. I.M. Singer and J.A. Thorpe, "Lecture Notes on Elementary Topology and Geometry", Scott, Foresman, Glenview, Ill. (1967).
8. L. Susskind, Phys. Rev. D16, 3031 (1977).
9. H.S. Sharatchandra, H.J. Thun, and P. Weisz, Nucl. Phys. B192, 205 (1981).
10. N. Kawamoto and J. Smit, Nucl. Phys. B192, 100 (1981).
11. M. Göckeler, Z. Physik C - Particles and Fields 18, 323 (1983).
12. A. Chodos and J.B. Healy, Nucl. Phys. B127, 426 (1977).
13. W. Celmaster and F. Krausz, Nucl. Phys. B220 FS8 , 434 (1983); W. Celmaster and F. Krausz, Phys. Rev. D28, 1527 (1983).
14. J. Kogut, M. Stone, H.W. Wyld, S.H. Shenker, J. Shigemitsu, and D.K. Sinclair, Nucl. Phys. B225 FS9 , 326 (1983).
15. B.K. Vainshtein, "Modern Crystallography I", Springer-Verlag, Berlin, Heidelberg, New York (1981).
16. B. Berg and A. Billoire, Nucl. Phys. B221, 109 (1983).
17. G. Parisi and Zhang Yi-Cheng, I.N.F.N. - Sezione di Roma preprint n. 357 (1983).
18. I.M. Benn and R.W. Tucker, Phys. Lett. 125B, 47 (1983).
19. S. Elitzur, E. Rabinovici, and A. Schwimmer, Phys. Lett. 119B, 165 (1982);
    H. Aratyn and A.H. Zimerman, preprint DESY 83-075 (1983).
20. P. Becher, in: "Proceedings of the 7th Johns Hopkins Workshop on Current Problems in High Energy Particle Theory" (Bad Honnef, 1983), to appear.
21. H.K. Nickerson, D.C. Spencer, and N.E. Steenrod, "Advanced calculus", van Nostrand Comp., Princeton, N.J. (1959).
22. I.M. Benn and R.W. Tucker, Phys. Lett. 119B, 348 (1982).
23. T. Banks, S. Raby, L. Susskind, J. Kogut, D.R.T. Jones, P.N. Scharbach, and D. Sinclair, Phys. Rev. D15, 1111 (1976).
24. P. Becher and H. Joos, Lett. Nuovo Cim. 38, 293 (1983).
25. A.N. Burkitt, A. Kenway, and R.D. Kenway, Phys. Lett. 128B, 83 (1983);
    P. Mitra, Phys. Lett. 123B, 77 (1983).
26. P. Mitra, Nucl. Phys. B227, 349 (1983).

27. P. Mitra and P. Weisz, Phys. Lett. 126B, 355 (1983).
28. O. Napoly, CEN-SACLAY preprint SPh. T/83/77 (1983).
29. M. Göckeler, Nucl. Phys. B224, 508 (1983).

PLANAR DIAGRAM FIELD THEORIES

Gerard 't Hooft

Institute for Theoretical Physics
Princetonplein, 5 P.O. Box 80.006
3508 TA Utrecht, The Netherlands

ABSTRACT

In this compilation of lectures field theories are considered which consist of N component fields q_i interacting with $N \times N$ component matrix fields A_{ij} with internal (local or global) symmetry group SU(N) or SO(N). The double expansion in 1/N and $\tilde{g}^2 = Ng^2$ can be formulated in terms of Feynman diagrams with a planarity structure. If the mass is sufficiently large and $\tilde{g}^2$ sufficiently small then the(extremely non-trivial) expansion in $\tilde{g}^2$ at lowest order in 1/N is Borel summable. Exact limits on the behavior of the Borel integrand for the $\tilde{g}^2$ expansion are derived.

1. INTRODUCTION

In spite of considerable efforts it is still not known how to compute physical quantities reliably and accurately in any four-dimensional field theory with strong interactions. It seems quite likely that if any strong interaction field theory exists in which accurate calculations can be done, then that must be an asymptotically free non-Abelian gauge theory. In such theories the small-distance structure is completely described by solutions of the renormalization group equations[1]; and there are reasons to believe that the continuum theory can be uniquely defined as a limit of a lattice gauge theory[2], when the size of the meshes of the lattice tends to zero, together with the coupling constant, in a way prescribed by this renormalization group[3]. Indeed, one can prove using this formalism[4] that this limit exists up to any finite order in the perturbation expansion for small coupling.

However, this result has not been extended beyond perturbation expansion. It is important to realize that this might imply that theories such as "quantum chromodynamics" (QCD) are not based on solid mathematics, and indeed, it could be that physical numbers such as the ratio between the proton mass and the string constant do not follow unambiguously from QCD alone. In view of the qualitative successes of the recent Monte-Carlo computation techniques[5] the idea that hadronic properties could be shaped by forces other than QCD alone seems to be far-fetched, but it would be extremely important if this happened to be the case. More likely, we may simply have to improve our mathematics to show that QCD is indeed an unambiguous theory. Either way, it will be important to extend our understanding of the summability aspects of higher order perturbation theory as well as we can. The following constitutes just such an attempt.

There are two categories of divergences when one attempts to sum or resum perturbation expansion for a field theory in four space-time dimensions. One is simply the divergence due to the increasingly large numbers of Feynman diagrams to consider at higher orders. They grow roughly as $n!$ at order g^{2n}. This is a kind of divergence that already occurs if the functional integral is replaced by some ordinary finite-dimensional integral of similar type:

$$I(g^2) = \int d\vec{\phi} \;\; e^{-S(\vec{\phi})} \;,$$

$$S(\vec{\phi}) = \tfrac{1}{2}(\vec{\phi},M\vec{\phi}) + g \sum A_{ijk}\phi_i\phi_j\phi_k + g^2 \sum B_{ijk\ell}\phi_i\phi_j\phi_k\phi_\ell \;. \tag{1.1}$$

Here the diagrams themselves are bounded by geometric expressions but the numbers $K(n)$ of diagrams of order n are such that the expansion is only asymptotic:

$$I(g^2) \to \sum_n K(n)\, C^n\, g^{2n} \;;$$

$$K(n) \to a^n\, n! \;, \tag{1.2}$$

where a is determined by one of the stationary points of the action S, called "instantons".

However in four-dimensional field theories the diagrams themselves are not geometrically bounded. In some theories it has been shown[6] that diagrams of the nth order that required k ultraviolet subtractions (with essentially $k \lesssim n$) can be bounded at best by

$$b^n \, k! \, g^{2n} \; . \tag{1.3}$$

But since the total number of such diagrams grow at most as $n!/k!$ we still get bounds of the form (1.2) for the total amplitude, however with a replaced by a different coefficient. This is a different kind of divergence sometimes referred to as ultraviolet "renormalons"[4,7].

If a field theory is asymptotically free the corresponding coefficient b is negative and one might hope that the ultraviolet renormalons are relatively harmless. But in massless theories a similar kind of divergence will then be difficult to cope with: the infrared renormalons, which, as the word suggests, are due to a build-up of infrared divergences at very high orders: individual diagrams may still be convergent but their sum diverges again with n! The theories we will study more closely are governed by planar diagrams only. Their numbers grow only geometrically[8] so that divergences due to instantons are absent. These diagrams are akin to but more complicated than Bethe-Salpeter ladder diagrams and by trying to sum them we intend to learn much about the renormalon divergences.

Infinite color quantum chromodynamics is of course the most interesting example of a planar field theory but unfortunately our analysis cannot yet be carried out completely there. We do get bounds on the behavior of its Borel functions however (sect. 17).

Examples of large N field theories that we can handle our way are given in sect. 3 and appendix A.

2. FEYNMAN RULES FOR ARBITRARY N

In order to show that the set of planar Feynman diagrams becomes dominant at large N values we first formulate a generic theory at arbitrary N, with a coupling constant g as expansion parameter in the usual sense.

Let us express the fields as a finite number K of N-component vectors $\psi^a_i(x)$ $(a=1,\ldots,K;\ i=1,\ldots,N)$, and a small number D of $N \times N$ matrices $A^b{}_i{}^j(x)$, where $b=1,\ldots,D$ and $i,j=1,\ldots,N$. Usually we will take ψ to be complex and $A_i{}^j$ to be Hermitean, in which case thesymmetry group will be U(N) or SU(N). The case that ψ_i are real and $A_i{}^j$ real and symmetric can easily be included after a few changes, in which case the symmetry group would be O(N) or SO(N), but the complex case seems to be more interesting from a physical point of view. In quantum chromodynamics K would be proportional to the number of flavors and D is the number of space-time dimensions plus two for the (non-Hermitean) ghost field.

In general then the Lagrangian has the form

$$\mathcal{L}(A,\psi,\psi^*) = - \sum_{a,\dots} \psi^{*a}\left(M_0^{ab} + gM_1^{abc}A^c + \tfrac{1}{2}g^2M_2^{abcd}A^cA^d\right)\psi^b$$

$$- \mathrm{Tr}\left(\tfrac{1}{2}R_0^{ab}A^aA^b + \frac{1}{3}gR_1^{abc}A^aA^bA^c + \frac{1}{4}g^2R_2^{abcd}A^aA^bA^cA^d\right) , \qquad (2.1)$$

with

$$A^{a}{}_{i}{}^{j} = \left(A^{a}{}_{j}{}^{i}\right)^* . \qquad (2.2)$$

Here the usual matrix multiplication rule with respect to the indices i,j,... is implied, and Tr stands for trace with respect to these indices. The objects M and R carry no indices i,j but only "flavor" indices a,b,... . Furthermore $M_{0,1}$ and $R_{0,1}$ may contain the derivatives of $\psi(x)$ or $A(x)$. So the case that ψ are fermions is included: then a,b,... may include spinor indices. The coupling constant g has been put in (2.1) in such a way that it is a handy expansion parameter.

In order to keep track of the indices i,j,... it is convenient to split the fields $A^a{}_i{}^j$ into complex fields for $i > j$ and real fields for $i = j$. One can then denote an upper index by an incoming arrow and a lower index by an outgoing arrow. The propagator is then denoted by a double line. In fig. 1, the A propagator stands for an $A_i{}^j$ propagator to the right if $i > j$; an $A_j{}^i$ propagator to the left if $i < j$ and a real propagator if $i = j$.

It is crucial now that the coefficients M and P in the Lagrangian respect the U(N) (or O(N)) symmetry: they carry no indices i,j,... . Hence the vertices in the Feynman graphs only depend on these indices *via* Kronecker deltas. We indicate such a Kronecker delta in a vertex by connecting the corresponding index lines. Since we have a unitary invariance group these Kronecker deltas only connect upper indices with lower indices, therefore the index lines carry an *orientation* which is preserved at the vertices. This is where the unitary case differs from the real orthogonal groups: the restriction to real fields with O(N) symmetry corresponds to dropping the arrows in Fig. 1: the index lines then carry no orientation.

As for the rest the Feynman rules for computing a diagram are as usual. For instance, fermionic and ghost loops are associated with extra minus signs.

In some theories (such as SU(N) gauge theories) we have an extra constraint:

$$\mathrm{Tr}\, A^a = 0 . \qquad (2.3)$$

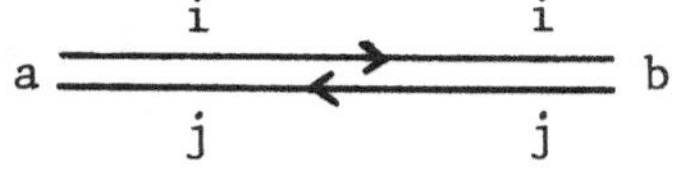

$P_{ab}(k) = \{R_o\}^{-1}_{ab}$ (N x N matrix propagator)

$P'_{ba}(k) = (M_o^{-1})_{ba}$ (N-vector propagator)

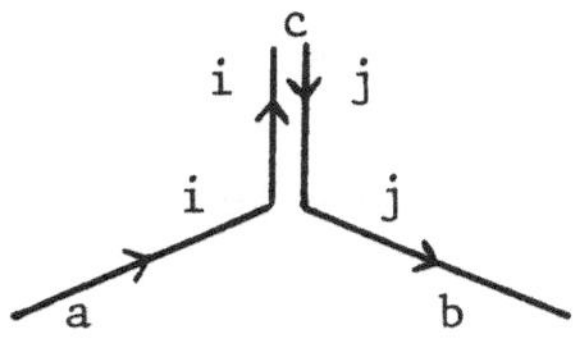

$-g\ M_1^{bac}$

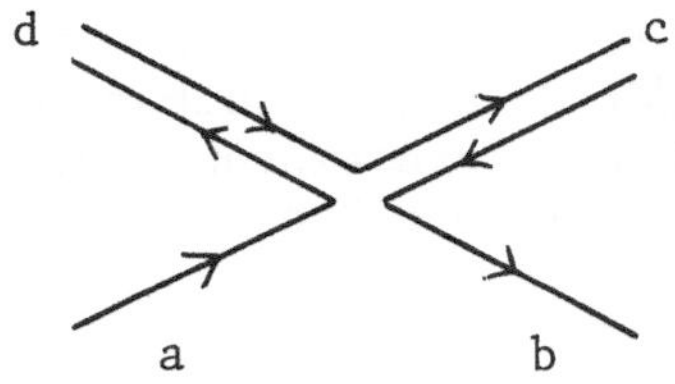

$-g^2\ M_2^{ba\ cd}$

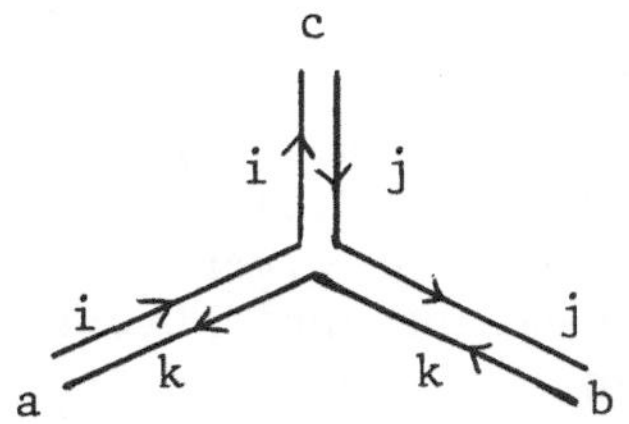

$-g\ R_1^{\{abc\}}$

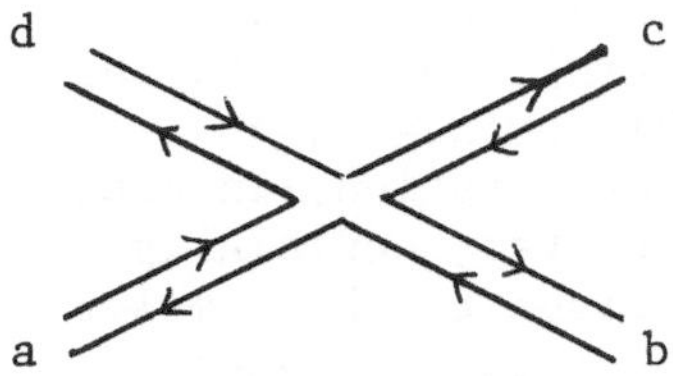

$-g^2\ R_2^{\{abcd\}}$

Fig. 1: Feynman Rules at arbitrary N. The accolades { } stand for cyclic symmetrization with respect to the indices a,b,...

In that case an extra projection operator is required in the propagator:

$$\delta_i^k \delta_\ell^j - \frac{1}{N} \delta_i^j \delta_\ell^k \ . \qquad (2.4)$$

The second term in (2.4) corresponds to an extra piece in the propagator, as given in Fig. 2. We will see later that such terms are relatively unimportant as $N \to \infty$.

For defining amplitudes it is often useful to consider source terms that preserve the (global) symmetry:

$$\mathcal{L}^{\text{source}} = J_\psi^{ab}(x)\psi^{*a}(x)\psi^b(x) + \tfrac{1}{2}J_A^{ab}(x)\,\mathrm{Tr}\,A^a(x)A^b(x) \ . \qquad (2.5)$$

The corresponding notation in Feynman graphs is shown in Fig. 3.

3. THE $N \to \infty$ LIMIT AND PLANARITY

As usual, amplitudes and Green's functions are obtained by adding all possible (planar and non-planar) diagrams with their appropriate combinatorial factors. Note that, apart from the optional correction term in (2.4), the number N does not occur in Fig. 1. But, of course, the number N will enter into expressions for the amplitudes, and that is when an index-line closes. Such an index-loop gives rise to a factor

$$\sum_i \delta_i^i = N \ . \qquad (3.1)$$

We are now in a position that we can classify the diagrams (with only gauge invariant sources as given by eq. (2.5)) according to their order in g and in N. Let there be given a connected diagram. First we consider the two-dimensional structure obtained by considering all closed index loops as the edges of little (simply connected) surface elements. All $N \times N$ matrix-propagators connect these surface elements into a bigger surface, whereas the N-vector-propagators form a natural boundary to the total surface. In the complex case the total surface is an oriented one; in the real case there is no orientation. In both cases the total surface may be multiply connected, containing "worm holes". For convenience we limit ourselves to the complex (oriented) case, and we close the surface by attaching extra surface elements to all N-vector-loops.

Let that surface have F faces (surface elements), P lines (propagators) and V vertices. We have $F = L + I + P_t$, where L is the number of N-vector-loops*) and I the number of index-loops; and we write (footnote: see next page)

$$a \quad \overset{i}{\underset{i}{\longrightarrow}} \quad \overset{\ell}{\underset{\ell}{\longleftarrow}} \quad b \qquad -\frac{1}{N}\{R_o\}^{-1}_{ab}$$

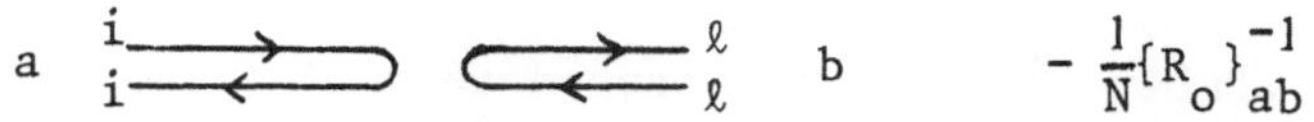

Fig. 2: Extra term in the propagator if Tr A is to be projected
out.

J^{ab}_ψ (N-vector source)

J^{ab}_A (matrix source)

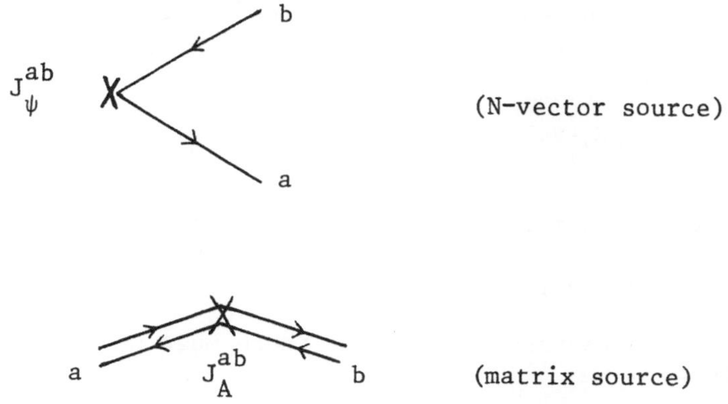

Fig. 3: Invariant source insertions.

$$V = \sum_n V_n ,$$

where V_n is the number of n-point vertices. (V_2 is the number of source insertions). The diagram is now associated with a factor

$$r = g^{V_3+2V_4} N^{I-P_t} . \tag{3.2}$$

Here P_t is the number of times the second term of (2.4) has been inserted to obtain traceless propagators. By drawing a dot at each end of each propagator we find that the total number of dots is

$$2P = \sum_n n V_n , \tag{3.3}$$

and eq. (3.2) can be written as

$$r = g^{2P-2V} N^{F-L-2P_t} . \tag{3.4}$$

Now we apply a well-known theorem of Euler:

$$F - P + V = 2 - 2H , \tag{3.5}$$

where H counts the number of "wormholes" in the surface and is therefore always positive (a sphere has H = o, a torus H = 1, etc.). And so,

$$r = (g^2N)^{\frac{1}{2}V_3+V_4} N^{2-2H-L-2P_t} . \tag{3.6}$$

Suppose we take the limit

$$N \to \infty , \ g \to o , \ g^2N = \tilde{g}^2 \text{ (fixed)} . \tag{3.7}$$

If there are N-vector-sources then there must be at least one N-vector-loop:

$$L \geqslant 1 .$$

The leading diagrams in this limit have H = o, P_t = o and L = 1. They have one overall multiplicative factor N, and they are all planar: an open plane with the N-vector line at its edge (Fig. 4a).

---

*) "N-vector" here stands for N-component vector in U(N) space, so the N-vector-loops are the quark loops in quantum-chromo-dynamics.

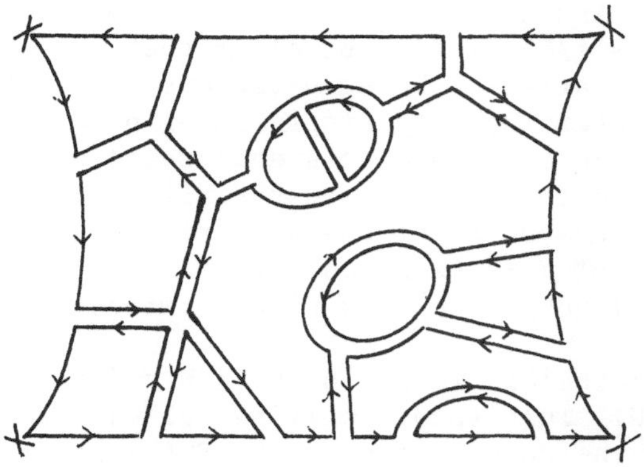

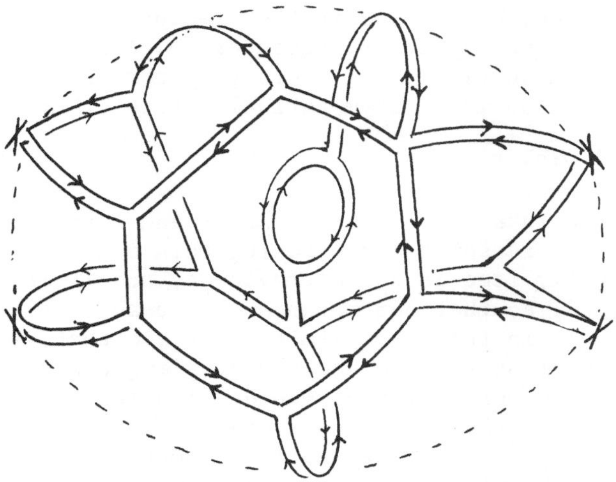

Fig. 4: Elements of the class of leading diagrams in the $N \to \infty$ limit. a) If vector sources are present. b) In the absence of vector sources (e.g. pure gauge theory).

If there are only matrix sources then $L \geqslant o$. The leading diagrams all have the topology of a sphere and carry an overall factor N^2 (see Fig. 4b). We read off from eq. (3.6) that next to leading graphs are down by a factor $1/N$ for each additional N-vector-loop (= quark loop in quantum chromodynamics) and a factor $1/N^2$ for each "wormhole". Also the difference between U(N) and SU(N) theories disappears as $1/N^2$. It will be clear that this result depends only on the field variables being N-vectors and $N \times N$ matrices and the Lagrangian containing only single inner products or traces (not the *products* of inner products and/or traces). Diagrams with $L = 1$ and $H = o$ are the easiest to visualize. In the sequel we discuss convergence aspects of the summation of those diagrams in all orders of $\tilde{g}$. Our main examples are

1) U(N) (or SU(N)) gauge theories with fermions in the N representation;
2) purely Lorentz scalar fields, both in N and in $N \times N$ representations of U(N). That theory will be called $\mathrm{Tr}\lambda\phi^4$, or $-\mathrm{Tr}\lambda\phi^4$, if λ is given the unusual sign. Both SU(N) gauge theory and $-\mathrm{Tr}\lambda\phi^4$ are asymptotically free[9]. The latter has the advantage that one may add a mass term, so that it is also infrared convergent. However the fact that λ has the wrong sign implies that that theory only exists in the $N \to \infty$ limit, not for finite N. A model that combines all "good" features of the previous model is:
3) an SU(N) Higgs theory with N Higgs fields in the elementary representation, N fermions in the elementary representation and a fermion in the adjoint representation. A global SU(N) symmetry then survives. All vector, spinor and scalar particles are massive, and it is asymptotically free if $\tilde{h}^2/\tilde{g}^2 = 1$;

$$\tilde{\lambda}/\tilde{g}^2 = \frac{1}{8}(\sqrt{129}-5) \ , \tag{3.8}$$

where h is a Yukawa coupling constant and λ the Higgs self coupling. The reason for mentioning this model is that it is asymptotically free in the ultraviolet, and it is convergent in the infrared, so that our methods will enable us to construct it rigorously in the $N \to \infty$ limit (provided that masses are chosen sufficiently large and the coupling constant sufficiently small), and positivity of the Hamiltonian is guaranteed also for finite N so that there is every reason for hope that the theory makes sense also at finite N, contrary to the $-\lambda\mathrm{Tr}\phi^4$ theory. This model is described in appendix A.

4. THE SKELETON EXPANSION

From now on we consider diagrams of the type pictured in figure 4a ($H = o$, $L = 1$). They all have the same N dependence, so once we restricted ourselves to these planar diagrams only we may drop the indices i,j,... and replace the double-line propagators by single lines. Often we will forget the tilde (~) on g^2 because the factor N is always understood. Only the (few) indices a,b,... of eq. (2.1), as far as they do not refer to the SU(N) group(s), are kept. The details of this surviving index structure are not important for what follows, as long as the Feynman rules (Fig. 1) are of the general renormalizable type.

Our first concern will be the isolation of the ultraviolet divergent parts of the diagrams. For this we use an ancient device[10] called "skeleton-expansion" *). It can be applied to any graph, planar or not, but for the planar case it is particularly useful.

Consider a graph with at least five external lines. A one-particle irreducible subgraph is a subset of more than one vertices with the internal lines that connect these vertices, that is such that if one of the internal lines is cut through then the subgraph still remains connected. We now draw boxes around all one-particle irreducible subgraphs that have four of fewer external lines. In general one may get boxes that are partially overlapping. A box is *maximal* if it is not entirely contained inside a larger box.

Theorem: All maximal boxes are not-overlapping. This means that two different maximal boxes have no vertex in common.

Proof: If two maximal boxes A and B would overlap then at least one vertex x_1 would be both in A and B. There must be a vertex x_2 in A but not in B, otherwise A would not be maximal. Similarly there is an x_3 in B but not in A. Now A was irreducible, so that at least two lines connect x_1 with x_2. These are external lines of B but not of A ∪ B. Now B may not have more than 4 external lines. So not more than two external lines of A ∪ B are also external lines of B. The others may be external lines of A. But there can also be not more than two of those. So A ∪ B has not more than four external lines and is also irreducible since A and B are, and they have a vertex in common. So we should draw a box around A ∪ B. But then neither A nor B would be maximal, contrary to our assumption. No planarity was needed in this proof.

The skeleton graph of the diagram is now defined by replacing all maximal boxes by single "dressed" vertices. Any diagram can now be decomposed into its "skeleton" and the "meat", which is the collection of all vertex and self-energy insertions at every two-, three- and four-leg irreducible subgraph. In particular the self-

---

*) The method described here differs from Bjorken and Drell[10] in that we do not distinguish fermions from bosons, so that also subgraphs with four external lines are contracted.

energy insertions build up the so-called dressed propagator. We call the dressed three- and four-vertices and propagators the "basic Green functions" of the theory. They contain all ultra-violet divergences of the theory. The rest of the diagram, the "skeleton" built out of these basic Green functions is entirely void of ultraviolet divergencies because there are no further (sub)graphs with four of fewer external lines, which could be divergent.

The skeleton expansion is an important tool that will enable us to construct in a rigorous way the planar field theory. For, under fairly mild assumptions concerning the behavior of the basic Green functions we are able to prove that, given these basic Green functions, the sum of all skeleton graphs contributing to a certain amplitude *in Euclidean space* is absolutely convergent (not only Borel summable). This proof is produced in the next 6 sections. Clearly, this leaves us to construct the basic Green functions themselves. A recursive procedure for doing just that will be given in sects. 12-15. Indeed we will see that our original assumptions concerning these Green functions can be verified provided the masses are big and the coupling constants small, with one exception: in the scalar $-\lambda \mathrm{Tr}\phi^4$ theory the skeleton expansion always converges even if the bare (minimally subtracted) mass vanishes! (sect. 18)

5. TYPE IV PLANAR FEYNMAN RULES

We wish to prove the theorem mentioned in the previous section: given certain bounds for the basic Green functions, then the sum of all skeleton graphs containing these basic Green functions inside their "boxes" converges in the absolute. In fact we want a little more than that. In sects. 13-15 we will also require bounds on the total sum. Those in turn will give us the basic Green functions. We have to anticipate what bounds those will satisfy. In general one will find that the basic Green functions will behave much like the bare propagators and vertices, with deviations that are not worse than small powers of ratios of the various momenta. Note that all our amplitudes are *Euclidean*.

First we must know how the dressed propagators behave at high and low momenta. The following bounds are required:

$$|P_{ab}(k)| \geqslant \frac{Z(k)}{k^2+m^2}\,, \qquad \text{if } k^2 \geqslant o\,. \tag{5.1}$$

Here $P_{ab}(k)$ is the propagator. From now on we use the absolute value symbol for momenta to mean: $|p| = \sqrt{p^2+m^2}$. Then the field renormalization factor $Z(k)$ is approximately:

$$Z(k) \cong \left[\log\left(1 + \frac{|k|}{m}\right)\right]^{\sigma} , \tag{5.2}$$

where σ is a coefficient that can be computed from perturbation expansion. The mass term in (5.1) is not crucial for our procedure but m in (5.2) can of course not be removed easily.

To write down the bounds on the three- and four-point Green functions in Euclidean space we introduce a convenient notation to indicate which external momenta are large and which are small.

For any planar Green function we label not the external momenta but the spaces in between two external lines by indices 1,2,3,... which have a cyclic ordering. An external line has momentum

$$p_{i,i+1} \overset{=}{\text{def}} p_i - p_{i+1} . \tag{5.3}$$

We have automatically momentum conservation,

$$\sum_i p_{i,i+1} = o , \tag{5.4}$$

and the p_i are defined up to an overall translation,

$$p_i \to p_i + q , \quad \text{all } i . \tag{5.5}$$

A channel (any in which possibly a resonance can occur) is given by a pair of indices, and the momentum through the channel is given by

$$p_{i,j} = p_i - p_j . \tag{5.6}$$

So we can look at the p_i as dots in Euclidean momentum space, and the distance between any pair of dots is the momentum through some channel. If we write

$$(((12)_{A_1}\ 3)_{A_2}\ 4)_{A_3} , \tag{5.7}$$

or simply $(((12)_1\ 3)_2\ 4)_3$, then this means:

$$\begin{aligned} |p_1 - p_2| &= A_1 , \\ |p_1 - p_3| &= A_2 \gg A_1 , \\ |p_1 - p_4| &= A_3 \gg A_2 . \end{aligned} \tag{5.8}$$

So the brackets are around momenta that form close clusters.

Our bounds for the three- and four-point functions are now defined in table 1.

Table 1

Bounds for the 3- and 4-point dressed Green functions. Z_{ij} stands for $Z(p_i-p_j)$. All other exceptional momentum configurations can be obtained by cyclic rotations and reflections of these. K_i are coefficients close to one.

$((12)_1\ 3)_2$	$K_1(Z_{12}Z_{23}Z_{31})^{-\frac{1}{2}}\ A_2(A_2/A_1)^{\alpha}\ g(A_2)$
$(((12)_1\ 3)_2\ 4)_3$	$K_2^2(Z_{12}Z_{23}Z_{34}Z_{41})^{-\frac{1}{2}}\ (A_2/A_1)^{\alpha}\ (A_3/A_2)^{\beta}\ g^2(A_3)$
$((12)_1\ (34)_2)_3$	$K_3^2(Z_{12}Z_{23}Z_{34}Z_{41})^{-\frac{1}{2}}\ (A_3^2/A_1A_2)^{\alpha}\ g^2(A_3)$
$(((13)_1\ 2)_2\ 4)_3$	$K_4^2(Z_{12}Z_{23}Z_{34}Z_{41})^{-\frac{1}{2}}\ (A_2A_3/A_1^2)^{\beta}\ g^2(A_3)$
$((13)_1\ (24)_2)_3$	$K_5^2(Z_{12}Z_{23}Z_{34}Z_{41})^{-\frac{1}{2}}\ (A_3/A_1)^{2\beta}\ g^2(A_3)$
	if $A_1 \leqslant A_2$
$((123)_1\ 4)_2$	$K_6^2(Z_{12}Z_{23}Z_{34}Z_{41})^{-\frac{1}{2}}\ (A_2/A_1)^{\beta}\ g^2(A_2)$
$((12)_1\ 34)_2$	$K_7^2(Z_{12}Z_{23}Z_{34}Z_{41})^{-\frac{1}{2}}\ (A_2/A_1)^{\alpha}\ g^2(A_2)$

Here α and β are small positive coefficients. $g(x)$ is a slowly varying running coupling constant. For the time being all we need is some g with

$$\max_i K_i\ g(x) \leqslant g \qquad \text{for all } x\ , \tag{5.9}$$

where we also assume that possible summation over indices a,b,... is included in the K coefficients. Clearly the bare vertices would satisfy the bounds with $\alpha = \beta = 0$. Having positive α and β allows us to have any of the typical logarithmic expressions coming from the radiative corrections in these dressed Green functions. Indeed we will see later (sect. 13) that those logarithms will never surpass our power-laws.

Table 1 has been carefully designed such that it can be re-obtained in constructing the basic Green functions as we will see in sects. 12-14. First we notice that the field renormalization factors $Z(p_i-p_j)^{-\frac{1}{2}}$ cancel against corresponding factors in our bounds for the propagator (5.1). The power-laws of Table 1 can be conveniently expressed in terms of a revised set of Feynman rules. These are given in Fig. 5. We call them type IV Feynman rules after a fourth attempt to reformulate our bounds (types I, II and

$\dfrac{1}{(k^2+m^2)^{1+\alpha}}$ (dressed elementary propagator)

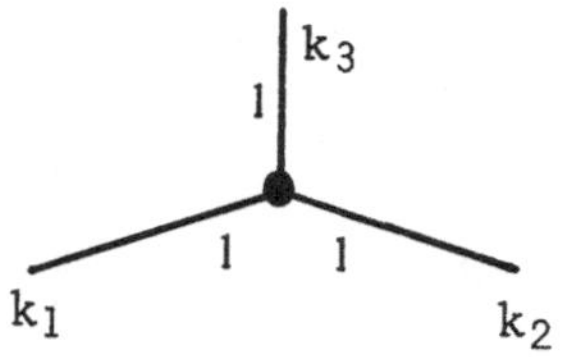

$g[\max(|k_1|, |k_2|, |k_3|)]^{1+3\alpha}$ (dressed 3-vertex)

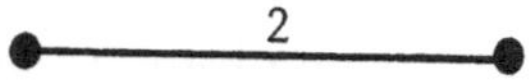

$\dfrac{1}{|k|^{2\beta}}$ (composite propagator)

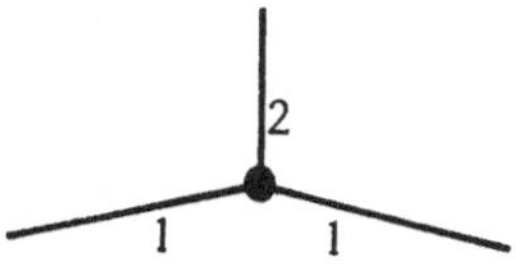

$g[\max(|k_1|, |k_2|, |k_3|)]^{2\alpha+\beta}$

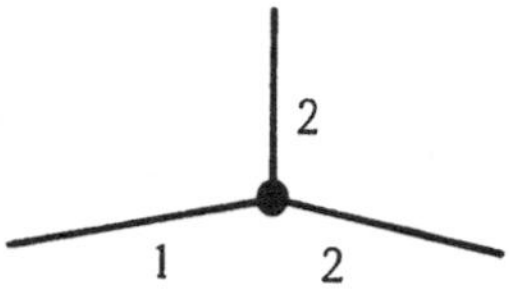

$g[\max(|k_1|, |k_2|, |k_3|)]^{\alpha+2\beta-1}$ (generalized vertices)

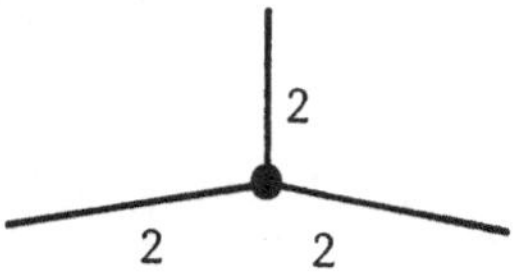

$g[\max(|k_1|, |k_2|, |k_3|)]^{3\beta-2}$

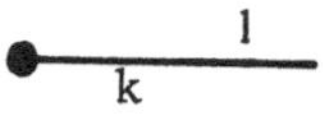

$|k|^{-\alpha}$ (external line)

Fig. 5: Type IV Feynman Rules. $|k|$ stands for $\sqrt{k^2+m^2}$.

III occur in refs. 11, 12 and are not needed here). The trick is to introduce a new kind of propagator, $\bullet\!\overset{2}{\rule{1.5em}{0.4pt}}\!\bullet$, that represents an exchange of two or more of the original particles in the diagram we started off with.

The procedure adapted in these lectures deviates from earlier work[11] in particular by the introduction of the last two vertices in Fig. 5. Notice that they decrease whenever two of the three external momenta become large.

It is now a simple exercise to check that indeed any diagram built from basic Green functions that satisfy the bounds of Table 1 can also be bounded by corresponding diagram(s) built from type IV Feynman rules. The four-vertex is simply considered as a sum of two contributions both made by connecting two three-point vertices with a composite propagator, and the factors $|k|^{-\alpha}$ from the propagators in Fig. 5 are considered parts of the vertex functions (the mass term of the propagator may be left out; it is needed at a later stage).

Elementary power counting now tells us that the superficial degree of convergence, Z, of any (sub)graph with E_1 external single lines and E_2 external composite lines is given by

$$Z = (1-\alpha)E_1 + (2-\beta)E_2 - 4 . \qquad (5.10)$$

Since we consider only skeleton graphs, all our graphs and subgraphs have

$$E_1 + 2E_2 \geqslant 5 . \qquad (5.11)$$

Thus, Z is guaranteed to be positive if we restrict our coefficients by

$$o < \alpha < 1/5 ;$$

$$o < \beta < 2/5 . \qquad (5.12)$$

(Infrared convergence would merely require $\alpha < 1$; $\beta < 2$, and is therefore guaranteed also.) So we know that with (5.12) all graphs and subgraphs are ultraviolet and infrared convergent. The theorem we now wish to prove is: the sum of all convergent type IV diagrams contributing to any given amplitude with 5 (or more) external lines converges in Euclidean space. It is bounded by the sum of all type IV tree graphs (graphs without closed loops) multiplied with a fixed finite coefficient.

A further restriction on the coefficients α and β will be necessary (eq. (8.15)).

6. NUMBER OF TYPE IV DIAGRAMS

The total number $G(E,L)$ of connected or irreducible planar diagrams with E external lines and L closed loops in any finite set of Feynman rules, is bounded by a power law (in contrast with the non-planar diagrams that contribute for instance to the Lth order term in the expansion such as (1.1) for a simple functional integral):

$$G(E,L) \leqslant C_1^E C_2^L \ , \tag{6.1}$$

for some C_1 and C_2.

In some cases C_1 and C_2 can be computed exactly and even closed expressions for $G(E,L)$ exist[8]. These mathematical exercises are beautiful but rather complicated and give us much more than we really need. In order to make these lectures reasonably self-sustained we will here derive a crude but simple derivation of ineq. (6.1) yielding C coefficients that can be much improved on, with a little more effort.

Let us ignore the distinction between the two types of propagators and just count the total number $G(E,L)$ of connected planar ϕ^3 diagrams with a given configuration of E external lines and L closed loops. We have (see Fig. 6)

$$G(E+1,L) = G(E+2,L-1) + \sum_{n,L_1} G(n+1,L_1)G(E+1-n,L-L_1) \ . \tag{6.2}$$

$$G(E,L) = o \quad \text{if} \quad E < 2 \quad \text{or if} \quad L < o \ ;$$

$$G(2,o) = 1 \ . \tag{6.3}$$

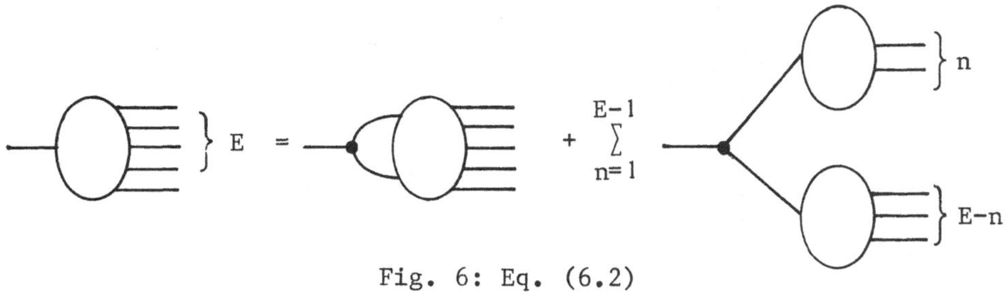

Fig. 6: Eq. (6.2)

We wish to solve, or at least find bounds for, $G(E,L)$ from (6.2) with boundary condition (6.3). A good guess is to try

$$G(E,L) \leqslant \frac{C_0 C_1^E C_2^L}{(E-1)^2 (L+1)^2}\ , \quad E \geqslant 2\ ,\ L \geqslant 0\ , \tag{6.4}$$

which is compatible with (6.3) if

$$C_0 C_1^2 \geqslant 1\ . \tag{6.5}$$

Using the inequality

$$\sum_{n=1}^{k} \frac{1}{n^2 (k-n)^2} \leqslant \frac{4}{k^2}\ , \tag{6.6}$$

we find that the r.h.s. of (6.2) will be bounded by

$$\frac{C_0 C_1^{E+2} C_2^{L-1}}{(E+1)^2\ L^2} + \frac{16 C_0^2 C_1^{E+2} C_2^L}{(E-1)^2 (L+1)^2}\ , \tag{6.7}$$

which is smaller than

$$\frac{C_0 C_1^{E+1} C_2^L}{E^2 (L+1)^2}\ , \tag{6.8}$$

if

$$4C_1/C_2 + 64 C_0 C_1 \leqslant 1\ . \tag{6.9}$$

This is not incompatible with (6.5) although the best "solution" to these two inequalities is an uncomfortably large set of values for C_0, C_1 and C_2. But we proved that they are finite.

The exact solution to eq. (6.2) is

$$G(E,L) = \frac{2^L (2E-2)!\,(2E+3L-4)!}{L!\,(E-1)!\,(E-2)!\,(2E+2L-2)!}\ , \tag{6.10}$$

which we will not derive here. Using

$$\frac{(A+B)!}{A!\,B!} \leqslant 2^{A+B-1} \tag{6.11}$$

we find that in (6.1),

$$C_1 \leqslant 16\ ;\ C_2 \leqslant 16\ . \tag{6.12}$$

For fixed E, in the limit of large L,

$$C_2 \to 27/2 \ . \qquad (6.13)$$

Similar expressions can be found for the set of irreducible diagrams. Since they are a subset of the connected diagrams we expect C coefficients equal to or smaller than the ones of eqs. (6.12) and (6.13). Limiting oneself to only convergent skeleton graphs will reduce these coefficients even further.

We have for the number of vertices V

$$V = E + 2L - 2 \ , \qquad (6.14)$$

and the number of propagators P:

$$P = V + L - 1 \ . \qquad (6.15)$$

So, if different kinds of vertices and propagators are counted separately then the number of diagrams is multiplied with

$$C_V^V C_P^P \ , \qquad (6.16)$$

which does not alter our result qualitatively. Also if there are elementary 4-vertices then these can be considered as pairs of 3-vertices connected by a new kind of propagators, as we in fact did. So also in that case the numbers of diagrams are bounded by expressions in the form of eq. (6.1).

7. THE SMALLEST FACETS

We now wish to show that every planar type IV diagram with L loops is bounded by a coefficient C^L times a (set of) type IV tree graph(s), with the same momentum values at the E external lines. This will be done by complete induction. We will choose a closed loop somewhere in the diagram and bound it by a tree insertion. Now even in a planar diagram some closed loops can become quite large (i.e. have many vertices) and it will not be easy to write down general bounds for those. Can we always find a "small" loop somewhere?

We call the elementary loops of a planar diagram facets. Now Euler's theorem for planar graphs is:

$$V - P + L = 1 \ . \qquad (7.1)$$

Take an irriducible diagram. Write

$$L = \sum_n F_n \; , \tag{7.2}$$

where F_n are the number of facets with exactly n vertices (or "corners"). Let

$$P = P_i + P_e \; , \tag{7.3}$$

where P_i is the number of internal propagators and P_e is the number of propagators at the edges of the diagram. Then, by putting a dot at every edge of each facet and counting the number of dots we get

$$\sum_n n\, F_n = 2P_i + P_e \; . \tag{7.4}$$

For the numbers V_n of n-point vertices we have similarly

$$\sum_n n\, V_n = 2P + E \; , \tag{7.5}$$

but in ourcase we only consider 3-point vertices (compare eqs. (6.14) and (6.15)):

$$3V = 2P + E \; . \tag{7.6}$$

Combining eqs. (7.1) - (7.6) we find

$$\sum_n (n-6)F_n = 2E - P_e - 6 \; . \tag{7.7}$$

This equation tells us that if a diagram has

$$L \geqslant 2E - 8 \tag{7.8}$$

then either it is a "seagull graph" ($P_e \leqslant 1$) which we usually are not interested in, or there must be at least one subloop with 6 or fewer external lines:

$$F_n > 0 \quad \text{for some} \quad n \leqslant 6 \; . \tag{7.9}$$

So diagrams with given E and large enough L must always contain facets that are either hexagons or even smaller.

In fact we can go further:
theorem: if a planar graph (with only 3-vertices) and all its irreducible subgraphs have $2E - P_e \geqslant 6$ then the entire graph obeys

$$L \leqslant \frac{E^2}{12} - \frac{E}{2} + 1 \; . \tag{7.10}$$

This simple theorem together with eq. (7.7) tells us that any diagram with a number of loops L exceeding the bound of (7.10) must have at least one elementary facet with 5 of fewer lines attached to it. Although we could do without it, it is a convenient theorem and now we devote the rest of this section to its proof (it could be skipped at first reading).

First we remark that if we have the theorem proven for all *irreducible* graphs up to a certain order, then it must also hold for reducible graphs up to the same order. This is because if we connect two graphs with one line we get a graph 3 with

$$L_3 = L_1 + L_2 \,,$$

$$P_{e3} = P_{e1} + P_{e2} + 2 \,, \qquad (7.11)$$

$$E_3 = E_1 + E_2 - 2 \,.$$

If L_1, E_1 and L_2, E_2 satisfy (7.10) then so do L_3 and E_3 (remember that E and L are integers and the smallest graph with $L > o$ has $E = 6$; propagators that form two edges of a diagram are counted twice in P_e).

Forthe irreducible graphs we prove (7.10) by a rather unusual induction procedure for planar graphs. We consider the outer rim of an irreducible graph and all the (in general not irreducible) graphs inside it (see Fig. 7). Let the entire graph have E external lines and P_e propagators at its sides. The subgraphs i inside the rim have e_i external lines and P_{ei} propagators at their sides. We count:

$$P_e = E + \sum_i e_i \,, \qquad (7.12)$$

and the number of loops L of the entire diagram is

$$L = 1 + \sum_i (L_i + e_i - 1) \,. \qquad (7.13)$$

Now each facet between the subgraphs and the rim must have at least 6 propagators as supposed, therefore

$$P_e + \sum P_{ei} + 2 \sum e_i \geqslant 6 \sum_i (e_i - 1) + 6 \,, \qquad (7.14)$$

but if some of the subgraphs are single propagators we need to be more precise

$$P_e + \sum P_{ei} + 2 \sum e_i \geqslant 6 \sum_i (e_i - 1) + 6 + 2N_2 \,, \qquad (7.15)$$

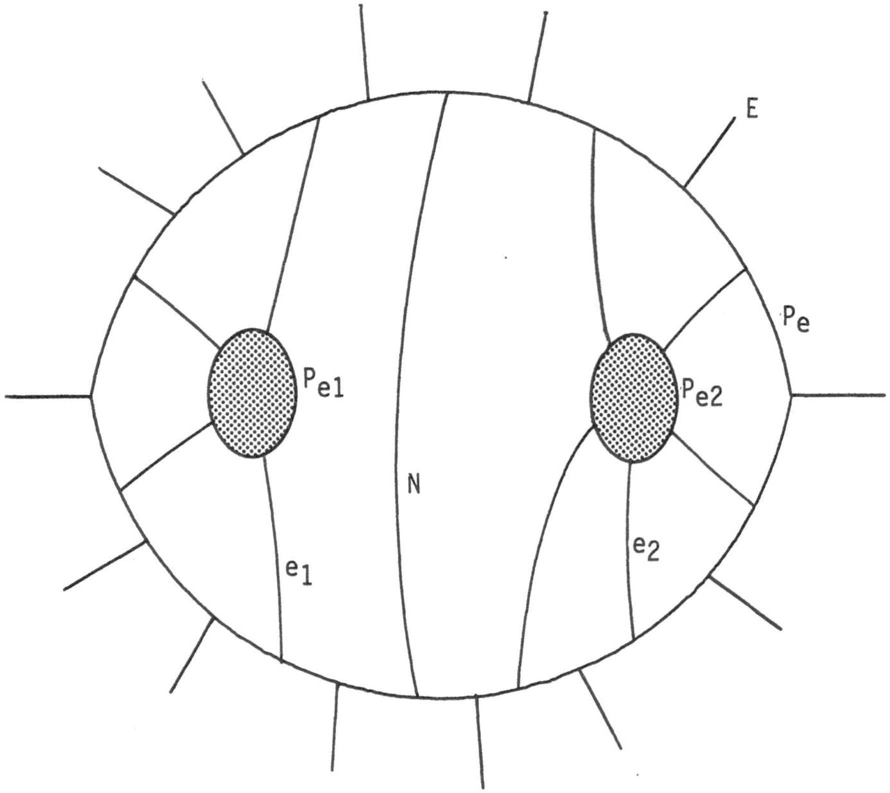

Fig. 7: Proving the theorem of sect. 7. The number E counts the external lines of the entire graph. P_e the number of sides and e_i and P_{ei} do the same for the subgraphs 1 and 2. N is the number of single propagators.

where N_2 is the number of single propagators, each of which contributes with $e = 2$ in eq. (7.13), and have $P_{ei} = o$. Now we use (7.12), and

$$P_{ei} \leqslant 2e_i - 6 \,, \tag{7.16}$$

as required, whereas $\Sigma(2e-6) + 2N_2 = o$ for the single propagators, to arrive at

$$E \geqslant \sum_i e_i + 6 \,. \tag{7.17}$$

From the assumption that all subraphs already satisfy (7.10) we get, writing $L_1 = \sum_i L_i$ and $E_1 = \sum_i e_i$:

$$L_1 \leqslant \frac{E_1^2}{12} - \frac{E_1}{2} + 1 , \tag{7.18}$$

and from (7.13)

$$L \leqslant L_1 + E_1 . \tag{7.19}$$

With (7.17) which reads $E \geqslant E_1 + 6$ we now see that (7.10) again holds for the entire diagram. The quadratic expression in (7.10) is the sharpest that can be derived from (7.17)-(7.19), and indeed large diagrams that saturate the inequality can be found, by joining hexagons into circular patterns.

Our conclusion is that if we wish to use an induction procedure to express a bound for diagrams with type IV Feynman rules and L loops in terms of one with a smaller number L' loops we can try to do that by replacing successively triangles, qudrangles and/or pentagons by type IV tree insertions, until the bound (7.10) is reached. In particular if $E = 5$ this leads us to a tree diagram. The next three sections show how this procedure works in detail.

8. TRIANGLES

Consider a (large) diagram with type IV Feynman rules. We had already decreed that it and all its subgraphs are ultraviolet and infrared convergent (divergent subgraphs had been absorbed into the vertices and propagators before). With eqs. (5.10)-(5.12) this means that each subgraph has

$$E_1 + 2E_2 \geqslant 5 , \tag{8.1}$$

so, in particular, there are no self-energy blobs. First we use the inequality of Fig. 8 to replace composite propagators by ordinary dressed propagators one by one until ineq. (8.1) forbids any further such replacements. The inequality is readily proven: we write for the propagators with its vertices:

Fig. 8. A composite propagator is smaller than an elementary one.

$$|p_1|^{1+\alpha_1+\alpha_2+\gamma}\;|p_2|^{1+\alpha_3+\alpha_4+\gamma}\;|k^2|^{-1-\gamma}\;, \tag{8.2}$$

where k is the momentum through the propagator, $|p_1| \geqslant |k|$. $|p_2| \geqslant |k|$, and $\alpha_i = \alpha$ for a dressed propagator, and $\alpha_i = \beta-1$ for a composite propagator. At the left hand side $\gamma = \beta-1$, and at the right hand side $\gamma = \alpha$. Clearly we have

$$\left(\frac{|p_1||p_2|}{k^2}\right)^{\beta-1} \leqslant \left(\frac{|p_1||p_2|}{k^2}\right)^{\alpha}\;, \tag{8.3}$$

α and β being both close to zero due to (5.12).

Now consider all elementary triangle loops in our diagram. Under what conditions can we replace them by type IV 3-vertices (Fig. 9)? Due to (8.1) there can be at most one elementary (dressed external line, the others are composite: E_2 = 2 or 3;

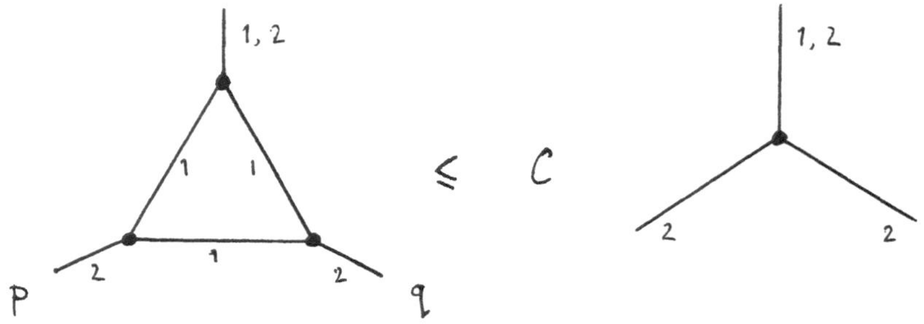

Fig. 9: Removal of elementary triangle facets.

E_1 = 1 or o. We write

$$\alpha_1 = \alpha \quad \text{or} \quad \beta - 1 \tag{8.4}$$

to cover both cases. Now let us replace the vertex functions by bounds that depend only on the momenta of the internal lines:

$$\Big(\max(|k|,|k+p|,|p|)\Big)^{\gamma} \leqslant R(\gamma)\Big(|k|^{\gamma} + |k+p|^{\gamma}\Big)\;, \tag{8.5}$$

with $\gamma = 1+3\alpha$; $R(\gamma) = 2^{3\alpha}$,

or $\gamma = \beta+2\alpha \ ; \ R(\gamma) = 1 \ .$ (8.6)

The integral over the loop momentum is then bounded by 8 terms all of the form

$$R \int d^4k \ (k^2)^{-\delta_1} (k+p)^{-2\delta_2} \ (k-q)^{-2\delta_3} \ , \tag{8.7}$$

where for a moment we ignored the mass term. It can be added easily later. We have convergence for all integrals:

$$Z = 2 \sum \delta - 4 = 1 - 2\beta - \alpha_1 > o \ . \tag{8.8}$$

and

$$\delta_1 \geqslant 1 + \alpha - \tfrac{1}{2}(2\alpha+\beta) - \tfrac{1}{2}(1+2\alpha+\alpha_1) = \tfrac{1}{2}(1-2\alpha-\alpha_1-\beta) \ ; \tag{8.9}$$

The integral (8.7) can be done using Feynman multiplicators:

$$\frac{R\pi^2\Gamma(\Sigma\delta-2)}{\Gamma(\delta_1)\Gamma(\delta_2)\Gamma(\delta_3)} \int_o^1 dx_1 dx_2 dx_3 \frac{\delta(\Sigma x-1) \ x_1^{\delta_1-1} \ x_2^{\delta_2-1} \ x_3^{\delta_3-1}}{\left(p^2x_2x_3+q^2x_3x_1+(p+q)^2x_1x_2\right)^{\Sigma\delta_i-2}} \tag{8.10}$$

Now if $|p| \geqslant |q| \geqslant |p+q|$ (the other cases can be obtained by permutation) then $|q| \geqslant \frac{1}{2}|p|$, so our integrals are bounded by

$$C\left[\max(|p|,|q|,|p+q|)\right]^{-\frac{1}{2}Z} \tag{8.11}$$

where C is the sum of integrals of the type

$$\frac{\pi^2\Gamma(\Sigma\delta-2)}{\Gamma(\delta_1)\Gamma(\delta_2)\Gamma(\delta_3)} \int_o^1 dx_1 dx_2 dx_3 \frac{\delta(\Sigma x-1) \ x_1^{\delta_1-1} \ x_2^{\delta_2-1} \ x_3^{\delta_3-1}}{(x_1x_2+\frac{1}{2}x_1x_3)^{\frac{1}{2}Z}} \tag{8.12}$$

which can be further bounded (replacing x_1x_2 by $\frac{1}{2}x_1x_2$) by

$$C \leqslant \sum \frac{2^{\Sigma\delta-2} \ \pi^2 \ \Gamma(\Sigma\delta-2)\Gamma(2-\delta_1)\Gamma(2-\delta_2-\delta_3)}{\Gamma(\delta_1) \ \Gamma(\delta_2+\delta_3)} \ , \tag{8.13}$$

if all integrals converge, of course. All entries in the Γ

functions must be positive. In particular, we must have

$$2 - \delta_2 - \delta_3 = \delta_1 - \tfrac{1}{2}Z > o \ . \tag{8.14}$$

Now with (8.8) and (8.9) this corresponds to the condition:

$$\beta > 2\alpha \ , \tag{8.15}$$

this is the extra restriction on the coefficients α and β to be combined with (5.12), and which we already alluded to in the end of sect. 5. A good choice may be

$$\alpha = 0.1 \ , \ \beta = 0.3 \ . \tag{8.16}$$

We conclude that we proved the bound of Fig. 9, if α and β have values such as (8.16), and the number C in Fig. 9 is bounded by the sum of eight finite numbers in the form of eq. (8.13).

9. QUADRANGLES

We continue removing triangular facets from our diagram, replacing then by single 3-vertices, following the prescriptions of the previous sections. We get fewer and fewer loops, at the cost of at most a factor C for each loop. Either we end up with a tree diagram, in which case our argument is completed, or we may end up with a diagram that can still be arbitrarily large but only contains larger facets. According to sect. 7 there must be quadrangles and/or pentagons among these.

Before concentrating on the quadrangles we must realize that there still may be larger subgraphs with only three external lines. In that case we consider those first: a minimal triangular subgraph is a triangular subgraph that contains no further triangular subgraphs. If our diagram contains triangular subgraphs then we first consider a minimal triangular subgraph and attack quadrangles (later pentagons) in these. Otherwise we consider the quadrangles inside the entire diagram.

Let us again replace as many composite propagators ($\overset{2}{\bullet\!\!-\!\!-\!\!\bullet}$) by single dressed propagators ($\overset{1}{\bullet\!\!-\!\!-\!\!\bullet}$) as allowed by ineq. (8.1) for each subgraph. Then one can argue that as a result we must get at least one quadrangle somewhere whose own propagators are all of the elementary type ($\overset{1}{\bullet\!\!-\!\!-\!\!\bullet}$), not composite ($\overset{2}{\bullet\!\!-\!\!-\!\!\bullet}$). This is because facets with composite propagators now must be adjacent to 4-leg subgraphs (elementary facets or more complicated), and then these in turn must have facet(s) with elementary propagators. Also (although we will not really need this) one may argue that there will be quadrangles with not more than one external composite propagator, the others elementary (the one exception is the case when one of the adjacent quadranglular subgraphs has itself only

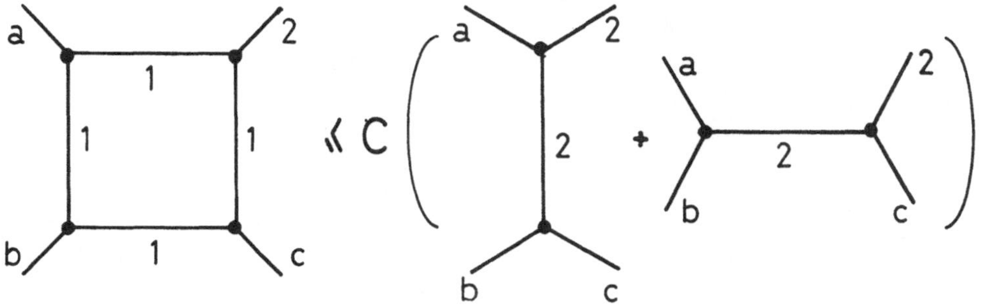

Fig. 10. Inequality for quadrangles. a, b and c may each be 1 or 2.

pentagons, but that case will be treated in the next section). As a result of these arguments, of all inequalities of the type given in Fig. 10 we only need to check the case that only one external propagator is composite, a = b = c = 1.

But in fact they hold quite generally, also in the other cases. This is essentially because of the careful construction of the effective Feynman rules of type IV in Fig. 5.

Rather than presenting the complete proof of the inequalities of Fig. 10 (5 different configurations) we will just present a simple algorithm that the reader can use to prove and understand these inequalities himself. In general we have integrals of the form

$$\int d^4k \prod_i \frac{1}{\left|k-p_i\right|^{2\delta_i}} . \tag{9.1}$$

We could write this as a diagram in Fig. 11, where the δ_i at the propagators now indicate their respective powers. The vertices are here ordinary point-vertices, not the type IV rules. Now write

$$\frac{1}{|k-p_1|^{\omega_2}|k-p_2|^{\omega_2}} \leqslant \frac{A(\omega_1,\omega_2)}{|p_1-p_2|^{\omega_2}|k-p_1|^{\omega_1}} + \frac{A(\omega_2,\omega_1)}{|p_1-p_2|^{\omega_1}|k-p_2|^{\omega_2}} \tag{9.2}$$

with

$$A(\omega_1,\omega_2) = \max\left[1,\left(\left(\frac{\omega_1}{\omega_1+\omega_2}\right)^{\omega_1} + \left(\frac{\omega_2}{\omega_1+\omega_2}\right)^{\omega_2}\right)^{-1}\right] . \tag{9.3}$$

Inserted in a diagram, this is the inequality pictured in Fig. 11.

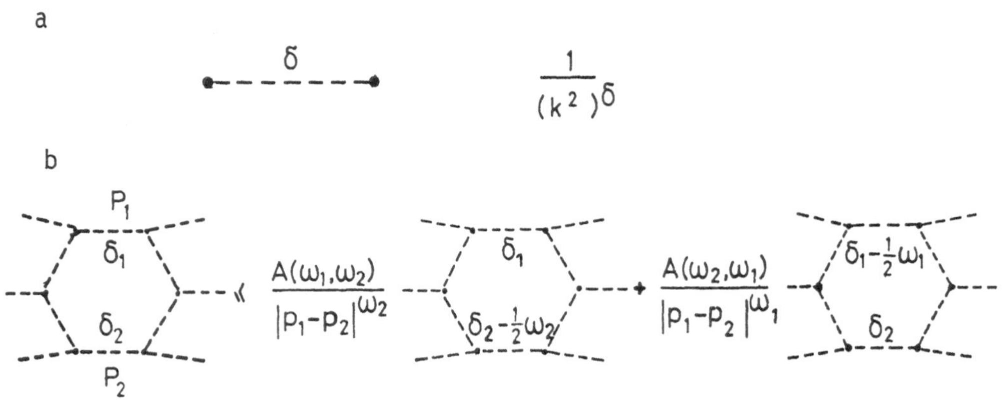

Fig. 11. a) Definition of a dotted propagator. Its vertices are just 1.
b) Extraction of a power of an external momentum.

We use it for instance when p_1-p_2 is the largest momentum of all channels, and if

$$\omega_1 \leqslant 2\delta_1 \;;\; \omega_2 \leqslant 2\delta_2 \;;\; \omega_i < Z \,, \tag{9.4}$$

where Z is the degree of convergence of the diagram: $Z = 2\Sigma\delta_i - 4$. We continue making such insertions, everytime reducing the diagrams to a convenient momentum dependent factor times a less convergent diagram. Finally we may have

$$Z < 2\delta_1 \;,\; Z < 2\delta_2 \;, \tag{9.5}$$

for two of its propagators. Then we use the inequality pictured in Fig. 12:

$$\int d^4k \prod_i (k-p_i)^{-2\delta_i} \leqslant C|p_1-p_2|^{-Z} \,. \tag{9.6}$$

Notice that in (9.5) we have a strictly unequal sign, contrary to ineq. (9.4). This C can be computed using Feynman multiplicators,

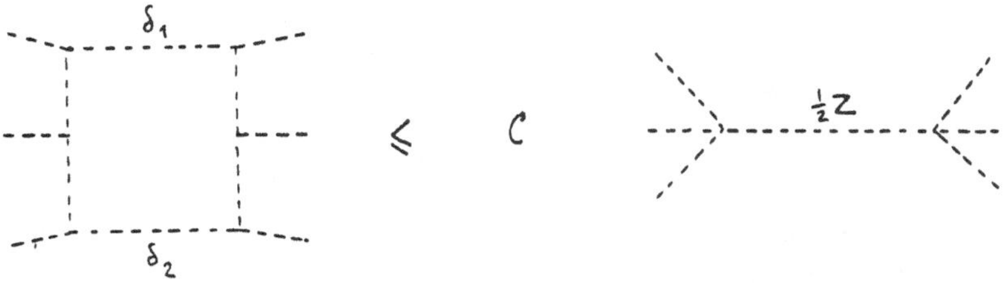

Fig. 12. Inequality holding if $\delta_{1,2}$ are strictly larger than $\frac{1}{2}Z$ (ineqs. (9.4), (9.5)).

much like in the previous section. We get

$$C \leqslant \frac{\pi^2\Gamma(\frac{1}{2}Z)}{\Gamma(2-\frac{1}{2}Z)} \prod_{i=1,2} \frac{\Gamma(\delta_i-\frac{1}{2}Z)}{\Gamma(\delta_i)} \; . \tag{9.7}$$

We see that ineqs. (9.2) and (9.6) have essentially the same effect: if two propagators have a power larger than a certain coefficient they allow us to obtain as a factor a corresponding "propagator" for the momentum in that particular channel. This is how proving the ineqs. of Fig. 10 can be reduced to purely algebraic manipulations. We discovered that ineq. (8.15) is again crucial. We notice that if the internal propagators of the quadrangle were elementary ones then the superficial degree of convergence of any of the other subgraphs of our diagram may change slightly, since in eq. (5.10) $\beta > 2\alpha$, but the left hand side of eq. (5.11) remains unchanged, because

$$\Delta E_1 = -2\Delta E_2 \; , \tag{9.8}$$

so our condition that all subgraphs be convergent remains fulfilled after the substitution of the inequalities of Fig. 10.

However, if one of the internal lines of the quadrangle had been a composite one ($\overset{2}{\bullet\!\!-\!\!-\!\!\bullet}$), then a subgraph would become more divergent, because we are unable to continue our scheme with something like a three-particle composite propagator ($\overset{3}{\bullet\!\!-\!\!-\!\!\bullet}$). A crucial point of our argument is that we will never really need such a thing, if we attack the quadrangle subgraphs in the right order.

10. PENTAGONS. CONVERGENCE OF THE SKELETON EXPANSION

As stated before, the order in which we reduce our diagram

into a tree diagram is:

1) remove triangular facets;
2) remove triangular subgraphs if any. By complete induction we prove this to be possible. Take a minimal triangular subgraph and go to 3;
3) remove quadrangular facets as far as possible. If any cannot be removed because of a crucial composite propagator in them, then
4) remove qudrangular subgraphs. After that we only have to
5) remove the pentagons.
6) If we happened to be dealing with a subgraph by branching at point 2 or 3, then by now that will have become a tree graph, because of the theorem in sect. 7. Go back to 1.

We still must verify point 5. If indeed our whole diagram contains pentagons then we can replace all propagators by elementary ones. But if we had branched at steps 2 or 3 then the subgraphs we are dealing with may still have composite external propagator(s). In that case it is easy to verify that there will be enough pentagons buried inside our subgraphs that do not need composite external lines. In that case we apply directly the inequality of Fig. 13. The procedure for proving Fig. 13 is

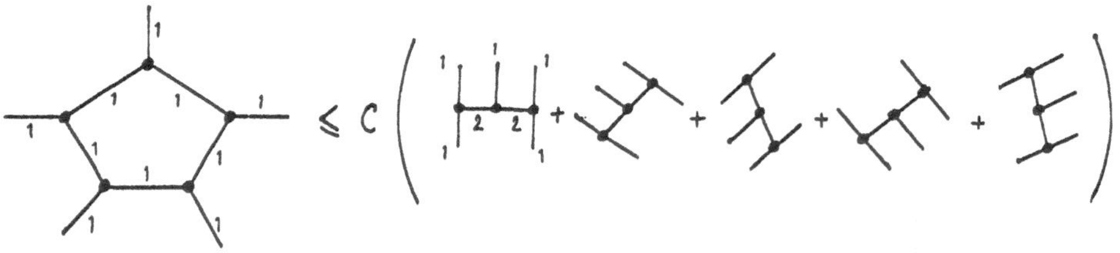

Fig. 13. Inequality for pentagons.

exactly as described for the quadrangles in the previous section. Again the degree of divergence of any of the adjacent subgraphs has not changed significantly. This now completes our proof by induction that any planar skeleton diagram with 5 external lines is equal to C^L times a diagram with type IV Feynman rules, where C is limited to fixed bounds. Since also the number of diagrams is an exponential function of L we see that for this set of graphs perturbation expansion in g has a finite radius of convergence. The proof given here is slightly more elegant than in Ref. 11, and also leads to tree graph expressions that are more useful for our manipulations.

If the diagram has 6 or more external lines then still a number of facets may be left, limited by ineq. (7.10), all having 6 or more propagators. If we wish we can still continue our procedure for these but that would be rather pointless: having a limited number of loops the diagram is finite anyhow. The difficulty would not so much be that no inequalities for hexagons etc. could be written down; they certainly exist, but our problem would be that the corresponding number C would not obviously be bounded by one universal constant. This is why our procedure would not work for nonplanar theories where ineq. (7.10) does not hold. In the non-planar case however similar theorems as ours have been derived[6].

11. BASIC GREEN FUNCTIONS

The conclusion of the previous section is that if we know the "basic Green functions", with which we mean the two- three- and four-point functions, and if these fall within the bounds given in Table 1, then all other Green functions are uniquely determined by a convergent sum. Clearly we take the value for the bound g^2 for the coupling constant (ineq. (5.9)) as determined by the inverse product of the coefficient C_2 found in sect. 6 and the maximum of the coefficients in the ineqs. pictured in Figs. 9, 10 and 13, times a combinatorial factor.

Now we wish not only to verify whether these bounds are indeed satisfied, but also we would like to have a convergent calculational scheme to obtain these basic Green functions. One way of doing this would be to use the Dyson-Schwinger equations. After all, the reason why those equations are usually unsoluble is that they contain all higher Green functions for which some rather unsatisfactory cut-off would be needed. Now here we are able to re-express these higher Green functions in terms of the basic ones and thus obtain a closed set of equations.

These Dyson-Schwinger equations however contain the bare coupling constants and therefore require subtractions. It is then hard to derive bounds for the results which depend on the difference between two (or more) divergent quantities. We decided to do these subtractions in a different way, such that only the finite, renormalized basic Green functions enter in our equations, not the bare coupling constants, in a way not unlike the old "bootstrap" models. Our equations, to be called "difference equations" will be solved iteratively and we will show that our iteration procedure converges. So we start with some *Ansatz* for the basic Green functions and derive from that an improved set of values using the difference equations. Actually this will be done in various steps. We start with assuming some function $g(x)$ for the *floating coupling constant*, where x is the momentum in the maximal channel (see Table 1):

$$x = \max_{i,j} |p_i - p_j| \ , \tag{11.1}$$

and a set of functions $g_{(i)}(x)$ with

$$g(x) \underset{\text{def}}{=} \max_i |g_{(i)}(x)| \ . \tag{11.2}$$

Here $g_{(i)}(x)$ is the set of independent numbers that determined the basic Green function at their "symmetry point":

$$|p_i - p_j| = x \quad \text{for all } i,j \ . \tag{11.3}$$

The index i in $g_{(i)}$ then simply counts all configurations in (11.3). With "independent" we mean that in some gauge theories we assume that the various Ward-Slavnov-Taylor[13] identities among the basic Green functions are fulfilled. This is not a very crucial point of our argument so we will skip any further discussion of these Ward or Slavnov-Taylor identities.

If the values of the basic n-point Green functions (n = 3 or 4) at their symmetry points are $A_i(x)$, then the relation between A_i and g_i is:

$$\begin{aligned} A_{3i}(x) &= \kappa^j_{i\mu}\, p_{j\mu}\, Z^{-3/2}(x)\, g_{ij}(x) \\ A_{4i}(x) &= Z^{-2}(x)\, g^2_{4i}(x) \ , \end{aligned} \tag{11.4}$$

where $\kappa^j_{i\mu}$ are coefficients of order one, and Z(x) is defined in (5.1) and (5.2). (We ignore for a moment the case of superrenormalizable couplings.) Our first Ansatz for $g_i(x)$ is a set of functions that is bounded by (11.2), with g(x) decreasing asymptotically to zero for large x as dictated by the lowest order term(s) of the renormalization group equations. We will find better equations for $g_i(x)$ as we go along. In any case we will require

$$\left| \frac{xd}{dx}\, g_i(x) \right| \leqslant \tilde{\beta}\, g^3(x) \tag{11.5}$$

for some finite coefficient $\tilde{\beta}$.

Our first *Ansatz* for the basic Green functions away from the symmetry points will be even more crude. All we know now is that they must satisfy the bounds of Table 1. In general one may start with choosing (11.4) to hold evenaway from the symmetry points, and

$$x = \max_{i,j} |p_i - p_j| \ . \tag{11.6}$$

After a few iterations we will get values still obeying the bounds of Table 1, and with *uncertainties* also given by Table 1 but K_i replaced by coefficients δK_i. Thus we start with

$$\delta K_i^{(0)} = K_i \ . \tag{11.7}$$

We will spiral towards improved *Ansätze* for the basic Green functions in two movements:

i) the "small spiral" is the use of difference equations to obtain improved values at exceptional momenta, *given the values $g_i(x)$ at the symmetry points*. These difference equations will be given in the next section.
ii) The "second spiral" is the use of a variant of the Gell-Mann-Low equation to obtain improved functions $g_i(x)$ from previous Ansätze for $g_i(x)$, making use of the convergent "small spiral" at every step. What is also needed at every step here is a set of integration constants determining the boundary condition of this Gell-Mann-Low equation. It must be ensured that these are always such that $g(x)$ in ineq. (11.2) remains bounded:

$$g(x) < g_0 \ ,$$

where g_0 is limited by the coefficients K_i and the various coefficients C from sects. 6, 8, 9 and 10, as in ineq. (5.9).

12. DIFFERENCE EQUATIONS FOR BASIC GREEN FUNCTIONS

The Feynman rules of our set of theories must follow from a Lagrangian, as usual.
For brevity we ignore the Lorentz indices and such, because those details are not of much concern to us. Let the dressed propagator be

$$P(p) = -G_2^{-1}(p) \ , \tag{12.1}$$

and let the corresponding zeroth order expressions be indicated by adding a superscript o. In massive theories:

$$P^0(p) = (p^2+m^2)^{-1} = -G_2^{0-1}(p) \ . \tag{12.2}$$

Define

$$G_2(p+k) - G_2(p) = G_{2\mu}(p|k)k_\mu \ , \tag{12.3}$$

so that

$$P(p+k) - P(k) = P(p+k)G_{2\mu}(p|k)k_\mu P(p) \ . \tag{12.4}$$

This gives us the "Feynman rule" for the difference of two dressed

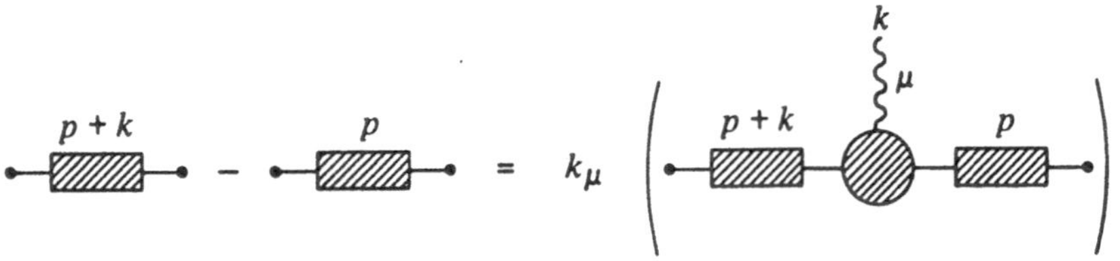

Fig. 14. Feynman rule for the difference of two dressed propagators. The 3-vertex at the right is the function $G_{2\mu}(p|k)$.

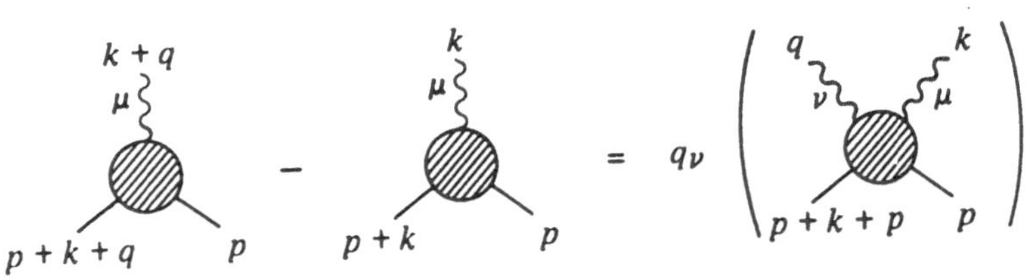

Fig. 15. Difference equation (12.6) for $G_{2\mu}$.

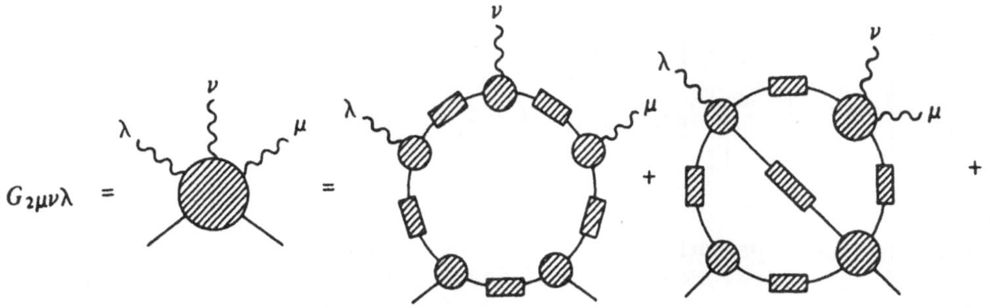

Fig. 16. Some arbitrarily chosen terms in the skeleton expansion for $G_{2\mu\nu\lambda}$.

propagators, depicted in Fig. 14. (Note that, in this section only, p and k denote external line momenta, not external loop momenta.)

We have also this Feynman rule for bare propagators. There $G^0_{2\mu}$ follows directly from the Lagrangian:

$$G^0_{2\mu} = -2p_\mu - k_\mu \ . \qquad (12.5)$$

Continuing this way we define

$$G_{2\mu}(p|k+q) - G_{2\mu}(p|k) = G_{2\mu\nu}(p|k|q)q_\nu \ , \qquad (12.6)$$

with

$$G^0_{2\mu\nu} = -\delta_{\mu\nu} \ . \qquad (12.7)$$

In Feynman graphs this is sketched in Fig. 15. Differentiating once more we get

$$G_{2\mu\nu}(p|k|q+r) - G_{2\mu\nu}(p|k|q) = G_{2\mu\nu\lambda}(p|k|q|r)r_\lambda \ . \qquad (12.8)$$

Of course $G_{2\mu\nu\lambda}$ can be computed formally in perturbation expansion. The rules for computing the new Green functions G_μ, $G_{\mu\nu}$, $G_{\mu\nu\lambda}$ are easy to establish. Let p_1 be one of the external loop momenta as defined in eq. (5.3). For a Green function $G(p_1)$ we have

$$G(p_1) = \int \dots \int dq_i \ f_1(p_1+q_1)f_2(p_1+q_2) \ \dots \ f_t(p_1+q_t).F \ , \qquad (12.9)$$

where $f_i(q_i)$ are bare vertex and/or propagator functions adjacent to the external facet labeled by 1. The remainder F is independent of p_1. We write

$$G(p_1+k) - G(p_1) = \int \ldots \int F\,dq \left(\prod_1^t f(p_1+q_i+k) - \prod_1^t f(p_1+q_i) \right)$$

$$= \int \ldots \int F\,dq \sum_{s=1}^{t} \prod_1^{s-1} f(p_1+q_i+k) \left(f(p_1+q_s+k) - f(p_1+q_s) \right) \prod_{s+1}^{t} f(p_1+q_i), \tag{12.10}$$

which is just the rule for taking the difference of two products. We find the difference of two *dressed* Green functions G in terms of the difference of *bare* functions f. Therefore the "Feynman rules" for the diagrams at the right hand sides of Figs. 14, 15 and the l.h.s. of Fig. 16 consist of the usual combinatorial rules with new bare vertices given by the eqs. (12.5) and (12.7). These bare vertices occur only at the edge of the diagram.

We see that the power counting rules for divergences in $G_{2\mu\nu\lambda}$ are just as in 5-point functions in gauge theories. Since the global degree of divergence is negative we can expand in skeleton graphs. See Fig. 16, in which the blobs represent ordinary dressed propagators and dressed vertices or dressed functions G_μ and $G_{\mu\nu}$.

Notice that one might also need $G_{3\mu}(p_1,p_2|k)$ defined by

$$G_3(p_1,p_2+k) - G_3(p_1,p_2) = G_{3\mu}(p_1,p_2|k).k_\mu \ . \tag{12.11}$$

In short, the skeleton expansion expresses $G_{2\mu\nu\lambda}$ but also $G_{3\mu\nu}$ etc. in terms of the few basic functions $G_{2\mu}$, $G_{2\mu\nu}$, $G_{3\mu}$ and the basic Green functions $G_{2,3,4}$. Also the function $G_{4\mu}$, defined similarly, can thus be expressed. The corresponding Feynman rules should be clear and straightforward.

We conclude that the basic Green functions can in turn be expressed in terms of skeleton expansions, and, up to overall constants, these equations, if convergent, determine the Green functions completely. Notice that we never refer to the bare Lagrangian of the theory, so, perhaps surprisingly, these sets of equations are the same for all field theories. The difference between different field theories only comes about by choosing the integration constants differently.

Planarity however was crucial for this chapter, because only planar diagrams have well defined "edges": the new vertices only occur at the edge of a diagram.

13. FINDING THE BASIC GREEN FUNCTIONS AT EXCEPTIONAL MOMENTA (THE "SMALL SPIRAL")

In this section we regard the basic Green functions at their symmetry points as given, and use the difference equations of

Sect. 12 to express the values at exceptional momenta in terms of these. If p_i-p_j is the momentum flowing through the planar channel ij, then in our difference equations we decide to keep

$$\mu = \max_{i,j} |p_i-p_j| \tag{13.1}$$

fixed. So the left hand side of our difference equations will show two Green functions with the same value for μ, one of which may be exceptional and the other at its symmetry point, and therefore known. (We use the concept of "exceptional momenta" as in ref. 14.)

Now the right hand side of these difference equations show a skeleton expansion of diagrams which of course again contain basic Green functions, also at exceptional momenta. But these only come in combinations of higher order, and the effect of exceptional momenta is relatively small, so at this point one might already suspect that when these equations are used recursively to determine the exceptional basic Green functions then this recursion might converge. This will indeed be the case under certain conditions as we will show in sect. 15.

Our iterative procedure must be such that after every step the bounds of Table 1 again be satisfied. This will be our guide to define the procedure. First we take the 4-point functions, and consider all cases of Table 1 separately.

The right hand side of our difference equations (Fig. 16) contains a skeleton expansion to which we apply the theorem mentioned in the end of sect. 5 and proven in sects. 5-10: the skeleton expansion for any 5-point Green function converges and is bound by tree diagrams constructed with type IV Feynman rules. Since the 5-point functions in Fig. 16 are irreducible, the internal lines in the resulting tree graph will always be composite propagators, as in the r.h.s. of Fig. 13. So we simply apply the type IV Feynman rules for 5 tree graphs to obtain bounds on the 5-point function in various exceptional regions of momentum space. Table 2 lists the results.

The power of $g^2(A_3)$ in the table applies where we consider the function $G_{4\mu}$. The other functions $G_{3\mu\nu}$ and $G_{2\mu\nu\lambda}$ have one and zero powers of $g(A_3)$, respectively. In front of all this comes a power series of the form

$$\sum_{n=1}^{\infty} C^n g_0^{2n} = Cg_0^2(1-Cg_0^2)^{-1} , \tag{13.2}$$

which converges provided that

Table 2

Bounds for the irreducible 5-point function at some exceptional momentum values.

$(((12)_1\ 53)_2\ 4)_3$	$\prod_i Z_{ii+1}^{-\frac{1}{2}}\ A_1^{-\alpha}\ A_2^{-1+\alpha-\beta}\ A_3^{\beta}\ g^2(A_3)$
$((12)_1\ 5\ (34)_2)_3$	$\prod_i Z_{ii+1}^{-\frac{1}{2}}\ A_1^{-\alpha}\ A_2^{-\alpha}\ A_3^{2\alpha-1}\ g^2(A_3)$
$(((513)_1\ 2)_2\ 4)_3$	$\prod_i Z_{ii+1}^{-\frac{1}{2}}\ A_1^{-1-2\beta}\ A_2^{\beta}\ A_3^{\beta}\ g^2(A_3)$
$((135)_1\ (24)_2)_3$	$\prod_i Z_{ii+1}^{-\frac{1}{2}}\ A_1^{-1-2\beta}\ A_3^{2\beta}\ g^2(A_3)$ if $A_2 > A_1$
$((1235)_1\ 4)_2$	$\prod_i Z_{ii+1}^{-\frac{1}{2}}\ A_1^{-1-\beta}\ A_2^{\beta}\ g^2(A_3)$

$$g_0 = \max_\mu |g(\mu)| < C^{-\frac{1}{2}} . \qquad (13.3)$$

We now write an equation such as (12.8) as follows:

$$\tilde{G}_4\{(((12)_1\ 3)_2\ 4)_3\} = G_4\{((523)_2\ 4)_3\} + $$
$$+ (p_1-p_5)_\mu\ G_{4\mu}\{(((12)_1\ 53)_2\ 4)_3\} , \qquad (13.4)$$

where $r = p_1-p_5$. In this and following expressions the tilde (~) indicates which quantities are being replaced by new ones in the iteration procedure.

If the Ansatz holds for $G_4\{((523)_2\ 4)_3\}$ then the new exceptional function will obey

$$|\tilde{G}_4\{(((12)_1\ 3)_2\ 4)_3| \leqslant K_6^2(Z_{45}Z_{52}Z_{23}Z_{34})^{-\frac{1}{2}} \left(\frac{A_3}{A_2}\right)^\beta g^2(A_3) +$$
$$+ (Z_{45}Z_{12}Z_{23}Z_{34})^{-\frac{1}{2}} \frac{Cg_0^2}{1-Cg_0^2}\ g^2(A_3) \left(\frac{A_2}{A_1}\right)^\alpha \left(\frac{A_3}{A_2}\right)^\beta . \qquad (13.5)$$

Choosing
$$\frac{Cg_0^2}{1-Cg_0^2} \equiv \gamma \qquad (13.6)$$

and considering that to a good approximation (since $|p_1-p_5| << |p_4-p_5|$):

$$Z_{45} = Z_{41} , \qquad (13.7)$$

we find

$$|\tilde{G}_4\{(((12)_1\ 3)_2\ 4)_3\}| \leqslant (Z_{12}Z_{23}Z_{34}Z_{41})^{-\frac{1}{2}}\ g^2(A_3)\left(\frac{A_2}{A_1}\right)^{\alpha}\left(\frac{A_3}{A_2}\right)^{\beta}\ .$$

$$.\left(\gamma + \left(\frac{Z_{12}}{Z_{52}}\right)^{\frac{1}{2}}\left(\frac{A_1}{A_2}\right)^{\alpha} K_6^2\right)\ . \tag{13.8}$$

What is now needed is a bound for the last term in (13.8). Let

$$x_{12} = |p_{12}|/m \geqslant 1 \tag{13.9}$$

(remember that $|p|$ stands for $\sqrt{p^2+m^2}$),
and

$$f(x_{12}) = (\log(1+x_{12})^{\sigma/2}\ .\ x_{12}^{\alpha}\ , \tag{13.10}$$

where σ is defined in (5.2).
When

$$x_{12} > \exp\ (-\sigma/2\alpha) = x_0 \tag{13.11}$$

this f is an increasing function, so that if

$$x_0 m \leqslant A_1 < A_2 \tag{13.12}$$

then

$$\frac{f(x_{12})}{f(x_{52})} < 1\ . \tag{13.13}$$

The range $1 \leqslant x \leqslant x_0$ is compact, so there exists a finite number L such that

$$\frac{f(x_{12})}{f(x_{52})} \leqslant L \tag{13.14}$$

as soon as

$$x_{12} < x_{52}\ . \tag{13.15}$$

So we find that after one iteration given by (13.4), the new K_2 coefficient satisfies

$$K_2^2 \leqslant \gamma + K_6^2 L\ . \tag{13.16}$$

Similarly we derive

$$K_7^2 \leqslant \gamma + L \;, \tag{13.17}$$

when the difference equation is used to express $\tilde{G}\{((12)_1\ 34)_2\}$ in terms of $G\{(5234)_2\}$. Also we use

$$\tilde{G}_4\{((12)_1\ (34)_2)_3\} = G_4\{(52(34)_2)_3\} + \\ + (p_1-p_5)_\mu\ G_{4\mu}\{((12)_1\ 5(34)_2)_3\} \;, \tag{13.18}$$

to find that after one step

$$K_3^2 \leqslant \gamma + K_7^2 L \leqslant \gamma + \gamma L + L^2 \;; \tag{13.19}$$

and for the three-point function

$$K_1 \leqslant K_7^2 + L \leqslant \gamma + 2L \;. \tag{13.20}$$

The remaining coefficients K_{4-6} must be computed in a slightly different way. Consider K_4. We replace p_1 by p_5 now in such a way that

$$|p_5-p_3| \simeq 2|p_1-p_3| \;; \qquad A_1 \rightarrow 2A_1 \tag{13.21}$$

and work with induction. Write

$$\tilde{\tilde{G}}_4\{(((13)_1\ 2)_2\ 4)_3\} = \tilde{G}_4\{(((53)_1\ 2)_2\ 4)_3\} \\ + (p_1-p_5)_\mu\ G_{4\mu}\{(((513)_1\ 2)_2\ 4)_3\} \;. \tag{13.22}$$

Inspecting Tables 1 and 2 we find now

$$K_4^2 = \max\left(K_6^2 \;, \frac{\gamma}{1-2^{-2\beta}}\right) \;. \tag{13.23}$$

Applying the same technique we compute the fifth exceptional configuration of Table 1. We separate p_1 from p_3 until $A_1 \rightarrow A_2$. Then we separate alternatively p_2 from p_4 and p_1 from p_3 keeping $A_1 \simeq A_2$. This makes the rate of convergence slightly slower:

$$K_5^2 = \frac{2\gamma}{1-2^{-2\beta}} \;. \tag{13.24}$$

Finally K_6 is found by widening the separation between p_1, p_2 and p_3 in successive steps of factors of 2:

$$\tilde{\tilde{G}}_4\{((123)_1\ 4)_2\} = \tilde{G}_4\{((563)_1\ 4)_2\} \\ + (p_5-p_1)_\mu\ G_{4\mu}\{((1235)_1\ 4)_2\} + (p_6-p_2)_\mu\ G_{4\mu}\{((2356)_1\ 4)_2\} \;, \tag{13.25}$$

where

$$|p_5-p_6| = |p_3-p_6| = |p_5-p_3| \simeq 2|p_1-p_2| \ . \tag{13.26}$$

We find

$$|\tilde{G}_4\{((123)_1\ 4)_2|\} \leqslant \gamma Z(A_1)^{-1}\ Z(A_2)^{-1}\left(\frac{A_2}{A_1}\right)^{\beta} \ .$$

$$\sum_{n=1}^{^2\log\,(A_2/A_1)} \left(2^{-\beta n}\ .\ 2\,\frac{Z(A_1)}{Z(2^n A_1)}\right) + |G_4\{(1234)_2\}| \ . \tag{13.27}$$

The sum can certainly be bounded:

$$\sum \leqslant L' \leqslant 2L(1-2^{\alpha-\beta})^{-1} \ . \tag{13.28}$$

Therefore

$$K_6^2 \leqslant \max(1,\gamma L') \ . \tag{13.29}$$

Thus all coefficients K_i have bounds that will be obeyed everywhere in the "small spiral" induction procedure. Note that these coefficients would blow up if $\alpha,\beta \to o$. In particular in (13.11) we need $\alpha > o$. Only if $g_0^2 \to o$ we can let $\alpha,\beta \to o$. It will be clear from the above arguments that our bounds are only very crude. Our present aim was to establish their existence and not to find optimal bounds.

In sect. 15 we show that the "small spiral" of iterations for the exceptional Green functions, given the non-exceptional ones in (13.27), actually converges geometrically.

14. NON-EXCEPTIONAL MOMENTA (THE "SECOND SPIRAL")

In order to formulate the complete recursion procedure for determining the basic Green functions we need relations that link these Green functions at different symmetry points. Again the difference equations are used:

$$G_4(p_1\ldots p_4) = G_4(2p_1,p_2p_3p_4) - p_\lambda G_{4\lambda}(p_1,2p_1,p_2p_3p_4)$$

$$= \ldots = G_4(2p_1,\ldots,2p_4) - \sum_{i=1}^{4} p_{i\lambda}G_{4\lambda}(p_1^{(i)},\ldots,p_5^{(i)}). \tag{14.1}$$

Here p_i and $p_j^{(i)}$ are external loop momenta. They are non-exceptional. We use a shorthand notation for (14.1). Writing $p_i^2 \simeq (p_i-p_j)^2 \simeq \mu^2$:

$$G_4(\mu) - G_4(2\mu) = \mu \sum_{i=1}^{4} G_{4\lambda}^{(i)}(2\mu,\mu) \ . \tag{14.2}$$

Similarly we have

$$G_{2,3}(\mu) - G_{2,3}(2\mu) = - \mu \sum_{i} G_{2,3\lambda}^{(i)}(2\mu,\mu) \ . \tag{14.3}$$

These are just discrete versions of the renormalization group equations. The right hand side of (14.2), [not (14.3)!] is to be expanded in a skeleton expansion which contains all basic Green functions at all μ, also away from their symmetry points. There we insert the values obtained after a previous iteration. It is our aim to derive from eq. (14.2) a Gell-Mann-Low equation[1] of the form

$$\mu\frac{\partial}{\partial\mu} g_i(\mu) = - \sum_{\ell=2}^{k} \beta_{ij_1\ldots j_\ell}^{(\ell)} g_{j_1}(\mu)\ldots g_{j_\ell}(\mu)$$

$$+ |g(\mu)|^N \rho_i(\mu) \ , \tag{14.4}$$

where $\beta^{(\ell)}$ are the first k coefficients of the β function, and they must coincide with the perturbatively computed β coefficients. Often (depending on the dimension of the coupling constant) only odd powers occur so that $k = N-2$ is odd. The rest function ρ must satisfy

$$|\rho_i(\mu)| \leqslant Q_N \tag{14.5}$$

for some constant $Q_N < \infty$. This inequality must hold in the sense that $|g(\mu)|^N \rho(\mu)$ must be a convergent expansion in the functions $g(\mu')$, with

$$\mu' \geqq m \tag{14.6}$$

(so that μ' may be smaller than μ), in such a way that the absolute value of each diagram contributes to Q_N and their total sum remains finite.

Now clearly eq. (14.2) is a difference equation, not a differential equation such as (14.4). Up till now differential equations were avoided because of infrared divergences. Just for ease of notation we have put (14.4) in differential form because the mathematical convergence questions that we are to consider now are insensitive to this simplification.

Consider the skeleton expansion of $G_{4\lambda}^{(i)}$ in (14.2). At each of the four external particle lines a factor $g(\mu_i)$ occurs with $\mu_i \geqq \mu$, so it may seem easy to prove (14.4) from (14.2) with N = 3 or 4. However, we find it more convenient* to have an equation of the form (14.4) with $N \leq 7$, and our problem is that the internal vertices of the $G_{4\lambda}^{(i)}$ might have momenta which are less than μ. We will return to this question later in this section.

In proving the difference equation variant of (14.4) from (14.2) we have to make the transition from G_4 to g^2 and G_3 to g, and this involves the coefficients $Z(\mu)$, associated to the functions G_2, by equations of the form

$$
\begin{aligned}
G_2(\mu) &= -\mu^2 Z^{-1}(\mu) \quad ; \\
G_3(\mu) &= \mu Z^{-3/2}(\mu) g_3(\mu) \quad ; \\
G_4(\mu) &= Z^{-2}(\mu) g_4^2(\mu) \quad .
\end{aligned}
\tag{14.7}
$$

where g_3, g_4 are just various components of the coupling constant g_i. In the following expressions we suppress these indices i when we are primarily interested in the dependence on μ (= $|p|$ at the symmetry point). Now from (14.2) and (14.3) we find not first order but third order differential equations for G_2, basically of the form

$$
\frac{\partial^3}{\partial\mu^3} G_2 = G_{2,\lambda\lambda\lambda} = \mathcal{O}(g^2(\mu)Z^{-1}(\mu)/\mu) \ , \tag{14.8}
$$

where $G_{2,\lambda\lambda\lambda}$ is just a shorthand notation for the combination of expandable functions $G_{2,\lambda\mu\nu}$ obtained after taking differences three times. Write

$$
U_2(\mu) = -\frac{\partial^2}{\partial\mu^2} G_2(\mu) = -G_{2\lambda\lambda}(\mu) \ , \tag{14.9}
$$

then

$$
\frac{\mu\partial}{\partial\mu} U_2(\mu) = -\mu G_{2,\lambda\lambda\lambda}(\mu) \ , \tag{14.10}
$$

and

$$
\mu^2 Z^{-1}(\mu) = \int_m^n (\mu-\mu_1) U_2(\mu_1) d\mu_1 + A\mu + B \ . \tag{14.11}
$$

---

* Closer analysis shows that actually N = 3 or 4 is sufficient to prove unique solubility. Only if we wish an exact, non-perturbative definition of the free parameters we need the higher N values. Note that not only Q_N but also g_0 may deteriorate as N increases.

Here A and B are free integration constants; A is usually determined by Lorentz invariance and B by the mass, fixed to be equal to m. In lowest order:

$$A = mU_2(m) \; ; \qquad B = -\tfrac{1}{2}m^2U_2(m) \; . \tag{14.12}$$

This strange-looking form of the integration constants is an artifact coming from our substitution of difference equations by differential equations. Using difference equations we can impose Lorentz invariance by symmetrization in momentum space, so that only one (for each particle) integration constant is left: the mass term. We choose at all stages $\frac{1}{2}U_2(m) = Z(m) = 1$.

A convenient way to implement eq. (14.12) is to formally define $U_2(\mu) = 2$ if $o \leqq \mu \leqq m$, and replace the lower bound of the integral in (14.11) by zero. Then after symmetrization: $A = B = o$.

Equation (14.11) has a linearly convergent integral, whereas (14.10) is logarithmic. Together they determine the next iterative approximation to G . In fact we have

$$\mu G_{2,\lambda\lambda\lambda}(\mu) = Z^{-1}(\mu)f(\{g\}) \; , \tag{14.13}$$

and in $f(\{g\})$, Z occurs only indirectly. So the iteration converges fastest if we replace (14.10) by

$$\frac{\mu\partial}{\partial\mu}\tilde{U}_2(\mu) = -\frac{Z(\mu)}{\tilde{Z}(\mu)}G_{2,\lambda\lambda\lambda}(\mu) \; , \tag{14.14}$$

where the tilde denotes the new function $U_2(\mu)$.

One can however also use (14.9) with U_2 replaced by $\tilde{U}_2$.

We find

$$\frac{\mu\partial}{\partial\mu}Z^{-1} = -\int_{m/\mu}^{1} d\tau(1-\tau)\mu G_{2,\lambda\lambda\lambda}(\tau\mu) \; . \tag{14.15}$$

As stated before, the $\mathcal{O}\left(\frac{m}{\mu}\right)$ terms have been removed by symmetrization.

This equation allows us to remove the Z factors from the functions $G_{3,4}$ and arrive at first order renormalization group integrodifferential equations for $g_i(\mu)$.

For the 3-point functions we must write

$$U_3(\mu) = G_{3,\lambda}(\mu) = \frac{\partial G_3}{\partial\mu} \; ,$$

$$\frac{\mu\partial}{\partial\mu} U_3(\mu) = \mu G_{3,\lambda\lambda}(\mu) \; , \qquad (14.16)$$

$$G_3(\mu) = \int_m^\mu \mu U_3(\mu) d\mu + C_3 \; , \qquad (14.17)$$

$$\frac{\mu\partial G_3(\mu)}{\partial\mu} = \int_{m/\mu}^1 d\tau \; \mu G_{3,\lambda\lambda}(\tau\mu) \; . \qquad (14.18)$$

A potential difficulty in writing down the renormalization group equation even for N = 4 is the convolutions in (14.15) and (14.18) which contain Green functions at lower μ values, and so they depend on $g(\mu')$ with $\mu' < \mu$. So a further trick is needed to derive (14.4). This is accomplished by realizing that the integrals in (14.15) and (14.18) converge linearly in μ. Suppose we require at every iteration step (see eq. 11.5):

$$\left|\frac{\mu\partial}{\partial\mu} g(\mu)\right| \leqq \tilde{\beta}|g(\mu)|^3 \quad \text{and} \quad |g(\mu)| \leqq g_0 \qquad (14.19)$$

for some $\beta < \infty$, $g_o < \infty$. Then it is easy to show that if $\mu_1 \leqq \mu$, then

$$|g(\mu_1)| \leqq |g(\mu)| + C\left(\frac{\mu}{\mu_1}\right)^\varepsilon |g^3(\mu)| \; , \qquad (14.20)$$

if

$$\varepsilon \geqq 3\tilde{\beta}g_0^2 + \tilde{\beta}/C \; . \qquad (14.21)$$

So with C large enough and g_0 small enough we can make ε as small as we like. Inequality (14.20) is proven by differentiating with μ. This enables us to replace $g(\tau\mu)$ by $g(\mu)$ in (14.15) and (14.18) while the factor $\tau^{-\varepsilon}$ does no harm to our integrals. So we find bounds for $\frac{\mu\partial}{\partial\mu}\tilde{Z}^{-1}$ and $\frac{\mu\partial}{\partial\mu}\tilde{G}_3$ in terms of a power series of $g(\mu)$. We must terminate the series as soon as the factors $\tau^{-\varepsilon}$ accumulate to give τ^{-1}. This implies that N must be kept finite, otherwise $g_0 \to o$.

The same inequality (14.16) is used to go from N = 4 to N = 7 in these equations. If in a skeleton diagram a vertex is not associated with any external line, then it may be proportional to a factor $g(\mu')$ with $\mu' < \mu$. But using (14.16) we see that it may be replaced by $g(\mu)$ at the cost of a factor $(\mu/\mu_1)^\varepsilon$. At most three of these extra factors are needed. If the three corresponding vertices are chosen not to be too far away from one of the external vertices of the diagram (which we can always arrange), then this just corresponds to inserting an extra factor

$\left(\frac{p^{ext}}{p_1}\right)^{\varepsilon}$ at an external vertex. We now note that such factors still leave our integrals convergent. In the ultraviolet of course the diagrams converge even better than they already did, and in the infrared our degree of convergence was at least 1-α (or 2-β) as can easily be read off from eq. (5.10): adding the external propagators to any diagram one demands

$$Z + (2+2\alpha)E_1 + 2\beta E_2 < 4(E_1+E_2-1) \ . \tag{14.22}$$

Thus infrared convergence requires

$$1 - \alpha - T\varepsilon > o \tag{14.23}$$

where $T \lesssim 5$ is the number of times our inequality (14.20) was applied.

From the above considerations we conclude that an equation of the form (14.4) can be written down for any finite N, such that Q_N in inequality (14.5) remains finite. We do expect of course that Q_N might increase rapidly with N, but then we only want the equation for $N \leqq 7$. We are now in a position to formulate completely our recursive definition of the Green functions G_2, G_3, G_4 of the theory:

1) We start with a given set of trial functions $G_2(\mu)$, $G_3(\mu)$, $G_4(\mu)$ for the basic Green functions at their symmetry points. They determine our initial choice for the floating coupling constants $g_i(\mu)$ and the functions $Z_i(\mu)$. We require their asymptotic behaviour to satisfy (5.2), (5.9) and (11.5) (= eq. (14.19)).

2) We also start with an *Ansatz* for the exceptional Green functions that must obey the bounds of Table 1.

3) Use the difference equations of sect. 13 to improve the exceptional Green functions (the new values are indicated by a tilde (~)). These will again obey Table 1 as was shown in sect. 13. Repeat the procedure. It will converge towards fixed values for the exceptional Green functions (as we will argue in sect. 15). This we call the "small spiral".

4) With these values for the exceptional Green functions we are now able to compute the right hand side of the renormalization group equation for G_2, or rather Z^{-1}, from (14.15), using (14.20):

$$\frac{\mu\partial}{\partial\mu}\tilde{Z}_i^{-1}(\mu) = \tilde{Z}_i^{-1}(\mu)\left(\gamma_{ijk}g_j(\mu)g_k(\mu) + g^4(\mu)\textstyle\sum(\mu)\right) , \tag{14.24}$$

where $\sum(\mu)$ is again bounded. Here γ_{ijk} are the one-loop γ coefficients[14] This gives us *improved* propagators. See sect. (15.b).

5) Now we can compute the right hand side of eq. (14.4). Before integrating eq. (14.4) it is advisable to apply Ward identities (if we were dealing with a gauge theory) in order to reduce the number of independent degrees of freedom at each μ. As is well known, in gauge theories one can determine all subtraction constants this way except those corresponding to the usual free coupling constants and gauge fixing parameters[15]. So the number of unknown functions $g_i(\mu)$ need not exceed the number of "independent" coupling constants of the theory*.

6) Eq. (14.4) is now integrated, giving improved expressions for $g_i(\mu)$. Now go back to 2. This is the "second spiral", which will be seen to converge towards fixed values of $g_i(\mu)$.

The question of convergence of these two spirals is now discussed in the following section.

15. CONVERGENCE OF THE PROCEDURE

a) Exceptional Momenta

In sect. 13 a procedure is outlined to obtain the Green functions at exceptional momenta, if the Green functions at the symmetry point are given. That procedure is recursive because eqs. (13.4), (13.18), (13.22) and (13.25) determine the Green functions $G_{2,3,4}$ in terms of the symmetry ones, and $G_{4\mu}$, $G_{3\mu\nu}$, $G_{2\mu\nu\lambda}$. But the latter still contain the previous ansatz for $G_{2,3,4}$. Fortunately it is easy to show that any error $\delta G_{2,3,4}$ will reduce in size, so that here the recursive procedure converges:

Let us indicate the bounds discussed in sects. 5 and 13 as

$$|G_n(p_1,\ldots,p_n)| \leqq B_n(p_1,\ldots,p_n) \;, \tag{15.1}$$

and assume that a first trial $G_n^{(1)}$ has an error

$$|\delta G_n^{(1)}| \leqq \varepsilon^{(1)} B_n \;, \tag{15.2}$$

with some $\varepsilon^{(1)} \leqq 2$.

Now $G_{4\mu}$, $G_{3\mu\nu}$, $G_{2\mu\nu\lambda}$ also satisfy inequalities of the form (15.1). Furthermore they were one order higher in g^2. So we have

---

* We put "independent" between quotation marks because our requirement of asymptotic freedom usually gives further relations among various running coupling constants, see appendix A

$$|\delta G_{4\mu}| \leqslant \varepsilon^{(1)} B_{4\mu} \sum_{n=4}^{\infty} n\, C^{n-2} g^{n-2} , \tag{5.3}$$

when the function $G_{4\mu}$ itself converges like

$$\sum C^n g^n ,$$

and $B_{4\mu}$ is the bound for $G_{4\mu}$ itself, as given by Table 2.

The procedure of sect. 13 can be applied unaltered to the error δG_n in the Green functions. But there is a factor in front,

$$\varepsilon^{(1)} \sum_{n=2}^{\infty} (n+2)\, C^n g^n = \varepsilon^{(2)} . \tag{15.4}$$

This gives for the newly obtained exceptional Green functions an error

$$|\delta G_n^{(2)}| \leqslant \varepsilon^{(2)} B \tag{15.5}$$

and $\varepsilon^{(2)} < \varepsilon^{(1)}$ if we reduce the maximally allowed value for g, as given by (5.9), somewhat more:

$$g_{max} \rightarrow 0.6527\, g_{max} . \tag{15.6}$$

We stress that the above argument is only valid as long as the Green functions at their symmetry points were kept fixed and are determined by $g(\mu)$, bounded by (15.6).

b) The Z Factors

Knowing that at any stage $g(\mu)$ satisfies ineq. (14.9), we find that the solution of (14.24) is

$$\log \tilde{Z}_i(\mu) = \int^{\mu} d \log \mu_1 (\gamma_{ijk} g_j(\mu_1) g_k(\mu_1) + g^4(\mu_1) \textstyle\sum_i(\mu_1)) ; \tag{15.7}$$

$$\tilde{Z}_i(\mu) = \left(\log \frac{\mu}{\Lambda}\right)^{\sigma_i} (1+\mathcal{O}(g^2)) , \tag{15.8}$$

where the $\mathcal{O}(g^2)$ terms are again bounded by a coefficient times $g^2(\mu)$. These equations must be solved iteratively, because the right hand side of (15.7) contains skeleton expansions that again contain $Z(\mu)$, hidden in the function $\sum(\mu_1)$. It is not hard to convince oneself that such iterations converge. A change

$$|\delta Z| \leqq \varepsilon^{(1)} g^2 Z \tag{15.9}$$

yields a change in the function $\sum(\mu_1)$ bounded by

$$|\delta\textstyle\sum| \leqq \varepsilon^{(1)} g_0^2 \sum , \tag{15.10}$$

so that

$$\frac{\delta\widetilde{Z}(\mu)}{\widetilde{Z}(\mu)} \leqq C\varepsilon^{(1)} g_0^2 g^2(\mu) \leqq \varepsilon^{(2)} g^2(\mu) \tag{15.11}$$

with $\varepsilon^{(2)} < \varepsilon^{(1)}$ if g_0 is small enough.

c) The coupling constants

We now consider the integro-differential equation (14.4). The solution is constructed iteratively by solving

$$\frac{\mu\partial}{\partial\mu} \widetilde{g}_i(\mu) + \sum_{\ell=2}^{k} \beta^{(\ell)}_{ij_1\ldots j_\ell} \widetilde{g}_{j_1}(\mu)\ldots\widetilde{g}_{j_\ell}(\mu) = |g(\mu)|^N \rho_i(\mu) , \tag{15.12}$$

where the tilde denotes the next "improved" function $g_i(\mu)$. Our first Ansatz will be a solution of (15.12) with $\rho_i\{g(\mu),\mu\}$ replaced by zero. This certainly exists because the β coefficients are determined by perturbation expansion and therefore finite. The integration constants must be chosen such that for all $\mu \geqslant m$ we have

$$|g(\mu)| \leqslant \kappa\, g_{max} \tag{15.13}$$

where g_{max} is the previously determined maximally allowed value of $g(\mu)$ and κ is again a constant smaller than 1 to be determined later. In practice this requirement implies asymptotic freedom[3]:

$$\lim_{\mu\to\infty} g(\mu) = o . \tag{15.14}$$

(It is constructive to consider also complex solutions.)

If we now substitute this $g(\mu)$ in the right hand side of eq. (15.12) we may find a correction:

$$g(\mu) \to g(\mu) = g(\mu) + \delta g(\mu) , \tag{15.15}$$

for which we may require

$$|\delta g(\mu)| \leqslant \varepsilon\, g_{max} \quad \text{for all } \mu . \tag{15.16}$$

We must start with:

$$\varepsilon+\kappa < 1 \tag{15.17}$$

Will a recursive application of eq. (15.12) converge to a solution? Let the first Ansatz produce a change (15.16). The next correction is then, up to higher orders in δg, given by

$$\frac{\mu\partial}{\partial\mu}\,\delta\tilde{g}_i(\mu) + M_{ij}(\mu)\delta\tilde{g}_j(\mu) = \delta f_i(\mu) \tag{15.18}$$

(where M_{ij} is determined by differentiation of (15.12) with respect to $g_i(\mu)$.

To estimate $\delta f(\mu)$ we must find a limit for the change in ρ. Our argument that $|\rho| < Q_N$ came from adding the absolute values of all diagrams contributing to ρ, possibly after application of (14.20) several times. Replacing (14.20) then by

$$|\delta g(\mu_1)| \leqslant \delta g(\mu) + 3c\left(\frac{\mu}{\mu_1}\right)^{\varepsilon} g^2(\mu)\delta g(\mu) \tag{15.19}$$

which indeed is true if g satisfies (15.18), or

$$\left|\frac{\mu\partial}{\partial\mu}\,\delta g\right| \leqslant 3\tilde{\beta}\, g^2\,\delta g\ , \tag{15.20}$$

as can be derived from (14.20) and (14.21), we find that we can write

$$|\delta\rho| < \varepsilon C'\ , \tag{15.21}$$

with C' slightly larger than C, and

$$|\delta f(\mu)| \lesssim \varepsilon(N+1)C'\ g(\mu)^N\ . \tag{15.22}$$

Now asymptotically,

$$M_{ij}(\mu) \rightarrow M^0_{ij}/\log\mu\ , \tag{15.23}$$

where M^0_{ij} is determined by one-loop perturbation theory. If there is only one coupling constant it is the number 3/2. In the more general case we now assume it to be diagonalized:

$$M^0_{ij} = M(i)\delta_{ij}\ , \tag{15.24}$$

with one eigenvalue equal to 3/2. (Our arguments can easily be extended to the special situation when M^0_{ij} cannot be diagonalized, in which case the standard triangle form must be used.) The asymptotic form of the solution to (15.18) is

$$\delta g_i(\mu) = (\log\mu)^{-M(i)} \int_{\mu(i)}^{\mu} d\log\mu(\log\mu)^{M(i)}\,\delta f_i(\mu) , \qquad (15.25)$$

where $\mu(i)$ are integration constants. If $M(i) < \frac{N}{2} - 1$ then we choose $\mu(i) = \infty$. If $M(i) > \frac{N}{2} - 1$ we set $\mu(i) = m$. Then in both cases we get

$$|\delta\tilde{g}_i(\mu)| < \frac{\varepsilon(N+1)C''}{\left|\frac{N}{2} - M(i) - 1\right|}\,|g(\mu)|^{N-2} , \qquad (15.26)$$

where C'' is related to C' and the first β coefficient. In a compact set of μ values where the deviation from (15.25) is appreciable we of course also have an inequality of the form (15.16).

If $M(i) = N/2 - 1$ then we simply pick another N value (which needs not be integer here), raising or lowering it by one unit. We see that we only need to consider $N \leqslant 4$. Comparing (15.26) with (15.16), noting that C'' is independent of A, we see that if

$$\frac{(N+1)C''}{\left|\frac{N}{2} - M(i) - 1\right|}\, g_{max}^{N-3} < 1 \qquad (15.27)$$

then our procedure converges. Since C'' stays constant or decreases with decreasing g_{max}, we find that a finite g_{max} will satisfy (15.27).

Also we should check whether $\tilde{g}(\mu)$ satisfies the *Ansatz* (14.19), with unchanged $\tilde{\beta}$. This however is obvious from the construction of $\tilde{g}$ through eq. (15.18).
Notice that the masses are adjusted in every step of the iteration for the Z functions, by choosing A and B in section 14. They are necessary now because we wish to confine the integrals (15.25) at $\mu \geqslant m$, limiting the solutions $g(\mu)$ to satisfy $|g(\mu)| \leqslant g$.

16. BOREL SUMMABILITY[16]

The fact that we obtained eq. (14.4) holding for $m \leqslant \mu < \infty$ is our central result. For simplicity of the following discussions we ignore the *next-to-leading* terms of the coefficients β. Let us take the case that the leading ones have $\ell = 3$. We find that

$$g_i(\mu) = \frac{b_i^2}{\log\mu^2 + C} + \mathcal{O}(g^4(\mu)) \qquad (16.1)$$

(the next-to-leading β coefficients give an unimportant correction in the denominator of the form $\log\log\mu^2$). The coefficients b_i are fixed by the leading renormalization group β coefficients. We must choose C such that

$$|g^2(\mu)| \leqslant g^2_{max} \quad \text{if} \quad \mu \geqslant m \ . \tag{16.2}$$

This is guaranteed if either

$$\mathrm{Re}\ C \gtrsim \frac{\max(b_i^2)}{g^2_{max}} - \log m^2 \equiv R - \log m^2 \tag{16.3a}$$

or

$$|\mathrm{Im}\ C| \gtrsim \frac{\max(b_i^2)}{g^2_{max}} \equiv R \ . \tag{16.3b}$$

Now the requirement of asymptotic freedom is usually so stringent that the constant C is the only free parameter besides the masses and possible dimension 1 coupling constants (a situation corresponding to the necessity of choosing all $\mu(i) = \infty$ in eq. (15.25)). But still we can do ordinary perturbation expansion, writing

$$g_i(m) = b_i g + \mathcal{O}(g^3) \ ; \tag{16.4}$$

$$C = \frac{1}{g^2} - \log m^2 \quad . \tag{16.5}$$

where g is now a regular expansion parameter. The $\mathcal{O}(g^3)$ terms are fixed by the asymptotic freedom requirement and can be computed perturbatively. We now claim that perturbation expansion in g is indeed Borel-summable:

$$G(g^2) = \int_0^\infty F(z) e^{-z/g^2}\, dz \ ; \tag{16.6a}$$

$$F(z) = \frac{1}{2\pi i} \int_C G(g^2) e^{+z/g}\, d(g^{-2}) \ , \tag{16.6b}$$

where the path C must be choosen to lie entirely in the region limited by eqs. (16.3) and (16.5) (see Fig. 17). Since in that region the Green functions $G(g^2)$ are approximately given by their perturbative values the integral (16.6b) will converge rapidly along this path, if

$$\mathrm{Re}\ z > 0 \ , \tag{16.7}$$

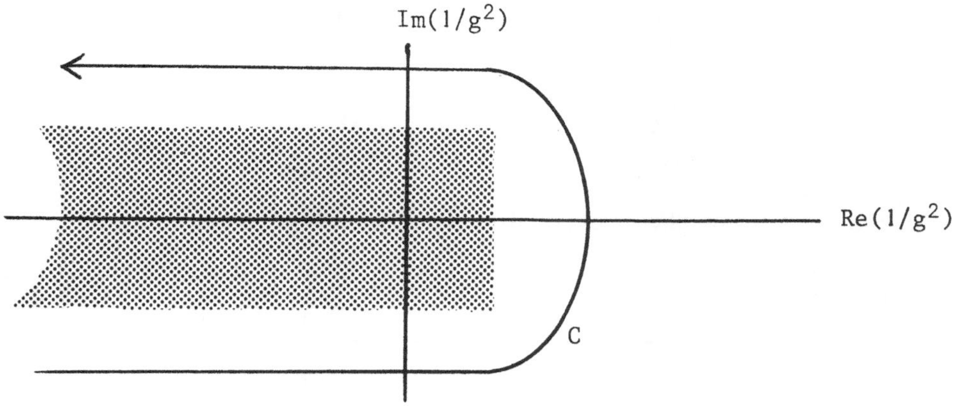

Fig. 17. The integration path C in eq. (16.7).

and F(z) will be bounded by

$$|F(z)| < A \exp(|z|/g^2_{max}) \ , \qquad (16.8)$$

where g_{max} is the allowed limit for g as derived in the previous sections.

However, eqs. (16.7) and (16.8) are not quite sufficient to prove Borel summability because we also want to have analyticity of F(z) for an open region around the origin. That this requirement is met can be seen as follows. Let us solve a variant of equation (14.4) of the following form:

$$\frac{\mu\partial}{\partial\mu} g_i(\mu,\Lambda) = \Big[- \sum_{\ell} \beta^{(\ell)}_{ij_1\ldots j_\ell} \, g_{j_1}(\mu,\Lambda)\ldots g_{j_\ell}(\mu,\Lambda) + |g(\mu,\Lambda)|^N \rho_i(\mu,\{g\}) \Big] \theta(\Lambda-\mu) \ , \qquad (16.9)$$

where $\theta(x)$ is the step function. The coefficients $\beta^{(\ell)}$ and the functional ρ are the same as before (constructed the same way via difference equations of sects. 12 and 13). Clearly we have, if $\mu > \Lambda$: $g(\mu,\Lambda) = g(\Lambda,\Lambda)$. And eq. (16.1) now reads

$$g_i(\mu) = \frac{b_i^2}{\log \mu^2 + C} + \mathcal{O}(g^4(\mu)) \ ; \ m \leqslant \mu \leqslant \Lambda \ . \qquad (16.10)$$

Our point is that this solution exists not only in the region (16.3), but also if

$$\mathrm{Re}\ C \leqslant -\frac{\max(b_i)}{g^2_{max}} - \log \Lambda^2 \equiv -R-\log \Lambda^2 \ . \tag{16.11}$$

We can now close the contour C (see Fig. 18).

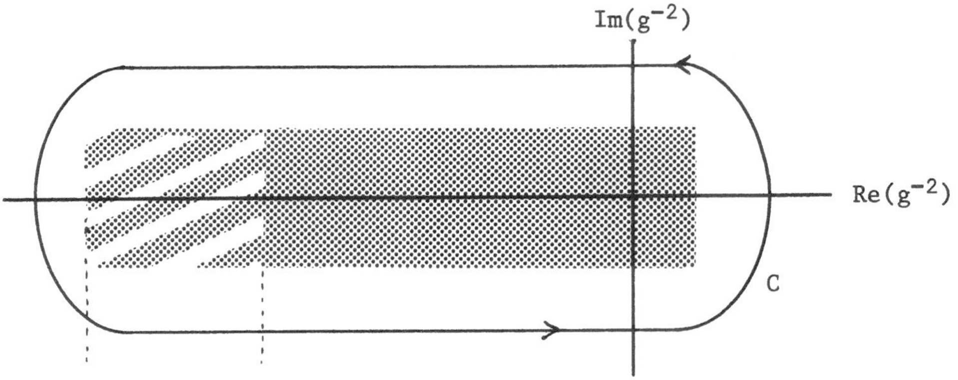

Fig. 18. The contour C of eq. (16.13); the forbidden regions of $G(g,\Lambda_i)$ are shaded.

$$-R-\log\left(\frac{\Lambda_2^2}{m^2}\right) \qquad -R-\log\left(\frac{\Lambda_1^2}{m^2}\right)$$

Now compare two different Λ values: Λ_1 and Λ_2, and compare the Green functions computed with these two Λ values, both as a function of g, taking

$$g_i(m,\Lambda) = b_i g + \mathcal{O}(g^3) \ . \tag{16.12}$$

We take these Green functions at (possibly exceptional) momentum values p, but always such that $\Lambda \gg |p|$. If they are computed directly following our algorithm then a slight Λ-dependence may still exist, coming from two sources: one is the fact that $g_i(\mu,\Lambda)$ depends on Λ because $\rho(\mu,\{g\})$ as a *functional* of g, may depend on $g(\mu')$ with $\mu' > \Lambda$. But clearly, since all integrals involved in the construction of ρ converge we expect this dependence to go like

$$|\delta G| \lesssim \Lambda^{\varepsilon(g)-1} \tag{6.13a}$$

or probably

$$|\delta G| \lesssim \Lambda^{\varepsilon(g)-2} \tag{16.13b}$$

(because linearly convergent equations can often be made quadratically convergent by symmetrization); here $\varepsilon(g) \downarrow o$ if $g \to o$. The second source of Λ-dependence comes from the application of difference equations to compute Green functions away from the symmetry point (sects. 12,13). Again, this error must behave like (16.13), because of convergence of the integrals involved.

The change in the Borel function F(z) can be read off from eq. (16.6b)

$$\left|F(z,\Lambda_2)-F(z,\Lambda_1)\right| \lesssim \frac{1}{2\pi i} \oint_C \Lambda_1^{\varepsilon(g)-2} e^{z/g^2}\, d(g^{-2}) \tag{16.14}$$

if $\Lambda_2 \gtrsim \Lambda_1$.
Under what conditions does this vanish in the limit $\Lambda_1 \to \infty$? In our integral g^{-2} ranges from $-R-\log\left(\frac{\Lambda_2^2}{m^2}\right)$ to R. So if Re $z \leqslant o$ then

$$\left|e^{z/g^2}\right| \lesssim \left(\frac{\Lambda_2^2}{m^2}\, e^R\right)^{-\mathrm{Re}\, z} . \tag{16.15}$$

If $|g^{-2}| \to \infty$ then $\varepsilon(g) \downarrow o$. This we may use on the far left of the curve C. Therefore ineq. (16.14) can be written as

$$|F(z,\Lambda_2) - F(z,\Lambda_1)| \lesssim e^{R}\left(\frac{\Lambda_2^2}{\Lambda_1^2}\right) \Lambda_2^{-2\,\mathrm{Re}\, z-2} . \tag{16.16}$$

By comparing a series of Λ values of the form $\Lambda_n = 2^n$, one easily sees that (16.16) guarantees convergence as soon as

$$\mathrm{Re}\ z > -1 , \tag{16.17}$$

which is a quite large region of analyticity of F(z). Note that $z = -1$ is the location of the first renormalon singularity[4,7]. Eq. (16.17) together with (16.8) implies complete Borel summability of this theory.

17. THE MASSLESS THEORY

In the sections 14-16 the mass m^2 had to be non-zero. This was the only way we could obtain the necessary ineq. (16.3a), allowing us to draw the contour C. What happens if we take the limit $m^2 \downarrow o$, considering not $g(m^2)$ but $g(\mu^2)$ at some fixed μ as expansion parameter? Our method to construct the entire theory then fails, but for some z values F(z) will still exist. The argument is analogous to the one of the previous section where we dealt with an ultraviolet problem. Now our difficulty is in the infrared. Comparing two different small values for m, we have

$$|\delta G| \lesssim (m^2)^{1-\varepsilon(g)} \;; \tag{17.1}$$

$$|F(z,m_2) - F(z,m_1)| \lesssim \frac{1}{2\pi i} \oint_C (m_1^2)^{2-\varepsilon(g)}\, e^{z/g^2}\, d(g^{-2}) \;, \tag{17.2}$$

if $m_2 \lesssim m_1$.
Now if Re z > o then

$$\left|e^{z/g^2}\right| \lesssim \left(\frac{m^2}{\mu^2}\, e^R\right)^{\mathrm{Re}z} \;. \tag{17.3}$$

Thus

$$|F(z,m_2) - F(z,m_1)| \lesssim e^R\left(\frac{m_1^2}{m_2^2}\right) m_2^{2-2\mathrm{Re}z} \;. \tag{17.4}$$

Now we have convergence as $m^2 \downarrow o$ as long as

$$\mathrm{Re}\, z < 1 \;, \tag{17.5}$$

and, indeed, z = 1 is a point where an infrared renormalon singularity is to be expected. Thus, the Borel transform F(z) of the massless theory is analytic in the region

$$-1 < \mathrm{Re}z < 1 \;. \tag{17.6}$$

This result guarantees that perturbation expansion in g^2 diverges not worse than

$$g^{2n}\, n!$$

but is clearly not enough for Borel summability.

18. THE $-\lambda\,\mathrm{Tr}\,\phi^4$ MODEL

A special case is the pure scalar planar field theory, with just one coupling constant λ and a mass m. If m is sufficiently large and λ sufficiently small then our analysis applies, and we find that all planar Green functions are uniquely determined. However, in this special case there is more: the Green functions can be uniquely determined as long as the masses in *all* channels are non-negative, and λ is negative. This was discovered by comparing with the much simpler "spherical model" (appendix B) which shows the same property.

The argument is fairly simple. Let us first take the theory at m^2 values which are so large that everything is well-defined. Now decrease the "bare mass" m_B^2 continuously.
The change in the dressed propagator

$$P(k,m^2) = - G_2^{-1}(k,m^2) \tag{18.1}$$

is determined by

$$\frac{\partial}{\partial m^2} G_2(k,m^2) \equiv G'(k,m^2) \tag{18.2}$$

which we take to start out at sufficiently large m^2.

Now the Feynman rules for

$$G''(k,m^2) \equiv \frac{\partial}{\partial m^2} G'(k,m^2) \ , \tag{18.3}$$

can easily be written down, just like those for

$$\frac{\partial}{\partial m^2} G_4(k_1,\ldots,k_4,m^2) \equiv G_4' \ . \tag{18.4}$$

Since G_2'' and G_4' are superficially convergent we can again express them in terms of a skeleton expansion containing only the functions G_4 and $-G_2'$ and the propagators P. *They are all positive* (remember that G_4 starts out as $-\lambda$, with $\lambda < 0$), *and all integrals and summations converge.* Only G_4' has one surviving minus sign from differentiating one propagator with m . Thus:

$$G_2'' > 0 \ ; \tag{18.5}$$

$$G_4' < 0 \ . \tag{18.6}$$

If we let m^2 decrease then clearly G_2 will stay positive and G_2' negative. Their absolute values grow however, until a point is reached where either the sum of all diagrams will no longer converge, or the two-point function G_2 becomes zero. As soon as this happens the theory will be ill-defined. A tachyonic pole tends to develop, followed by catastrophes in all channels.
The point we wish to make in this section however is that as long as this does not happen, indeed all summations and integrals converge, so that our iterative procedure to produce the Green functions will also converge. For all those values of λ and m^2 this theory will be Borel summable.

This result only holds for the special case considered here, namely $-\lambda$ Tr ϕ^4 theory, because all skeleton diagrams that con-

tribute to some Green function, carry the same sign. They can never interfere destructively.
Notice that what also was needed here was convergence of the diagram expansion. Now we know that at finite N the non-planar graphs give a divergent contribution. Thus the "tachyons" will develop already at infinite m^2: the theory is fundamentally unstable. Of course we knew this already: λ after all has the wrong sign. The instantons that bring about the decay of our "false vacuum" carry on action S proportional to $-N/\tilde{\lambda}$ which is finite for finite N.

19. OUTLOOK

Apart from the model of sect. 18, the models we are able to construct explicitly now lack any appreciable structure, so they are physically not very interesting. Two (extremely difficult) things should clearly be tried to be done: one is the massless planar theories such as SU(∞) QCD. Clearly that theory should show an enormously intricate structure, including several possible phase-transitions. We still believe that more and better understanding of the infrared renormalons that limited analyticity of our borel functions in sect. 17 could help us to go beyond those singular points and may possibly "solve" that model (i.e. yield a demonstrably convergent calculational scheme).
Secondly one would try to use the same or similar skeleton techniques at finite N (non-planar diagrams). Of course now the skeleton expansion does not converge, but, in Borel-summing the skeleton expansion there should be no renormalons, and all divergences may be due entirely to instantonlike structures. More understanding of resummation techniques for these diagrams by saddle point methods could help us out. If such a program could work then that would enable us to write down SU(3) QCD in a finite (but small) box. QCD in the real world could then perhaps be obtained by gluing boxes together, as in lattice gauge theories.

Another thing yet to be done is to repeat our procedure now in Minkowsky space instead of Euclidean space. Singling out the obvious singularities in Minkowsky space may well be not so difficult, so perhaps this is a more reasonable challenge that we can leave for the interested student.

APPENDIX A. ASYMPTOTICALLY FREE INFRARED CONVERGENT PLANAR HIGGS MODEL

In discussing examples of planar field theories for which our analysis is applicable we found that pure SU(∞) gauge theory (with possibly a limited number of fermions) is asymptotically free as required, but unbounded in the infrared - so that even the ultraviolet limit cannot be treated exactly (see sects. 14 and 15).

$-\lambda$ Tr φ^4 is asymptotically free and can be given a mass term so that infrared convergence is also guaranteed. At $N \to \infty$ this is a fine planar theory, but at finite N the vacuum is unstable. The only theory that suffers from none of these defects is an SU(∞) gauge theory in which all bosons get a mass due to the Higgs mechanism. But then a new scalar self-coupling occurs that tends to be not asymptotically free. Asymptotic freedom is only secured if, curiously enough, several kinds of fermionic degrees of freedom are added. The following model is an example (similar examples can also be constructed at finite N, such as SU(2)).

In general a renormalizable model can be written as

$$\mathcal{L} = -\tfrac{1}{4}G^a_{\mu\nu}G^a_{\mu\nu} - \tfrac{1}{2}(D_\mu\phi_i)^2 - V(\phi_i) - \bar{\psi}(\gamma D + W(\phi))\psi \,, \tag{A.1}$$

where G is the covariant curl, ϕ_i is a set of scalar fields and ψ a set of spinors. V is a quartic and W a linear polynomial in ϕ. We write

$$\begin{aligned}
G^a_{\mu\nu} &= \partial_\mu A^a_\nu - \partial_\nu A^a_\mu + g^{abc}A^b_\mu A^c_\nu \\
D_\mu\phi_i &= \partial_\mu\phi_i + T^a_{ij}A^a_\mu\phi_j \\
D_\mu\psi_i &= \partial_\mu\psi_i + U^a_{ij}A^a_\mu\psi_j \\
W &= S + iP\gamma_5 \;;\; \hat{W} = S - iP\gamma_5 \\
U &= U_s + U_p\gamma_5 \;;\; \hat{U} = U_s - U_p\gamma_5 \\
C^{ab}_1 &= g^{apq}g^{bpq} \;;\; C^{ab}_2 = -\mathrm{Tr}\,T^aT^b \\
C^{ab}_3 &= -\mathrm{Tr}\left(U^a_LU^b_L + U^a_RU^b_R\right) = -2\mathrm{Tr}\left(U^a_sU^b_s + U^a_pU^b_p\right) .
\end{aligned} \tag{A.2}$$

The most compact way to write the complete set of one-loop β functions is to express them in terms of the one-loop counter-Lagrangian, $[8\pi^2(4-n)]^{-1}\Delta\mathcal{L}$, where $\Delta\mathcal{L}$ has been found to be[17], after performing the necessary field renormalizations,

$$\begin{aligned}
\Delta\mathcal{L} = {} & G^a_{\mu\nu}G^b_{\mu\nu}\left[\frac{11}{12}C^{ab}_1 - \frac{1}{24}C^{ab}_2 - \frac{1}{6}C^{ab}_3\right] \\
& + \frac{1}{4}V^2_{ij} + \frac{3}{2}V_i(T^2\phi)_i + \frac{3}{4}(\phi T^aT^b\phi)^2 \\
& + \bar{\psi}\left\{\frac{1}{4}W_i\hat{W}_iW + \frac{1}{4}W\hat{W}_iW_i + W_i\hat{W}W_i\right\}\psi \\
& + \frac{3}{2}\bar{\psi}(\hat{U}^2W + WU^2)\psi + \phi_i(V_j + \bar{\psi}W_j\psi)\mathrm{Tr}(S_iS_j + P_iP_j) \\
& - \mathrm{Tr}(S^2+P^2) + \mathrm{Tr}[S,P]^2 \,.
\end{aligned} \tag{A.3}$$

Here V_i stands for $\partial V/\partial\varphi_i$, etc.
The scaling behavior of the coupling constants is then determined by

$$\frac{\mu\partial\mathcal{L}}{\partial\mu} = -\frac{\Delta\mathcal{L}}{8\pi^2}\ . \tag{A.4}$$

We now choose a model with $U(N)_{\text{local}} \times SU(N)_{\text{global}}$ symmetry. Besides the gauge field we have a scalar ϕ_i^s in the $N_{\text{local}} \times N_{\text{global}}$ representation, and two kinds of fermions:

$\psi_{(1)i}{}^j$ in $N_{\text{local}} \times N_{\text{local}}$ and $\psi_{(2)a}{}^i$ in $N_{\text{local}} \times N_{\text{global}}$. We choose

$$V(\phi) = \frac{\lambda}{2}\,\phi^{*s}_i\,\phi^t_i\,\phi^{*t}_j\,\phi^s_j\ , \tag{A.5}$$

and

$$\bar\psi W\psi = h\left[\bar\psi_{(2)}{}^a{}_i\,\phi^{*j}_a\,\psi_{(1)j}{}^i + \text{h.c.}\right]\ . \tag{A.6}$$

Writing

$$C_1^{ab} = C_2^{ab} = \tilde g^2\,\delta^{ab}\ ;\ C_3^{ab} = 3\tilde g^2\,\delta^{ab}\ , \tag{A.7}$$

we find:

$$-8\pi^2\,\frac{\mu\partial\tilde g^2}{\partial\mu} \equiv \Delta\tilde g^2 = \frac{3}{2}\,\tilde g^4\ ; \tag{A.8}$$

and with $\tilde\lambda \equiv N\lambda$; $\tilde h^2 = Nh^2$, by substituting (A.5) and (A.6) in the expression (A.3) after some algebra, and in the limit $N \to \infty$:

$$\Delta\tilde h^2 = -3\tilde h^4 + \frac{9}{2}\,\tilde h^2\tilde g^2\ ; \tag{A.9}$$

$$\Delta\tilde\lambda = -2\tilde\lambda + 3\tilde g^2\tilde\lambda - \frac{3}{4}\,\tilde g^4 - 4\tilde h^2\tilde\lambda + 4\tilde h^4\ . \tag{A.10}$$

These equations (A.8)-(A.10) are ordinary differential equations whose solutions we can study. The signs in all terms are typical for any such models with three coupling constants g, λ and h. Only the relative magnitudes of the various terms differ from one model to another. For asymptotic freedom we need that the second and last terms of (A.10) and the last of (A.9) are sufficiently large. Usually this implies that the fermions must be in a sufficiently large representation of the gauge group, which explains our choice for the fermionic representations. Our model has an asymptotically free solution if all coupling constants stay in a fixed ratio with respect to each other:

$$\tilde\lambda = \hat\lambda\,\tilde g^2\quad ;\quad \tilde h = \hat h\,g\ , \tag{A.11}$$

and then, from (A.8)-(A.10) we see:

$$\hat{h} = 1 \;;$$

$$\hat{\lambda} = \frac{1}{8}\,(\sqrt{129}-5) \;. \tag{A.12}$$

So indeed we have a solution with positive λ.

It now must be shown that in this model all particles can be made massive via the Higgs mechanism. We consider spontaneous breakdown of $SU(N)_{local} \times SU(N)_{global}$ into the diagonal $SU(N)_{global}$ subgroup. Take as a mass term

$$-\mu\, \phi_i^{*s}\, \phi_i^s \;. \tag{A.13}$$

We can write V as

$$V = \frac{\lambda}{2}\left|\phi_i^{*s}\,\phi_i^t - F^2\delta^{st}\right|^2 + \text{const.} \tag{A.14}$$

Clearly this is minimal if

$$\phi_i^s = F\delta_i^s \;, \tag{A.15}$$

or a gauge rotation thereof.
All vector bosons get an equal mass:

$$-D^*\phi\, D\phi \Rightarrow -\, g^2F^2A_\mu^2 \;; \tag{A.16}$$

$$M_A^2 = 2g^2F^2 \;, \tag{A.17}$$

and of course the scalars get a mass:

$$M_H^2 = \lambda F^2 \;. \tag{A.18}$$

Thus the mass ratio is given by

$$\sqrt{\hat{\lambda}/2} = 0.6303\ \ 6778\ldots \;. \tag{A.19}$$

This is a fixed number of this theory, but it will be affected by higher order corrections. The fermions can each be given a mass term:

$$-\, m_1\bar{\psi}_{(1)}\psi_{(1)} - m_2\bar{\psi}_{(2)}\psi_{(2)} \;, \tag{A.20}$$

and the Yukawa force will give a mixing of a definite strength. The model described in this Appendix is probably the simplest completely convergent planar field theory with absolutely stable vacuum. It is unlikely however that it would have a direct physical significance.

APPENDIX B. THE N-VECTOR MODEL IN THE N → ∞ LIMIT (SPHERICAL MODEL). SPONTANEOUS MASS GENERATION

When only N-vector fields are present (rather than N×N tensors) then the N → ∞ limit is easily obtained analytically. This is the quite illustrative spherical model. Let the bare Lagrangian be

$$\mathcal{L} = -\tfrac{1}{2}(\partial_\mu \vec{\phi})^2 - \tfrac{1}{2}m_B^2\vec{\phi}^2 - \frac{\lambda_B}{8}(\vec{\phi}^2)^2 \ . \tag{B.1}$$

The only diagrams that dominate in the N → ∞ limit are the chains of bubbles (Fig. 19).

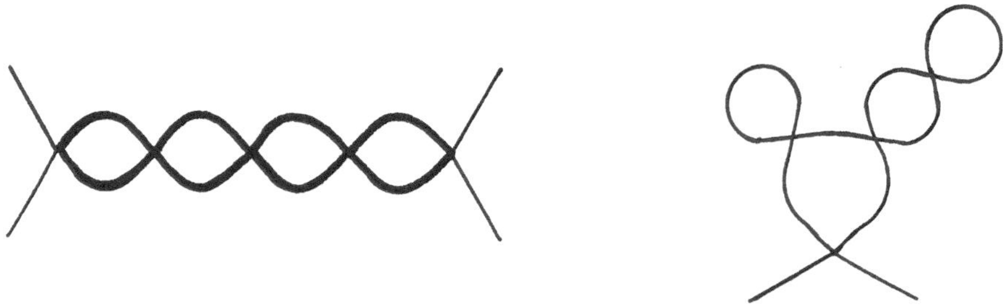

a b

Fig. 19. a) Dominating diagrams for the 4-point function.
b) Mass renormalization.

Let us remove some factors π^2 by defining

$$\tilde{\lambda}_B = N\lambda_B/16\pi^2 \ . \tag{B.2}$$

The diagrams of Fig. 19 are easily summed. Mass and coupling constant need to be renormalized. Dimensional renormalization is appropriate here. In terms of the finite constants $\tilde{\lambda}_R(\mu)$ and $m_R(\mu)$, chosen at some subtraction point μ, and the infinitesimal $\varepsilon = 4-n$, where n is the number of space-time dimensions, one finds:

$$\tilde{\lambda}_B = -\varepsilon\left(1 + \frac{\varepsilon}{\tilde{\lambda}_R(\mu)}\right)\mu^\varepsilon \ , \tag{B.3}$$

and

$$m_B = \mu m_R(\mu)\sqrt{\frac{-\varepsilon}{\tilde{\lambda}_R(\mu)}} \ . \tag{B.4}$$

The sum of all diagrams of type 19a gives an effective propagator of the form

$$F(q) = \frac{32\pi^2/N}{\gamma - \frac{2}{\tilde{\gamma}_R(\mu)} + \log\frac{\pi m^2}{\mu^2} + f(\frac{q^2}{m^2}) - i\varepsilon} , \qquad (B.5)$$

where q is the exchanged momentum, γ is Euler's constant, and

$$f(\frac{q^2}{m^2}) = \int_o^1 dx \log\left|1 + \frac{x(1-x)q^2}{m^2}\right| - \pi i\theta(-q^2-4m^2)\sqrt{1+\frac{4m^2}{q^2}} , \qquad (B.6)$$

and m is the *physical* mass in the propagators of Fig. 19a; that is because these should include the renormalizations of the form of Fig. 19b.

Fig. 20.

Fig. 20 shows how m follows from m_R:

$$2m_R^2\mu^2 = \tilde{\lambda}_R m^2 \log\left(\frac{\pi m^2}{\mu^2} + \gamma - 1 - \frac{2}{\tilde{\lambda}_R}\right) = o . \qquad (B.7)$$

From (B.3) we see that the renormalization-group invariant combination is

$$\frac{1}{\tilde{\lambda}_R(\mu)} + \log\mu , \qquad (B.8)$$

so that inevitably $\tilde{\lambda}_R < o$ at large μ. Indeed, in this model the vacuum would become unstable as soon as N is made finite. In the limit $N \to \infty$ however everything is still fine.

Now in Fig. 21 we plot both m_R^2 and the composite mass M^2, determined by the pole of F(q), as a function of the physical mass m^2. We see that at negative m_R^2 there are two solutions for m^2, but one should be rejected because M^2 would be negative, an indication for an unstable choice of vacuum.
The observation we wish to make in this appendix is that in the allowed region for m_R^2 we get an entirely positive 4-point function in Euclidean space ($\tilde{F}(q) > o$). If we chose m_R^2 to be fixed and vary $\tilde{\lambda}_R$ (or rather vary $\tilde{\lambda}_B$) then at $m_R^2 \geqslant o$ *all* $\tilde{\lambda}$ values are allowed, at negative m_R^2 only sufficiently small values. At $m_R^2 = o$ we see a

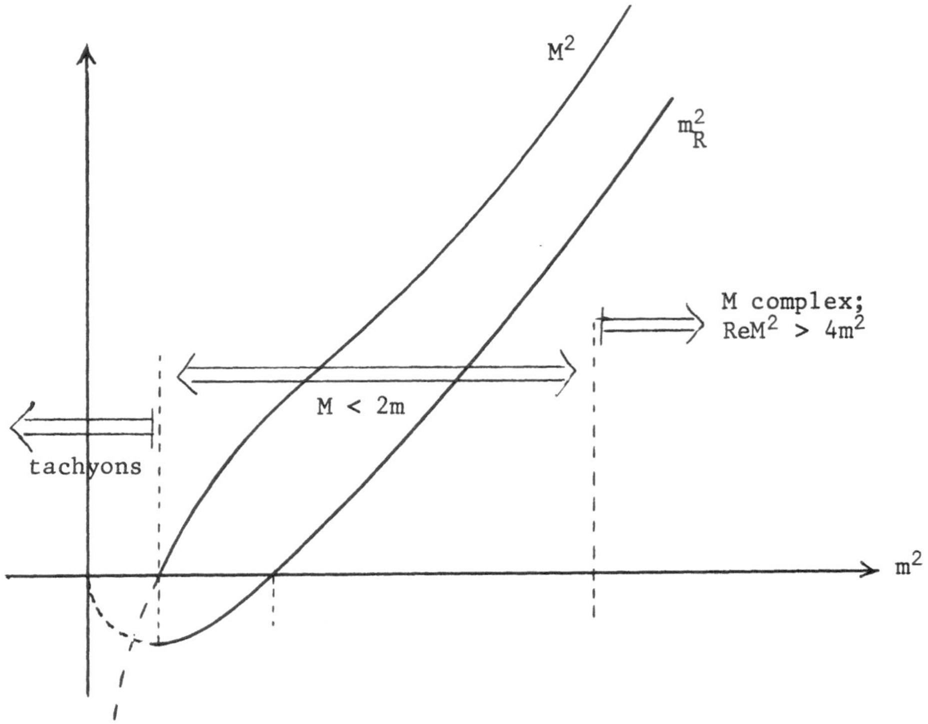

Fig. 21. Mass ratios at given value for $\tilde{\lambda}_R(\mu)$.

"spontaneous" generation of a finite value for m . Perturbation expansion in $\tilde{\lambda}_R$ would show the "infrared renormalon" difficulty. Apparently here the difficulty solves itself via this spontaneous mass generation.

REFERENCES

1. E.C.G. Stueckelberg and A. Peterman, Helv. Phys. Acta 26 (1953), 499; M. Gell-Mann and F. Low, Phys. Rev. 95 (1954) 1300.
2. K.G. Wilson, Phys. Rev. D10, 2445 (1974).
   K.G. Wilson, in "Recent Developments in Gauge Theories", ed. by G. 't Hooft et al., Plenum Press, New York and London, 1980, p. 363.
3. G. 't Hooft, Marseille Conference on Renormalization of Yang-Mills Fields and Applications to Particle Physics, June 1972, unpublished; H.D. Politzer, Phys. Rev. Lett. 30, 1346 (1973); D.J. Gross and F. Wilczek, Phys. Rev. Lett. 30, 1343 (1973).
4. G. 't Hooft, in "The Whys of Subnuclear Physics", Erice 1977, A. Zichichi ed. Plenum Press, New York and London 1979, p. 943.
5. M. Creutz, L. Jacobs and C. Rebbi, Phys. Rev. Lett. 42, 1390 (1979).
6. C. de Calan and V. Rivasseau, Comm. Math. Phys. 82, 69 (1981); 91, 265 (1983).

7) G. Parisi, Phys. Lett. 76B, 65 (1978) and Phys. Rep. 49, 215 (1979).
8) J. Koplik, A. Neveu and S. Nussinov, Nucl. Phys. B123, 109 (1977).
W.T. Tuttle, Can. J. Math. 14, 21 (1962).
9) See ref. 3.
10) J.D. Bjorken and S.D. Drell, Relativistic Quantum Mechanics (McGraw-Hill, New York, 1964).
11) G. 't Hooft, Commun. Math. Phys. 86, 449 (1982).
12) G. 't Hooft, Commun. Math. Phys. 88, 1 (1983).
13) J.C. Ward, Phys. Rev. 78 (1950) 1824.
Y. Takahashi, Nuovo Cimento 6 (1957) 370.
A. Slavnov, Theor. and Math. Phys. 10 (1972) 153 (in Russian). English translation: Theor. and Math. Phys. 10, p. 99.
J.C. Taylor, Nucl. Phys. B33 (1971) 436.
14) K. Symanzik, Comm. Math. Physics 18 (1970) 227; 23 (1971) 49; Lett. Nuovo Cim. 6 (1973) 77; "Small-Distance Behaviour in Field Theory", Springer Tracts in Modern Physics vol. 57 (G. Höhler ed., 1971) p. 221.
15) G. 't Hooft, Nucl. Phys. B33 (1971) 173.
16) G. 't Hooft, Phys. Lett. 119B, 369 (1982).
17) G. 't Hooft, unpublished.
R. van Damme, Phys. Lett. 110B (1982) 239.

FIELDS ON A RANDOM LATTICE

C. Itzykson

Service de Physique Théorique

Cen-Saclay, 91191 Gif sur Yvette, Cedex

I. INTRODUCTION

Elegant as it is, the proposal by Christ Friedberg and Lee[1] of studying (Euclidean) quantum field theory on a random -as opposed to regular- lattice has not yet attracted a spectacular interest. This is most likely due to the fact that numerical simulations or analytical (for instance strong coupling) expansions seem much more difficult than in the regular lattice case. Spurious localization effects are also involved and not easy to master. These unhappy circumstances should not however hide the merits of a very fascinating subject which comes as close as possible to a cutoff but still translational and rotational invariant field theory fulfilling all reasonable criteria. On the other hand it looks like a first step towards promoting geometry to a dynamical coupled system. Indeed in the realm of lattice models the random one, corresponds to the use of arbitrary coordinate systems, even though the underlying geometry is kept euclidean, this being in principle not necessary.

We present a review based on the work of reference [1], as well as earlier contributions such as a classical article of Meijering [2], of the formulation of field theory or equivalently statistical mechanics on the standard Poissonian random lattice. Here and there we report what seem to be new results.

In part II we discuss random geometry and reproduce various

average of local quantities. The domain of correlations seems to be largely unexplored.

In part III we recall the construction of Lagrangians and classical field equations. The natural framework is simplicial cohomology. This yields very natural (abelian) gauge invariant equations for arbitrary antisymetric tensor fields.

Section IV discusses some elementary conjectures on the spectrum of the random Laplacian in conjunction with the question of localization. In the final section V this is illustrated on the simple but instructive one dimensional case [3].

It is also worth mentioning the intimate relationship of this topic with work done in the context of random media. A general presentation, is given in a book by Ziman [4] where one will find numerous relations and references to work in condensed matter physics. See also the articles by Collins [5] and Zallen [6].

During the preparation of this lecture I had the privilege of collaborating with E. Gardner, B. Derrida and J.M. Drouffe. I am happy to take this opportunity to thank them as well as the organizers of the Cargèse Summer Institute.

II. RANDOM GEOMETRY

We describe here the Poissonian random lattice, the Dirichlet-Voronoï cell construction and present the computation of various local averages. The mathematical context of geometric probability is presented in the book of Santalo [7].

Let Ω be a large regular volume inbedded in Euclidean metric d-dimensional space. N points are chosen at random, independently with uniform probability. The finite density limit is always understood, with $||\Omega||$ the measure of the volume Ω

$$N \to \infty \quad ||\Omega|| \to \infty \quad \frac{N}{||\Omega||} = \rho = a^{-d} \tag{1}$$

Hence a plays the role of elementary length scale-proportional to the average spacing between points. For simplicity we set a=1, so that ρ is also one, meaning one point per unit volume.

If the dimension is equal to one, i.e. on the line, we can order the points and denote by $\ell_{i,i+1}$ the length of the interval between neighbors. It is readily checked that these variables are also independent in the infinite volume limit (1), with a distribution equal to

$$p(\ell)\, d\ell = e^{-\ell}\, d\ell \tag{2}$$

This justifies the name Poissonian lattice.

A) Two dimensions

To exemplify the construction of this lattice in higher dimension let us look first at the case d=2. The generalization to higher dimension will then be straightforward. The following construction is attributed to Dirichlet and Voronoï in the mathematical literature. In a solid state physics context one would call it by the name of Wigner and Seitz.

Let $\{M_i\}$ be the N points with coordinates $\vec{x}_i$. To each M_i we assign a closed cell C_i of all those points of the plane which are nearer to M_i than to any other M_j, $j \neq i$. Clearly C_i is a convex polygon (the intersection of half planes) and for $j \neq i$ the intersection $C_i \cap C_j$ is either empty or a segment of the bisecting line between M_i and M_j. In this last case we say that (M_i, M_j) form a pair of neighbors. Let q_i be the number of neighbors of M_i, or equivalently, the number of sides of the cell C_i (local coordination number). Similarly for (ijk) distinct, if $C_i \cap C_j \cap C_j$ is not empty, it is a vertex of each of the cells C_i, C_j, C_k and we call the triangle an elementary 2-simplex (triangle here) as $(M_i M_j)$ was an elementary 1-simplex. The plane region Ω is now paved by elementary 2-simplices as it is paved by elementary cells C_i. Obviously the two sets of objects are (geometrically) dual.

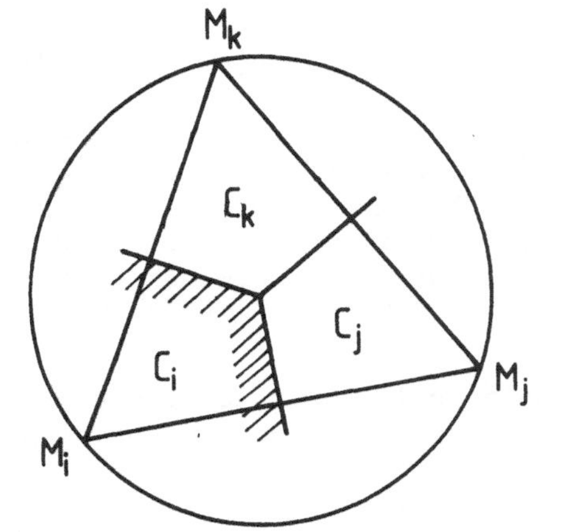

Fig.1 : The Dirichlet Voronoï of 1- and 2- simplices in the 2 dimensional case.

Let

$N_o \equiv N$ = number of points = number of cells

N_1 = number of links (between neighbors) = number of cell edges

N_2 = number of 2-simplices = number of cell vertices.

Among these numbers we have the Euler relation

$$N_o - N_1 + N_2 = \chi \qquad (3)$$

where the Euler characteristic (2 for the sphere, 0 for the torns ...) is not extensive, i.e. in the infinite volume limit (1) $\chi/N \to 0$. Cutting the plane along the edges of the cells we get N_o pieces (the cells) with $2N_1 = \sum_i q_i$ edges and $3N_2$ vertices. Since each cell has as many vertices as edges (Euler's relation in dimension 1) we find

$$2N_1 = 3N_2 = \sum_i q_i \qquad N_o = \sum_i 1 \tag{4}$$

Combining (3) and (4) in the limit $N \to \infty$ we get, with

$$q = \lim_{N\to\infty} \frac{\sum_i q_i}{\sum_i 1} \tag{5}$$

the average coordination

$$q = 6 \qquad \frac{N_1}{N_o} = 3 \qquad \frac{N_2}{N_o} = 2 \tag{6}$$

Thus, on average, each cell has six sides and there are twice as many triangles than cells. Equivalently if $n_{i,j}$ $i > j$ is the average number of i-simplices incident on the j simplex we have

$$n_{1,o} = n_{2,o} = 6 \tag{7}$$

A remarkable property of the two-dimensional case is the existence of a regular lattice (namely the triangular one) where (6) is true locally instead as an average. The random lattice may thus be described as a deformation of a regular triangular lattice with defects (points at which $q_i \neq 6$). Those points where $q_i - 6$ is different from zero may be thought as carrying a "charge" of "curvature", while globally the system is neutral.

Up to now we have only used topological aspects of the construction. Let us turn to metrical properties. We use the notations

ℓ_{ij} = distance between the neighbors (M_iM_j) length of 1-simplex (ij) $\qquad \langle \ell_{ij} \rangle = \ell_{(1)}$

ℓ_{ijk} = area of 2-simplex (ijk) $\qquad \langle \ell_{ijk} \rangle = \ell_{(2)}$

Similarly

σ_i = area of cell C_i $\qquad \langle \sigma_i \rangle = \sigma_{(2)}$

σ_{ij} = length of edge perpendicular to the link (ij) $\qquad \langle \sigma_{ij} \rangle = \sigma_{(1)}$

Since the density is unity, we have

$$\sigma_{(2)} = 1 \tag{8}$$

and since there are twice as many simplices as there are cells on the average

$$\ell_{(2)} = \frac{1}{2} \tag{9}$$

To compute $\ell_{(1)}$ and $\sigma_{(1)}$ we need more information. One makes the following simple but critical observation. Let (ijk) be a 2-simplex and O the dual vertex common to the cells C_i, C_j, C_k. The circle Γ or radius R circumscribed to the triangle (ijk) has O as its center. Then there are no lattice points inside Γ. Indeed the distance $|OM_\ell|$ for $\ell \neq \{i,jk\}$ is greater or equal to $|OM_i|$, $|OM_j|$, $|OM_k|$ by definition. Therefore such points M_ℓ lie outside or on the circle Γ. This last possibility occurs with vanishing probability.

Given a point M_o of the lattice, the probability that M_1 and M_2 lying within d^2x_1 and d^2x_2 respectively altogether form a 2-simplex is therefore

$$dp = n_{2,o}^{-1} \times C_{N-1}^{2} \times \frac{d^2x_1 d^2x_2}{\Omega^2} \times \frac{(\Omega-\pi R^2)^{N-3}}{\Omega^{N-3}}$$

$n_{2,o}^{-1}$	C_{N-1}^{2}	$\frac{d^2x_1 d^2x_2}{\Omega^2}$	$\frac{(\Omega-\pi R^2)^{N-3}}{\Omega^{N-3}}$
≠ of 2-simplices/ point	choice of 2 out of N-1 points	uniform probabilities for M_1 and M_2	N-3 other points required to lie outside Γ

in the infinite limit volume this reads

$$dp = \frac{e^{-\pi R^2}}{12} d^2x_1 d^2x_2 \tag{10}$$

Instead of the four parameters $(\vec{x}_1, \vec{x}_2)$ we can use the radius R of the circumscribed circle and the polar angles φ_o, φ_1, φ_2 of OM_o, OM_1, OM_2. Then

$$d^2x_1 d^2x_2 = R^3 |\sin(\varphi_1-\varphi_2) + \sin(\varphi_2-\varphi_o) + \sin(\varphi_o-\varphi_1)| \, dR \, d\varphi_o d\varphi_1 d\varphi_2$$

with a Jacobian, equal to 2R × area of the 2-simplex, vanishing as the triangle degenerates, as it should, and obviously invariant under permutations. We can check that $\int dp = 1$ as follows

$$\int dp = \frac{1}{12} \int_o^\infty dR\, R^3\, e^{-\pi R^2} \times 2\pi \times 2 \int_o^{2\pi} d\varphi_2 \int_o^{\varphi_2} d\varphi_1 \left\{ \sin(\varphi_2-\varphi_1) - \sin\varphi_2 + \sin\varphi_1 \right\}$$

$$= 2\pi^2 \int_o^\infty dR\, R^3\, e^{-\pi R^2} = 1$$

Similarly the average separation between two neighbors $\ell_{(1)}$ is the mean of say

$$|M_oM_1| = 2R\ |\sin \frac{(\varphi_1-\varphi_o)}{2}| \text{ over } dp, \text{ i.e.}$$

$$\ell_{(1)} = \frac{1}{12}\int_o^\infty dR\ R^3\ e^{-\pi R^2} \times 2\pi\times 2 \int_o^{2\pi} d\varphi_2 \int_o^{\varphi_2} d\varphi_1\ 2R \sin \frac{1}{2}\varphi_1$$

$$\left\{\sin(\varphi_2-\varphi_1) - \sin\varphi_2 + \sin\varphi_1\right\}$$

$$= \frac{32}{9\pi} = 1.1317684... \tag{11}$$

It is worth to point out that, as remarked by the authors of Ref.[1], this is justified in dimension 2 since each link belongs to a constant number of triangles, namely two, so that

$$\ell_{(1)} = \frac{1}{N_1}\sum_{\text{links}} \ell_{ij} = \frac{1}{2N_1}\sum_{\text{triangles}}\ \sum_{\ell_{ij} \text{ belongs to a given triangle}} \ell_{ij} =$$

$$\frac{3N_2}{2N_1} \times \text{average of a triangle side}$$

and $3N_2/2N_1 = 1$. In three dimensions, the number of tetrahedra sharing a link is not constant through the lattice so the average length of a tetrahedron side is not really the average of a link of the lattice. But again the procedure is correct for the average area of a triangular face. For lack of a better procedure the quantities referred to as $\ell_{(i)}$ in dimension d are in fact computed as the averages of the corresponding elements of a typical d-simplex.

One easily verifies that the average area of the triangle is 1/2 as indicated in (9) and computes various other quantities. For instance the relative variance of the area of the simplex is

$$\frac{\delta\ell_{(2)}}{\ell_2} = \left[\frac{<\ell_{ijk}^2>}{<\ell_{ijk}>^2} - 1\right]^{1/2} = \left(\frac{35}{2\pi^2} - 1\right)^{1/2} = 0.8793 \tag{12}$$

The average perimeter of the cells has been found by Meijering using a similar very nice argument. Let M be a lattice point and consider for any other lattice point M_k the bisecting line Δ_k of the segment MM_k. The number of such lines at a distance ρ, up to $d\rho$, is the number of points M_k at distance 2ρ up to $d2\rho$, i.e.

$$N\ \frac{8\pi\rho d\rho}{\Omega} = 8\pi\rho d\rho$$

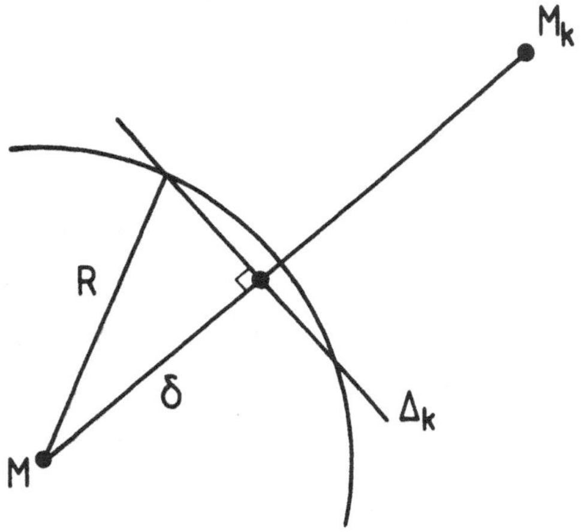

Fig. 2 : Geometric construction involved in Meijering's argument.

The fraction of length of such a line Δ_k at a distance between R and R + dR from M is

$$\frac{2RdR}{\sqrt{R^2-\rho^2}}$$

This means that the total amount of would be cell boundary length between R and R + dR is obtained by integrating in ρ from zero to R the product of these factors

$$2RdR \int_0^R \frac{8\pi\rho d\rho}{\sqrt{R^2-\rho^2}} = 16\pi\ R^2 dR$$

An element of this length will belong to the boundary of the cell around M if the circle of radius R centered at the length element and passing through M does not contain any interior lattice point. This occurs with probability $e^{-\pi R^2}$. Thus the perimeter of the cell boundary at a distance between R and R + dR is given by

$$dP = 16\ \pi\ R^2\ dR\ e^{-\pi R^2} \tag{13}$$

Integrating over R gives the mean perimeter

$$P = 16\pi \int_0^\infty R^2\ e^{-\pi R^2}\ dR = 4 \tag{14}$$

Since on average a cell has six sides, we find

$$\sigma_{(1)} = \frac{2}{3}$$

Table I summarizes the results and compares them with the regular triangular lattice (numbers in brackets) of equal density

Table 1 : Averages in two dimensions

	Direct lattice			dual lattice	
	number	average number incident on a point	Mean length or area	Mean length or area	
 point	$N_o = N$	$n_{o,o} = 1$		$\sigma_{(2)} = 1$ [1]	
 1-simplex	$N_1 = 3N$	$n_{1,o} = 6$	$\ell_{(1)} = \frac{32}{9\pi}\left[\frac{2^{1/2}}{3^{1/4}}\right]$	$\sigma_{(1)} = \frac{2}{3}\left[\frac{2^{1/2}}{3^{3/4}}\right]$	
 2-simplex	$N_2 = 2N$	$n_{2,o} = 6$	$\ell_{(2)} = \frac{1}{2}$		

$\ell_{(1)}$ random $= \frac{32}{9\pi} = 1.1318$ $\quad\sigma_{(1)}$ random $= \frac{2}{3} = 0.6667$

$\ell_{(1)}$ regular $= \frac{2^{1/2}}{3^{1/4}} = 1.0746$ $\quad\sigma_{(1)}$ regular $= \frac{2^{1/2}}{3^{3/4}} = 0.6204$

It may be amusing to report here some findings obtained in collaboration with J.M. Drouffe [8]. Following earlier investigators [9], but using different methods, we have computed several quantities pertaining to the distribution of cells of n sides (recall $\bar{n}=q=6$) in particular the probability p_n of the occurrence of such a cell up to very large n (of order 50 where $p_n \sim 10^{-75}$!!) using an analytical (and untractable) integral representation for p_n. Some examples are given in table II. A striking feature is that n=6 only occurs in approximately 30% of the cases with 4, 5, 7 and 8 sided faces still in appreciable amounts

Table II : Probability distribution for n-sided cells

n	p_n in %	n	p_n in %
3	1.13 ± 0.01	9	2.88 ± 0.07
4	10.70 ± 0.2	10	0.69 ± 0.02
5	25.7 ± 0.77	20	(1.5 ± 0.8) 10^{-11}
6	29.4 ± 0.3	35	(3.6 ± 2.7) 10^{-38}
7	19.8 ± 0.3	50	(1.5 ± 1.5) 10^{-73}
8	9.3 ± 0.8		

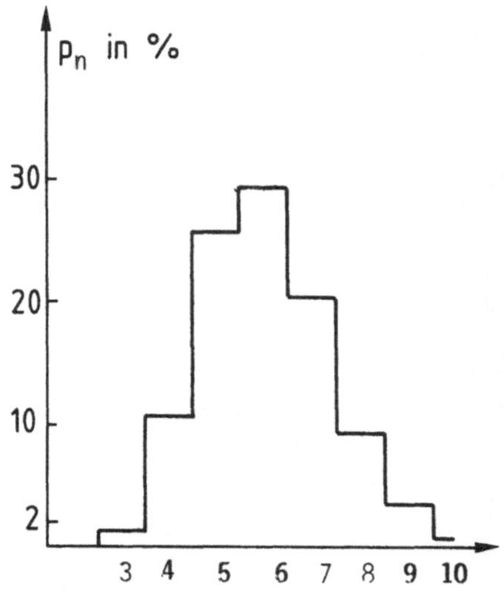

Fig. 3.

We were also able to show that there exists two constants A, B such that $\frac{A}{n^{2n}} \leq p_n \leq \frac{B}{n^n}$, while the data seem to favor a n^{-2n} behavior corresponding to the lower bound.

B) Three dimensions

The Voronoi-Dirichlet construction goes through with $N=N_o$, the number of points, N_1 of links, N_2 of triangles, N_3 of tetrahedra. Euler's relation is

$$\chi = N_o - N_1 + N_2 - N_3 \tag{15}$$

and of course $\chi/N \to 0$. Each triangle is shared by two tetrahedra and each tetrahedron has four triangles, hence

$$2N_2 = 4N_3 \tag{16}$$

If we define as above $n_{k,o}$ as the number of k-simplices incident on a point, then

$$n_{k,o} = (k+1)\,\frac{N_k}{N} \tag{17}$$

It follows therefore from topology alone that

$$n_{1,o} = 2 + \frac{1}{2}\,n_{3,o} \tag{18}$$

$$n_{2,o} = \frac{3}{2}\,n_{3,o} \tag{19}$$

leaving as unknown $n_{3,o}$.

As above we choose the units such that the density is one. The number of tetrahedra N_3 is given by

$$N_3 = \frac{N!}{(N-4)!4!||\Omega||^4} \int \prod_{o}^{3} d^3x_i \, e^{-\frac{4\pi R^3}{3}} \simeq \frac{1}{4!} ||\Omega|| \int \prod_{1}^{3} d^3x_i \, e^{-\frac{4\pi R^3}{3}} \tag{20}$$

where R is the radius of the circumscribed sphere.

Now $n_{3,o} = 4 \frac{N_3}{N}$ and $\frac{N}{||\Omega||} = 1$ yielding

$$n_{3,o} = \frac{1}{3!} \int \prod_{1}^{3} d^3x_i \, e^{-\frac{4\pi R^3}{3}} \tag{21}$$

It follows that

$$dp = \frac{1}{3!n_{3,o}} \prod_{1}^{3} d^3x_i \, e^{-\frac{4\pi R^3}{3}} \tag{22}$$

is the probability to find a tetrahedron with one vertex M_o at the origin and the three other ones M_i, i = 1,2,3, at $\vec{x}_i$ within d^3x_i. If O is the center of the tetrahedron, $\hat{n}_o$ $\hat{n}_1$ $\hat{n}_2$ $\hat{n}_3$ unit vectors along OM_o, OM_1, OM_2, OM_3 and R^3w the volume of the tetrahedron.

$$w = \frac{1}{3!} \left| \det \begin{bmatrix} 1 & 1 & 1 & 1 \\ \hat{n}_o & \hat{n}_1 & \hat{n}_2 & \hat{n}_3 \end{bmatrix} \right| \tag{23}$$

then one shows that

$$\frac{1}{3!} d^3x_1 d^3x_2 d^3x_3 = R^8 dR \, w \, d^2\hat{n}_o d^2\hat{n}_1 d^2\hat{n}_2 d^2\hat{n}_3 \tag{24}$$

a formula which generalises in d-dimensions to a d simplex with a vertex at the origin as

$$\frac{1}{d!} d^dx_1 \dots d^dx_d = R^{d^2-1} dR \, w \prod_{o}^{d} d^{d-1}\hat{n}_i \tag{25}$$

where $d^{d-1}\hat{n}$ stands for the rotational invariant measure on the unit sphere in d dimension $d^{d-1}\hat{n} = 2\delta(n^2-1)\, d^dn$.

Returning to dimension 3 with <w> denoting the rotational average of w, i.e.

$$\langle w\rangle = \int \prod_{o}^{3} \frac{d^2\hat{n}_i}{4\pi} w \tag{26}$$

we find

$$n_{3,o} = 72\pi \langle w\rangle \tag{27}$$

It remains to compute <w>. We shall present below a general formula which restricted to d=3 yields

$$n_{3,o} = \frac{96}{35}\pi^2 = 27.0709... \tag{28}$$

From (19) we then deduce $n_{1,o}$ and $n_{2,o}$. These numbers are irrational. They do not correspond to any regular lattice as they did in two dimensions. The relative number of tetrahedra is

$$\frac{N_3}{N} = \frac{n_{3,o}}{4} = \frac{24}{35}\pi^2 \tag{29}$$

It then follows that

$$n_{1,o} = 2 + \frac{48}{35}\pi^2 = 15.53546... \qquad \frac{N_1}{N} = \left(1 + \frac{24}{35}\pi^2\right) \tag{30}$$

and

$$n_{2,o} = \frac{144}{35}\pi^2 = 40.6064... \qquad \frac{N_2}{N} = \frac{48}{35}\pi^2 \tag{31}$$

The average volume of a tetrahedron in natural units is the reciprocal of N_3/N i.e.

$$\ell_{(3)} = \frac{35\pi^2}{24} = 0.1478... \tag{32}$$

From (22) one also computes the average area of a 2-simplex, a triangle

$$\ell_{(2)} = \left(\frac{3}{4\pi}\right)^{2/3} \frac{875}{243}\frac{1}{\pi}\Gamma\left(\frac{2}{3}\right) = 0.5973... \tag{33}$$

and the mean length of a one simplex, i.e. the average distance between neighbors

$$\ell_{(1)} = \left(\frac{3}{4\pi}\right)^{1/3} \frac{1,715}{2,304} \Gamma\left(\frac{1}{3}\right) = 1.2371 \tag{34}$$

with the limitations explained above.

For the dual lattice we have of course

$$\sigma_{(3)} = 1 \tag{35}$$

The average area is obtained by following the same steps as in two dimensions with the result that

$$\text{average cell area} = \left(\frac{4\pi}{3}\right)^{1/3} \frac{8}{3} \Gamma\left(\frac{2}{3}\right) = 5.821 \tag{36}$$

. The average area of a face $\sigma_{(2)}$ is obtained by dividing by the connectivity $n_{1,o}$, hence

$$\sigma_{(2)} = \frac{\left(\frac{4\pi}{3}\right)^{1/3} \frac{4}{3} \Gamma(2/3)}{1 + \frac{24}{35}\pi^2} = 0.3747\ldots \tag{37}$$

For three dimension one gets the following table

Table III : cell data in 3 dimensions

Simplices	Number N_i	Number incident on a point $n_{i,o}$	average length/area/volume	dual lattice average volume/ area/length	mean number percell	
	$N_o=N$	$n_{oo}=1$		1	1	cell
	$N_1=\left(1+\frac{24}{35}\pi^2\right)N$	$n_{1,o}=2+\frac{48}{35}\pi^2$	$\ell_{(1)}=\left(\frac{3}{4\pi}\right)^{1/3}\frac{1,715}{2304}\Gamma\left(\frac{1}{3}\right)$	$\sigma_{(2)}=\frac{\left(\frac{4\pi}{3}\right)^{1/3}\frac{4}{3}\Gamma\left(\frac{2}{3}\right)}{1+\frac{24}{35}\pi^2}$	n_{10}	face
	$N_2=\frac{48}{35}\pi^2 N$	$n_{2,o}=\frac{144}{35}\pi^2$	$\ell_{(2)}=\left(\frac{3}{4\pi}\right)^{2/3}\frac{875}{243}\frac{1}{\pi}\Gamma\left(\frac{2}{3}\right)$	$\sigma_{(1)}=$	n_{20}	edge
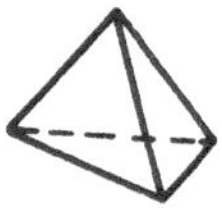	$N_3=\frac{24}{35}\pi^2 N$	$n_{3,o}=\frac{96\pi^2}{35}$	$\ell_{(3)}=\frac{35\pi^2}{24}$		$n_{3,o}$	vertices

C) Four dimensions

Similar calculations in four dimensions yield Table IV

Table IV : cell data in 4 dimensions

	Number	Number incident on a point	Average length/area/volume/hyperfine
0-simplex	$N_o = N$	$n_{o,o} = 1$	
1-simplex	$N_1 = \frac{170}{9} N$	$n_{1,o} = \frac{340}{9}$	$\ell_{(1)} = \left(\frac{2}{\pi^2}\right)^{1/4} \frac{3^3}{7\times 11} \frac{16!!}{15!!} \frac{1}{\pi} \Gamma\left(\frac{1}{4}\right) = 1.3225...$
2-simplex	$N_2 = \frac{590}{9} N$	$n_{2,o} = \frac{590}{3}$	$\ell_{(2)} = 2^{1/2} \frac{2^7}{11^2 \times 13} \frac{17!!}{16!!} \Gamma\left(\frac{1}{2}\right) = 0.6809...$
3-simplex	$N_3 = \frac{715}{9} N$	$n_{3,o} = \frac{2860}{9}$	$\ell_{(3)} = \left(\frac{2}{\pi^2}\right)^{3/4} \frac{2^7 7}{3\times 5\times 11\times 13^2} \frac{18!!}{17!!} \Gamma\left(\frac{3}{4}\right) = 0.1923...$
4-simplex	$N_4 = \frac{286}{9} N$	$n_{4,o} = \frac{1430}{9}$	$\ell_{(4)} = \frac{9}{286} = 0.0315...$

Note that the average coordination grows very fast beeing already $\frac{340}{9} = 37.778$ in dimension four as opposed to $4\times 2 = 8$ on a regular hypercubic lattice. The average volume of the boundary of a cell is according to Meijering

$$\text{volume of cell boundary} = (2\pi^2)^{1/4} \frac{4}{5} \frac{1}{\pi} \Gamma\left(\frac{3}{4}\right) \tag{38}$$

so that the average volume of an hyperface is

$$\sigma_{(3)} = \frac{9}{425} (2\pi^2)^{1/4} \frac{1}{\pi} \Gamma\left(\frac{3}{4}\right) \tag{39}$$

D) Some results for arbitrary d

Let us first quote some topological constraints derived in [1]. Since each d-1 simplex belongs to 2 d-simplices, and each d simplex has d+1 d-1-simplices, we have the relation

$$2N_{d-1} = (d+1) N_d \quad \text{or} \quad n_{d-1,o} = \frac{d}{2} n_{d,o} \tag{40}$$

(see for example (19)).

In general, Euler's relation applied to the surface of a Voronoi cell homeomorphic to a shpere gives in d-dimensions

$$+ n_{1,o} - n_{2,o} + \ldots - (-1)^{d-1} n_{d,o} = +1 - (-1)^d = \begin{array}{ll} d=2 & 0 \\ d=3 & 2 \\ d=4 & 0,\ldots \end{array} \tag{41}$$

More generally one has from similar reasoning

$$\sum_{k=m+1}^{d} (-1)^{k-m-1} n_{k,m} = 1 - (-1)^{d-m} \tag{42}$$

or equivalently since

$$n_{k,m} = \frac{N_k}{N_m} C_{k+1}^{m+1} \qquad d \geq k \geq m \tag{43}$$

$$\sum_{k=m}^{d} C_{k+1}^{m+1} (-1)^{d-k} N_k = N_m \tag{44}$$

The case m=d is a triviality, m=d-1 and m=d-2 give equivalent information, etc... Hence (42) or (44) leave $\frac{d}{2}-1$ (d even) or $\frac{d-1}{2}$ (d-odd) unknown. This means no unknown in two dimension, 1 unknown in three and four dimensions, etc..., in agreement with the previous sections.

Next we turn to the computation of the average number of d-simplices incident on a point $n_{d,o}$, given by

$$n_{d,o} = \frac{1}{d!} \int d^d x_1 \ldots d^d x_d \; e^{-v} \tag{45}$$

v = (hyper) volume of circumscribed sphere to simplex $(M_o, M_1, \ldots, M_d)$ with coordinates $(\vec{x}_o=0, \vec{x}_1, \ldots, \vec{x}_d)$, of radius R, i.e. $(R^d/d)^1 S_d$, where S_d is the "area" of the unit sphere in d-dimensional space

$$S_d = \frac{2\pi^{d/2}}{\Gamma(d/2)} \tag{46}$$

Taking the origin at the center of the sphere with $\hat{n}_o, \ldots, \hat{n}_d$ n+1 unit vectors along $OM_o, OM_1, \ldots, OM_d$

$$n_{d,o} = \int_o^\infty R^{d^2-1} \, dR \; e^{-\frac{R^d}{d} S_d} \; S_d^{d+1} \; \langle w \rangle$$

$$= S_d \, d^{-2} \, d! \, \langle w \rangle \tag{47}$$

$\langle w \rangle$ = average of the volume of simplex when the radius is unity

$$= \int_0^d \prod \frac{d^{d-1} n_k}{S_d} \frac{1}{d!} \left| \det_{d+1} \begin{bmatrix} 1 & 1 & & 1 \\ \hat{n}_o & \hat{n}_2 & \cdots & \hat{n}_d \end{bmatrix} \right|$$

$$= \int_0^d \prod \frac{d^{d-1} n_k}{S_d} \frac{1}{d!} \det_d (1+\hat{n}_\alpha . \hat{n}_\beta)^{1/2} \tag{48}$$

It is easy to see that were the simplex regular, i.e.

$$\hat{n}_\alpha . \hat{n}_\beta = \begin{Bmatrix} \alpha=\beta & 1 \\ \alpha\neq\beta & -\frac{1}{d} \end{Bmatrix} \text{ then,}$$

$$w_{reg} = \frac{1}{d!} \frac{(1+d)^{\frac{1+d}{2}}}{d^{d/2}} \tag{49}$$

Using the method of reference [1] we write $w = \frac{1}{d} h \Delta$ where Δ is the "volume" of d-1 simplex out of d points $(\hat{n}_1 \ldots \hat{n}_d)$ and h is the height. The picture is difficult to draw in d dimensions but the idea is clear. We want to use as variables, instead of $\hat{n}_1 \ldots \hat{n}_d$, the direction of the unit normal $\hat{u}$ to the hyperplane containing the d points where $\hat{n}_1 \ldots \hat{n}_d$ pierce the unit sphere. This normal is defined up to a sign. If we integrate it over the unit sphere then

$$\int f(\hat{n}_1 \ldots \hat{n}_d) \prod_1^d d^{d-1} n_k = \int \frac{d\hat{u}}{2} (d-1)! \; \Delta \prod_1^d d^{d-1} n_k$$

$$\delta(\hat{u}.[\hat{n}_2 - \hat{n}_1]) \ldots \delta(\hat{u}.[\hat{n}_d - \hat{n}_1]\; f(\hat{n}_1 \ldots \hat{n}_d)$$

For an arbitrary vector u we can write each $\hat{n}_k (1 \leq k \leq d)$ as $\hat{n}_k = \cos\theta_k \, \hat{u} + \sin\theta_k \, \hat{m}_k$ defining then d unit vectors $\hat{m}_k$ on the unit sphere in a d-1 dimensional space orthogonal to $\hat{u}$. Then $d^{d-1} \hat{n}_k = \sin\theta_k^{d-3} \, d\cos\theta_k \, d^{d-2} \hat{m}_k$. The arguments of the δ functions are $\cos\theta_k - \cos\theta_1 \quad k \geq 2$, and enforce the fact that $\hat{u}$ is orthogonal to the above hyperplane. One now integrates $f = \frac{h\Delta}{d}$ and

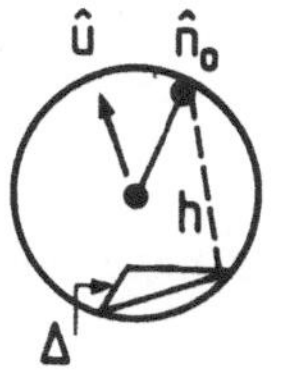

Fig. 4: Recursive computation of <w>

notes that $\Delta = \sin\theta^{d-1}\, w_{d-1}$. Thus

$$\langle w_d \rangle = \frac{1}{d}\left(\frac{S_{d-1}}{S_d}\right)^d \frac{(d-1)!}{2}\, S_d\, \langle w_{d-1}^2 \rangle \int_{\theta=0}^{\theta=\pi} d\cos\theta\, \sin\theta^{d(d-3)+2(d-1)}$$
$$\times \frac{\int_0^\pi d\varphi\, \sin\varphi^{d-2} |\cos\varphi - \cos\theta|}{\int_0^\pi d\varphi\, \sin\varphi^{d-2}}$$

The last ratio of integrals comes from the average of the height h. The integrals are readily performed and yield

$$\langle w_d \rangle = \frac{(d-2)!}{d^2}\, d+1\, \frac{(S_{d-1})^{d+1}}{S_d^d} \frac{\Gamma\left(\frac{1}{2}\right)\Gamma\left(\frac{d^2-1}{2}\right)}{\Gamma\left(\frac{d^2}{2}\right)} \langle w_{d-1}^2 \rangle \qquad (50)$$

It remains to obtain $\langle w_{d-1}^2 \rangle$ as a spherical average over d unit vectors $\hat{m}_1 \ldots \hat{m}_k$ in d-1 space

$$\langle w_{d-1}^2 \rangle = \frac{1}{(d-1)!^2} \langle \det(1 + \hat{m}_a . \hat{m}_b) \rangle$$

We expand the d×d determinant as a sum of permutations

$$\eta_d = \langle \det(1+\hat{m}_a . \hat{m}_b) \rangle = \sum_Q (-1)^Q \langle \pi(1+\hat{m}_a . \hat{m}_{Qa}) \rangle$$

Grouping the cycles in a permutation we observe that the average factorizes. Since

$$\langle \hat{m}_a{}^i \hat{m}_b{}^j \rangle = \delta^{ij}\, \delta_{ab} \frac{1}{d-1}$$

we get for a cycle of length ℓ

$$\langle (1+\hat{m}_1 . \hat{m}_2)(1+\hat{m}_2 . \hat{m}_3) \ldots (1+\hat{m}_\ell . \hat{m}_1) \rangle =$$
$$1 + \langle \hat{m}_1 . \hat{m}_2\; \hat{m}_2 . \hat{m}_3 \ldots \hat{m}_\ell . \hat{m}_1 \rangle = 1 + \left(\frac{1}{d-1}\right)^{\ell-1}$$

and

$$\eta_d \equiv \langle \det(1+\hat{m}_a . \hat{m}_b) \rangle = \frac{1}{2} \sum_{\text{permutations}} (-1)^Q \prod_{\substack{\text{cyclic decomposition} \\ \text{of permutation } Q}} \left[1 + \frac{1}{(d-1)}\, \ell_{p-1}\right] \qquad (51)$$

Therefore

$$\langle w_d \rangle = \frac{d+1}{d^2(d-1)^2(d-2)!} \frac{(S_{d-1})^{d+1}}{(S_d)^d} \frac{\Gamma\left(\frac{1}{2}\right)\Gamma\left(\frac{d^2-1}{2}\right)}{\Gamma\left(\frac{d^2}{2}\right)} \eta_d \tag{52}$$

In η_d we regroup terms corresponding to the same cyclic decomposition. Let $\alpha_p \geq 0$ be the number of cycles of length p, $\sum_p p\alpha_p = d$, the number of permutations having this cyclic decomposition is

$$\frac{d!}{1^{\alpha_1}\alpha_1!2^{\alpha_2}\alpha_2!\ldots d^{\alpha_d}\alpha_d!}$$

and the signature of the permutation is $(-1)^{\sum p\alpha p(p-1)}$. Therefore

$$\frac{1}{d!}\eta_d = \sum_{\{\alpha_p\}} \frac{1}{1^{\alpha_1}\alpha_1!\ldots d^{\alpha_d}\alpha_d!} \prod_p (-1)^{\alpha_p(p-1)}\left[1+\left(\frac{1}{d-1}\right)^{p-1}\right]^{\alpha_p}$$

$$= \text{coefficient of } t^d \text{ in } \exp \sum_{p=1}^{\infty} (-1)^{p-1}\left[1+\left(\frac{1}{d-1}\right)^{p-1}\right]\frac{t^p}{p}$$

$$= \text{coefficient of } t^d \text{ in } (1+t)\left(1+\frac{t}{d-1}\right)^{d-1}$$

$$= \left(\frac{1}{d-1}\right)^{d-1}$$

Inserting this remarkably simple result in (52) yields

$$\langle w_d \rangle = \frac{d+1}{d} \frac{S_{d-1}^{d+1}}{((d-1)S_d)^d} \frac{\Gamma\left(\frac{1}{2}\right)\Gamma\left(\frac{d^2-1}{2}\right)}{\Gamma\left(\frac{d^2}{2}\right)} \tag{53}$$

and finally

$$n_{d,o} = \frac{2}{d} \frac{(d-1)!}{\Gamma\left(\frac{d+1}{2}\right)^2} \left[\frac{\Gamma\left(\frac{1}{2}\right)\Gamma\left(\frac{d}{2}+1\right)}{\Gamma\left(\frac{d+1}{2}\right)}\right]^{d-1} \frac{\Gamma\left(\frac{1}{2}\right)\Gamma\left(\frac{d^2+1}{2}\right)}{\Gamma\left(\frac{d^2}{2}\right)} \tag{54}$$

Asymptotically $n_{d,o}$ grows like a

$$n_{d,o} \sim \frac{2}{d^{1/2}} e^{1/4} (2\pi d)^{\frac{d-1}{2}} \tag{55}$$

Formula (54) can be compared with the results quoted for d=2,3,4. From it we derive the average d-simplex (hyper-volume)

$$\ell_{(d)} = \frac{d+1}{n_{d,o}} \tag{56}$$

Similarly we obtain the average total "area" of a Voronoi cell, call it P,

$$P = 2^{d+1} \frac{1}{d} \Gamma\left(2 - \frac{1}{d}\right) \frac{1}{\Gamma\left(d - \frac{1}{2}\right)} \left(\Gamma\left(\frac{d}{2} + 1\right)\right)^{2 - \frac{1}{d}} \tag{57}$$

and the average "area" of a d-1 simplex

$$\ell_{(d-1)} = \frac{d^2}{\pi^{1/2}} \frac{\Gamma\left(d + \frac{d-1}{d}\right)}{(\Gamma(d))^2} \left(\frac{\Gamma\left(\frac{d}{2} + 1\right)}{\pi^{d/2}}\right)^{\frac{d-1}{d}} \frac{\Gamma\left(\frac{d+1}{2}\right)^{d}}{\Gamma\left(\frac{d}{2} + 1\right)^{d-1}} \left(\frac{\Gamma\left(\frac{d^2}{2}\right)}{\Gamma\left(\frac{d^2+1}{2}\right)}\right)^2 \frac{\Gamma\left(\frac{d^2+d}{2}\right)}{\Gamma\left(\frac{d^2+d-1}{2}\right)} \tag{58}$$

I believe that formulas (54) and (57) and (58) are new. Of course much remains to be done, not only in general dimension, but also with respect to correlations [10].

III. ACTIONS AND LAPLACIANS

The next step is to formulate classical and quantum field theory on these random lattices, and to analyse functions defined on various geometric objects. Two lattices have in fact been defined. In the direct lattice L we have 0, 1, 2... simplices. The zero simplices are points i, we set $\ell_i \equiv 1$. The one-simplices are segments joining two "neighboring" points (ij) i.e. segments that cross a face of a cell, and $\ell_{ij} = \ell_{ji}$ is there length and so on... The dual lattice $\tilde{L}$ is made of d, d-1,... cells we call i the d-cell to which the point i belongs and σ_i its volume, (ij) is a d-1 cell common to two neighboring cells i and j and $\sigma_{ij} = \sigma_{ji}$ its "area" etc...

Define L_o as the set of o-forms, i.e. functions defined on sites $i \to \varphi_i$. Similarly L_1 is the set of antisymmetric 1 forms $\varphi_{ij} \equiv \varphi_{ji}$ defined on 1-simplices; L_2 antisymmetric 2 forms $\varphi_{ijk} = -\varphi_{ijk} = \ldots$ defined on 2 simplices and so on. Furthermore let us call d-densities the functions ψ_i defined on the (positively) oriented d-cells and the set of such densities $\tilde{L}_d$. The d-1 densities ψ_{ij} defined on oriented d-1 cells build L_{d-1} and so on. The orientation on the cell (ij...) and on the simplex (ij...) are compatible, i.e. their product is unity.

The various φ_i, φ_{ij},φ_{ijk} ... may be thought of as restrictions on the random lattice of scalar, vector, antisymmetric tensor... fields defined in the continuum, the correspondence being

$$\begin{aligned}
\varphi_i &\leftrightarrow \varphi(x_i) \\
\varphi_{ij} &\leftrightarrow \frac{1}{\ell_{ij}} \int_i^j dx^\mu \; \varphi_\mu(x) \\
\varphi_{ijk} &\leftrightarrow \frac{1}{ijk} \iint_{\substack{\text{oriented}\\ \text{triangle ijk}}} dx^\mu \wedge dx^\nu \; \varphi_{\mu\nu}(x)
\end{aligned} \qquad (1)$$

Between L_p and $\tilde{L}_{d-p}$ there exists a natural scalar product, in other terms $\tilde{L}_{d-p}$ is the dual of the vector space L_p. We call this scalar product $\langle\varphi|\psi\rangle$ with

$$\begin{aligned}
p=0 \qquad & \langle\varphi|\psi\rangle = \frac{1}{1!} \sum_i \varphi_i \psi_i \\
p=1 \qquad & \langle\varphi|\psi\rangle = \frac{1}{2!} \sum_{\substack{ij \\ \{ij\}\text{ neighbors}}} \varphi_{ij} \psi_{ij} \\
p=2 \qquad & \langle\varphi|\psi\rangle = \frac{1}{3!} \sum_{\{ijk\}\text{ neighbors}} \varphi_{ijk} \psi_{ijk} \\
& \ldots
\end{aligned} \qquad (2)$$

Moreover we can define a one to one correspondence between L_p and $\tilde{L}_{d-p}$ which we call duality in such a way that each one of them becomes a (real) Hilbert space as follows

$$\begin{aligned}
&\varphi \leftrightarrow \psi \\
&L_p \leftrightarrow \tilde{L}_{d-p} \\
&\ell_{ijk\ldots} \varphi_{ijk\ldots} = \frac{\psi_{ijk\ldots}}{\sigma_{ijk\ldots}} \quad \text{(no summation)}
\end{aligned} \qquad (3)$$

Note that for any p this entails in terms of dimensions

$$[\psi] = [\varphi]\,[\text{length}]^d$$

the volume $[\text{length}]^d$ is the volume of a cell for p=0. For p=1 it is d times the sum of the volumes of two pyramides of apex i and j with base σ_{ij} and so on. To denote the correspondence (3) we set $\psi=\tilde{\varphi}$ or $\varphi=\tilde{\psi}$.

We define next on L_p a potential (or mass term) as

$$V(\varphi) = \frac{1}{2} \langle\varphi|\tilde{\varphi}\rangle \tag{4}$$

i.e.

$$\begin{aligned}
&p=0 \qquad V(\varphi) = \frac{1}{2} \sum_i \sigma_i \varphi_i^2 \\
&p=1 \qquad V(\varphi) = \frac{1}{2} \sum_{(ij)} \ell_{ij} \sigma_{ij}\, \varphi_{ij}^2 \\
&p=2 \qquad V(\varphi) = \frac{1}{2} \sum_{(ijk)} \ell_{ijk} \sigma_{ijk}\, \varphi_{ijk}^2
\end{aligned} \tag{5}$$

To proceed we need kinetic terms. This will be obtained when we know the analogs of gradient and divergence. The gradient (or exterior derivative) is an operator d from L_p to L_{p+1} (giving o on L_d) such that

$$L_p \overset{d}{\rightarrow} L_{p+1}$$

$$\begin{aligned}
(d\varphi)_{ij} &= \frac{\varphi_i - \varphi_j}{\ell_{ij}} \\
(d\varphi)_{ijk} &= \frac{\ell_{ij}\varphi_{ij} + \ell_{jk}\varphi_{kj} + \ell_{ki}\varphi_{ki}}{\ell_{ijk}} \\
(d\varphi)_{i_o \ldots i_q} &= \sum_{s=0}^{q} \frac{(-1)^s\, \ell_{i_o \ldots \hat{i}_s \ldots i_q}\, \varphi_{i_o \ldots \hat{i}_s \ldots i_q}}{\ell_{i_o \ldots i_q}}
\end{aligned} \tag{6}$$

Obviously

$$d^2 = 0 \tag{7}$$

Using the scalar product (2) we can transpose d into d^T which operates as

$$\tilde{L}_p \xrightarrow{d^T} \tilde{L}_{p+1} \tag{8}$$

and satisfies

$$<\varphi|d^T\psi> = <d\varphi|\psi> \tag{9}$$

Explicitly

$$\begin{aligned} &\tilde{L}_{d-1} \rightarrow \tilde{L}_d && (d^T\psi)_i = \sum_{j(i)} \frac{\psi_{ij}}{\ell_{ij}} \\ &\tilde{L}_{d-2} \rightarrow \tilde{L}_{d-1} && (d^T\psi)_{ij} = \sum_{k(ij)} \frac{\ell_{ij}\psi_{ijk}}{\ell_{ijk}} \\ &\qquad\dots \end{aligned} \tag{10}$$

Clearly

$$d^{T^2} = 0 \tag{11}$$

The divergence operator d^* is then obtained by shifting back d^T using the duality map (3) according to the diagram

$$\begin{array}{ccc} L_p & \xrightarrow{\text{duality}} & \tilde{L}_{d-p} \\ d^* \Downarrow & & \downarrow d^T \\ L_{p-1} & \xleftarrow[\text{duality}]{} & \tilde{L}_{d-p+1} \end{array} \tag{12}$$

For instance on 1-forms

$$L_1 \xrightarrow{d^*} L_o$$

We start with φ_{ij}, get from duality $\tilde{\varphi}_{ij} = \ell_{ij}\sigma_{ij}\varphi_{ij}$, apply d^T, to obtain

$$(d^T\psi)_i = \sum_{j(i)} \frac{\tilde{\varphi}_{ij}}{\ell_{ij}} = \sum_{j(i)} \sigma_{ij}\varphi_{ij}.$$

Returning by duality to L_o yields (with $\ell_i \equiv 1$)

$$(d^*\varphi)_i = \frac{1}{\sigma_i} \sum_{j(i)} \sigma_{ij}\varphi_{ij} \tag{13}$$

which is seen to be the discrete analog of the divergence. Again

$$d^{*^2} = 0 \tag{14}$$

and we note that d^* annihilates L_o.

Obviously we can also define d^{T*} with square equal to zero. We are now in a position to define a kinetic term as

$$\begin{aligned} K(\varphi) &= V(d\varphi) = \frac{1}{2} \langle d\varphi | \tilde{d}\varphi \rangle \\ &= \frac{1}{2} \sum_{i_o,\ldots,i_p} \frac{\sigma_{i_o \ldots i_p}}{\ell_{i_o \ldots i_p}} \left(\sum_{s=0}^{p} (-1)^s \ell_{i_o \ldots \hat{i}_s \ldots i_p} \varphi_{i_o \ldots \hat{i}_s \ldots i_p} \right)^2 \end{aligned} \tag{15}$$

For instance for scalars or vectors

$$\begin{aligned} &\varphi \in L_o \qquad K(\varphi) = \frac{1}{2} \sum_{(ij)} \frac{\sigma_{ij}}{\ell_{ij}} (\varphi_i - \varphi_j)^2 \\ &\varphi \in L_1 \qquad K(\varphi) = \frac{1}{2} \sum_{(ijk)} \frac{\sigma_{ijk}}{\ell_{ijk}} (\ell_{ij}\varphi_{ij} + \ell_{jk}\varphi_{jk} + \ell_{ki}\varphi_{ki})^2 \end{aligned} \tag{16}$$

We note that the dimension of K is

$$[K] = [\varphi]^2 \, [\text{length}]^{d-2} \tag{17}$$

as in the continuum theory.

To obtain (free massless) field equations we minimize the kinetic term K considered as an action

$$\delta K = \langle . | \delta\tilde{\varphi} \rangle = 0$$

Let ψ denote $\tilde{\varphi}$ and ψ_1 denote $\widetilde{d\varphi} = d^{T^*} \psi$ according to the diagram

$$\begin{array}{ccc} \varphi & \xleftarrow{\text{duality}} & \psi \\ d \downarrow & & \\ d\varphi & \xrightarrow[\text{duality}]{} & \psi_1 = d^{T^*}\psi \end{array}$$

Now in detail

$$\delta K(\varphi) = \frac{1}{2}\,\delta\langle d\varphi|\psi_1\rangle = \frac{1}{2}\langle d\delta\varphi|\psi_1\rangle + \frac{1}{2}\langle d\varphi|d^{T^*}\delta\psi\rangle$$

$$= \frac{1}{2}\langle \delta\varphi|d^T\psi_1\rangle + \frac{1}{2}\langle d^* d\varphi|\delta\psi\rangle$$

But $d^T\psi_1 = d^T d^{T*}\psi$ so both terms are equal. Hence

$$K(\varphi) = \langle d^* d\varphi|\delta\psi\rangle$$

and the field equations(*) are

$$d^* d\varphi = 0 \tag{18}$$

The dimension of the operator d^*d is $[\text{length}]^{-2}$ and it transforms L_p into L_p.

A) Scalar case

To see the content of (18) we consider first the case of scalars. The operator d^*d can then be identified with the Laplacian Δ, up to a sign, $(d^*d + dd^*) = (d+d^*)^2 = -\Delta$, and

$$(-\Delta\varphi) = (d^* d\varphi)_i = \frac{1}{\sigma_i}\sum_{j(i)} \sigma_{ij}\cdot\frac{\varphi_i-\varphi_j}{\ell_{ij}} \tag{19}$$

Observe that the right hand side can be rewritten $(2d)\sum_{j(i)} \frac{\sigma_{ij}\ell_{ij}}{2d\sigma_i}\cdot\frac{(\varphi_i-\varphi_j)}{\ell_{ij}{}^2}$. For fixed i, $p_{j,i} = \frac{\sigma_{ij}\ell_{ij}}{2d\sigma_i}$ when summed over j adds up to one and can identified with hopping probabilities $p_{j,i}$ from site i to site j

$$\left[-\frac{1}{2d}\,\Delta\varphi\right]_i = \sum_{j(i)} p_{j,i}\,\frac{(\varphi_i-\varphi_j)}{\ell_{ij}{}^2} \tag{20}$$

Because ℓ_{ij} is not bounded from below, as on a regular lattice, we may expect that the spectrum of $-\Delta$ is not bounded. As emphasized by the authors of Ref.[1] harmonic functions i.e. solutions of

------

(*) Addition of a mass term would replace this equation by $(d^*d+m^2)\,\varphi=0$. This would however spoil the "gauge" invariance of (18), i.e. its invariance under the replacement $\varphi \to \varphi + d\varphi$, or $\varphi \to \varphi + \text{cst}$ for p=0.

$\Delta\varphi=0$ are of the form

$$\varphi_i = a + \vec{k}.\vec{x}_i \tag{21}$$

where $\vec{k}$ is a constant vector as in the continuum. Indeed since $\ell_{ij} = |\vec{x}_i-\vec{x}_j|$

$$\sum_{j(i)} \sigma_{ij} \frac{\vec{k}.(\vec{x}_i-\vec{x}_j)}{\ell_{ij}} = \vec{k}. \sum_{j(i)} \sigma_{ij} \frac{\vec{x}_i-\vec{x}_j}{|\vec{x}_i-\vec{x}_j|} = -\int_{\substack{\text{surface}\\ \text{of cell}}} d\sigma\, \vec{k}.\vec{n} = 0 \tag{22}$$

from Gauss's theorem. If one studies a diffusion equation of the form (D a diffusion constant)

$$\left(\frac{1}{D}\frac{\partial}{\partial t}\varphi + d^* d\varphi\right) = 0 \tag{23}$$

then the integral over φ i.e. $Q(\varphi) = \langle\varphi|1\rangle = \sum_i \sigma_i\varphi_i$ is a conserved quantity, since

$$\frac{1}{D}\dot{Q} = -\sum_i \sum_{j(i)} \frac{\sigma_{ij}}{\ell_{ij}} (\varphi_i-\varphi_j) = 0$$

The solution of the diffusion equation (23) can be obtained as a limit of a sum over paths

$$\begin{aligned} \varphi_j(t) &= \sum_i (j|e^{-Dtd^*d}|i) \\ &= \lim_{n\to\infty} \left(j\left|\left(1-\frac{Dt}{n}d^*d\right)^n\right|i\right) \\ &= \lim_{n\to\infty} \sum_{\substack{\text{path of}\\ n \text{ steps}}} C_{path} \end{aligned} \tag{24}$$

A path of n steps is a set $\omega_{ij} = \{i\ k_1k_2\ldots k_{n-1}\}$ with the constraint that $k_{a+1} \equiv k_a$ or k_{a+1} is a neighbor of k_a. Correspondingly the contribution C_{path} is a product of factors, one for each step such that

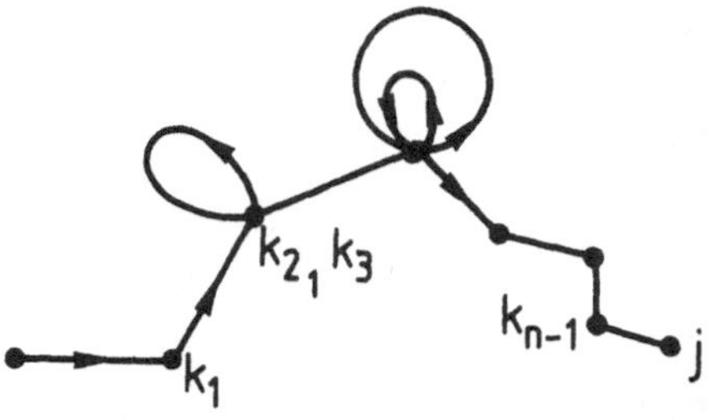

Fig. 5 A path form i to j.

$$\text{if } k_{a+1} = k_a \quad \text{factor} \left(1 - \frac{Dt}{n} \sum_{q(k_a)} \frac{\sigma_{qk_a}}{\sigma_{k_a} \ell_{qk_a}}\right)$$

$$\text{if } k_{a+1} \neq k_a \quad \text{factor} \ \frac{Dt}{n} \frac{\sigma_{k_{a+1},k_a}}{\sigma_{k_a} \ell_{k_{a+1},k_a}} \tag{25}$$

For n going to infinity since all the factors are positive barring a very improbable situation where some $\sum_{j(i)} \frac{\sigma_{ji}}{\sigma_i \ell_{ij}}$ would be arbitrarily large, the kernal $(j|e^{-Dtd^x d}|i)$ is a positive number as in the continuum. Thus if φ is initially positive (or zero) it remains so and if its total "charge" $Q(\varphi)$ is originally normalized it will remain so at any time. It is then possible to interpret it as a probability density in the cell i with $\sigma_i \varphi_i$ the total probability in the same cell i.

The links a b where $\frac{\sigma_{ab}}{\sigma_a \ell_{ab}}$ gets abnormally large as compared to its mean value, play a particular role in the sum over paths. They will increase the tendancy for the paths to leave a and to visit b. We shall have to return to these considerations when we study the average spectrum of the Laplacian.

Vector case

Returning to equations (18) we observe that for 1,2,... forms d^*d is not the Laplacian anymore, very much as Maxwell's equations (the case p=1) do not entail without a special choice of gauge (the Lorentz gauge) that the vector potential is harmonic. When p=1 we can write (18) as

$$(d^* d\varphi)_{ij} = \frac{1}{\sigma_{ij}} \sum_{k(ij)} \sigma_{ijk} \frac{\ell_{ij}\varphi_{ij} + \ell_{jk}\varphi_{jk} + \ell_{ki}\varphi_{ki}}{\ell_{ijk}} \tag{26}$$

To get the meaning of (26) assume for instance that the dimension is three and that φ_{ij} results from an underlying vector field $A_\mu(x)$ defined in the continuum, then

$$\varphi_{ij} \leftrightarrow \frac{1}{\ell_{ij}} \int_i^j dx^\mu \, A_\mu(x)$$

$$\frac{\ell_{ij}\varphi_{ij}+\ell_{jk}\varphi_{jk}+\ell_{ki}\varphi_{ki}}{ijk} \leftrightarrow \frac{1}{\ell_{ijk}} \oint_{\substack{\text{boundary}\\ \text{of } ijk}} dx^\mu \, A_\mu(x)$$

$$= \frac{1}{\ell_{ijk}} \iint_{(ijk)} dx^\mu \wedge dx^\nu (\partial_\mu A_\nu - \partial_\nu A_\mu)$$

In three dimensions we have on the r.h.s. the average flux of the corresponding magnetic field B through the (oriented) triangle ijk i.e. $\vec{B}.\vec{n}$ where $\vec{n}$ is the unit normal. The dual of (ijk) is a cell edge with length σ_{ijk}. The oriented edge is represented by the vector $\sigma_{ijk}\vec{n} = \vec{\sigma}_{ijk}$. In (26) we have then

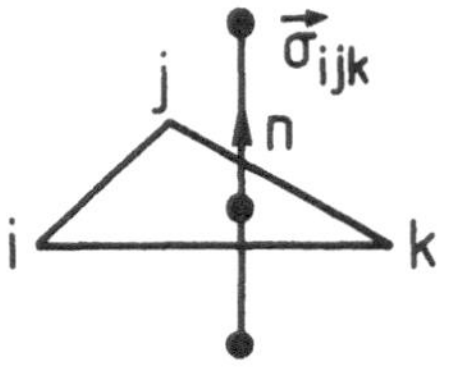

$$(d^*d\varphi)_{ij} \Longleftrightarrow \frac{1}{\sigma_{ij}} \sum_{k(ij)} \vec{B}.\vec{\sigma}_{ijk}$$

on the r.h.s. we find the circulation of B around the face (ij) with area σ_{ij}

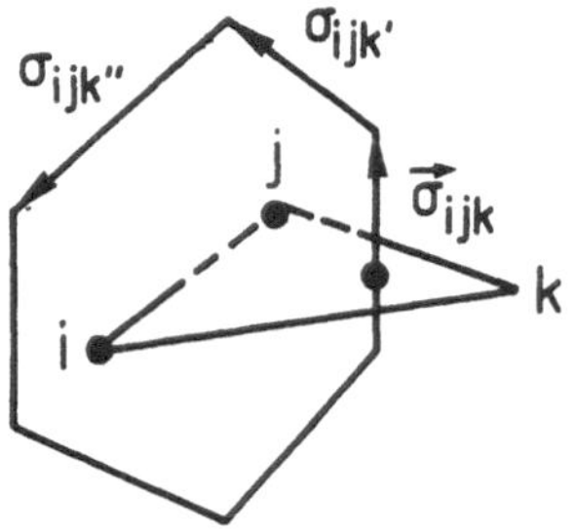

$$\frac{1}{\sigma_{ij}} \sum_{k(ij)} \vec{B}.\sigma_{ijk} = \frac{1}{\sigma_{ij}} \oint_{(ij)} \vec{B}.d\vec{\ell}$$

$$= \frac{\iint \vec{\nabla}\times\vec{B}.\vec{m}d^2\sigma}{\iint d^2\sigma} \sim \vec{\nabla}\times\vec{B}.\vec{m}_{ij}$$

with $\vec{m}_{ij}$ the unit normal to the face (ij).

Fig. 6: Geometry for Maxwell's equation.

In general dimension we find similarly with $F_{\mu\nu}=\partial_\mu A_\nu - \partial_\nu A_\mu$, a discrete version of Maxwell's equations $\partial^\mu F_{\mu\nu} = \Delta A_\nu - \partial_\nu(\partial .A)$.

As emphasized before this is not the Laplacian on vector fields. Indeed from the expression for $\partial^\mu F_{\mu\nu}$ we get $\Delta A_\nu = \partial^\mu(\partial_\mu A_\nu - \partial_\nu A_\mu) + \partial_\nu(\partial . A)$. In general terms this reads div × grad + grad × div or more abstractly

$$\underline{o} = d^*d + dd^* = (d+d^*)^2 \tag{27}$$

As a result of (27) we can formulate a Dirac-Kähler [11] equation on a collection $\Phi \equiv \{\varphi_i, \varphi_{ij}, \varphi_{ijk}, \ldots\}$ of 0,1,...d forms of the type

$$(d+d^*)\Phi = 0 \tag{28}$$

Further generalizations to include interactions or gauge fields can similarly be handled.

IV. SPECTRAL PROPERTIES - ELEMENTARY CONSIDERATIONS

On a random lattice we do not have, strictly speaking, translational or rotational invariance. The latter only hold on average. There are various approaches, none of them entirely satisfactory to replace Fourier analysis in terms of plane waves. If we study for instance a scalar function φ_i we might expand it in terms of eigenfunctions of the Laplace operator labelled by an index α (which replaces momentum). Assume this is done in a large box. The eigenfunctions are normalized according to

$$<\varphi^{(\alpha)}|\tilde{\varphi}^{(\beta)}> \equiv \sum_i \sigma_i \, \varphi_i^{(\alpha)} \, \varphi_i^{(\beta)} = \delta^{\alpha\beta} \tag{1}$$

this being the correct scalar product with respect to which the Laplacian is symmetric (up to boundary terms). Then we would analyze an arbitrary φ as

$$\varphi_i = \sum_\alpha \gamma_\alpha \, \varphi_i^{(\alpha)}$$

$$\gamma_\alpha = <\varphi|\tilde{\varphi}^{(\alpha)}> \equiv \sum_i \sigma_i \, \varphi_i \, \varphi_i^{(\alpha)} \tag{2}$$

In this way the Laplacian is diagonal

$$(-\Delta\varphi)_i = \sum_\alpha \gamma_\alpha \, \omega_\alpha \, \varphi_i^{(\alpha)} \tag{3}$$

and the $\varphi^{(\alpha)}$ replace the plane waves. In (3) ω_α is of course the eigenvalue corresponding to the mode $\varphi(\alpha)$, both of which are still stochastic variables.

This leads to the much harder problem of understanding the most likely distribution of eigenvalues and corresponding structure of eigenfunctions.

On a large lattice of N sites, α runs over N values. As N goes to infinity we shall denote by N $\rho(\omega)\, d\omega$ the number of eigenvalues between ω and $\omega+d\omega$. By definition

$$\int_0^\infty d\omega\ \rho(\omega) = 1 \tag{4}$$

We may expect that the thermodynamic limit ensures that $\rho(\omega)$ is already the most likely distribution. For $\omega \to 0$ we expect that plane waves are almost eigenstates of the Laplacians. This suggests that as $\omega\to 0$ $d\omega\rho(\omega) = \frac{d^d k}{(2\pi)^d}$, $\omega = k^2$, i.e.

$$\rho(\omega) \underset{\omega\to 0}{\to} \frac{\pi^{d/2}}{\Gamma(d/2)} \frac{\omega^{\frac{d}{2}-1}}{(2\pi)^d} = \frac{\omega^{\frac{d}{2}-1}}{(4\pi)^{d/2}\Gamma(d/2)} \tag{5}$$

To lend credence to (5) we can perform the following exercise. Consider

$$\varphi_i(\vec{k}) = e^{i\vec{k}.\vec{x}_i} \tag{6}$$

and the approximate "eigenvalue", still a function of the site L

$$\omega_i(\vec{k}) = \varphi_i(k)^* (-\Delta\varphi(\vec{k}))_i = \frac{1}{\sigma_i} \sum_{j(i)} \frac{\sigma_{ij}}{\ell_{ij}} \left[1 - e^{i\vec{k}.(\vec{x}_j - \vec{x}_i)}\right] \tag{8}$$

From (III.22) $\omega_i(k)$ is of order k^2. Consider then the following average

$$\overline{\omega}(\vec{k}) = \frac{\Sigma_i \sigma_i \omega_i(\vec{k})}{\Sigma_i \sigma_i} \tag{9}$$

which can be interpreted in a complex Hilbert space as the average of $(-\Delta)$ in the state $\varphi(\vec{k})$ i.e. the diagonal elements in the plane wave basis. For $N\to\infty$ we have no prefered direction in the lattice so (9) can be averaged over the directions of $\vec{k}$. When this is done we get

$$\overline{\omega}(k) = \sum_{p=1}^{\infty} (-1)^{p-1} \frac{k^{2p}}{2^{2p}} \frac{1}{p!} \frac{\Gamma(\frac{d}{2})}{\Gamma(\frac{d}{2}+p)} < \int d\sigma\ \ell^{2p-1} > \tag{10}$$

The meaning of the last bracket is

$$<\int d\sigma\ \ell^{2p-1}> = \frac{1}{N}\sum_{i}\sum_{j(i)} \sigma_{ij}\ \ell_{ij}^{2p-1} \tag{11}$$

It can be computed following Meijering's arguments to yield

$$<\int d\sigma\ \ell^{2p-1}> = 2^{2(p-1)+d}\ \frac{S_{d-1}}{\Omega_d^{1+\frac{2(p-1)}{d}}}\ \frac{\Gamma(p+\frac{d-1}{2})\,\Gamma(\frac{d-1}{2})}{\Gamma(p+d-1)} \times \Gamma\left[2+\frac{2(p-1)}{d}\right] \tag{12}$$

This yields

$$\bar{\omega}(k) = \frac{2^d}{d\pi^{1/2}}\,\Gamma\left(1+\frac{d}{2}\right)^2 \sum_{p=1}^{\infty}\left[\Gamma\left(1+\frac{d}{2}\right)^{\frac{d}{2}}\right]^{p-1} \frac{\Gamma\left(p+\frac{d-1}{2}\right)\Gamma\left(2+\frac{2(p-1)}{d}\right)}{\Gamma(p+\frac{d}{2})\,\Gamma(p+d-1)} \times \frac{k^{2p}}{p!} \tag{13}$$

when d=1,2 this reduces to (recall that the unit of length has been set equal to unity)

$$d=1 \qquad \bar{\omega}(k) = \ln(1+k^2) \tag{14}$$

$$d=2 \qquad \bar{\omega}(k) = 2\pi\left[1 - e^{-\frac{k^2}{2\pi}}\, I_o\!\left(\frac{k^2}{2\pi}\right)\right] \tag{15}$$

with $I_o(z)$ the modified Bessel function

$$I_o(z) = \sum_{n=0}^{\infty}\left(\frac{z^2}{4}\right)^n \frac{1}{(n!)^2} \underset{z\to\infty}{\sim} \frac{e^z}{\sqrt{2\pi z}}\left(1+\frac{1}{8z}+\dots\right) \tag{16}$$

The general formula as well as the explicit forms (14) and (15) show that independently of d, as k goes to zero $\bar{\omega}(k) \to k^2$, a fact which can readily be checked directly from the definition (9).

The function $\bar{\omega}(k)$ being the average diagonal term of the Laplacian in the plane wave basis is appropriate for an estimate of the lower end of the spectrum. Its meaning is uncertain as k increases. Surprisingly the dimension d=1 is marginal. Up to d=1, $\bar{\omega}(k)$ increases without bound as $k\to\infty$ while for d>1 it is bounded. For instance for d=2, $\overline{\omega}(k) \to 2\pi(1 - \frac{1}{k} + \ldots)$ as $k\to\infty$ and in general by dropping the oscillatory term in (8)

$$\lim_{k\to\infty} \overline{\omega}(k) \equiv \omega_\infty = \pi d \frac{\Gamma\left[2\left(1 - \frac{1}{d}\right)\right]}{\left[\Gamma\left(1 + \frac{d}{2}\right)\right]^{2/d}} \qquad d > 1 \tag{17}$$

At any rate this shows that the spectrum of $(-\Delta)$ contains at least the interval $[0,\omega_\infty]$ so that in one dimension we have a proof that it covers all the positive axis.

We can also estimate the large ω behavior of $\rho(\omega)$ on the basis of the paths expansion for the diffusion equation (section III). What is suggested is that for short times i.e. ω large, we can think of ultralocalized wave functions corresponding to links where σ_{ij}/ℓ_{ij} was exceptionally large. For those links a bold approximation is that the wave function φ^α is essentially concentrated on the two neighboring points i and j. As a result

$$\begin{aligned} \omega\varphi_i &= \frac{1}{\sigma_i}\frac{\ell_{ij}}{\sigma_{ij}}(\varphi_i - \varphi_j) \\ \omega\varphi_j &= \frac{1}{\sigma_j}\frac{\ell_{ij}}{\sigma_{ij}}(\varphi_j - \varphi_i) \\ \omega &= \frac{\sigma_{ij}}{\ell_{ij}}\left(\frac{1}{\sigma_i} + \frac{1}{\sigma_j}\right) \end{aligned} \tag{18}$$

It is assumed of course that the right hand side is (with very small probability) very large. A natural assumption is that this results from the fact that i and j are exceptionnally close hence ℓ_{ij} very small. If this is the case we may think of the two neighboring cells as a unique cell, call it α cut by the (ij) bissecting plane up to minute corrections. This plane is very close

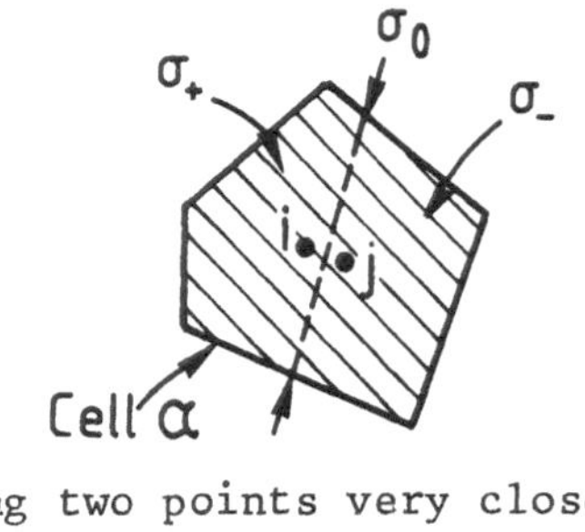

Fig. 7: Bringing two points very close to each other.

to a random plane through the center of the cell α cutting it in two parts of volumes σ_+ and σ_-, having and "area" σ_o in the cell. Then for $\omega\to\infty$ we expect

$$\rho(\omega) \underset{\omega\to\infty}{\sim} \left\langle \left[\sigma_o\left(\frac{1}{\sigma_+} + \frac{1}{\sigma_-}\right)\right]^d \right\rangle \frac{2\pi^{d/2}}{\Gamma(\frac{d}{2})} \frac{1}{\omega^{d+1}} \tag{19}$$

where the average is both on the direction of the plane and the various cells. This fast decrease with ω is of course in agreement with the fact that the integral of $\rho(\omega)$ is one. There is to be contrasted with the continuum case, where there are infinitely many modes per unit volume and therefore the integral of ρ is infinite.

A tail in the spectrum of the type (14) does not occur on a regular lattice and our argument shows that it is related to localization effects. This leads into the much more delicate question of the structure of wave functions. At $\omega\to\infty$ wave functions are localized while as $\omega\to 0$, this is not so. We shall show that in one dimension we have a finite localization length no matter how small ω is. Current wisdom would suggest that d=2 might be a lower critical dimension for this phenomenon (how the wave functions decrease at large distances is yet another question). For d>2 there might exist a range $[0,\omega_c]$ in which wave functions extend to infinity, ω_c being a localization threshold. This is of course a very difficult problem to analyse in the present case. A quotation from reference [4] is perhaps appropriate: "the theory of excitations of topologically disordered systems is particularly intractable".

It is nevertheless to get some understanding of this point in order ro proceed to interacting systems where the interplay of two length scales of localization and of ordering might lead to complex results.

To conclude this section let us add two remarks. From (8) and (17) one can derive an exact sum rule on the spectrum

$$\langle\omega\rangle = \int_0^\infty d\omega\ \omega\rho(\omega) = \omega_\infty = \pi d \frac{\Gamma\left(2 - \frac{2}{d}\right)}{\Gamma\left(1 + \frac{d}{2}\right)^{2/d}} \tag{20}$$

This is of course infinite for d=1, but eventually grows like d for large d due to the suppression of small and large frequencies.

Furthermore one can also derive the following inequality

$$\frac{\langle\omega^2\rangle}{\langle\omega\rangle 2} - 1 \geq \frac{1}{q} \tag{21}$$

where q is the average coordination. This is in fact an equality on any regular lattice. It could perhaps be that (21) becomes an equality as d tends to infinity.

To prove it we use Schwartz's inequality as follows. First

$$<\omega> = \int_0^\infty d\omega\, \omega\rho(\omega) = \overline{\sum_{j(i)} \frac{\sigma_{ij}}{\ell_{ij}}} = \frac{1}{N} \overline{\sum_i \sum_j \frac{\sigma_{ij}}{\ell_{ij}}} = q\, \overline{\frac{\sigma_{ij}}{\ell_{ij}}}$$

$$<\omega^2> = \int_0^\infty d\omega\, \omega^2\rho(\omega) = \overline{\frac{1}{\sigma_i} \sum_{j(i)} \left[\frac{\sigma_{ij}^2}{\ell_{ij}^2} + \left(\sum_{j(i)} \frac{\sigma_{ij}}{\ell_{ij}}\right)^2\right]}$$

Observe then that for two positive random variables $\sigma \equiv \sigma_i$, such that $<\sigma> = 1$, and $x \equiv \sum_{j(i)} \frac{\sigma_{ij}}{\ell_{ij}}$ one has the inequality

$$<x>^2 \leq <\frac{x^2}{\sigma}>$$

and therefore

$$<\omega>^2 \leq <\omega^2> - \overline{\frac{1}{\sigma_i} \sum_{j(i)} \frac{\sigma_{ij}^2}{\ell_{ij}^2}}$$

The fact average is, applying again the same inequality,

$$\overline{\frac{1}{\sigma_i} \sum_{j(i)} \frac{\sigma_{ij}^2}{\ell_{ij}^2}} = q\, \overline{\frac{\sigma_{ij}^2}{\sigma_i \ell_{ij}^2}} \geq q \left(\overline{\frac{\sigma_{ij}}{\ell_{ij}}}\right)^2 = \frac{<\omega>^2}{q}$$

So that we have indeed (21). Of course with harder work one could compute the successive moments in terms of averages of local quantities, with the limitations implied by equation (19).

V. ONE DIMENSIONAL EXAMPLE

With B. Derrida and E. Gardner we have studied [3] the case d=1 in some detail using a simplified version of the Laplacian, namely

$$\omega\varphi_n = - \left\{\frac{\varphi_{n+1} - \varphi_n}{\ell_n} - \frac{\varphi_n - \varphi_{n-1}}{\ell_{n-1}}\right\} \qquad (1)$$

The random points have been ordered and ℓ_n denotes the successive independent Poissonian distributed intervals

$$p(\ell)\, d\ell = e^{-\ell}\, d\ell \tag{2}$$

The cell volume $\frac{1}{2}(\ell_n+\ell_n+1)$ was replaced by its average, namely one. Our results were given in terms of small and large ω expansions for both $\rho(\omega)$, the density of eigenvalues and $L(\omega)$ the localization length. For small ω we found

$$\rho(\omega) = \frac{1}{2\pi\omega^{1/2}}\left(1 - \frac{1}{128}\,\omega + 0(\omega^2)\right) \tag{3}$$

$$L(\omega) = \frac{8}{\omega} - \frac{1}{8} + 0(\omega) \tag{4}$$

in agreement with the discussion of section IV. No matter how small ω is, $L(\omega)$ is finite, but of course diverges as ω approaches zero. Also the leading correction to the behaviour of $\rho(\omega)$ is very small. In the large ω regime, again in agreement with expectations, $\rho(\omega)$ behaves as ω^{-2} and $L(\omega)$ tends to vanish as (C is related to Euler's constant γ through $C = e^{\gamma}$)

$$\rho(\omega) = \frac{2}{\omega^2} - \frac{4}{\omega^3}\left(\ell n\,\frac{\omega}{2C} - \gamma - \frac{3}{2}\right) + \ldots \tag{5}$$

$$\frac{1}{L(\omega)} = \ell n\,\frac{\omega}{C} - \frac{2}{\omega}\left(\ell n\,\frac{\omega}{2C} + 1\right) - \frac{1}{\omega^2}\left[\left(\ell n\,\frac{\omega}{2C} - 1\right)^2 + \eta - 4 - \pi^2\right] + \ldots$$

$$\eta = 0.13070\ldots \tag{6}$$

The appearance of logarithms is characteristic of a Poissonian distribution for the ℓ_n with no cutoff at small values.

These expressions were found to be in good agreement with a numerical simulation as shown on the following figure

One can also study Green's functions at arbitrary separations and observe that since averages of products are different from product of averages the "free field" theory is in effect an interacting one with in particular a short range attraction. In any case the one dimension case illustrates the points made in the general discussion.

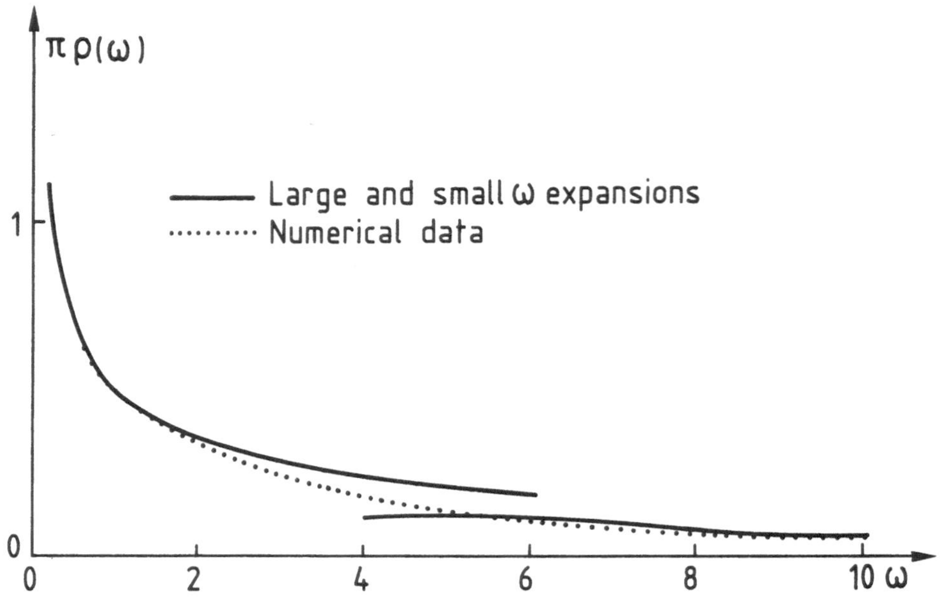

Fig. 7: The density of states $\rho(\omega)$ for the random one dimensional Laplacian showing agreement between numerical simulations and analytical expansion at small and large ω.

REFERENCES

[1] N.H. Christ, R. Friedberg, T.D. Lee, Nucl. Phys. B202, 89 (1982), B210 (FS6), 310, 337 (1982)

T.D. Lee, talk given at the Shelter Island II Conference (1983)

[2] J.L. Meijering, Philips Research Report 8, 270 (1953)

[3] C. Itzykson, in "Non-Perturbative Field Theory and QCD" Proceedings of the Trieste Workshop, R. Lengo, A. Neveu, P. Olesen, G. Parisi editors, World Scientific, Singapore (1983)

E.J. Gardner, C. Itzykson, B. Derrida, submitted to J. Phys. A

[4] J.M. Ziman, Models of disorder, Cambridge University Press, 1979

[5] R. Collins in "Phase Transitions and Critical Phenomena" VoL.2, C. Domb and M.S. Green editors, Academic Press, London (1972)

[6] R. Zallen in "Fluctuation Phenomena", E.W. Montroll, J.L. Lebowitz editors, North Holland (1979)

[7] L.A. Santalo, Integral Geometry and Geometrical probability. Addison-Wesley, Reading, Mass (1976)

[8] J.M. Drouffe, C. Itzykson "Random Geometry and the Statistics of Two Dimensional Cells" Saclay preprint 1983

[9] A. Rahman, J. Chem. Phys. 45, 2585 (1966).

[10] H.G. Hanson, Journal of Statistical Physics 30, 591 (1983) and references therein

[11] P. Becher, H. Joos, Z. Phys. C, Particles and Fields 15, 343 (1982)

DEFECT MEDIATED PHASE TRANSITIONS IN SUPERFLUIDS, SOLIDS, AND THEIR RELATION TO LATTICE GAUGE THEORIES

H. Kleinert
Institut für Theorie der Elementarteilchen
Freie Universität Berlin
Arnimallee 14, 1000 Berlin 33
Germany

ABSTRACT

We compare the defects in different physical systems, exhibit their relevant properties for phase transitions, and point out the similarity of the lattice field theories by which their ensembles can be studied.

In recent years it has become clear that phase transitions of many non-linear systems are caused by the proliferation of topological excitations.[1-4] These are characterized by the failure of certain fields to satisfy integrability conditions and will therefore be called defects.

For example, in superfluid Helium II, there are vortex lines along which the phase of the order parameter has non-commuting derivatives (see Fig. 1).

$$\varepsilon_{kij}\, \partial_i \partial_j\, \gamma(\underset{\sim}{x}) = \alpha_k(\underset{\sim}{x}) \neq 0 \tag{1}$$

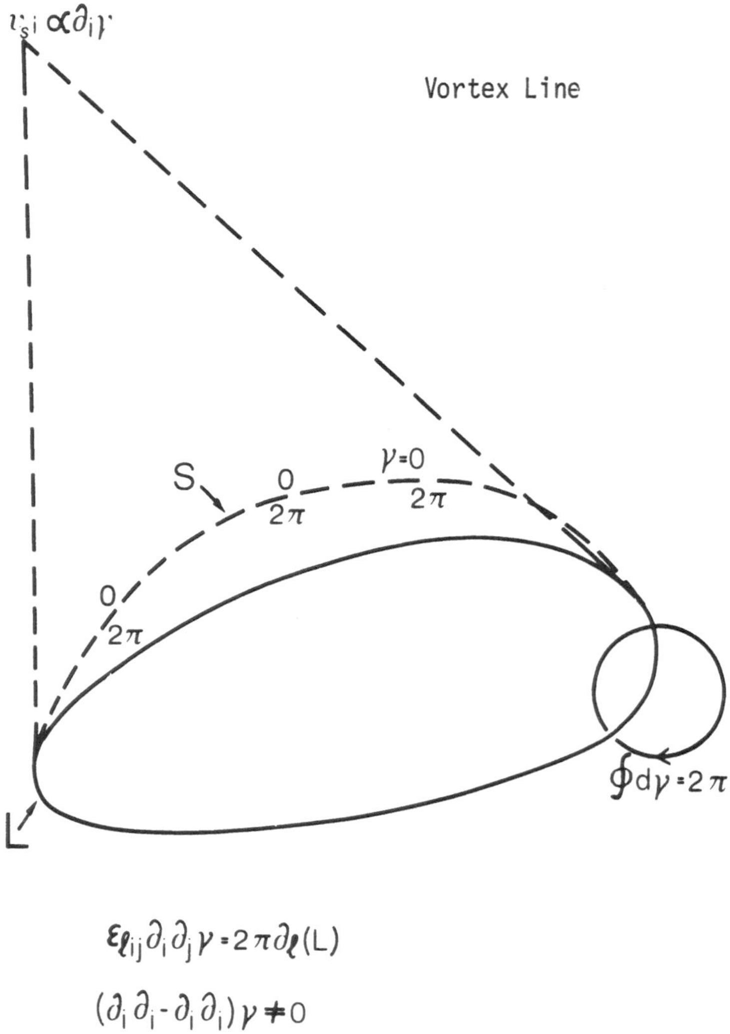

Fig. 1. A vortex line with its non-integrable phase angle which satisfies $\epsilon_{\ell ij}(\partial_i\partial_j - \partial_j\partial_i)\gamma = 2\pi\delta_\ell(L)$

If

$$\delta_k(L) \equiv \int ds \, \frac{dx_k(s)}{ds} \, \delta^{(3)}(\underset{\sim}{x} - \underset{\sim}{x}(s)) \tag{2}$$

denotes the delta function singular along a vortex line L, parametrized by x(s), the vortex density is

$$\alpha_k(\underset{\sim}{x}) = 2\pi n \, \delta_k(L) \tag{3}$$

In a crystal, the displacement vector $u_i(\underset{\sim}{x})$ may be non-integrable due to dislocation lines (see Fig. 2-4)[5]

$$\varepsilon_{kij} \partial_i \partial_j u_\ell(\underset{\sim}{x}) = \alpha_{\ell k}(\underset{\sim}{x}) \neq 0 \tag{4}$$

where $\alpha_{lk}(\underset{\sim}{x})$ is called the dislocation density. A single line along L is characterized by a Burgers vector b_i (which is a multiple of a lattice vector) and $\alpha_{lk}(\underset{\sim}{x})$ reads

$$\alpha_{\ell k}(\underset{\sim}{x}) = b_\ell \, \delta_k(L) \tag{5}$$

Dislocations may pile up to form disclinations (see Fig. 5) in which case the local rotation field $\omega_i \equiv \frac{1}{2}\varepsilon_{ijk}\partial_j u_k$ may become non-integrable with[6]

$$\varepsilon_{kij} \partial_i \partial_j \omega_\ell(\underset{\sim}{x}) = \Theta_{\ell k}(\underset{\sim}{x}) \neq 0 \tag{6}$$

For a single disclination line one has (see Fig. 6,7,8)

$$\Theta_{\ell k}(\underset{\sim}{x}) = \Omega_\ell \, \delta_k(L) \tag{7}$$

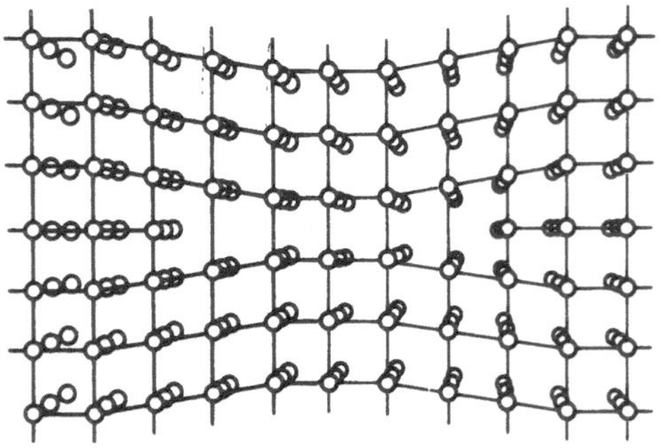

edge dislocation

Fig. 2. An edge dislocation as the boundary line of a missing section of a lattice plane.

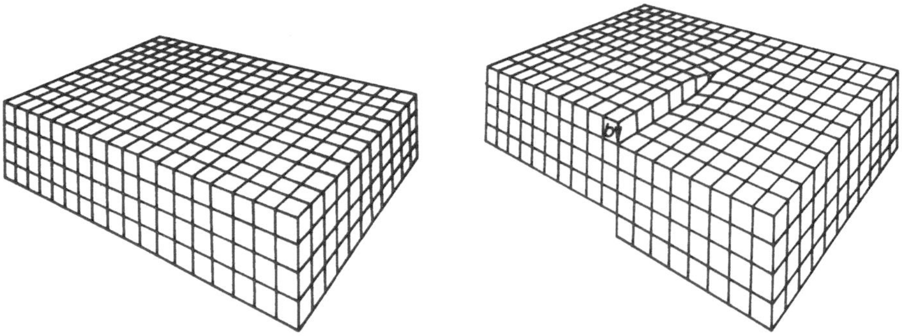

screw dislocation

Fig. 3. A screw dislocation as the boundary line of a torn part of the crystal.

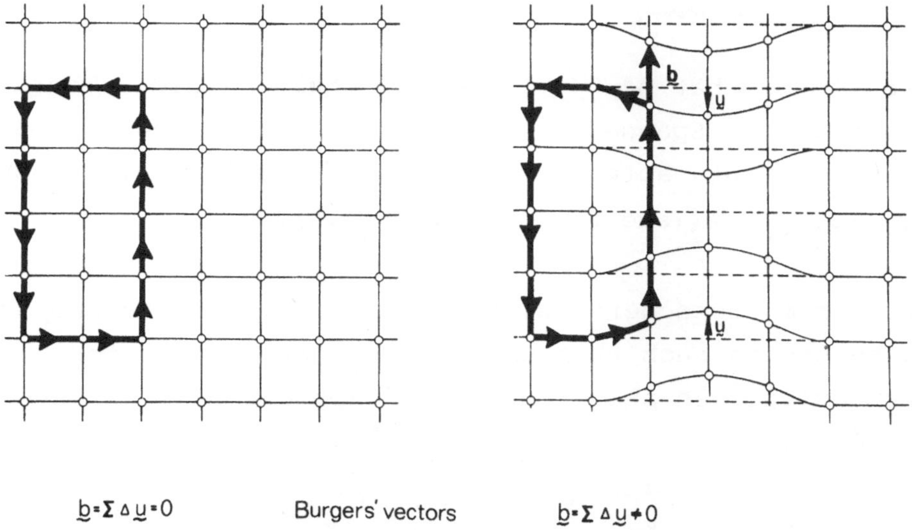

Fig. 4. The non-integrability of the displacement field at a dislocation line.

and Ω_l is called the Frank vector of the line. Instead of a differential characterization, one can state (1),(4),(6) also in the form of circuit integrals around the defect line L

$$\oint d\gamma = 2\pi n \tag{8}$$

$$\oint du_i = b_i \tag{9}$$

$$\oint d\omega_i = \Omega_i \tag{10}$$

These show that γ, u_i, ω_i are not single valued. There is a surface S Spanned by the line L, whose precise position is irrelevant, across which these variables have a jump by $2\pi n, b_i, \Omega_i$ (see Figs. 1,2,3,6,7,8).

Apart from these defects, the fields γ, u_i, ω_i are supposed to be smooth such that the physical quantities formed from its derivatives are integrable. For example, the superfluid velocity is given by $v_{s_k} \propto \partial_k \gamma$ and satisfies

$$(\partial_i\partial_j - \partial_j\partial_i)\partial_k \gamma(\underset{\sim}{x}) = 0 \tag{11}$$

The strain is given by $u_{ln} = \frac{1}{2}(\partial_l u_n + \partial_n u_l)$ and satisfies

$$(\partial_i\partial_j - \partial_j\partial_i)\, u_{ln}(\underset{\sim}{x}) = 0 \tag{12}$$

Moreover, also the derivative of strain and rotation field are supposed to be integrable

$$(\partial_i\partial_j - \partial_j\partial_i)\partial_k\, u_{ln}(\underset{\sim}{x}) = 0 \tag{13}$$

$$(\partial_i\partial_j - \partial_j\partial_i)\partial_k\, \omega_l(\underset{\sim}{x}) = 0 \tag{14}$$

These integrability conditions imply conservation laws.[7] Contracting (11) with ε_{ijk} shows that $\alpha_k(x)$ is divergenceless

$$\partial_k \alpha_k(\underset{\sim}{x}) = 0 \tag{15}$$

such that vortex lines can never end. Similarly, writing the dislocation density (4) as

$$\begin{aligned} \alpha_{\ell k}(\underset{\sim}{x}) &= \varepsilon_{kij}\, \partial_i \partial_j u_\ell = \varepsilon_{kij}\, \partial_i (u_{j\ell} + \varepsilon_{j\ell n}\,\omega_n) \\ &= \varepsilon_{kij}\, \partial_i\, u_{j\ell} + \delta_{k\ell}\, \partial_i \omega_i - \partial_\ell \omega_k \end{aligned} \tag{16}$$

and differentiating with respect to ∂_k, equ. (12) leads to

$$\partial_\ell \alpha_{k\ell}(\underset{\sim}{x}) = \varepsilon_{kij}\, \Theta_{ij} \tag{17}$$

which says that dislocations can end only on disclinations. For disclinations themselves, a few manipulations lead from (9),(10) and (11) to the conservation law

$$\partial_\ell \Theta_{k\ell}(\underset{\sim}{x}) = 0 \tag{18}$$

which says that disclination cannot end.

The conservation laws (17) and (18) can be used to construct a defect tensor

$$\eta_{ij} = \Theta_{ij} - \varepsilon_{jnk}\, \partial_k \left(\alpha_{ni} - \tfrac{1}{2}\delta_{ni}\, \alpha_{\ell\ell}\right) \tag{19}$$

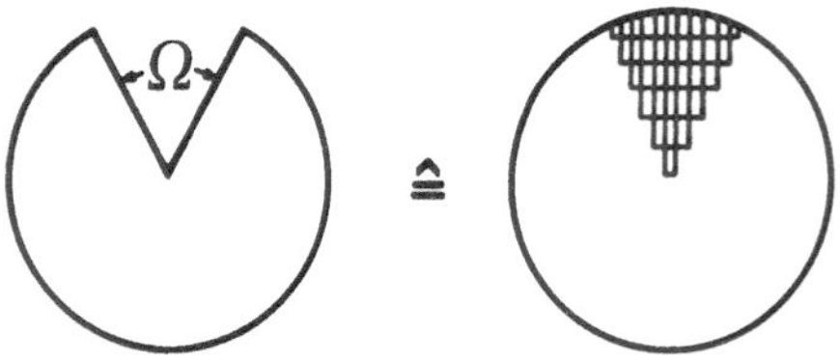

Fig. 5. Dislocation lines can stack up to form a rotational defect, a disclination.

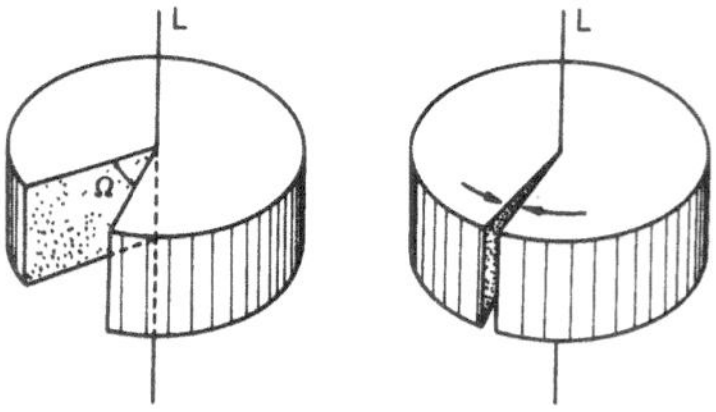

Fig. 6. A wedge disclination with Frank vector parallel to the line.

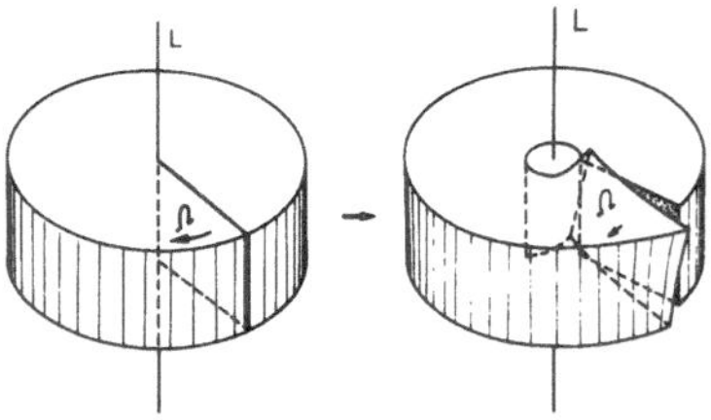

Fig. 7. A twist disclination with Frank vector orthogonal to the line and the cutting interface.

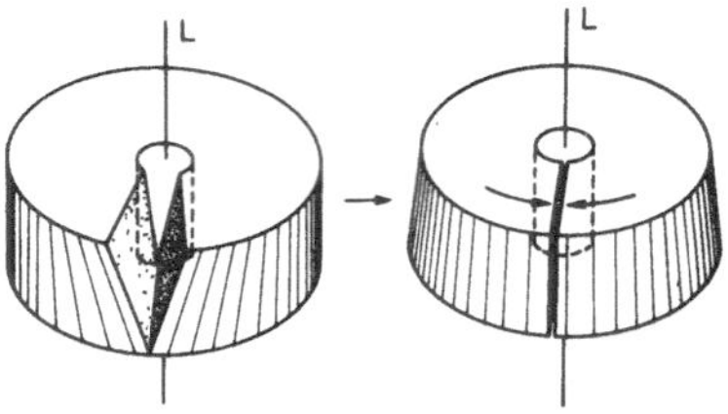

Fig. 8. A splay disclination with Frank vector orthogonal to the line but parallel to the cutting interface.

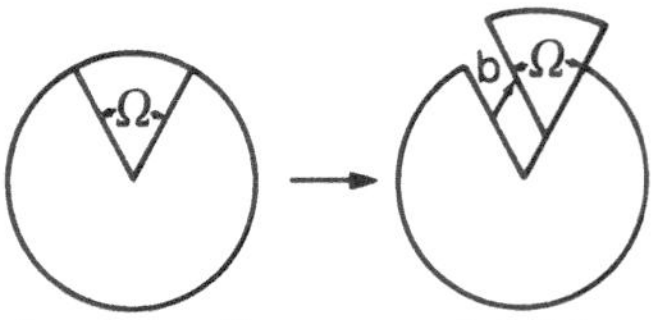

Fig. 9. A disclination and an antidisclination spaced a distance $\underset{\sim}{b}$ apart from a dislocation of Burgers' vector $\underset{\sim}{b}$.

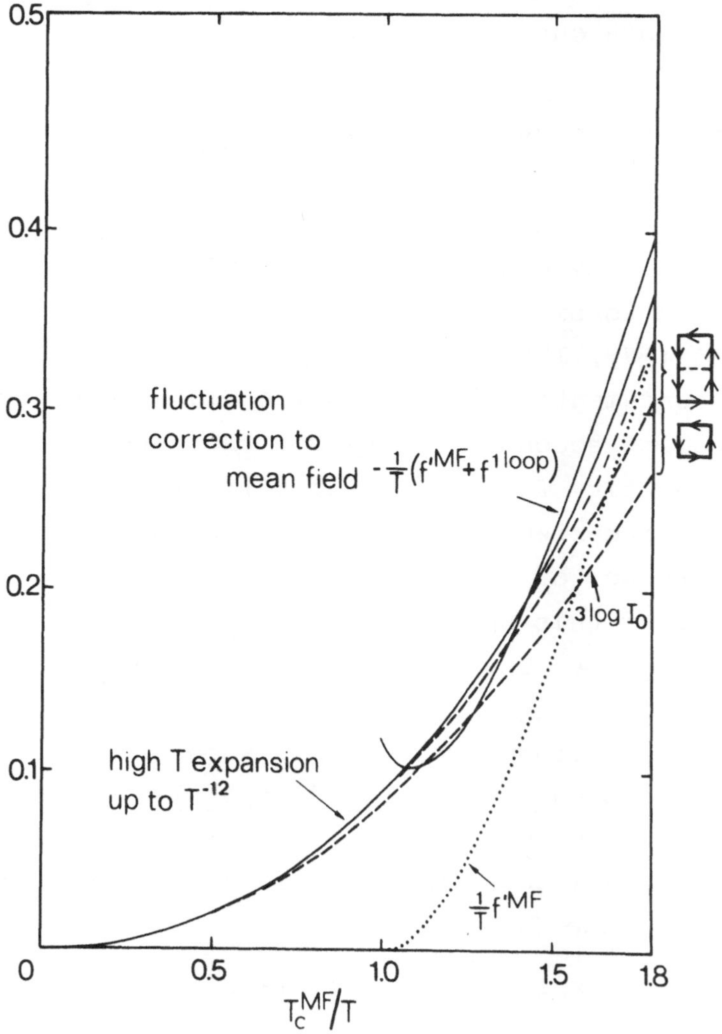

Fig. 10. The mean field potential $-\beta v = -\beta f$ for the XY model plus the one loop correction as compared with the high temperature expansion up to β^{12}. The intercept lies at $\beta_c/\beta_{MF} \sim 1.4$. The exact critical value is 1.35.

which is symmetric, due to (17), and divergenceless, due to (18). Notice that this equals Belinfante's construction of a symmetric energy momentum tensor from a canonical one Θ_{ij} and a spin current density $S_{ij,k} = \varepsilon_{ij\ell}\alpha_{\ell k}$ which reads

$$\eta_{ij} = \Theta_{ij} - \tfrac{1}{2}\partial_k\left(S_{ij,k} - S_{jk,i} + S_{ki,j}\right) \tag{20}$$

This is no accident since disclination density and dislocation density can be identified with Einstein curvature tensor[8] and torsion tensor of a general affine (metric) space and (17),(18) are the linearized manifestations of the two fundamental identities (in Schouten's[9] convention the so-called second identity for the torsion and the Bianchi identity for the curvature). The expression (20) is a linearized version of the symmetric energy momentum tensor of a gravitational Lagrangian $\int dx\sqrt{g}\,R$. The metric of this space is $g_{ij} = 2u_{ij}$ and the connection $\Gamma_{ijk} = \partial_i\partial_j u_k$. The geometry of this space has a simple operational meaning: Imagine a distorted crystal to be embedded in Euclidean space. Distances are measured by counting atoms in the distorted crystal. Vectors are parallel if they correspond to parallel vectors before the distortion (remaining attached to the crystalline atoms during the distortion).

Inserting the explicit forms (16),(6) into (19) we find

$$\eta_{ij} = \varepsilon_{ik\ell}\varepsilon_{jmn}\partial_k\partial_m u_{\ell n} \tag{21}$$

This double curl operation is called incompatibility.[5] It plays the same role for symmetric tensors as the single curl does for vectors. If a curl of a vector vanishes everywhere, the vector can be written as a gradient of an integrable scalar field. Similarly, if the double curl vanishes, the strain tensor can be written as $\frac{1}{2}(\partial_i u_j + \partial_j u_i)$ with integrable u_i fields! Hence we can conclude: A crystal is free of defects if $\eta_{ij}=0$.

The phenomenon of quark confinement arises if a gas of magnetic monopoles squeezes electric flux lines into thin tubes. In the usual description of magnetism it is customary to use a vector potential A_i whose curl is the magnetic field

$$B_i = \varepsilon_{ijk} \partial_j A_k \tag{22}$$

A magnetic monopole of charge m is a source of B_i field lines, defined by the condition

$$\partial_i B_i = \varepsilon_{ijk} \partial_i \partial_j A_k = m\, \delta^{(3)}(\underset{\sim}{x}) \tag{23}$$

Thus, in the presence of monopoles, the vector potential fails to satisfy the integrability condition

$$(\partial_i \partial_j - \partial_j \partial_i) A_k(\underset{\sim}{x}) = 0 \tag{24}$$

Monopoles are <u>defects</u> in the vector potential!

Notice that in a scalar description

$$B_i = \partial_i u \tag{25}$$

this would not be the case. The scalar field of a monopole $u(\underset{\sim}{x}) = (\partial^2)^{-1} m\, \delta(\underset{\sim}{x})$ is singular but integrable. However, in the scalar description, electric currents would cause defects since along them

$$\varepsilon_{ijk}\,\partial_i B_k = \varepsilon_{ijk}\,\partial_i \partial_k u = I\,\delta_i(L) \tag{26}$$

which is a relation analogous to (1),(3) for vortex lines in superfluid ^{4}He.

Thus, in a system with electric currents <u>and</u> magnetic monopoles there are always defects. With the traditional choice of a vector potential, these are the monopoles.

Lattice formulations of partition functions permit a simple study of ensembles of defects. If these are line-like, there is usually a certain temperature where the entropy overwhelms the energy and causes a proliferation of lines. On a lattice in 3 dimensions, a line involving n links has $(2D)^n$ configurations. If the energy per link is ε , the partition function

$$Z = \sum_n (2D)^n e^{-\frac{\varepsilon}{T}n}$$

indicates a proliferation for[2] $T > T_c = \varepsilon/\log 2D$. Vortex lines destroy the superfluid order[10], dislocation lines the translational order (creating liquid crystals)[11] and disclination lines the rotational order.[12]

The magnetic monopoles in three dimensions behave like a classical Coulomb gas. Since Debye, it has been known that there is screening which changes the propagation of fields for all temperatures, i.e. independent of how dilute the gas is. This Debye screening causes quark confinement, as was first observed by Polyakov.[4]

It is interesting to discover a great similarity between the three models describing such different systems:

Superfluid He is studied by means of an XY model on a simple cubic lattice with sites $\underset{\sim}{x}$.[13]

$$Z_{XY} = \prod_{\underset{\sim}{x}} \int_{-\pi}^{\pi} \frac{d\gamma(\underset{\sim}{x})}{2\pi} \, e^{\beta \sum_{x,i} (\cos \nabla_i \gamma(\underset{\sim}{x}) - 1)} \tag{27}$$

where $\nabla_i \gamma = \gamma(\underset{\sim}{x}+\underset{\sim}{i}) - \gamma(\underset{\sim}{x})$ is the lattice gradient across the links $\underset{\sim}{i}=(1,0,0),(0,1,0),(0,0,1)$ connecting $\underset{\sim}{x}$ and $\underset{\sim}{x}+\underset{\sim}{i}$. The Villain approximation,[14]

$$Z_{XY} \simeq Z_V = R_V^N \sum_{\{n_i(x)\}} \prod_{\underset{\sim}{x}} \int_{-\pi}^{\pi} \frac{d\gamma(x)}{2\pi} \, e^{-\frac{\beta_V}{2} \sum_{x,i} (\nabla_i \gamma - 2\pi n_i)^2} \tag{28}$$

with $R_V = (I_0(\beta) e^{-\beta} \sqrt{2\pi\beta_V})$ and $\beta_V = -\left[2 \log(I_1(\beta)/I_0(\beta))\right]^{-1}$ displays the surfaces over which γ jumps ($n_3(o)=1$ means that γ jumps when passing the XY plane for 0 to 1). Instead of integrating from $-\pi$ to π we can integrate over the entire γ axis, if we remove from n_i the gradient of an integer field, i.e.

$$n_i \to n_i - \nabla_i n \tag{29}$$

By choosing $n = \nabla_3^{-1} n_i$, the remaining n_i can be taken to satisfy $n_3=o$. Hence

$$Z_V = R_V^N \sum_{\{n_i(\underset{\sim}{x})\}} \delta_{n_3,0} \prod_{\underset{\sim}{x}} \int_{-\infty}^{\infty} \frac{d\gamma(x)}{2\pi} e^{-\frac{\beta_V}{2} \sum_{x,i} (\nabla_i \gamma - 2\pi n_i)^2} \tag{30}$$

Introducing a conjugate variable B_i, which is the superfluid velocity and plays the same role as a magnetic field for electric currents, in the gradient representation (25), we can rewrite this as

$$Z_V = R_V^N \sum_{\{n_i(\underset{\sim}{x})\}} \delta_{n_3,0} \prod_{\underset{\sim}{x},i} \int_{-\infty}^{\infty} \frac{dB_i}{\sqrt{2\pi\beta}} \prod_{\underset{\sim}{x}} \int_{-\infty}^{\infty} \frac{d\gamma}{2\pi} e^{-\frac{1}{2\beta_V}\sum_{\underset{\sim}{x},i} B_i^2 + i\sum_{\underset{\sim}{x},i} B_i(\nabla_i\gamma - 2\pi n_i)} \tag{31}$$

The integrals over γ force B_i to be divergenceless. Hence we can express B_i in terms of a vector potential

$$B_i = \varepsilon_{ijk}\partial_j A_k \tag{32}$$

and write

$$Z_V = R_V^N \sum_{\{n_i(\underset{\sim}{x})\}} \delta_{n_3,0} \int_{-\infty}^{\infty} \frac{dA_i}{\sqrt{2\pi\beta}} \delta(\nabla_i A_i)\, e^{-\frac{1}{2\beta_V}\sum_{\underset{\sim}{x}}(\nabla\times A)^2 + 2\pi i \sum_{\underset{\sim}{x},i} A_i l_i} \tag{33}$$

where we have set

$$l_i(\underset{\sim}{x}) = (\bar{\nabla}\times n)(\underset{\sim}{x}) \tag{34}$$

This is an integer field satisfying $\bar{\nabla}_i l_i = 0$. It can be pictured as superposition of closed integer lines and is the lattice version of the vortex density $\alpha_k(\underset{\sim}{x})$ (recall (3)). Due to (33), the vortex lines interact in the same way as electric currents (Biot-Savart forces). From (34) we see that the l_i form the boundary lines of the surfaces across which the phase γ jumps. Hence l_i are the vortex lines. The smallest is given by a single $n_i(x)$ being equal to one which gives a loop around one plaquette. This can be identified with a roton. In fact, its energy turns out to be about the same as that of a roton. Rotons have dipole forces. These do not change the field propagators as long as they are dilute. Only when they proliferate does screening set in and there is a close relation to the Meissner screening of magnetic field lines in super conductors. There exists a disorder field theory of the Ginzburg-Landau type describing this proliferation of vortices and the Meissner like screening.[10]

In a crystal, we can construct a similar model containing dislocations and disclinations. The linear elastic energy is, in terms of the displacement vector

$$e(\underset{\sim}{x}) = \frac{\mu}{4} \left(\partial_i u_j + \partial_j u_i\right)^2 \tag{35}$$

if we neglect the Lamé constant λ , for brevity. Dislocation lines are characterized by jumps of u_i across certain surfaces. Their ensemble can be studied by[15,16]

$$Z_{1\,\mathrm{melt}} = \prod_{\underset{\sim}{x},i} \int_{-\pi}^{\pi} \frac{dA_i(\underset{\sim}{x})}{2\pi}\, e^{\beta \left[\sum_{\underset{\sim}{x},i>j} \left(\cos(\nabla_i A_j + \nabla_j A_i) - 1\right) + 2 \sum_{\underset{\sim}{x},i} \left(\cos \nabla_i A_i - 1\right)\right]} \tag{36}$$

which has a Villain approximation

$$Z_{1\,\mathrm{melt}} \approx Z_{\mathrm{melt},V} = R_V^{6N} \sum_{\{n_{ij}(\underset{\sim}{x})\}} \prod_{\underset{\sim}{x},i} \int_{-\pi}^{\pi} \frac{dA_i}{2\pi}\, e^{-\frac{\beta_V}{2}\left[\sum_{\underset{\sim}{x},i>j} \left(\nabla_i A_j + \nabla_j A_i - 2\pi n_{ij}\right)^2 + \frac{1}{2}\sum_{\underset{\sim}{x},i} \left(\nabla_i A_i - 2\pi n_{ii}\right)^2\right]} \tag{37}$$

We have renormalized u_i to $A_i = \frac{2\pi}{a} \cdot u_i$ such that it is periodic in 2π , and set $\beta \equiv \mu a^3/((2\pi)^2 T)$. We can again extend the A_i integrations over all real values by restricting n_{ij} to

$$n_{3i} = n_{i3} = 0 \tag{38}$$

Just as in (30) it is possible to introduce a conjugate stress variable σ_{ij} and write

$$Z_{\text{melt},V} = R_V^{(N} \prod_{\underset{\sim}{x},i\geq j} \int_{-\infty}^{\infty} \frac{d\sigma_{ij}}{\sqrt{2\pi\beta_V}}\, e^{\frac{1}{4\beta_V}\sum_{\underset{\sim}{x},i,j} \sigma_{ij}^2} \tag{39}$$

$$\cdot \sum_{\{n_{ij}(\underset{\sim}{x})\}} \delta_{n_{3i},0}\, e^{\frac{i}{2}\sum_{\underset{\sim}{x},i>j} \sigma_{ij}(\nabla_i A_j + \nabla_j A_i - 2\pi n_{ij}) + i\sum_{\underset{\sim}{x},i} \sigma_{ii}(\nabla_i A_i - 2\pi n_{ii})}$$

The integrals over A_i force

$$\bar{\nabla}_i \sigma_{ij} = 0 \tag{40}$$

which is the standard equation for linear elasticity. This can be taken advantage of to express σ_{ij} in terms of a symmetric gauge field

$$\sigma_{ij} = \varepsilon_{ik\ell}\, \varepsilon_{jmn}\, \bar{\nabla}_k \bar{\nabla}_m \chi_{\ell n} \tag{41}$$

In this way, (39) becomes

$$Z_{\text{melt},V} = R_V^{6N} \prod_{x,i\geq j} \int_{-\infty}^{\infty} \frac{d\chi_{ij}(\underset{\sim}{x})}{\sqrt{2\pi\beta_V}}\, \Phi[\chi] \tag{42}$$

$$\sum_{\{n_{ij}(\underset{\sim}{x})\}} \delta_{n_{3j},0}\, e^{-\frac{1}{4\beta_V}\sum_{\underset{\sim}{x}}(\varepsilon\varepsilon\bar{\nabla}\bar{\nabla}\chi)^2 + 2\pi i \sum_{\underset{\sim}{x},i,j} \chi_{ij}\eta_{ij}}$$

where $\Phi[\chi]$ is a gauge fixing factor (for example $\prod_{x,i} \delta(\chi_{3i}(x))$) and

$$\eta_{ij} = \varepsilon_{ik\ell}\, \varepsilon_{jmn}\, \bar{\nabla}_k \bar{\nabla}_m n_{\ell n} \tag{43}$$

is a symmetric integer tensor satisfying

$$\bar{\nabla}_i \eta_{ij}(\underset{\sim}{x}) = 0 \tag{44}$$

This is the lattice version of the defect density (21). integrating out the χ fields gives

$$Z_{melt,V} = R_V^{CN} \frac{1}{(4\pi\beta_V)^{\frac{3N}{2}}} e^{-\beta_V 4\pi^2 \sum_{\underset{\sim}{x},\underset{\sim}{x}'} \eta_{en}(\underset{\sim}{x}) v_4(x-x') \eta_{en}(\underset{\sim}{x})} \tag{45}$$

where

$$v_4(\underset{\sim}{x}) = \int \frac{d^3k}{(2\pi)^3} \left(2D - 2\sum_i \cos k_i\right)^{-2} \tag{46}$$

is the lattice version of the $1/k^4$ potential. It diverges which implies that only such defect configurations $\eta_{en}(\underset{\sim}{x})$ can contribute which are neutral

$$\sum_{\underset{\sim}{x}} \eta_{en}(\underset{\sim}{x}) = 0 \tag{47}$$

For these, in turn, we can use the finite subtracted potential

$$v_4'(\underset{\sim}{x}) \equiv v_4(\underset{\sim}{x}) - v_4(\underset{\sim}{0}) \tag{48}$$

Similar to the XY model, the melting model has a critical temperature at which dislocation loops proliferate. These cause a screening of the elastic forces, which can again be viewed as a Meissner effect, by going to a disorder field description.[17] Contrary to the XY model, the transition is now of first order. The reason for this is quite simple: Lattice defects allow for certain collective formations whose energy is much lower than the sum of the individual energies. An example was given before: Many dislocation lines can pile up on top of each other and form a disclination line (see Fig. 5). If two such disclination lines run through the crystal in opposite directions the energy between line elements grows like R. Hence they are permanently confined. In fact, one may consider the disclinations as fundamental defect lines as confined pairs (see Fig. 9). When the dislocation lines proliferate, they screen the elastic forces from R to 1/R.[18] This leads to the deconfinement of its constituents. The proliferation alone would be a second order transition, the ensuing deconfinement opens up an additional reservoir of entropy and this is what causes the discontinuity of the transition.

An ensemble of monopoles in a magnetic field can be studied with the model[4]

$$Z_{\text{gauge}} = \prod_{\underline{x},i} \int_{-\pi}^{\pi} \frac{dA_i(\underline{x})}{2\pi}\, e^{\beta \sum_{\underline{x},i>j} \left(\cos\left(\nabla_i A_j - \nabla_j A_i\right) - 1\right)} \tag{49}$$

called lattice QED. Its Villain approximation reads

$$Z_{\text{gauge},V} = R_V^{3N} \sum_{\{n_{ij}(\underline{x})\}} \prod_{\underline{x},i} \int_{-\pi}^{\pi} \frac{dA_i(\underline{x})}{2\pi}\, e^{-\frac{\beta_V}{2} \sum_{\underline{x},i>j} \left(\nabla_i A_j - \nabla_j A_i - 2\pi n_{ij}\right)^2} \tag{50}$$

Introducing a conjugate magnetic field gives

$$Z_{\text{gauge},V} = R_V^{3N} \sum_{\{n_{ij}(\underline{x})\}} \prod_{\underline{x},i>j} \int_{-\infty}^{\infty} \frac{df_{ij}(\underline{x})}{\sqrt{2\pi\beta_V}} \prod_{\underline{x},i} \int_{-\pi}^{\pi} \frac{dA_i(\underline{x})}{2\pi}\, e^{-\frac{1}{2\beta_V} \sum_{\underline{x},i>j} f_{ij}^2 + i \sum_{\underline{x},i>j} f_{ij}\left(\nabla_i A_j - \nabla_j A_i - 2\pi n_{ij}\right)} \tag{51}$$

Summing over the n_{ij} forces f_{ij} to be integer whereupon the integrations over A_i give

$$\bar{\nabla}_i f_{ij} = 0 \tag{52}$$

Therefore one can write

$$f_{ij} = \varepsilon_{ijk} \bar{\nabla}_k \varphi \tag{53}$$

with an integer scalar field $\varphi(\underset{\sim}{x})$. This, in turn, can be integrated over all real values if one couples it to an integer field m(x):

$$Z_{gauge,V} = \left(R_V^3 \frac{1}{2\pi\beta_V}\right)^N \sum_{\{m(\underset{\sim}{x})\}} \prod_{\underset{\sim}{x}} \int \frac{d\varphi(\underset{\sim}{x})}{\sqrt{2\pi\beta^V}} \, e^{-\frac{1}{2\beta_V}\sum_{\underset{\sim}{x}} (\bar{\nabla}\varphi)^2 + 2\pi i \sum_{\underset{\sim}{x}} \varphi(\underset{\sim}{x}) m(\underset{\sim}{x})} \tag{54}$$

Integrating out the φ field gives

$$Z_{gauge,V} = \left(R_V^3 \frac{1}{2\pi\beta_V}\right)^N \sum_{\{m(\underset{\sim}{x})\}} e^{-\frac{\beta_V}{2} 4\pi^2 \sum_{\underset{\sim}{x},\underset{\sim}{x}'} m(\underset{\sim}{x})\, v_2(\underset{\sim}{x}-\underset{\sim}{x}')\, m(\underset{\sim}{x}')} \tag{55}$$

where

$$v_2(\underset{\sim}{x}) = \int \frac{d^3k}{(2\pi)^3} e^{i\underset{\sim}{k}\underset{\sim}{x}} \left(2D - 2\sum_i \cos k_i\right)^{-1} \tag{56}$$

is the lattice Coulomb potential. The field $m(\underset{\sim}{x})$ parametrizes the ensemble of monopole charges. It is related to the jumps n_{ij} via

$$m(\underset{\sim}{x}) = \varepsilon_{ijk} \nabla_i n_{jk}(\underset{\sim}{x}) \tag{57}$$

under the constraint $n_{3j} = n_{j3} = 0$.

The Debye screening can be seen by separating the self-energy of the monopoles and writing

$$v_2(\underset{\sim}{x}) = v_2(\underset{\sim}{0}) + v_2'(\underset{\sim}{x}) = .2527 + v_2'(\underset{\sim}{x}) \tag{58}$$

which allows bringing (55) to the form

$$Z_{gauge,V} = \left(R_V^3 \frac{1}{2\pi\beta_V}\right)^N \prod_{\underset{\sim}{x}} \int_{-\infty}^{\infty} \frac{d\varphi}{\sqrt{2\pi\beta_V}} \, e^{-\frac{1}{2\beta_V} \sum_{\underset{\sim}{x},\underset{\sim}{x}'} \varphi(\underset{\sim}{x}) \frac{-\nabla_\ell \bar{\nabla}_\ell}{1+v(0)\nabla_\ell\bar{\nabla}_\ell}(\underset{\sim}{x},\underset{\sim}{x}')\varphi}$$

$$\sum_{\{m(\underset{\sim}{x})\}} e^{-\beta_V \frac{4\pi^2}{2} v_2(\underset{\sim}{0}) \sum_{\underset{\sim}{x}} m^2(x) + 2\pi i \sum_{\underset{\sim}{x}} \varphi(\underset{\sim}{x}) m(\underset{\sim}{x})} \tag{59}$$

For large β, only $m(\underset{\sim}{x}) = 0, \pm 1$ has to be summed and the last factor becomes

$$e^{2 e^{-\beta_V \frac{4\pi^2}{2} v_2(\underset{\sim}{0})} \cdot \sum_{\underset{\sim}{x}} \cos 2\pi \varphi(\underset{\sim}{x})} \tag{60}$$

The curvature of the cosine term induces a mass

$$m^2 = 8\pi^2 e^{-\beta_V \frac{4\pi^2}{2} v_2(\underset{\sim}{0})} \tag{61}$$

and this is responsible for confinement.[4]

Comparing the partition function (27), (36), and (49) we notice the following relation. The latter two cases can be considered as special constraint versions of an extended XY model for three angular variables

$$Z'_{XY} = \prod_{x,i} \int_{-\pi}^{\pi} \frac{d\gamma_i(x)}{2\pi} \, e^{\beta \sum_{x,i,j} (\cos \nabla_i \gamma_j - 1)} \tag{62}$$

As long as the angles are independent, this partition function is simply the cube of the individual XY models.

$$Z'_{XY} = Z_{XY}^3 \tag{63}$$

The melting model counts only the energy of symmetric combinations, the gauge model that of antisymmetric combinations of $\nabla_i \gamma_j$.

The XY model has a second order phase transition and so does (62). The antisymmetric combination has no phase transition (permanent confinement). The symmetric combination, on the other hand, describes melting which is a first order transition. Thus, symmetrization of the tensor $\nabla_i \gamma_j$ in (62) hardens the transition, antisymmetrization softens it.

The simplest way of studying the phase diagram of such theories is via a mean field approximation plus loop corrections.[19,20] This follows from the possibility of rewriting the partition functions in the following way

$$Z_{XY} = e^{-3N\beta} \prod_{x} \int_{-\pi}^{\pi} \frac{d\gamma(x)}{2\pi} \, e^{\beta \sum_{x,i} U(x) U^{+}(x+i)} \tag{64}$$

$$Z_{melt} = e^{9N\beta} \prod_{\underset{\sim}{x}, i} \int_{-\pi}^{\pi} \frac{dA_i}{2\pi} \, e^{\beta\left[\sum_{\underset{\sim}{x}, i>j} U_i(\underset{\sim}{x}) U_j^+(\underset{\sim}{x}+\underset{\sim}{i}) U_i^+(\underset{\sim}{x}+\underset{\sim}{j}) U_j(\underset{\sim}{x}) + 2\sum_{\underset{\sim}{x}, i} U_i(\underset{\sim}{x}) U_i^+(\underset{\sim}{x}+\underset{\sim}{i})\right]} \tag{65}$$

$$Z_{gauge} = e^{3N\beta} \prod_{x, i} \int_{-\pi}^{\pi} \frac{dA_i}{2\pi} \, e^{\beta \sum_{\underset{\sim}{x}, i>j} U_i(\underset{\sim}{x}) U_j(\underset{\sim}{x}+\underset{\sim}{i}) U_i^+(x+j) U_j^+(x)} \tag{66}$$

where $U(\underset{\sim}{x}) = e^{i\gamma(\underset{\sim}{x})}$, $U_i(\underset{\sim}{x}) = e^{iA_i(\underset{\sim}{x})}$. Melting and the first term in the gauge theory differ only by the position of a star on the U's which corresponds graphically to exchanging a rotation graph  by a distortion graph  .

The introduction of two auxiliary complex fields via the identity

$$\prod_{\underset{\sim}{x}} \iint_{-i\infty}^{i\infty} \frac{d\alpha(\underset{\sim}{x}) \, d\alpha^+(\underset{\sim}{x})}{(2\pi)^2} \iint_{-\infty}^{\infty} du(\underset{\sim}{x}) \, du^+(\underset{\sim}{x}) \, e^{\frac{i}{2}\sum_{\underset{\sim}{x}} \alpha^+(u - U) + c.c.} \tag{67}$$

permits a reformulation of the partition functions as integrals over the fluctuating energies[19,20]

$$-\beta F_{xy} = \sum_{\underset{\sim}{x}} \left[\beta \sum_i u(\underset{\sim}{x}) u^+(\underset{\sim}{x}+\underset{\sim}{i}) - \tfrac{1}{2}\left(\alpha^+(\underset{\sim}{x}) u(\underset{\sim}{x}) + c.c.\right) - \log I_0(|\alpha|)\right] \tag{68}$$

$$-\beta F_{melt} = \sum_{\underset{\sim}{x}} \left[\beta \sum_{i>j} u_i(\underset{\sim}{x}) u_j^+(\underset{\sim}{x}+\underset{\sim}{i}) u_i^+(\underset{\sim}{x}+\underset{\sim}{j}) u_j(\underset{\sim}{x}) + 2\beta \sum_i u_i(\underset{\sim}{x}) u_i(\underset{\sim}{x}+\underset{\sim}{i}) - \tfrac{1}{2}\sum_i \left(\alpha_i^+(x) u_i(x) + c.c.\right) - \sum_i \log I_0(|\alpha_i|)\right] \tag{69}$$

$$-\beta F_{gauge} = \sum_{\underline{x}} \Big[\beta \sum_{i>j} u_i(\underline{x}) u_j(\underline{x}+\underline{i}) u_i^+(\underline{x}+\underline{j}) u_j^+(\underline{x}) - \frac{1}{2} \sum_i (\alpha_i^+(\underline{x}) u_i(\underline{x}) + c.c.) - \sum_i \log I_0(|\alpha_i|) \Big] \tag{70}$$

The mean field equations are

$$|u| = \frac{I_1(|\alpha|)}{I_0(|\alpha|)} \tag{71}$$

in all three cases and

$$\alpha = 6\beta u \quad \text{for XY} \tag{72}$$

$$|\alpha_i| = 4\beta(|u_i|^3 + |u_i|) \quad \text{for melting} \quad \text{model} \tag{73}$$

$$|\alpha_i| = 4\beta|u_i| \quad \text{for gauge} \tag{74}$$

This gives a second order transition for the XY model at $\beta_c^{MF} = 1/3$. For melting[16] and guage models[20] one finds, on the other hand, a first order transition at

$$\beta_c^{MF} = .49 \qquad \text{melting}$$

$$\beta_c^{MF} = 1.52 \qquad \text{gauge}$$

For $\beta < \beta_c^{MF}$, the mean fields are zero, for $\beta > \beta_c^{MF}$ they grow from zero to $u \sim 1$, $\alpha_i \sim 6\beta, 8\beta, 4\beta$. Fluctuation corrections change the energy for low β by adding a power series in β which is known[21] up to β^{12} . Above β_c^{MF} , the loop correction gives the major contributions and shifts $-\beta f$ slightly upwards. For the XY model, this brings β_c up to .47 (see Fig. 11).

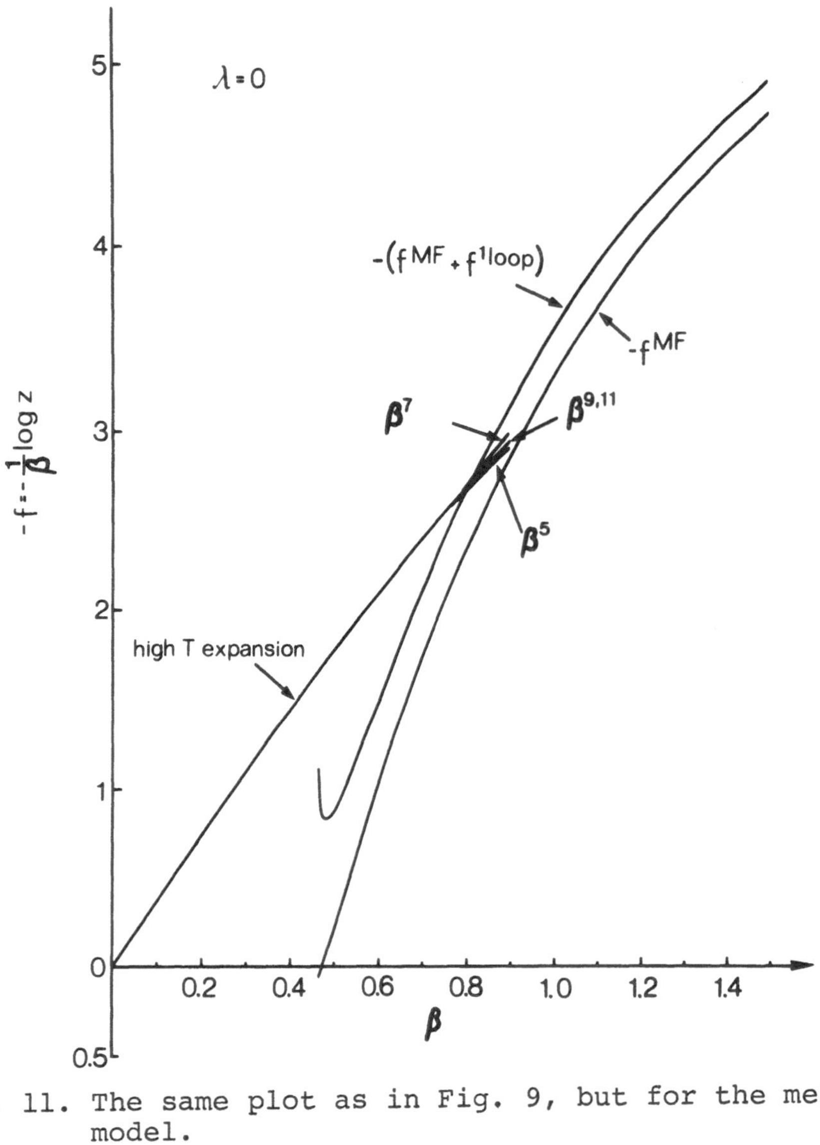

Fig. 11. The same plot as in Fig. 9, but for the melting model.

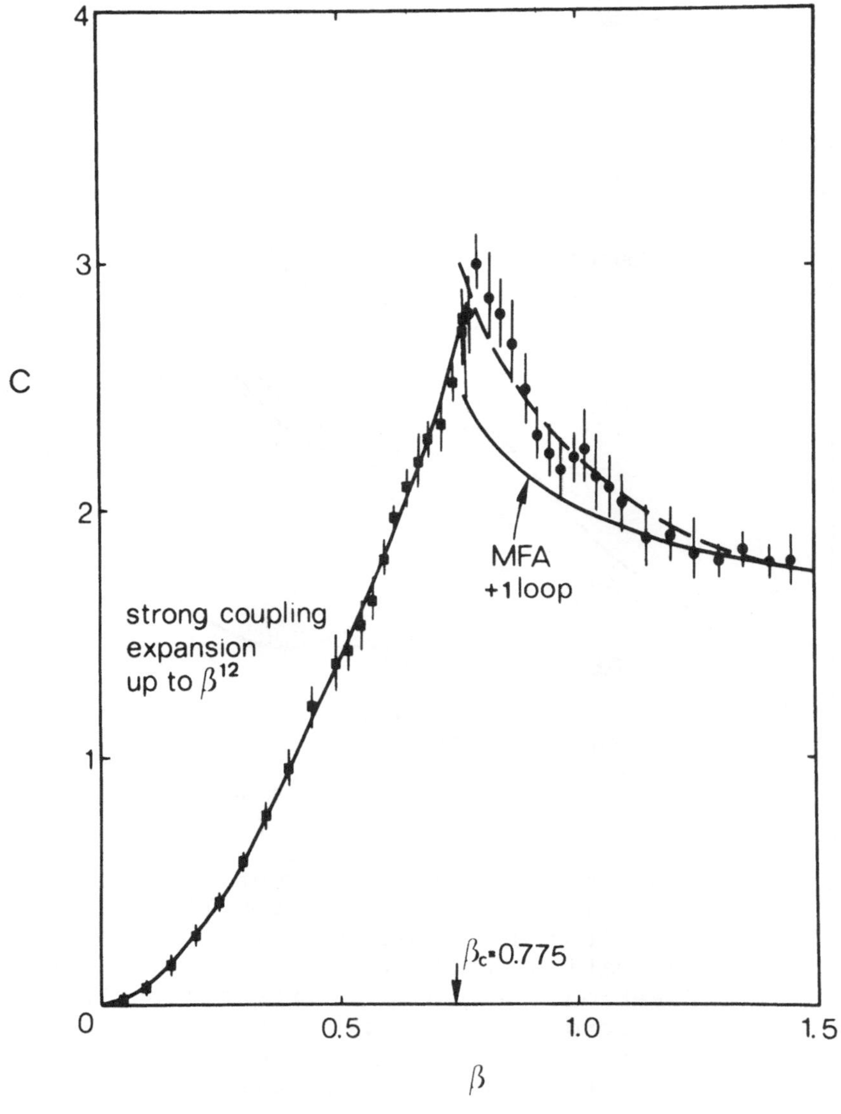

Fig. 12. The specific heat of the melting model as compared with mean field calculation plus one loop correction for $\beta > \beta_c$ and with a low temperature expansion up to β^{12} for $\beta < \beta_c$.

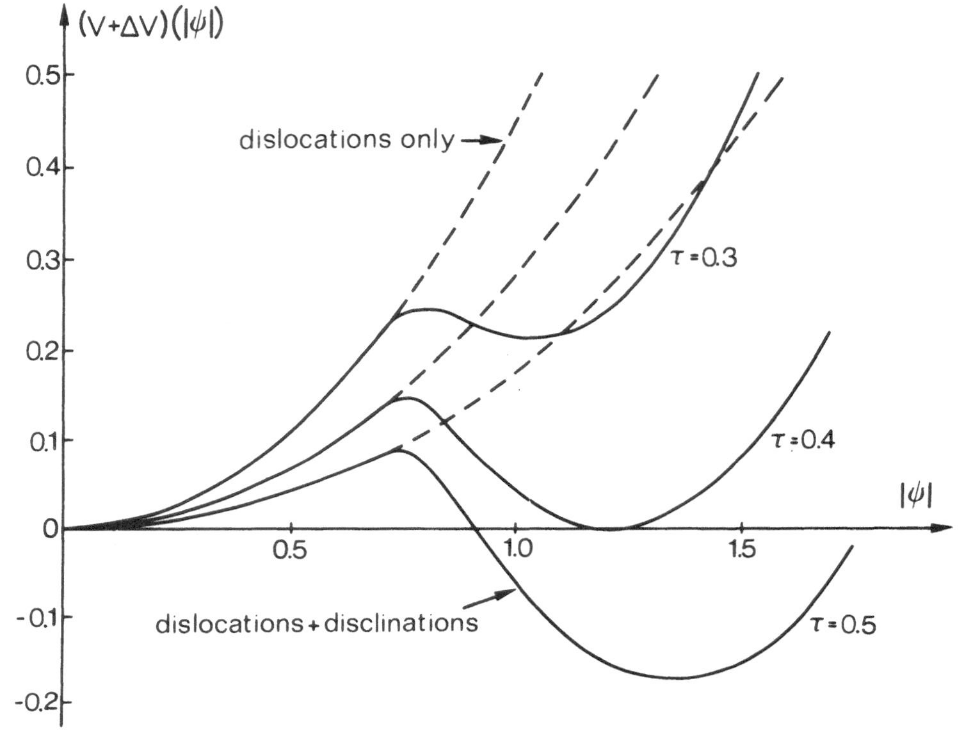

Fig. 13. The potential of the disorder field theory of dislocation plus disclination lines as a function of the dislocation field $|\Psi|$. The curve "dislocations only" corresponds to the pure proliferation (second order transitions). The break-up due to Meissner screening causes the additional lowering of the energy which causes the first order phase transition.

For the gauge theory, the corrections are so important that they wipe out the transition completely. The model always remains on the energy branch obtained by the low β expansion.

The melting theory, on the other hand, retains the first order of the mean field transition. The jump in entropy comes out $S=1.4k_B$ per site, in good agreement with experimental data.[22,23] The shape of the specific heat agrees reasonably with experimental curves (see Fig. 12.)

In conclusion we see that defects form the common basis for the understanding of many phase transitions. Their analysis can give an important clue concerning the order of the transition. Melting and U(1) lattice gauge theory are, in one respect, quite similar, namely by having the same mean field approximations. However, their defects are quite different, and this explains why melting is a first order transition while the gauge theory has confinement at all temperatures.

Closed defect lines have dipole interactions. For this reason, few of them cannot cause screening. Their proliferation is necessary to achieve this. Point defects, with Coulomb forces, on the other hand, screen for all temperatures in three dimensions, as is known from Debye's classical work. If closed defect lines can break up after screening the coupled transition "proliferation plus break up" is of first order (see Fig. 13.)

REFERENCES

1) In solids this was noticed in 1952 by W. Shockley in L'Etat Solid, Proc. Neuvieme Conseil de Physique, Brussels, ed. R. Stoops (Institut International du Physique, solvay, Brussels, 1952).

2) In superfluid He this was proposed by R.P. Feynman in 1955 in Progress in Low Temperature Physics, 1, ed. C.J. Gorter (North-Holland, Amsterdam, 1955) and later by V.N. Popov, Sov. Phys. JETP 37, 341(1973). In two dimensions, the study was performed in detail in He II by J.M. Kosterlitz and D.J. Thouless, J. Phys. C (Solid State Phys.) 6, 1181 (1973) and by J.V. Jorgé, L.P. Kadanoff et al., Phys. Rev. B16, 1217(1977). The two dimensional melting was attempted in complete analogy, by D.R. Nelson and B.I. Halperin, Phys. Rev. B19, 2457(1979), but these authors mutilated the physical properties of defects by giving them an unphysical core energy which led to strange phases. For a correct treatment see H. Kleinert, Phys. Lett. 95A, 381(1983).

3) In four dimensional QCD, a similar study was initiated by A. Polyakov et al., Phys. Lett. 59B, 85(1975) and G. t'Hooft, Phys. Rev. D14, 3432(1976) and continued by many others such as C. Callan, R.F. Dashen, and D.J. Gross, see Phys. Rev. Lett. 44D, 435(1980) and references.

4) The confinement in 3 dimensional compact QED was realized by A. Polyakov, Phys. Lett. 59B, 82(1975) and Nucl. Phys. B120, 429(1977). See also J.B. Bronzan and Ashok Das, Phys. Rev. D26, 1415(1982) for a recent study, and T. Banks, R. Myerson and J. Kogut, Nucl. Phys. B129, 493(1977).

5) E. Kröner, Lecture presented at the 1980 Les Houches Summer School on The Physics of Defects, ed. R. Balian (North-Holland, Amsterdam, 1981).

6) R. de Wit, Journal of Research (Nat. Bur. of Standards-A, Phys. and Chem.), 77A, 49(1973).

7) For more details see H. Kleinert, Gauge Theory of Defects and Stresses, Gordon and Breach, 1984.

8) H. Kleinert, Lett. Nuovo Cimento, 35, 41(1982).

9) J.A. Schouten, Ricci-Calculus, Springer, Berlin, 1954.

10) H. Kleinert, Phys. Lett. A93, 861(1982).

11) W. Helfrich, Proc. International Liquid Crystal Conference, Bangalore, Heyden and Son, London, 1980.

12) H. Kleinert, Phys. Lett. 95A, 381, 493(1983).

13) R.G. Bowers and G.S. Joyce, Phys. Lett. 19, 630(1967). See also D.D. Bettd in Phase Transitions and Critical Phenomena, Vol 3, eds. C. Domb and M. Green, Academic, New York, 1947.

14) J. Villain, J. de Physique (Paris), J. Phys. 36, 581 (1975). It is important to notice that the approximation is valid not only for large β (where it is obvious with $\beta_V \sim \beta$) but also for small β. In fact, the phase transition lies at $\beta \sim .46$, i.e. in the low β regime where $\beta_V \sim -(2\log /2 - {}^2/4 + 5\ {}^4/192 - \ldots)^{-1}\ .33$, not in the high β regime where $\beta_V \sim \beta(1 - {}^1/2\beta + \ldots)$. It corresponds to the approximation $I_n(\beta)/I_o(\beta) \sim \exp(n^2 \log /2)$ for $n = 0, \pm 1$ and small β, rather than $I_n(\beta)/I_o(\beta) \sim \exp(-n^2/2\beta)$ for large β. This was overlooked by T. Banks, R. Myerson, and J. Kogut, Nucl. Phys. B129, 493(1979) and many later papers.

15) H. Kleinert, Phys. Lett. A91, 295(1982).

16) H. Kleinert, Lett. Nuovo Cimento 37, 425(1983).

17) H. Kleinert, Phys. Lett. A89, 294(1982), and Lett. Nuovo Cimento 34, 464(1982).

18) H. Kleinert, Phys. Lett. A95, 493(1983).

19) E. Brezin and J.M. Drouffe, Nucl Phys. B200 FS4 , 93(1982), J.M. Drouffe, Nucl. Phys. B205, FS5 27(1982).

20) J. Greensite and B. Lautrup, Phys. Lett. 104B, 41(1981). H. Flyvbjerg, B. Lautrup, and J.B. Zuber, Phys. Lett. 110B, 279(1982), B. Lautrup, Mean Field Methods in Gauge Theories, Lecture presented at the 13th Symposium on High Energy Physics, Bad Schandau, DDR, 1982.

21) R. Balian et al., Phys. Rev. D11, 2104(1975).

22) S. Ami, T. Hofsäss and R. Horsley, Phys.Lett. 101A, 145 (1984).

23) L. Jacobs and H. Kleinert, J.Phys. A15, L 361 (1984).

LATTICE GAUGE THEORY

C.P. Korthals Altes

Centre de Physique Théorique
C.N.R.S. - Luminy - Case 907
F-13288 Marseille Cedex 9 (France)

INTRODUCTION

Lattice gauge theory[1] is now ten years old. Apart from the theoretical insight the lattice formulation gives, it is very well suited for computer simulations, as its inventor advocated already some five years ago at this school[2]. Since three years[3] this approach has extracted useful information out of lattice gauge theory and spurred many interesting questions.

In the first lecture I will assume there are no experts in the audience and explain some basic facts in quarkless quantumchromodynamics on a lattice (QCD). Then, in the second lecture, we shall review tests for the consistency of the numerical results so far obtained. The third lecture shall deal with a more esoteric subject : that of large N reduced models. The list of references is by no means meant to be exhaustive ; for that the reader is referred to ref. 27.

LECTURE I. SOME BASIC FACTS IN LATTICE QCD

This lecture is meant to be an elementary introduction ; educated readers can skip this section and go to lecture II and III.

I.1. QCD, Asymptotic Freedom and Confinement.

Quantum chromodynamics is the theory of quarks interacting with gluons. The gauge group is the SU(3) group but in these notes we shall consider the SU(N) gauge group, with the N^2-1 generators λ^a in the defining (NxN) representation. We will write the N^2-1 gauge potentials $A_\mu(x)$ in matrix form :

$$A_\mu(x) = \sum_a \frac{\lambda^a}{2} A_\mu^a(x) \tag{1.1}$$

with the normalisation

$$\mathrm{Tr}\, \lambda^a \lambda^b = 2\delta^{ab} \tag{1.2}$$

The gauge group SU(N) consists of NxN special unitary matrices $\Omega(x)$ and transforms the gauge potentials in the wellknown way :

$$A_\mu(x) \to A_\mu^\Omega(x) \equiv \Omega(x) A_\mu(x) \Omega^\dagger(x) + \frac{1}{i} \Omega(x) \partial_\mu \Omega^\dagger(x) \tag{1.3}$$

Then we can define the gauge covariant curl $G_{\mu\nu}(x)$ as follows :

$$G_{\mu\nu} = \partial_\mu A_\nu - \partial_\nu A_\mu - i\left[A_\mu, A_\nu\right] \tag{1.4}$$

The action of QCD is gauge invariant and reads

$$S = \frac{1}{2g^2} \mathrm{Tr}\, G_{\mu\nu} G_{\mu\nu} \tag{1.5}$$

The bare charge g is -apart from eventual massterms for the quarks and the θ parameter- the only parameter in the theory.

The QCD action (1.5) gives upon quantization rise to a theory that is thought to marry two aspects of hadron physics :

i) asymptotic freedom : a quark-antiquark pair feels at short distances $R \ll \Lambda^{-1}$ a mildly modified Coulomb force :

$$V(R) \cong -\frac{1}{4\pi R} \frac{1}{|\log \Lambda R|} \tag{1.6}$$

ii) confinement : at large distances a constant force acts between the quarks :

$$V(R) = \sigma R \tag{1.7}$$

The experimental values of Λ ($\sqrt{\sigma}$) are about 100 MeV (400 MeV) respectively.

The Coulomb force (1.6) is computed in terms of the small fluctuations of the potentials A, i.e. for small values of the bare charge g.

The constant force at large distance (1.7) has been verified from large fluctuations of the potentials and for large values of the bare coupling but with the unfortunate proviso that the cut-off in the theory is kept fixed.

Nobody has succeeded in a reliable computation of the ratio of the two length scales appearing in the force laws (1.6) and (1.7), based on theoretical understanding. It was Creutz's famous Monte-Carlo simulation[3] that produced an acceptable ratio and paved the way for more elaborate simulations producing mass spectra for the hadrons. Still, this novel way of approach has to be considered as exploratory. It is not only relevant for the hadron spectrum generated by the (u,c,b) families, but also for the more pretentious theory like technicolor, SU(5) and so on.

I.2. Lattice Version of QCD

The quantization of the theory necessitates a cut-off. Wilson[1] introduced the lattice cut-off by discretizing space in a four dimensional Euclidean world into a four dimensional hypercubic lattice with lattice points $n = (n_0 n_1 n_2 n_3)(n_\mu$ any integer for $\mu = 0,1,2,3)$.

Let $\hat{\mu}$ be the unit vector in the direction μ on such a lattice. Then on a link (n,μ) between the points n and $n+\hat{\mu}$ lives an SU(N) valued matrix

$$U_\mu(n) = \exp iag A_\mu(n) \tag{1.8}$$

where a is the lattice length.

The gauge transformations Ω live on the lattice points n and transform the link variables $U_\mu(n)$ as follows :

$$U_\mu(n) \to U_\mu^\Omega(n) = \Omega(n)\, U_\mu(n)\, \Omega^+(n+\hat{\mu}) \tag{1.9}$$

To lowest order in the lattice length a the transformation law (1.9) reduces to (1.3).

The analogue of the gauge covariant curl $G_{\mu\nu}$ in equ.(1.4) is constructed from the link variables on the border of a plaquette P given by a point n and two directions μ and ν :

$$U_{\mu\nu}(n) = U_\mu(n)\, U_\nu(n+\hat{\mu})\, U_\mu^+(n+\hat{\nu})\, U_\nu^+(n) \equiv \exp iga^2 \psi_{\mu\nu}(n) \tag{1.10}$$

The quantity $\psi_{\mu\nu}(n)$ is the analogue of $G_{\mu\nu}(n)$. The transformation property of $\psi_{\mu\nu}(n)$ under (1.9)

$$\psi_{\mu\nu}(n) \to \Omega(n)\, \psi_{\mu\nu}(n)\, \Omega^+(n) \tag{1.11}$$

and its formal limit to lowest order in a :

$$\psi_{\mu\nu}(n) = \frac{1}{a}\Delta_\mu^+ A_\nu(n) - \frac{1}{a}\Delta_\nu^+ A_\mu(n) - ig\left[A_\mu(n), A_\nu(n)\right] \tag{1.12}$$

justify the analogy. Δ_μ^+ is the lattice derivative :

$$\Delta_\mu^+ A_\nu(n) \equiv A_\nu(n+\hat{\mu}) - A_\nu(n) \tag{1.13}$$

Now we have to define the lattice action S_L. This can be done in many ways provided the continuum action (1.5) is retrieved when expanding S_L in terms of a.

The simplest choice is that of Wilson :

$$S_W = \frac{1}{g^2} \sum_{n;\mu,\nu} \mathrm{Tr}\, U_{\mu\nu}(n) \qquad (1.14)$$

Another choice is one where not only the defining representation of SU(N) but also the adjoint representation appear :

$$S_M = \frac{1}{g^2} \sum_{n;\mu,\nu} \mathrm{Tr}\, U_{\mu\nu}(n) + \frac{1}{g_A^2} \sum_{n;\mu,\nu} \left| \mathrm{Tr}\, U_{\mu\nu}(n) \right|^2 \qquad (1.15)$$

There are other choices discussed in the literature but we shall not be concerned with them in these lectures.

The action S_L determines the quantum average of observables $O(\{ U_\mu(n) \})$:

$$\langle O \rangle = \frac{1}{Z} \int \prod_{n,\mu} dU_\mu(n)\, O(\{U_\mu(n)\}) \exp S_L(\{U_\mu(n)\}) \qquad (1.16)$$

where the partition function Z is defined by

$$Z = \int \prod_{n,\mu} dU_\mu(n) \exp S_L(\{U_\mu(n)\}) \qquad (1.17)$$

Averages like (1.16) will depend on the form of S_L and the parameters (e.g. g^2, g_A^2) appearing in S_L.

1.3. The Coulomb Force in Lattice QCD

The force between a pair of quark-antiquarks is computed from a rectangular Wilson loop C (see Fig. 1) with fixed width R,

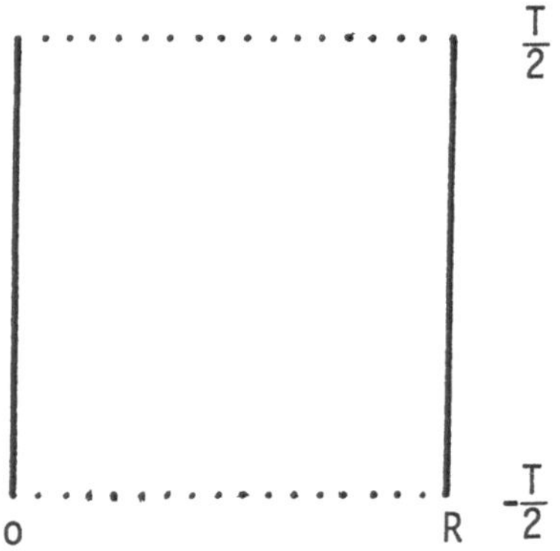

Fig. 1
Wilson loop.

the distance between the quarks, and a time span T which tends to infinity. At large $|T|$ the gauge fields are supposed to be pure gauge configurations :

$$U_\mu(n) = \Omega(n)\Omega^+(n+\hat{\mu}) \qquad na = (\vec{x},T) \; , |T| \to \infty \tag{1.18}$$

so that the Wilson loop on the dotted parts of loop reduces to the unit matrix.

Then we have for the average of the Wilson loop calculated with the Wilson action (1.14) :

$$\left\langle \frac{\mathrm{Tr}\, U(C)}{N} \right\rangle \underset{|T|\to\infty}{\simeq} \exp\left[- V(R)T - \varepsilon T - \ldots\right] \tag{1.19}$$

where $V(R)$ is the potential we are interested in. ε is the self-energy of the quarks, the dots indicate terms non-leading in T.

The perturbative calculation follows the usual rules[6] : expand the action of Wilson, equ.(1.14), in terms of the vector potential equ.(1.8). The resulting bilinear part in the potentials is only invertible when we add a gauge fixing term, and the concomitant Faddeev-Popov ghosts. In Feynman gauge the propagator takes the simple form :

$$\frac{1}{4\sum_\mu \sin^2 \frac{a}{2} p_\mu} \quad \delta_{\rho\sigma}\, \delta^{ab} \tag{1.20}$$

where $-\pi \leq ap_\mu \leq \pi$, $\mu = 0,1,2,3$. As the lattice length a shrinks to zero the propagator takes the continuum Euclidean form $1/p^2$.

When calculating the average (1.19) we take connected graphs ; some of them are shown in fig. 2. The lowest order graph fig. 2(a) is easily evaluated and gives in the limit of $a \to 0$

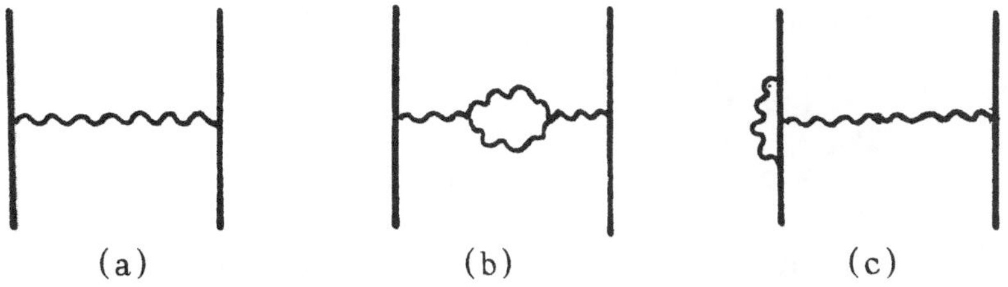

Fig. 2.
Tree (a) and one loop (b), (c) contributions to the heavy quark potential.

the Coulomb law :

$$V(R) = -\frac{1}{4\pi R} g^2 N (1 - N^{-2}) \tag{1.21}$$

The one loop graphs, only partially shown in Fig. 2(b) and (c), do contribute to the potential :

$$V(R)\Big|_{\text{one loop}} = \frac{1}{4\pi R} (g^2 N)^2 (1-N^{-2}) (b_0 \log a^2 R^{-2} + F_{oL}) \tag{1.22}$$

This calculation is complicated, due to the proliferation of vertices in the lattice action and Wilson loop. The coefficient in front of the divergence is universal ; it is obtained in any regularisation scheme, e.g. the dimensional one :

$$V(R)\Big|_{\text{one loop}} = \frac{1}{4\pi R} (g^2 N)^2 (1-N^{-2}) (b_0 \log(\mu R)^{-2} + F_{od}) \tag{1.23}$$

The arbitrary mass scale μ enters always the dimensional regularisation. It is the finite part F_{oL} (F_{od}) that can be different. F_{od} has been computed[18] and so has been b_0 :

$$b_0 = \frac{11}{3} \frac{1}{16\pi^2} \tag{1.23a}$$

Adding tree and one loop expressions gives

$$V(R) = -\frac{1}{4\pi R} g^2(R) N \tag{1.24}$$

where the renormalised coupling obeys

$$\tilde{g}(R)^2 \equiv g(R)^2 N = \tilde{g}^2 [1 - \tilde{g}^2 (b_0 \log a^2 R^{-2} + F_{oL}) + O(\tilde{g}^4)] \tag{1.24a}$$

Clearly, from its physical meaning, this definition of renormalised coupling is gauge choice independent. However it is hard to compute on the lattice. Other, less physical definitions, like that with a background field (see Lecture II) are much simpler computationally. From Callan and Symanzik's work[5] we know that $\tilde{g}^2(R)$ obeys to all orders in perturbation theory :

$$\begin{aligned} R \frac{\partial}{\partial R} \tilde{g}^2(R)\Big|_{\tilde{g}^2, a} &= 2\beta(\tilde{g}^2(R)) \\ &\equiv 2\beta_0 \tilde{g}^4(R) + 2\beta_1 \tilde{g}^6(R) + 2\beta_2 \tilde{g}^8(R) + \cdots \end{aligned} \tag{1.25}$$

Since $\tilde{g}^2(R)$ is the physical potential between heavy quarks we expect equ.(1.25) to be the same in whatever regularisation scheme we compute. We will make use of this property in Section I.6 and

Lecture II. From equ.(1.24a) (or equ.(1.23)) we find $b_0 = \beta_0$ and the two loop coefficient β_1 is known as well[27,28]:

$$\beta_1 = \frac{34}{3}\frac{1}{(16\pi^2)^2} \tag{1.26}$$

Solving equ.(1.25) for R in terms of $\tilde{g}^2(R)$ gives :

$$2\log\Lambda_{coul}R = \frac{-1}{\beta_0\tilde{g}^2(R)} - \frac{\beta_1}{\beta_0^2}\log\beta_0\tilde{g}^2(R) - \left(\frac{\beta_2}{\beta_0^2} - \frac{\beta_1^2}{\beta_0^3}\right)\tilde{g}^2(R) + \ldots \tag{1.27}$$

The parameter Λ_{coul} is an integration constant independent of R and $\tilde{g}^2(R)$. For R much smaller than Λ_{coul} the Coulomb potential becomes :

$$V(R) = \frac{1}{4\pi R}\left(1 - N^{-2}\right)\frac{1}{2\beta_0\log\Lambda_{coul}R} \tag{1.28}$$

so Λ_{coul} is in principle measurable.

For R larger than Λ_{coul} perturbation theory breaks down and we have to resort to strong coupling methods to determine the potential.

I.4. Strong Coupling on the Lattice

Let us suppose the physical distance R between the quarks much larger than Λ_{coul}. Lattice regularisation is ideally suited for strong coupling expansions[8,9,10]. Only the bare outlines will be sketched here. Consider Fig. 3 where a Wilson loop of size $R = ra \times T = ta$ is shown. Strong coupling with the Wilson action (1.8) consists of expanding the average (1.14) in terms of $1/g^2$ and integrating out the link variables. The two dominant graphs are shown in Fig. 3a and 3b. The result for the potential V(R) defined

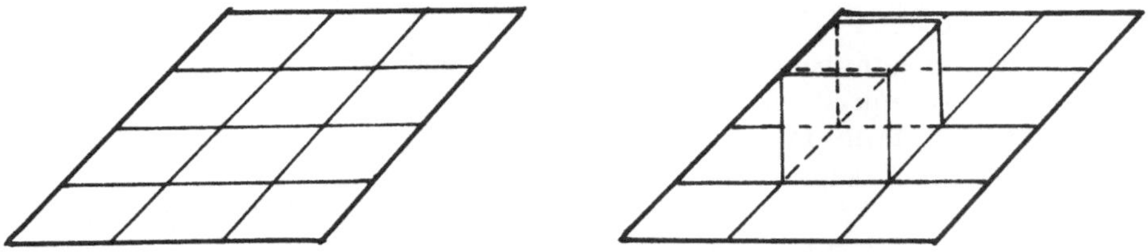

(a) (b)

Fig. 3
Dominant strong coupling graph (a) and cubic correction (b)

in (1.19) is then :

$$V(R) = \frac{1}{a^2}\,\rho(\tilde{g}^{-2})\,R \tag{1.29}$$

where $$\rho(\tilde{g}^{-2}) = \log \tilde{g}^2 - 4(\tilde{g}^{-2})^4 - \cdots \tag{1.30}$$

The lattice string tension falls off quite rapidly with decreasing $\tilde{g}^2$. It is known to have a non-zero radius of convergence[11]. So without further hypotheses the confining behaviour (1.29) is restricted to some finite range in $\tilde{g}^{-2}$.

I.5. The Missing Link

Let us hypothesize that ρ, the lattice string tension, is non-zero for any value of $\tilde{g}$, even in weak coupling. Then, keeping the physics fixed, when the lattice length a goes to zero we have from equ.(1.29) and (1.7) :

$$\frac{1}{a^2}\,\rho(\tilde{g}^{-2}) = \sigma \tag{1.31}$$

As we keep the physics fixed the bare coupling $\tilde{g}$ will become cut-off dependent, and we have the Gell-Mann-Low equation :

$$a\frac{\partial}{\partial a}\tilde{g}^2\Big|_{\tilde{g}^2(R)} = 2b_0\tilde{g}^4 + 2b_1\tilde{g}^6 + 2b_{2L}\tilde{g}^8 + \cdots \tag{1.32}$$

The coefficients b_0 and b_1 are the same as those appearing in equ. (1.25): the third coefficient is depending on the type of regularisation and lattice action used, hence the suffix L (see Section I.6).
Integrating equ.(1.32) to get the cut-off as a function of $\tilde{g}$ gives :

$$2\log a\Lambda_L = \frac{-1}{b_0\tilde{g}^2} - \frac{b_1}{b_0^2}\log b_0\tilde{g}^2 - \left(\frac{b_{2L}}{b_0^2} - \frac{b_1^2}{b_0^3}\right)\tilde{g}^2 + \cdots \tag{1.33}$$

The integration constant Λ_L does not depend on a and $\tilde{g}$. It follows from equ.(1.31) that the lattice string tension behaves in the weak coupling like

$$\log\rho = \log\frac{\sigma}{\Lambda_L^2} - \frac{1}{b_0\tilde{g}^2} - \frac{b_1}{b_0^2}\log b_0\tilde{g}^2 - \left(\frac{b_{2L}}{b_0^2} - \frac{b_1^2}{b_0^3}\right)\tilde{g}^2 + \cdots \tag{1.34}$$

The first two terms in (1.34) have been fitted from Montecarlo[38] simulations for SU(2)[3], SU(3)[29] and higher groups SU(N)[30] as shown in Fig. 4. Table 1 gives the values of the transition couplings where the scaling behaviour (1.34) sets in. The (still

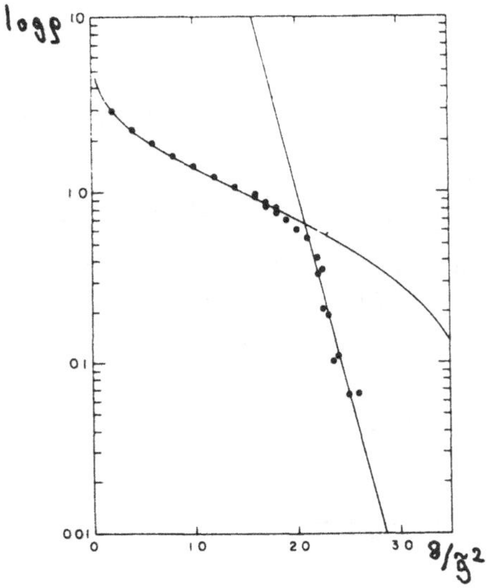

Fig. 4
Lattice string tension as a function of bare coupling for SU(2).

Transition couplings and Λ-scales for various SU(N) groups.

Table 1

N	$\tilde{g}^{-2}$	$\Lambda_L \sigma^{-\frac{1}{2}} \times 10^3$
2	.28	11 ± 2
3	.31	6 ± 1
4	.32	4.5 ± .5
5	.33	> 5
∞	.34-.38	> 4

changing) values of $\Lambda_L \sigma^{-\frac{1}{2}}$ have been put in as well. Their surprisingly small values are discussed in the next section. Obviously knowledge of the $O(\tilde{g}^2)$ term is needed, in view of the numbers in Table 1.

I.6. The Parameters in the Lattice String Tension

The asymptotic form of the string tension (1.34) contains the known universal parameters b_0 and b_1 (equ.(1.23a) and (1.26)). However Λ_L and b_{2L} will change as the lattice action changes ; we would like to know how they change. Also we would like to know their relationship to similar quantities in dimensional regularisation (Λ_{min} and b_{2d}). This relationship is basically due to the fact that the Coulomb potential $V(R)$ is independent of the regularisation used. To see this in detail we rewrite equs.(1.23) and (1.24) in their full glory :

$$\tilde{g}^{-2}(R) = \tilde{g}^{-2}(a) + b_0 \log a^2R^{-2} + F_{0L} + \tilde{g}^2(a)\left(b_1 \log a^2R^{-2} + F_{1L}\right) + \\ + \tilde{g}^4(a)\left\{-\tfrac{1}{2}b_0 b_1 \left(\log a^2R^{-2}\right)^2 + \left(b_{2L} - b_0 F_{1L}\right)\log a^2R^{-2} + F_{2L}\right\} \quad (1.35)$$

and

$$\tilde{g}^{-2}(R) = \tilde{g}^{-2}(\mu) + b_0 \log(\mu R)^{-2} + F_{0d} + \tilde{g}^2(\mu)\left(b_1 \log(\mu R)^{-2} + F_{1d}\right) +$$

$$+ \tilde{g}^4(\mu)\left\{-\tfrac{1}{2} b_0 b_1 \left(\log(\mu R)^{-2}\right)^2 + (b_{2d} - b_0 F_{1d}) \log(\mu R)^{-2} + F_{2d}\right\} \quad (1.36)$$

The coefficients of the logarithms in these equations are such that the unrenormalized couplings $\tilde{g}$ (a) ($\tilde{g}$ (μ)) do obey the Gell-Mann-Low equations

$$-\tfrac{1}{2} a \frac{\partial}{\partial a} \tilde{g}^{-2}(a)\Big|_{\tilde{g}^2(R)} = b_0 + b_1 \tilde{g}^2(a) + b_{2L} \tilde{g}^4(a) + \cdots \quad (1.37)$$

$$-\tfrac{1}{2} \mu \frac{\partial}{\partial \mu} \tilde{g}^{-2}(\mu)\Big|_{\tilde{g}^2(R)} = b_0 + b_1 \tilde{g}^2(\mu) + b_{2d} \tilde{g}^4(\mu) + \cdots \quad (1.38)$$

The requirement that the same Callan-Symanzik equation (1.25) comes out from (1.35) and (1.36) leads to :

$$b_0 = \beta_0 \quad (a)$$

$$b_1 = \beta_1 \quad (b)(1.39)$$

$$b_{2L} - b_0 F_{1L} + b_1 F_{0L} = b_{2d} - b_0 F_{1d} + b_1 F_{0d} = \beta_2 \quad (c)$$

The first two equalities we knew already. The third one relates b_{2L} to b_{2d} through knowledge of the two and one loop finite parts. Since b_{2d} is known[21] the computation of b_{2L} involves the latter one and two loop quantities. This is necessary to calculate the $O(\tilde{g}^2)$ term in the lattice string-tension.

The comparison between b_{2L} and $b_{2L'}$ (defined by two lattice actions S_L and $S_{L'}$) is much easier and has been done recently [14,15] (see Lecture II). The integration constants Λ_L and Λ_{min} appear by solving (1.37) and (1.38) (see e.g. (1.33) for Λ_L, Λ_{min} is defined analogously). It is easy to verify from the definition of Λ_{coul}, equ.(1.27), that both Λ_L and Λ_{min} are related to Λ_{coul} by :

$$2 \log \Lambda_{coul} \Lambda_L^{-1} = -\frac{1}{\beta_0} F_{0L} \quad 1.40(a)$$

$$2 \log \Lambda_{coul} \Lambda_{min}^{-1} = -\frac{1}{\beta_0} F_{0d} \quad 1.40(b)$$

Thus Λ_L can be related through knowledge of the one loop finite parts[4] to any other Λ -scale parameter[7] . This relationship was first examined in the pioneering work of A. and P. Hasenfratz[4] . Their result showed a large ratio between the Λ_{mom}[7] and the Wilson action lattice scale Λ_L, thus explaining the large ratio of $\sigma^{1/2}\Lambda_L^{-1}$ in Montecarlo simulations (see Table 1). To wit : the existence of Gell-Mann-Low equations for a given scheme (like in (1.37) or (1.38)) is equivalent to the existence of C-S-equations (like (1.25)). This is simple to prove. That the C-S-equations (1.25) are the same for all schemes is non-trivial. For lattice gauge theories this is known to be case up and including two loops (see end of Lecture II).

LECTURE II. THE MIXED ACTION

In this lecture we will study the mixed action (1.15), slightly rewritten :

$$N^{-2} S_M = \tilde{g}_F^{-2} \sum_{n,\mu,\nu} \frac{\mathrm{Tr}\, U_{\mu\nu}(n)}{N} + \frac{1}{2} g_A^{-2} \sum_{n,\mu,\nu} \left| \frac{\mathrm{Tr}\, U_{\mu\nu}(n)}{N} \right|^2 \qquad (2.1)$$

As usual the tilde on a coupling constant means

$$\tilde{g}_F^2 = g_F^2 N$$

The question we ask is : what are the curves of equal string tension in the $(\tilde{g}_F^{-2}, g_A^{-2})$ plane, supposing we are in the weak coupling régime ?

If these theoretical curves do match the Montecarlo curves beyond the transition points (see Table 1) then we have additional evidence that indeed we are in the scaling region.

II.1. Effective Coupling versus Bare Coupling.

The answer to the question is partly contained in the work of Section I.6. Let us introduce an effective coupling $\tilde{g}_{eff}$ which is a function of $\tilde{g}_F$ and g_A and describes curves of equal string tension :

$$\tilde{g}_{eff}^2 = \tilde{g}_{eff}^2 (\tilde{g}_F^2, g_A^2) \qquad (2.2)$$

Put this effective coupling in the pure Wilson action

($g_A^{-2} = 0$ in equ.(2.1)) to compute $\tilde{g}^2(R)$ appearing in the "Coulomb law" (1.24) :

$$\tilde{g}^{-2}(R) = \tilde{g}^{-2}_{eff} + b_0 \log a^2 R^{-2} + F_{0W} + \tilde{g}^{2}_{eff}\left(b_1 \log a^2 R^{-2} + F_{1W}\right)$$

The suffix W serves to remind the reader that the finite parts are computed from the pure Wilson action. A suffix A will indicate that the g_A^{-2} term in the mixed action is taken into account.

On the other hand we can take the mixed action S_M (2.1) and expand the link variables $U_\mu(n) = \exp i A_\mu(n)$. The quadratic part in A_μ will be multiplied by a coefficient which we will call the bare coupling and equals :

$$\tilde{g}_b^{-2} = \tilde{g}_F^{-2} + g_A^{-2} \tag{2.4}$$

Then the Coulomb force becomes :

$$\tilde{g}^{-2}(R) = \tilde{g}_b^{-2} + b_0 \log a^2 R^{-2} + F_{0A} + \tilde{g}_b^2\left(b_1 \log a^2 R^{-2} + F_{1A}\right) + \ldots \tag{2.5}$$

Compare now equations (2.3) and (2.5). We find easily a relation between $\tilde{g}^2_{eff}$ and $\tilde{g}^2_b$:

$$\tilde{g}_{eff}^{-2} = \tilde{g}_b^{-2} + (F_{0A} - F_{0W}) + \tilde{g}_b^2 (F_{1A} - F_{1W}) + \ldots . \tag{2.6}$$

To lowest order the effective coupling equals the bare coupling (2.4) : this gives a set of parallel straight lines in the ($\tilde{g}_F^{-2}, \tilde{g}_A^{-2}$) plane (see Fig. 5) and clearly necessitates corrections.

The one-loop correction is given by the second term in equ.(2.6). The actual calculation of this term and the two loop correction is discussed in Section 2.4. The one loop result is plotted in Fig. 5 for SU(2) and $\tilde{g}_{eff}^{-2} = .31$; a value which we would say from Table 1 is in the scaling region. However the prediction[15] fails dramatically.[44] The two loop result does improve somewhat the comparison with the data. Note that in the comparison between equ.(2.3) and (2.5) the logarithms drop out. This phenomenon persists on the three and higher loop level as well : it is a simple consequence of the Callan-Symanzik equation (1.25) being the same for any regularisation scheme (see Section I.6). That indeed the Callan-Symanzik equations are the same on the two loop level for pure Wilson and mixed action is also borne out by direct calculation (see Section II.4).

Accepting identical C-S equations we can easily understand the absence of logarithms in equ.(2.6).

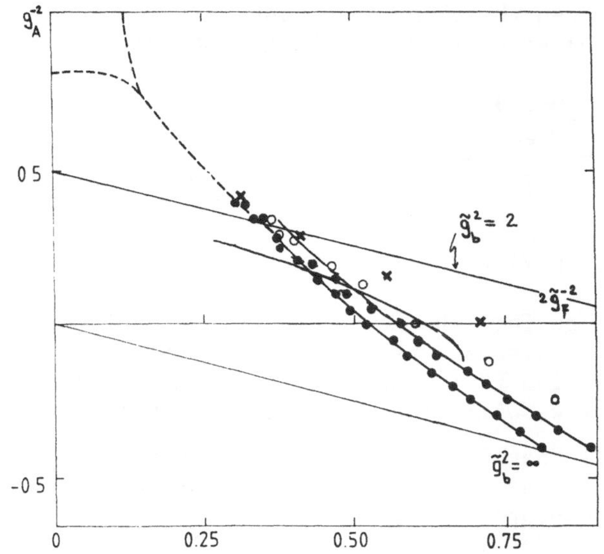

Figure 5
Failure of the one loop corrected effective coupling (drawn curve) to match the equal stringtension data (empty circles). Dots related by lines and crosses do also represent equal stringtension data for various values of ρ .

Call $L \equiv \log a^2 R^{-2}$; then we have from equations (2.3) and (2.5)

$$\frac{\partial}{\partial L} \tilde{g}^{-2}(R)\Big|_{\tilde{g}^2_{eff}} = \beta(\tilde{g}^2(R)) \tag{2.3'}$$

and

$$\frac{\partial}{\partial L} \tilde{g}^{-2}(R)\Big|_{\tilde{g}^2_b} = \beta(\tilde{g}^2(R)) \tag{2.5'}$$

The left hand side of equ. (2.5') can also be expressed as :

$$\frac{\partial}{\partial L} \tilde{g}^2(R)[\tilde{g}^2_{eff}(\tilde{g}^2_b, L), L] =$$

$$\frac{\partial \tilde{g}^2(R)}{\partial \tilde{g}^2_{eff}}\Big|_L \frac{\partial \tilde{g}_{eff}}{\partial L}\Big|_{\tilde{g}^2_b} + \frac{\partial}{\partial L} \tilde{g}^2(\tilde{g}^2_{eff}, L)\Big|_{\tilde{g}^2_{eff}} \tag{2.7}$$

The second term on the right hand side of (2.7) equals the term on the left hand side of (2.7) ; both equal $\beta(\tilde{g}^2(R))$, so indeed :

$$\frac{\partial}{\partial L}\tilde{g}^2_{eff}\Big|_{\tilde{g}^2_b} = 0$$

No logarithms are left when we express $\tilde{g}^2_{eff}$ in terms of $\tilde{g}^2_b$.

II.2. The Large N View Point

Before we go into the actual calculation of the effective coupling we will discuss its behaviour in the large N limit[22] of the mixed action (2.1). Apart from the numerical success of the large N limit we will also get some nice insights.

As is well known[23a] correlations between physical observables O_i do vanish in the limit of large N. This has been checked in strong and weak coupling expansions :

$$\langle O_1 O_2 \cdots \rangle = \langle O_1 \rangle \langle O_2 \rangle \cdots + O(N^{-2}) \tag{2.8}$$

Based on this factorisation we can show the following : in the large N limit the physics of the mixed action is described by a pure Wilson theory with an effective coupling $\tilde{g}^2_{eff}$ obeying the equation :

$$\tilde{g}^{-2}_{eff} = \tilde{g}_F^{-2} + g_A^{-2}\left\langle \frac{\mathrm{Tr}\, U_{\mu\nu}(n)}{N}\right\rangle_W (\tilde{g}^2_{eff}) \tag{2.9}$$

The average plaquette value in equ.(2.9) is calculated in the pure Wilson theory as indicated by the suffix W.

To prove this statement let us study the expectation value of some observable O in the mixed theory (2.1) :

$$\langle O \rangle (\tilde{g}_F^2, g_A^2) = Z^{-1}\int DU \; O \exp S_M \tag{2.10}$$

where Z is the partition function

$$Z = \int DU \exp S_M \tag{2.11}$$

We rearrange the mixed action in the following way :

$$N^{-2} S_M = \left(\tilde{g}_F^{-2} + g_A^{-2}\left\langle \frac{\mathrm{Tr}\, U_{\mu\nu}(n)}{N}\right\rangle\right) \sum_{n,\rho,\sigma} \frac{\mathrm{Tr}\, U_{\rho\sigma}(n)}{N} +$$

$$+ \frac{1}{2} g_A^{-2} \sum_{n,\rho,\sigma} \left| \frac{\mathrm{Tr}\, U_{\rho\sigma}(n)}{N} - \left\langle \frac{\mathrm{Tr}\, U_{\mu\nu}(n)}{N}\right\rangle \right|^2 + \tag{2.12}$$

$$- \frac{1}{2} g_A^{-2} \sum_{n,\mu,\nu} \left| \frac{\mathrm{Tr}\, U_{\mu\nu}(n)}{N} \right|^2 \tag{2.12}$$

and define $S_c \equiv \frac{1}{2} \sum_{n,\rho,\sigma} \left| \frac{\mathrm{Tr}\, U_{\rho\sigma}(n)}{N} - \left\langle \frac{\mathrm{Tr}\, U_{\mu\nu}(n)}{N} \right\rangle \right|^2$

The average plaquette value in (2.12) is computed with the action S_M. The last term can be omitted in averages. We call the coefficient of the first (pure Wilson) term the effective coupling $\hat{g}_{eff}^{-2}$.

Then we consider the average $\langle 0 \rangle$ in the large N limit ; expand the term $g_A^{-2}\, S_c$ in the mixed action (2.12) and find after some trivial algebra :

$$\langle 0 \rangle (\tilde{g}_F^2, g_A^2) = \langle 0 \rangle_w (\tilde{g}_{eff}^2) + g_A^{-2} \langle (0 - \langle 0 \rangle_w) S_c \rangle + \ldots\ldots \tag{2.13}$$

The second term in (2.13) is a correlation and is negligible with respect to the first term according to factorisation (2.8). So at $N = \infty$ we find

$$\langle 0 \rangle (\tilde{g}_F^2, g_A^2) = \langle 0 \rangle_w (\tilde{g}_{eff}^2)$$

which is in particular true for $0 = \frac{\mathrm{Tr}\, U_{\mu\nu}(n)}{N}$

Thus we have for the effective coupling at $N = \infty$ indeed equation (2.9).

This equation predicts constant physics lines in the $(\tilde{g}_F^{-2}, g_A^{-2})$ plane, with a slope given by the value of the mean action. This slope becomes more negative the bigger $\tilde{g}_{eff}^2$, and the pattern that emerges fits qualitatively amazingly well the data of SU(2)[17] !

Let us look at the weak coupling expansion of the mean action ; it is known up to two loops [31]:

$$\left\langle \frac{\mathrm{Tr}\, U_{\mu\nu}(n)}{N} \right\rangle (\tilde{g}_{eff}^2) = 1 - w_1\, \tilde{g}_{eff}^2 - w_2\, \tilde{g}_{eff}^4 - \ldots \tag{2.14}$$

with :

$$w_1 = \frac{1}{8} \frac{N^2-1}{N^2} \quad , \quad w_2 = \frac{N^2-1}{4N^2} \left(0{,}0203 - \frac{1}{32N^2} \right) \tag{2.14a}$$

From equations (2.9) and (2.14) we find easily that $\tilde{g}_{eff}^2$ and $\tilde{g}_b^2$ (equ.(2.4)) are related :

$$\tilde{g}_{eff}^{-2} = \tilde{g}_b^{-2} + r w_1 + \tilde{g}_b^2 \left(-r^2 w_1^2 + r w_2 \right) + \ldots \tag{2.15}$$

Here the parameter $r = -\tilde{g}_b^2 \, g_A^{-2}$ is constant along lines emanating from the origin in the $(\tilde{g}_F^{-2}, g_A^{-2})$ plane. The values of w_1 and w_2 are of course taken at $N = \infty$, from (2.14a). Thus at $N = \infty$ the computation of the one end two loop finite parts in equ.(2.6) is already done !

The corrections to the equation (2.14) for $\tilde{g}_{eff}^2$ are of order N^{-2}, N^{-4},... They have been evaluated up and including two loops [14,15]. How to do them will be explained in the next two sections.

II.3. Background Field Method and BRS Symmetry on the Lattice

We are now faced with computation of the finite parts appearing in the effective coupling.

First one should realize that the heavy quark-potential, though a physical quantity, is from a computational point of view not useful : there is a proliferation of diagrams already at the one loop level. Therefore we must look for another process which is computationally easier. The method of the background field is very well suited. One imagines a classical ("background") field on the lattice[12] with link variables $U_{c\mu}(n) = \exp ia\, B_\mu(n)$. $B_\mu(n)$ should vary slowly with respect to the lattice length a and should be a local minimum of the lattice action S_L. The background field is coupled to the quantum fields $V_\mu(n) = \exp i\, g\, Q_\mu(n)$ through the lattice action $S_L(U)$ with $U_\mu(n) = V_\mu(n)\, U_{c\mu}(n)$.

One computes in a loop expansion the fluctuations of the quantum field $V_\mu(n)$ aromund the classical field $U_{c\mu}(n)$. One has to introduce a gauge fixing term to get rid of the fluctuations in the direction of the gauge transforms of the background field :

$$G(n) = \sum_\mu \Delta^-_\mu(U_c) Q_\mu(n) \equiv \sum_\mu \left(U_{c\mu}(n-\hat{\mu})\, Q_\mu(n-\hat{\mu})\, U^+_{c\mu}(n-\hat{\mu}) - Q_\mu(n) \right)$$

The crucial point is that this gauge fixing respects a symmetry :

$$\begin{aligned} U_{c\mu}(n) &\to \Omega(n)\, U_{c\mu}(n)\, \Omega^+(n+\hat{\mu}) \\ Q_\mu(n) &\to \Omega(n)\, Q_\mu(n)\, \Omega^+(n) \end{aligned} \qquad (2.17)$$

This symmetry is a particular realisation of the gauge transformation $\Omega(n)$ acting on $U_\mu(n) = V_\mu(n)\, U_{c\mu}(n)$: it views the quantum field as a matter field, the background field as a true gauge field. To generate from the gauge fixing term the Faddeev-Popov term we introduce another realisation of the gauge transformation ; we keep the background field fixed :

$$\delta V_\mu(n) = \delta\Omega(n) V_\mu(n) - V_\mu(n) U_{c\mu}(n) \delta\Omega(n+\hat{\mu}) U^\dagger_{c\mu}(n)$$

$$\equiv -\left[\Delta^+_\mu (V U_c)\, \delta\Omega(n)\right] V_\mu(n) \tag{2.18a}$$

Define the anticommuting ghost field $\omega(n) = \omega^a(n)\frac{\lambda^a}{2}$ and an anticommuting variable $\delta\lambda$ such that :

$$\delta\Omega(n) = i\,\omega(n)\,\delta\lambda$$

and let

$$\delta\omega(n) \equiv -i\,\omega^2(n)\,\delta\lambda \tag{2.18b}$$

The anticommuting ghost field $\bar{\eta}(n)$ obeys :

$$\delta\bar{\eta}(n) = -\frac{1}{\alpha} G(n)\,\delta\lambda \tag{2.18c}$$

These are the B.R.S. transformations[24, 25] in the presence of a background field.

It is easy to see that the double variations of V and ω fields vanish :

$$\delta^2 V_\mu(n) = \delta^2\omega(n) = \delta^2 V^\dagger_\mu(n) = 0 \tag{2.19}$$

To see that the double variation of the field $Q_\mu(n)$ vanishes, form the hermitean field $H^a_\mu(n)$:

$$H^a_\mu(n) \equiv \frac{1}{2ig} \mathrm{Tr}\,\lambda^a \left(V_\mu(n) - V^\dagger_\mu(n)\right) = Q^a_\mu(n) + O(Q^3) \tag{2.20}$$

Clearly, from (2.19), the field $H^a_\mu(n)$ has double variation equal to zero. Let us invert (2.20) to express Q in terms of H. Then the double variation of Q becomes

$$\frac{\delta^2 Q^a_\mu(n)}{\delta\lambda^2} = -\frac{\delta^2 Q^a_\mu(n)}{\delta H^b_\rho(n)\,\delta H^c_\sigma(n)} \frac{\delta H^b_\rho(n)}{\delta\lambda} \frac{\delta H^c_\sigma(n)}{\delta\lambda} = 0 \tag{2.21}$$

since the last two factors in (2.21) anticommute.
Now we are on familiar ground ; the effective action[45] :

$$S_{eff} = S_L - \frac{1}{2\alpha} G.G + \bar{\eta}.\frac{\delta G}{\delta\lambda} + J.\frac{\delta Q}{\delta\lambda} + K.\frac{\delta\omega}{\delta\lambda} \tag{2.22}$$

is invariant under the BRS transformation (2.18).
Define the one particle irreducible generating functional Γ in the traditional[19,25] way, by adding source terms $j.Q$, $\bar{\xi}.\omega$ and $\bar{\eta}.\xi$ to S_{eff}. The resulting partition function Z is a functional of U_c, j, ξ and $\bar{\xi}$ and defines the connected generating functional W through

$$W \equiv \log Z$$

Then we have for Γ :

$$\Gamma = W - j\frac{\delta W}{\delta j} - \bar{\xi}\frac{\delta W}{\delta \bar{\xi}} - \frac{\delta W}{\delta \xi}.\xi \tag{2.23}$$

The BRS symmetry (2.18) implies that the generating functional $\hat{\Gamma} \equiv \Gamma + (2\alpha)^{-1} G^2$ obeys the well known BRS identities[24,25] since the measure $DU\, D\omega\, D\bar{\eta}$ is invariant :

$$\frac{\delta\hat{\Gamma}}{\delta Q}\cdot\frac{\delta\hat{\Gamma}}{\delta J} + \frac{\delta\hat{\Gamma}}{\delta\omega}\cdot\frac{\delta\hat{\Gamma}}{\delta K} = 0 \tag{a}$$

(2.24)

$$\left\langle\frac{\delta G}{\delta\lambda}\right\rangle - \xi = 0 \tag{b}$$

In these identities the background field is like an external parameter, since it is not varied by the BRS transformation (2.18). The gauge fixing term is that of equ.(2.16).

These identities are useful in two loop calculations when one wants to check properties of one loop subdiagrams.

II.4. The Renormalized Background Field Coupling

The process from which we want to calculate the renormalization of the coupling is the effective potential $\Gamma(U_c)$, which we get from Γ (equ.(2.23)) by putting all the quantum field expectation values equal to zero. So $\Gamma(U_c)$ has only external background field legs.

It is standard to derive from the symmetry (2.17) that

$$\Gamma(U_c) = \Gamma(U_c^{\Omega})$$

As we are interested in singular and finite parts only as the lattice length a shrinks to zero we find in this limit :

$$\Gamma(U_c) = -N\Gamma_0\, S(B) \tag{2.25}$$

where $S(B)$ is the continuum version 1.5 of the background field B_μ.

It is Γ_0 which replaces $\tilde{g}^2(R)$ in equations (2.3) and (2.5).

Equation (2.25) permits us to calculate Γ_0 from the two point function of the background field (see Fig. 6) : a great simplification with respect to the heavy quark potential computation.

Fig. 6
Two point function of the background field

All what one has to do to get the effective coupling in equation (2.6) is : compute the differences of finite parts in the background field propagator.

On the other hand the modification of equation (2.14) taking one and two loop effects into account is due to the analogue of equation (2.13) for the effective potential Γ (U_c) :

$$\Gamma(U_c) = \Gamma(U_c)\big|_{N=\infty} + g_A^{-2}\langle S_c\rangle + \frac{1}{2!}\, g_A^{-4}\langle S_c^2\rangle$$

Only one particle irreducible diagrams contribute to the averages. Some of them are shown in Fig. 7. The semi-circles in this figure

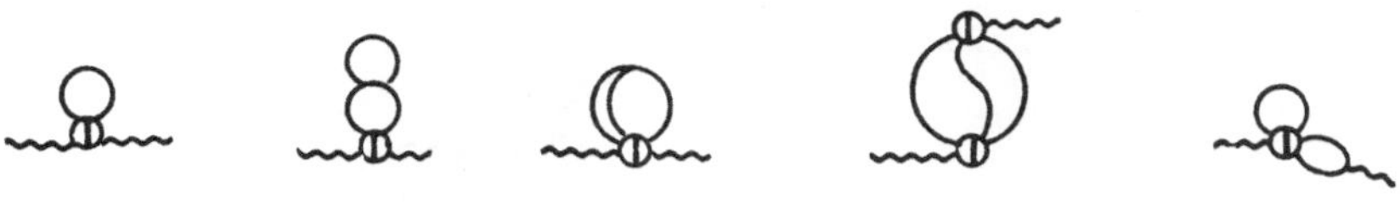

(a) (b) (c) (d) (e)

Fig. 7
Some diagrams contributing to the background field two point function

represent the operator $N^{-2}|\mathrm{Tr}\, U_{\mu\nu}|^2$ appearing in S_c as defined underneath equation (2.12). All other vertices are those appearing in the pure Wilson action. Wiggly lines represent the background field, internal lines the quantum field propagators in the gauge (2.16).

II.5. Results for the Effective Coupling

The two loop modification of Equation (2.14) becomes :

$$\tilde{g}_{eff}^{-2} = \tilde{g}_F^{-2} + g_A^{-2}\, w(\tilde{g}_{eff}^2) - g_A^{-2} \frac{1}{N^2-1}\left(1 - w^2(\tilde{g}_{eff}^2)\right) - g_A^{-2} \frac{N^2-3}{48N^4}\tilde{g}_{eff}^4 - g_A^{-4}\frac{N^2+1}{64N^4}\tilde{g}_{eff}^6 \qquad (2.26)$$

The first two terms are like in equation (2.14) ; the third term comes from diagrams like in Fig. 7(a) and (b) ; the fourth from diagrams like in Fig. 7(c) and the last from diagrams like in Fig. 7(d).

We used the expansion of the mean action $w(\tilde{g}_{eff}^2)$, equations (2.14) and (2.14 (a)), to plot equal string tension lines from equation (2.26) (Fig. 8). The one loop result in (2.26) still produces straight lines with the slope somewhat shifted, the two

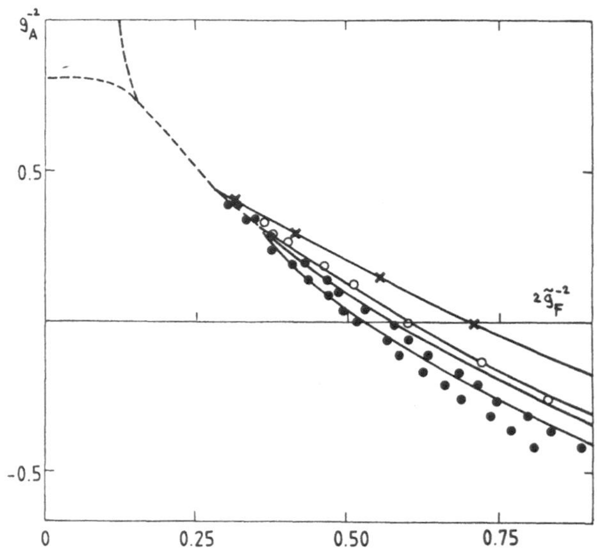

Fig. 8
Equal string tension curves from equ.(2.26).

loop result produces the slight bending of the curves. Comparison with the data is good.

Of course we can convert equation (2.26) in the corresponding equation (2.26). Some trivial algebra shows that :

$$F_{0A} - F_{0W} = \frac{1}{8} \cdot \frac{N^2+1}{N^2} \cdot r \qquad (2.27)$$

with r like in equation (2.15), which is the $N = \infty$ limit of (2.27), as it should.

The expression for $F_{1A} - F_{1W}$ is long, and since it does not improve drastically the bad result (2.27) plotted in Fig. 5 for $N = 2$, we don't quote it here[15]. A salient feature of the $1/N$ expansion of $\tilde{g}_{eff}$ is its rapid convergence, in contrast with the conventional expansion in terms of $\tilde{g}_b$.

Let us finally note that the BRS identities are instrumental in doing the computation on the two loop level : they permit us to relate the set of -individually singular - diagrams like in Fig. 7(e) to a simpler finite diagram[14]. This means absence of logarithms on the two loop level for the effective coupling, or, according to equation (2.7) equality of the two-loop coefficient in the C-S equations for Wilson and mixed theories.

For details, the reader is referred to the original papers[14,15].

LECTURE III. LARGE N REDUCED MODELS

Let us take a rest from the trying calculations in the preceding lectures and contemplate in this lecture the beautiful interplay between topology and the collapse of space into one single point for large N theories.

The idea of reduction is best illustrated by asking the following question : suppose we calculate by strong coupling methods the average of a Wilson loop sitting in a finite box with periodic boundary conditions in all directions. Then we ask : what are the effects on the average due to the finiteness of the box ?

In strong coupling a typical diagram contributing a finite size effect is one where the plaquettes are attached to the outside of the loop and connect through a surface to the walls of the periodic box. Due to the periodicity such a diagram is a torus with a window formed by the inside of the loop (see Fig. 9). Such a graph is generally of order $\exp(-\rho L)$ with L the length of the box. However due to its topology the graph is suppressed altogether at $N = \infty$ with respect to the planer graphs as depicted in Fig. 3. One can argue this way that a Wilson loop average does not feel the finiteness of the periodic box at $N = \infty$, at least in strong coupling !

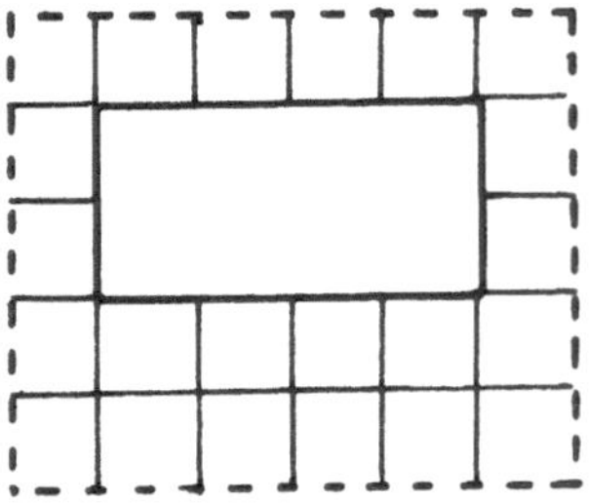

Fig. 9.
Wilson loop (drawn rectangle) connected to the (dashed) walls of the periodic box.

Eguchi and Kawai[32] pushed this phenomenon to its most extreme : they showed the following :

At $N = \infty$ the average of a Wilson loop on an infinite lattice $C(x, \mu, \nu, \rho, \ldots, \tau)$ obeys the same Schwinger Dyson equations as the Wilson loop $C(\mu, \nu, \rho, \ldots \tau)$ on a one point periodic lattice, provided the expectation value of any loop $C(\mu, \nu, \ldots)$ on the small lattice corresponding to an open loop on the big lattice is zero.

The notation for the loops is self-evident : the variable x is some point on the big lattice from which we start the loop in the μ direction, go on in ν, ρ ... direction and return to x through the τ direction. On the small lattice there is only one point x, only the directions $\mu, \nu, \ldots$ are present.

If the Schwinger-Dyson equations admit only one solution, then the theory on infinite lattice reduces to that on a one point lattice ! It soon turned out that the proviso in the Eguchi-Kawaï proposal was not valid at weak coupling [33]. This had been realized also in ref. 34 : it turns out that precisely those toroidal properties of the periodic box, that did suppress the finite size effects in strong coupling, do admit zero action solutions, which change perturbation theory radically at long wave lengths.
Above a critical dimension $d_c = 2N/(N-1)$ it is the quartic modes from the commutator term in the action that dominate and do change perturbation theory. For $N = \infty$ we have $d_c = 2$ and therefore the Eguchi-Kawaï proposal is not working in $d > 2$. Otherwise the idea is very attractive from a Montecarlo point of view (only $d(N^2-1)$ degrees of freedom) and intriguing from a theoretical point of view.

III.1. Curing the Eguchi Kawai Model

The disease at weak coupling can be cured[35] by a simple prescription : the action on the one point periodic lattice

$$S = g^{-2}\,\mathrm{Tr}\,(\mathbb{1} - U_\mu U_\nu U_\mu^\dagger U_\nu^\dagger) \tag{3.1}$$

has to be replaced by the twisted action

$$S_t = g^{-2}\,\mathrm{Tr}\,(\mathbb{1} - z_{\mu\nu} U_\mu U_\nu U_\mu^\dagger U_\nu^\dagger) \tag{3.2}$$

where the twist $z_{\mu\nu}$ is a phase factor from the center $Z(N)$ of the gauge group $SU(N)$:

$$z_{\mu\nu} \equiv \exp - i \frac{2\pi}{N} n_{\mu\nu} \tag{3.3}$$

The twist tensor $n_{\mu\nu}$ is antisymmetric and defined modulo N. The matrices U_μ ($\mu = 0, 1, 2, 3$) are in the group $SU(N)$.

The averages are now defined with the Boltzmann factor exp - S_t and the measure $\prod_\mu dU_\mu$.

The twisted action (3.2) will have zero action solutions[42] if and only if[39]

$$\frac{1}{4} n_{\mu\nu} \tilde{n}_{\mu\nu} = \sigma N, \; \tilde{n}_{\mu\nu} \equiv \varepsilon_{\mu\nu\rho} \frac{n_\rho}{2} \quad , \sigma \text{ integer} \tag{3.4}$$

This condition may at first sight seem strange. It is related to the Pontryagin index P as we will explain below. Equation (3.4) is necessary because zero action solution to (3.2) must obey :

$$U_\mu U_\nu = \exp\left(i \frac{2\pi}{N} n_{\mu\nu}\right) U_\nu U_\mu \tag{3.5}$$

This condition is mathematically the same as the condition that there exist zero action configurations in a box with twisted periodic boundary conditions[39] in the continuum. Since in the continuum the action 1.5 obeys:

$$S \geq 8\pi^2 |P| \; , \; P \doteq \frac{1}{32\pi^2} \int d^4x \, G^a_{\mu\nu} \tilde{G}^a_{\mu\nu} \tag{3.6}$$

and since[43]

$$P = \frac{N-1}{4N} \cdot \tilde{n}_{\mu\nu} n_{\mu\nu} + p \qquad p \text{ integer} \tag{3.7}$$

we have as necessary condition equ. (3.4). Its sufficiency has been shown in ref. 40 .

III.2. Conversion of Colour Degrees of Freedom into Momenta

Let us concentrate on a simple choice of $n_{\mu\nu}$:

$$n_{\mu\nu} = L \begin{pmatrix} 0 & 1 & 0 & 0 \\ -1 & 0 & 0 & 0 \\ 0 & 0 & 0 & 1 \\ 0 & 0 & -1 & 0 \end{pmatrix} \quad , \ L \text{ integer}, L^2 = N \tag{3.8}$$

With this choice one can show the following :

① The zero action solutions $\{\Gamma_\mu\}_0^3$ obeying (3.5) are up to unitary similarity transformation given by :

$$\Gamma_0 = P_L \times 1\!\!1_L \ , \Gamma_1 = Q_L \times 1\!\!1_L \ , \Gamma_2 = 1\!\!1_L \times P_L \ , \Gamma_3 = 1\!\!1_L \times Q_L \tag{3.9}$$

The L x L matrices P_L, Q_L are defined by [42]

$P_L e_k = e_{k-1}$ cyclically for any unit vector, k = 1, ..., L.

$Q_L = \text{diag}\,(1, \exp i\frac{2\pi}{L}, \ldots, \exp i\frac{2\pi}{L}(L-1))$

It follows that $P_L Q_L = \exp i\frac{2\pi}{L}\, Q_L P_L$, and with the direct product rule in (3.9) we recover (3.5). Any other set $\{\Gamma'_\mu\}_0^3$ obeying (3.5) is unitarily equivalent to $\{\Gamma_\mu\}_0^3$ in (3.9) :

$$\Gamma'_\mu = \Omega\, \Gamma_\mu\, \Omega^+ \qquad \mu = 0, 1, 2, 3 \tag{3.10}$$

This follows in the same way as Pauli's theorem for Dirac matrices[37].

② The matrices $\Gamma(p)$ defined by

$$\exp[i\frac{\pi}{N}(p_0 p_1 + p_2 p_3)]\,\Gamma_0^{-p_1}\,\Gamma_1^{p_0}\,\Gamma_2^{-p_3}\,\Gamma_3^{p_2} \tag{3.11}$$

(where the "momentum" components p_μ run from 0 to $L-1$) do span the Lie algebra of SU(N). This is due to two simple facts :

i) $\text{Tr}\, P_L^k = \text{Tr}\, Q_L^k = 0$ if $k \neq L\nu$, ν integer number

$\text{Tr}\,\Gamma(p) = 0$ $\qquad p \neq Lw$, w integer vector

ii) $\Gamma^+(p) = \Gamma(-p)$

$$\Gamma(p)\,\Gamma(p') = \exp\left(i\frac{\pi}{N}\, p_\mu n_{\mu\nu} p'_\nu\right)\Gamma(p+p')$$

These properties can easily checked by inspection, and it follows that any linear combination of the N^2 $\Gamma(p)$'s with

$$\sum_p a(p)\,\Gamma(p) = 0$$

implies after multiplication with $\Gamma^+(p')$ and taking the trace, that $a(p') = 0$ for any p'. So the $N^2\ \Gamma(p)$ are linearly independent and the N^2-1 traceless $\Gamma(p)$ do span the Lie algebra of SU(N).

③ Consider the fluctuations $V_\mu = e^{iQ_\mu}$ around a "background field" Γ_μ:

$$U_\mu = V_\mu \Gamma_\mu$$

By use of the commutationrelations (3.5) of the Γ_μ the action (3.2) can be rewritten as :

$$S_t = g^{-2} \sum_{\mu,\nu} \mathrm{Tr}\left(\mathbb{1} - e^{iQ_\mu}\, e^{i\Gamma_\mu Q_\nu \Gamma_\mu^\dagger}\, e^{-i\Gamma_\nu Q_\mu \Gamma_\nu^\dagger}\, e^{-iQ_\nu}\right) \tag{3.12}$$

Define from the fluctuation matrices Q_μ "momentum" components $\tilde{Q}_\mu(p)$ by using the property ② :

$$Q_\mu = \frac{1}{N^2} \sum_p \tilde{Q}(p)\, \Gamma(p)$$

The momenta p deserves their name because in :

$$\Gamma_\rho Q_\mu \Gamma_\rho^\dagger = \frac{1}{N^2} \sum_p \tilde{Q}_\mu(p)\, \Gamma_\rho \Gamma(p) \Gamma_\rho^\dagger$$

$$= \frac{1}{N^2} \sum_p \tilde{Q}_\mu(p) \exp i \frac{2\pi}{L} p_\rho\ \Gamma(p) \tag{3.13}$$

the momenta p come in in exactly the way one would expect from translating the field Q_μ over a unit step in the direction ρ . Thus the action (4.12) is nothing but a field theory on a periodic lattice of size L in disguise ! We converted the N^2-1 colour degrees of freedom into N^2-1 momenta. Only the zero momentum components are absent.

III.3. Weak Coupling Expansion of the Twisted Action

Let us now compare the diagrammatic expansion of the field theory (3.12) with a conventional SU(N) gauge theory in a periodic box of size L . The field theory (3.12) gives, after expansion in Q_μ, a propagator which is not defined. This is due to the symmetry (3.10), and can be resolved by introducing a gauge fixing term, e.g.:

$$G = \sum_\mu \left(\Gamma_\mu^\dagger Q_\mu \Gamma_\mu - Q_\mu\right) \tag{3.14}$$

The Faddeev-Popov ghost term follows from this gauge fixing term by mimicking the steps in equ.(2.18).

A gauge transformation on our 1 point lattice is just

$$U_\mu^\Omega = \Omega U_\mu \Omega^\dagger = \Omega V_\mu \Gamma_\mu \Omega^\dagger$$

Keep the "background" field Γ'_μ fixed and find

$$\delta U_\mu \doteq \delta V_\mu = \delta\Omega V_\mu - V_\mu \Gamma'_\mu \delta\Omega \Gamma'^{+}_\mu$$
$$= \omega V_\mu \delta\lambda - V_\mu \Gamma'_\mu \omega \Gamma'^{+}_\mu \delta\lambda \qquad (3.15)$$

In the last term of (3.15) appears the translate of the ghost field ω, $\Gamma'_\mu \omega \Gamma'^{+}_\mu$, as it should. So (3.14) and (3.15) have to be compared with a gauge choice $G = \Delta^-_\mu Q_\mu$ and a gauge variation (2.18) in the conventional theory without background field.

These remarks define the weak coupling expansions we are going to compare now. The propagators in the two theories are identical provided the gauge choice is done as described above. Any vertex with s outgoing legs contains a momentum factor $V(p_1, \ldots, p_s)$ which is identical for the two theories. The conventional theory has a trace after colour matrices :

$$\mathrm{Tr}\, \lambda^{a_1} \ldots\ldots \lambda^{a_s} \qquad (3.16)$$

whereas the action (3.12) gives a trace over Γ' matrices :

$$\mathrm{Tr}\, \Gamma'(p_1) \ldots\ldots \Gamma'(p_s) = N \delta_L(\Sigma p) \exp i \frac{\pi}{N} \sum_{r=1}^{s-2} \Delta_r(p) \qquad (3.17)$$

where the phase $\Delta_r(p) = (p_1 + p_2 + \cdots p_r)_\mu n_{\mu\nu} p_{r+1}$ as follows easily from properties i) and ii) in the preceding section (see also Fig. 10). Momentum is conserved modulo L as follows from i).

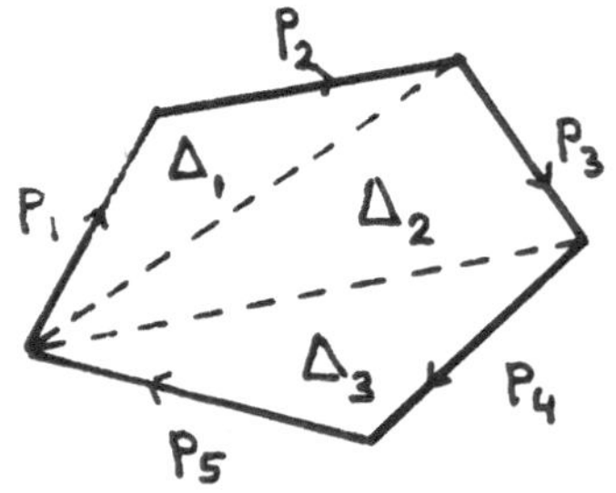

Fig. 10

Momentum polygon (s = 5) with phases obtained by combining subsequent Γ' (p) (equ.(3.17)).

The conventional theory can be analyzed in terms of the double line representation of 't Hooft[23a]. Then we have planar diagrams, diagrams with one handle, two handles, etc... In these diagrams, the colour traces (3.16) do combine to damping factors N^{-2H+2} where H is the number handles of the diagram ; in each s-vertex appears the momentum factor $V(p_1 \ldots p_s)$. In our theory (3.12) we can also draw such diagrams with the same vertex factors $V(p_1 \ldots p_s)$ but with the phases as in (3.17).

The phases can be seen to cancel in planar diagrams. To this end draw a planar diagram (see Fig. 11) and consider some propagator P in it with momentum p. Draw the two momentum polygons belonging to the two vertices at the endpoints of P. Clearly the momentum

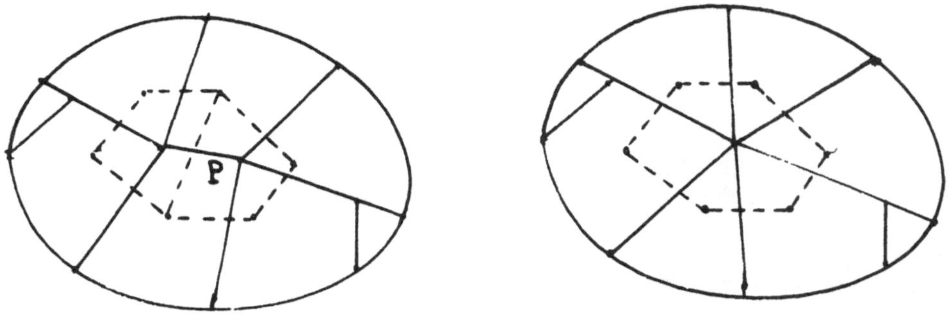

(a) (b)

Fig. 11
Contraction of propagator P in planar diagram (a) with merging momentum polygon (b).

polygons have one side in common namely the momentum of P. Thus we can merge the two into one big one by omitting the common side. This merger has the same phase as the sum of the two and corresponds to the planar diagram with P omitted. Repeat this process till only one vertex is left. The momentum polygon corresponding to this one single vertex is degenerate because of the planarity of the diagram, and has phase zero : planar diagrams in the two theories are identical.

Turn to the case with one handle : we can associate to it a planar diagram by having lines, that did jump each other in the original non-planar diagram (Fig. 12(a)), to cross each other in an extra vertex (Fig. 12(b)). Thus the one handle diagram has a phase equal to minus the phases of the extra vertices needed to make it planar, since we know that the resulting planar diagram has phase zero. The reader can easily check that the resulting phase equals (Fig. 12(c)) :

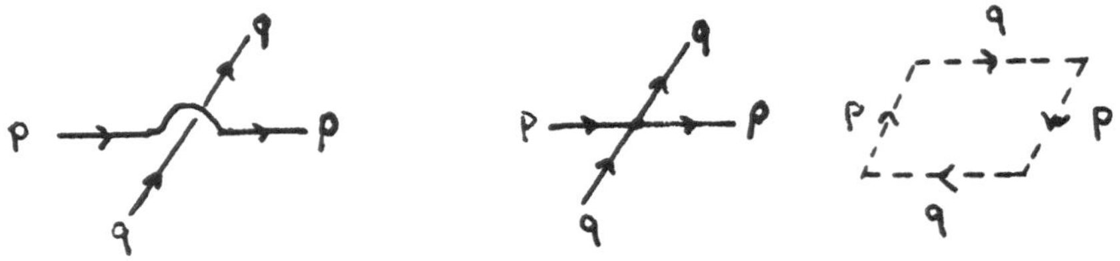

(a) (b) (c)

Fig. 12
Phase of a non planar diagram (a) relates to that of a planar diagram (b) with extra phase (c).

$$\exp -i\,\frac{2\pi}{N}\,P_{t\mu}\,n_{\mu\nu}\,Q_{t\mu\nu}$$

where P_t and Q_t are the total momentum flowing through the handle parallel to the two generators. The multihandle case is as easy.

III.4. Equality of Twisted and Conventional Theory at Large N

Do the non-planar diagrams vanish when $N = L^2 = \infty$? In this limit the phase factor can be written in terms of the continuum momentum

$$\hat{P}_t \equiv \frac{2\pi}{L}\,P_t \quad , \quad \hat{Q}_t \equiv \frac{2\pi}{L}\,Q_t$$

and becomes :

$$\exp i\,\frac{1}{2\pi}\,\hat{P}_{t\mu}\,n_{\mu\nu}\,\hat{Q}_{t\nu}$$

As $L \to \infty$ this phase factor starts to oscillate rapidly and one would expect the non-planar diagram to vanish. Indeed one can show this to be the case, but lack of space forbids us to do this in these notes. Thus as $N \to \infty$ the two theories become identical, and consist of only planar diagrams.

III.5. Outlook

The twisted Eguchi-Kawai model (4.12) has been tested by Montecarlo methods [38] and indeed the results for the string tension exhibit scaling. The model lends itself also to finite temperature physics for large N theories : by a judicious choice of the twist tensor[37, 41] one can build into the model a fixed temperature, whereas the three dimensional infrared cut-off is lifted as N becomes large.

A version of the model exists which avoids altogether the lattice[36] . The momenta p become continuous ; the fluctuation matrices Q_μ and the shift operators $\Gamma(p)$ (see equ.(3.13) become infinite dimensional right from the start :

$$\Gamma(p) = \exp i \frac{N}{2\pi}\left(p_\mu \frac{n_{\mu\nu}}{L} \gamma_\nu\right)$$

The infinite dimensional operators $\{\gamma_\mu\}_0^3$ obey the commutation relations

$$[\gamma_\mu , \gamma_\nu] = i \cdot \frac{n_{\mu\nu}}{L} \cdot \frac{2\pi}{N}$$

and are the generators of infinitesimal translations. The $\Gamma(p)$ obey the same algebraic rules as in Section III.2, and when we express the fluctuation matrices Q in terms of the $\Gamma(p)$ nothing changes in the analysis of Section III.3 and III.4.

Thus, apart from its practical use in Montecarlo simulations the model is a very compact formulation of the sum of all planar diagrams.

Epilogue

Lattice gauge theory clearly needs more theoretical backing in order to test the consistency of the Montecarlo methods. This was the message of Lectures I and II. A small inroad into the complicated two loop calculations has been made and the results are encouraging. They indicate the usefulness of large N expansions.

The large N theory itself can be formulated in a very compact way as we showed in the last lecture. The essential ingredient was the conversion of space degrees of freedom into colour degrees of freedom : "A la recherche de l'espace perdu" was our leitmotiv. Maybe, one day, we will realize that no time was lost in our efforts.

Acknowledgements

I wish to thank Guido Martinelli and Jurek Jurkiewicz for useful discussions. My gratitude goes to the organizers for giving me the opportunity to be at this school.

REFERENCES

1. K.G. Wilson, Phys. Rev. D14, 2455 (1974).
2. K.G. Wilson, Recent Developments in Gauge Theories, edited by G.'t Hooft (Plenum Press, New York, 1980).
3. M. Creutz, Phys. Rev. D21, 2308 (1980).
4. A. Hasenfratz, P. Hasenfratz, Phys. Lett. 93B, 165 (1980).
5. C.G. Callan, Phys. Rev. D2, 1541 (1970).
   K. Symanzik, Comm. Math. Phys. 49, 424 (1970).
6. B.E. Baaquie, Phys. Rev. D16, 2612 (1977).
7. W. Celmaster, R.J. Gonzalvez, Phys. Rev. D20, 1420 (1979).
8. G. Münster, Nucl. Phys. B180 FS 2 (1981), 23.
9. J.M. Drouffe, J.B. Zuber, Phys. Repts., to be published.
10. C.P. Korthals Altes, Proceedings of the Colloquium on Recent Progress in Lagrangian Field Theory, Marseille, 1974.
11. K. Osterwalder, E. Seiler, Ann. of Phys. (NY), 110 (1978), 440.
12. D. Gross, R. Dashen, Phys. Rev. D23 (1981), 2340.
   A. and P. Hasenfratz, Nucl. Phys. B193 (1981), 210.
13. A. Gonzalez-Arroyo, C.P. Korthals Altes, Nucl. Phys. B205 FS 5 (1982), 45.
14. J.W. Dash, J. Jurkiewicz, C.P. Korthals Altes, CERN TH, 3621, June 1983, to appear in Nucl. Phys.
15. R.K. Ellis, G. Martinelli, Rome Preprint, October 1983.
16. P. Weisz, S. Scharatchadran, DESY 81-083 (1981).
17. B. Grossmann, S. Samuel, Rockefeller preprint, RU 82/B/28 (1982).
18. W. Fischler, Nucl. Phys. B129, 157 (1977).
19. L.F. Abbott, Nucl. Phys. B185 (1981), 189.
20. G. 't Hooft, Acta Univ. Wratislavensis n° 38.
21. O. Tarasov et al., Phys. Lett. 93B (1980), 429.
22. Yu. Makeenko, M.I. Polikarpov, Nucl. Phys. B205 FS 5 (1982), 386.
23. (a) G.'t Hooft, Nucl. Phys. B52 (1973), 444.
   (b) G.'t Hooft, Nucl. Phys. B138 (1978), 1.
24. C. Becchi, A. Rouet, R. Stora, Comm. Math. Phys. 42 (1975),127.
25. J. Zinn-Justin, Trends in Elementary Particle Theory, Springer, Berlin, 1975.
26. M. Creutz, L. Jacobs, C. Rebbi, Phys. Repts., to be published.
27. W.E. Caswell, Phys. Rev. Lett. 33 (1974), 244.
28. D.R.J. Jones, Nucl. Phys. B75 (1974), 531.
29. E. Pieterinen, Nucl. Phys. B190 (1981), 349.
30. D. Barkai et al., BNL preprints, January, June, 1983.
31. A. di Giacomo, G.C. Rossi, Phys. Lett. 100B (1981), 481.
32. T. Eguchi, H. Kawai, Phys. Rev. Lett. 48 (1982), 1063.
33. G. Bhanot, U. Heller, H. Neuberger, Phys. Lett. 113B (1982), 147.
34. J. Jurkiewicz, A. Gonzalez-Arroyo, C.P. Korthals Altes, Proceedings Nato Advanced Summer Institute 1981, Plenum Press.
35. A. Gonzalez-Arroyo, M. Okawa, Phys. Lett. 120B (1983), 174 and Phys. Rev., to be published.

36. A. Gonzalez-Arroyo, C.P. Korthals Altes, Phys. Lett. 131B (1983), 296.
37. K. Fabricius, C.P. Korthals Altes, Wuppertal preprint, December 1983.
38. A. Gonzalez-Arroyo, M.Okawa, BNL 32 988 (1982).
K. Fabricius, OM. Haan, WU B 84-1.
39. G.'t Hooft, Comm. Math. Phys. 81 (1981), 267.
40. P. van Baal, Comm. Math. Phys. 85, 529 (1982), Utrecht Preprints, 1983.
41. P. van Baal, F. Klinkhamer, Utrecht preprints, (1983).
42. J. Groeneveld, J. Jurkiewicz, C.P. Korthals Altes, Physica Scripta 23 (1981), 1022.
43. This is perfectly consistent with the index theorem which states that the difference between the number of positive helicity zero modes and negative helicity zero modes for a Dirac field in the adjoint representation on the torus equals : - 2NP.
44. G. Bhanot, R. Dashen, Phys. Lett. 113B, (1982), 299.
45. Dots mean summation over all indices.

CONSTRUCTIVE THEORY OF CRITICAL PHENOMENA

Antti Kupiainen

Helsinki University of Technology
Department of Technical Physics
Espoo 15, Finland

K. Gawedzki*

CNRS, IHES,
Bures-sur-Yvette, France

1. BLOCKSPINS

The purpose of these lectures is to introduce some nonperturbative renormalization group (RG) methods for the study of certain problems of statistical mechanics and quantum field theory. Our aim is a nonperturbative construction of asymtotically free quantum field theories and non-gaussian critical points.

To be more specific, we study lattice spin systems, given by the Gibbs state in a finite box $\Lambda \subset Z^d$

$$\langle - \rangle_H = \frac{1}{Z} \int - \exp(-H(\vec{\phi}))D\vec{\phi}$$

$$D\vec{\phi} \equiv \prod_{x \in \Lambda} d^N \vec{\phi}_x \qquad (1)$$

where H is some Hamiltonian for the N-component field (spin) $\vec{\phi}$. L(1) may be viewed as a lattice approximation to <u>Euclidean field theory</u>, or a model of classical

---

* On leave from Department of Mathematical Methods of Physics, Warsaw University

statistical mechanics of its own right. There is a fairly complete theory of such models deep in the massive regime, where H is such that

$$\langle\vec{\phi}_x;\vec{\phi}_y\rangle \underset{|x-y|\to\infty}{\sim} e^{-m|x-y|} \qquad (2)$$

using high and low temperature expansions, see eg. [3] . However, the regime of interest for field theories and critical phenomena is the massless regime

$$\langle\vec{\phi}_x\cdot\vec{\phi}_y\rangle \sim \frac{1}{|x-y|^{\alpha}} \qquad (3)$$

with infinite correlation length. Approach to criticality gives us via scaling limits Euclidean field theories. The problem in the critical regime is the absence of any length scale which invalidates the usual expansions there. The well known remedy to this problem is the RG of Wilson [13]. Instead of integrating over all the degrees of freedom in (1) simultaneously, one performs an infinite (as $\Lambda\to Z^d$) sequence of fixed momentum scale integrations, i.e. reduces the problem to an iteration of a massive problem.

The problem with the RG has been that in practice in the process of iteration one invariably resorted to approximations and truncations due to the large spaces of interaction H's involved. Our approach will be to use the standard perturbative arguments as a qualitative input and then take care of the corrections exactly in a nonperturbative analysis. To define the setting of the RG, we intruduce the Kadanoff block spin variables [10] $\vec{\phi}'$, related to the average values of the original spins in LxLx...xL blocks (cubes) of the lattice by

$$\vec{\phi}'_x = L^{\frac{\alpha}{2}}\phi^{av}_{Lx} \equiv L^{\frac{\alpha}{2}}(L^{-d}\sum_{|y^\mu - Lx^\mu|<\frac{L}{2}}\vec{\phi}_y) \equiv (C\vec{\phi})_x \qquad (4)$$

(take L odd).

The block spin transformation R_L by definition

produces from the distribution of $\vec{\phi}$ the one of $\vec{\phi}'$, i.e. it fixes the block spins $\vec{\phi}'$ and integrates out the remaining degrees of freedom in $\vec{\phi}$:

$$\exp[-R_L H(\vec{\phi}')] = \text{const.} \int \exp[-H(\vec{\phi})]\,\delta(\vec{\phi}' - C\vec{\phi})\,D\vec{\phi} \tag{5}$$

(the overall normalization of $R_L H$ is irrelevant for us here). $R_L H$ is also called the effective (rescaled) interaction for distances $\geq L$. The exponent α in (4) is the same as in (3) and the reason for this choise is heuristically the following. Namely, from (3) it is plausible, that for $x \neq y$, the limit

$$\lim_{L\to\infty} L^{\alpha} \langle \vec{\phi}^{av}_{Lx}\, \vec{\phi}^{av}_{Ly} \rangle \tag{6}$$

exists. But by (4) and (5) this equals

$$\lim_{L\to\infty} \langle \vec{\phi}'_x\, \vec{\phi}'_x \rangle_{R_L H} \tag{7}$$

Thus we may hope, that in a theory with the scaling behaviour (3) in the infrared (IR), $R_L H$ converges as $L\to\infty$ to a fixed point H^* of R_L : R_L : $R_L H^* = H^*$. H^* exhibits a generalized scale invariance in the form

$$\langle \vec{\phi}_x \vec{\phi}_y \rangle_{H^*} = L^{\alpha} \langle \vec{\phi}^{av}_{Lx} \vec{\phi}^{av}_{Ly} \rangle_{H^*}$$

since $R_{L^n} = R_L{}^n \equiv R^n$ we may expect that the control of iterations of R will simplify as n increases.

2. Models

The models we consider will all be governed by a fixed point H^* of R. We take H generically as

$$H(\vec{\phi}) = \frac{1}{2} \sum_x (\nabla \vec{\phi}_x)^2 + V(\vec{\phi}). \tag{8}$$

The block spin transformation, (5) has a gaussian fixed point H_0^*, easily obtained from (8) with V=0 by iteration of R, with α having the canonical value $\alpha = d-2$(see e.g. [5]). The models governed by the gaussian fixed point will be called asymtotically free (AF), either in the IR or in the ultraviolet (UV). Let us discuss some of them.

(A) Dipole gas and the like

We take in (8) N=1 and

$$V(\phi) = \tilde{V}(\nabla\phi) = \sum_{x \in \Lambda} v(\nabla\phi_x) \tag{9}$$

with e.g.

$$v(\nabla\phi) = \begin{cases} z \int d\rho(\hat{n}) \cos \hat{n}\cdot\nabla\phi & \text{(dipole gas)} \\ \lambda \sum_{\mu} (\nabla_\mu \phi)^4 & \text{(anharmonic crystal)} \end{cases} \tag{10}$$

This representation of the dipole gas is obtained by the sine-Gordon transformation from the gas picture [4]. We prove the IR AF of (9) for very general even v (see section 3). That is, (the $\Lambda \to Z^d$ is taken here)

$$\langle \prod_i \phi_{x_i} \rangle_{R^n H} \underset{n\to\infty}{\to} \langle \prod_i \phi_{x_i} \rangle_{z^{-1} H_0^*} \tag{11}$$

(for $d \geq 3$, for d=2 analogous results hold for expectations of $\nabla\phi$'s). Stated for $\langle - \rangle_H$, we get eg.

$$\langle \phi_x \phi_y \rangle_H / (z(-\Delta)^{-1})_{xy} \xrightarrow[|x-y|\to\infty]{} 1. \tag{12}$$

In (11) and (12) z is the dielectric constant (wave-function renormalization) and Δ the Laplacian on Z^d (or R^d). As a byproduct we get detailed information on the models: analyticity in parameters, Borel summability etc. (see Sect. 3). $\lambda(\nabla\phi)^4$ model has also been controlled by a phase space cell expansion [11] , an alternative analysis of different length scales.

(B) ϕ^4_d $d \geq 4$

Now

$$V(\phi) = \lambda \sum_{x \in \Lambda} \phi_x^4 \tag{13}$$

and we expect again IR AF. In the d=4 case the approach of $R^n H$ to $z^{-1} H_0^*$ is logarithmic, $\lambda_n \sim \frac{1}{n} \sim (\log L^n)^{-1}$, and one has logarithmic corrections to scaling:

$$\langle \phi_x \phi_y \rangle_H \sim \frac{1}{|x-y|^2} (\log|x-y|)^{\gamma} \quad \text{as } |x-y| \to \infty \tag{14}$$

so z^{-1} above diverges . (14) is the result predicted by the perturbative β-function. We expect the analysis of (A) to extend to this case too. In a hierarchical approximation to the BST this has been shown in [6] and the full model is under investigation. The non-perturbative infrared renormalization of ϕ_4^4 carries some analogies with the corresponding ultraviolet problem in UV AF models to whose control we consider it as a useful prelude. It also would partly complement the discussion [1] of the triviality of the continuum ϕ_4^4.

(C) Non-linear σ-model

This is a renormalizable, AF field theory model. We wish to take the scaling limit of the N-component (N≥3) Heisenberg model in d=2:

$$H_g(\vec{\phi}) = \frac{1}{g^2} \sum_x (\nabla \vec{\phi}_x)^2 \qquad \vec{\phi}_x^2 = 1 \tag{15}$$

Here g^2 is the temperature, or, bare coupling constant. One expects the critical point of (15) to be $g^2 = 0$, so the scaling limit will be taken by approaching it in a specific way. The perturbative RG[2,12] predicts, that taking

$$g(\Lambda)^{-2} \underset{\Lambda \to \infty}{\sim} c_1 \log \Lambda + c_2 \log\log\Lambda + C_3 \tag{16}$$

the field theory Schwinger functions will be given by

$$\lim_{\Lambda\to\infty} Z(\Lambda)^{-n/2} \langle\vec{\phi}_{\Lambda x_1}\cdots\vec{\phi}_{\Lambda x_n}\rangle_{g^2(\Lambda)} = G(x_1\ldots x_n) \tag{17}$$

with a certain (computable) $Z(\Lambda)$. Stated in terms of R, we wish to show the effective unit cutoff theory

$$H_{eff}(\vec{\phi}) = \lim_{\Lambda\to\infty} R_\Lambda H_{g(\Lambda)}(Z(\Lambda)^{1/2}\vec{\phi}) \tag{18}$$

exists. We hope to carry out this construction for N large, using the $\frac{1}{N}$-expansion as a perturbative input in computing RH. As a by-product we would expect to get the existence of a massgap in the Heisenberg model for all temperatures. We are pursuing this program in the collective field approach to RG, see [7] .

For critical phenomena in d=3, the interesting fixed point is not the gaussian one, but a non-gaussian with non-canonical scaling. In d=3 such a fixed point may be computed perturbatively [14] in $\frac{1}{N}$-expansion. We have constructed it in a hierarchical version of the model. For details of our attempt to use the collective field method to the full model, see [7] .

3 A non-perturbative analysis of R

We will now consider in more detail how the analysis of R will procede in the simplest of the models introduced above, namely the dipole gas of (A). The basic idea will be the same also for the more complicated models. Our first step is to realize the transformation (5) explicitely as an integration over fluctuations within the blocks. To this end, we "invert" (4) (α=d-2 from now on)

$$\phi_x = L^{\frac{-d-2}{2}} \psi'(\phi')_{x/L} + Z_x \tag{19}$$

where $\psi'(\phi')$ is ϕ' (which lives on unit lattice) suitably smeared on $\frac{1}{L}$ lattice (which naturally arises from (41) whereas Z is a fluctuation field with zero average within blocks: $Cz = 0$. In fact, $\psi'(\phi')$ may be chosen as

$$\psi'(\phi')_x = \sum_y A_{xy}\phi'_y \tag{20}$$

in such a way, that ψ' and Z are independent when V is zero:

$$H_0 = \sum(\nabla\phi_x)^2 = \int(\nabla\psi'_x)^2 + \sum(\nabla Z_x)^2 \tag{21}$$

i.e. one chooses $\psi'(\phi')$ as the "classical minimum" of H_0 when $C\phi=\phi'$ is given . In (21) we use $\int_1$ and ∇ for Riemann sums and finite differances on $\frac{1}{2}$ lattice. $R_L H$ may now be written as (up to a constant)

$$\exp[-R_L H(\phi')] = \exp\left[-\tfrac{1}{2}\int(\nabla\psi'_x)^2\right]\int d\mu_\Gamma(Z) \exp[-V(L^{\frac{2-d}{2}}\psi!/_L+Z] \tag{22}$$

with $d\mu_\Gamma(z)$ gaussian

$$d\mu_\Gamma(Z) = N \exp\left[-\tfrac{1}{2}(Z,\Gamma^{-1}Z)\right] \equiv N'\delta(CZ)\, e^{-\frac{1}{2}\sum(\nabla Z)^2} \tag{23}$$

The point of (22) is, that Γ is a <u>massive</u> propagator

$$|\Gamma_{xy}| \leq c \exp\left[-m\ |x-y|\right] \tag{24}$$

Intuitively this is clear, since Z involves only fluctuations within the blocks i.e. momenta $\geq\frac{1}{L}$. (24) may be established by an explicit computation [5].

Writing (2) for the $\nabla\phi$-model (9), we are led to study the transformation

$$V\to V' : V'(\nabla\psi') = -\log\int d\mu_\Gamma(Z) \exp[-V(L^{\frac{d}{2}}\nabla\psi!/_L+Z)] \tag{25}$$

(25) is a small perturbation of a massive (high temperature) problem (if the couplings in V are small), were it not the "external" field ψ' which may take arbitary values. These large field contributions must be subjected to a nonperturbative treatment. Before going into that, let us look at the perturbative argument.

Perturbation theory

The linear approximation to (25) gives (with V as in (9) and use of $\int$ for $\sum$)

$$V'(\nabla\psi')_{LINEAR}=\int d\mu_\Gamma(Z)\int dx\, v(L^{-\frac{d}{2}}\nabla\psi'_{\frac{x}{L}}+\nabla Z_x). \tag{26}$$

Thus e.g.

$$\begin{aligned}&\int(\nabla\phi_x)^2\to\int(\nabla\psi_x)^2+\text{const.} &(27)\\ &\int(\nabla\phi_x)^4\to L^{-d}\int(\nabla\psi_x)^4+\text{const.}\int(\nabla\psi_x)^2+\text{const.}\end{aligned}$$

(27) show the familiar marginal and irrelevant behaviors determened by the (canonical) dimensions of the couplings. Thus, indeed, we expect the $(\nabla\psi)^4$-model (10) to be IR AF λ going to zero with repeated application of R and producing an $\mathcal{O}(\lambda)$ increment to the $(\nabla\psi)^2$-term in R^nH resulting in the wave function renormalization z of (11) and (12).

One might now attempt to compute V' perturbatively as a power series in v (for definitiveness the reader should consider the $\lambda(\nabla\psi)^4$ as a canonical example). As remarked already, one cannot expect this to be very reasonable for large $\nabla\psi'$. But even for small $\nabla\psi'$ we encounter the usual divergences of perturbation theory. Formally

$$V'(\nabla\psi')=\sum_{n=0}^{\infty}\frac{(-1)^n}{n!}\int dx_1\ldots dx_n\ \langle v(L^{-\frac{d}{2}}\nabla\psi'_{\frac{x_1}{L}}+\nabla Z_{x_1});$$

$$\ldots;\ v(L^{-\frac{d}{2}}\nabla\psi'_{\frac{x_n}{L}}+\nabla Z_{x_n})\rangle^T_\Gamma \tag{28}$$

The expectation in (28) is a connected one in $d\mu_\Gamma$. For $v = \lambda(\nabla\phi)^4$ it has $\sim(n!)^2$ terms leading to an n! divergence. The reason for this is clear too: perturbation theory should be good only when v is small, i.e. for not too big $\nabla\phi = L^{-\frac{d}{2}} \nabla\psi' + \nabla Z$. For $v = \lambda(\nabla\phi)^4$, taking e.g. $|\nabla\phi| < \mathcal{O}(|\log \lambda|^\nu)$ would guarantee smallness of v. But in (28) ∇Z takes arbitary values. Thus a careful study of small and large field regions is needed. This is complicated by the fact, that the propagator Γ even if massive is not totally local, thus mixing the regions. The non-locality of Γ is controlled by a high temperature expansion leading essentially to local expressions + small corrections.

Non-perturbative contribution

Before discussing the non-localities and in order to see clearly how the large fields are non-perturbatively taken into account, let us imagine performing (25) for a lattice of the size of one block. Consider first $\nabla\psi'$ small

$$|\nabla\psi'_x| \le |\log \lambda|^\nu \quad , \tag{29}$$

with $\nu > \frac{1}{2}$. Compute the large and small ∇Z-contributions to (25) separately. Let $\chi(Z)$ be the characteristic function for $|\nabla Z_x| < \frac{1}{2}|\log \lambda|^\nu$ for all x, and write

$$e^{-V'(\nabla\psi')} = \int d\mu_\Gamma(Z)\chi(Z)e^{-V(\dots)} + \int d\mu_\Gamma(Z) (1-\chi(Z))e^{-V(\dots)} \tag{30}$$

The point of (30) is, that the first term involves

$$|\nabla\phi_x| = |L^{-\frac{d}{2}} \nabla\psi_{x/L} + \nabla Z_x| \le (L^{-\frac{d}{2}} + \tfrac{1}{2}) \cdot |\log\lambda|^\nu < |\log \lambda|^\nu \tag{31}$$

for $L \ge 3$ say. Thus perturbation theory should (and will!) converge for that term, giving a contribution (schematically ,we ignore here the λ dependence of χ, for full discussion, see [8])

$$V_1'(\nu\psi') = \sum_n b_n(\nabla\psi')\lambda^n = \sum_n c_n^{\prime}(\lambda)(\nabla\psi')^n \tag{32}$$

analytic both in a small λ and in $\nabla\psi'$ for $|\nabla\psi_x| < |\log\lambda|^\nu$. The second term in (30) involves the non-perturbative region: it is non-analytic at $\lambda=0$ as its perturbative series carries the n! divergences mentioned above. But on the support of $1-\chi(Z)$, $|\nabla Z_x| > \frac{1}{2}|\log\lambda|^\nu$ for some x, which has a small probability in the gaussian measure $d\mu_\Gamma(Z)$ (see (23)). Thus, provided that e^{-V} is suitably bounded for large fields, we get from the second term a contribution

$$O(\exp[-c\ |\log\lambda|^{2\nu}]) < O(\lambda^n) \quad \text{for all } n \tag{33}$$

In fact it is easy to see, that a bound

$$|e^{-V(\nabla\phi)}| \le e^{\kappa\int(\nabla\phi_x)^2} \tag{34}$$

will suffice for the large fields, if $\kappa<\frac{1}{2}$. But this bound should hold for the model to be stable in the first place. A crucial observation now is that the second term besides being small, is also analytic in small $\nabla\psi'$ (all the time for $v(\nu\phi) = \lambda(\nabla\phi)^4$) whence (32) and (33) combine to give

$$V'(\nabla\psi') = \sum_{n=0}^{\infty} C_n(\lambda)(\nabla\psi')^n \tag{35}$$

with $C_n(\lambda) = C_n'(\lambda) + O(|\log\lambda|^{-\nu n} e^{-C|\log\lambda|^{2\nu}})$. The lesson to be learned from the above is, that we should expect $V'(\nabla\psi')$ to be analytic in the small $\nabla\psi'$ region, even if perturbation theory in λ diverges (this is reflected in the non-analyticity of $C_n(\lambda)$ in λ). To guarantee this for general V, we used besides the stability bound for e^{-V} in the large field region the analyticity of e^{-V} in a strip

$$|\text{Im}\ \nabla\phi_x| < |\log\lambda|^\nu, \tag{36}$$

see (25).

These are the only large field properties of V which enter the small $\nabla\psi'$ analysis of V'.

We would like to carry over these properties to V'. From (25) it is clear, that e^{-V} analytic in a strip (36) will in fact imply analyticity of $e^{-V'}$ in a strip $|\nabla\psi'_x| < L^{d/2}|\log\lambda|^{\nu}$ (provided we have proper bounds for the integral). From (31) it is also clear, that we may replace (29) by $|\nabla\psi'_x| < K|\log\lambda|^{\nu}$ for $K = 1+\varepsilon$ thus expanding the small field region. The stability found (34) will also produce an analogous bound for $e^{-V'(\nabla\psi')}$ due to (21).

We may now subject the small field information (35) and the large field bound to an iterative argument, the upshot being the expansion of the small field region, and contraction of the "couplings" $C_n(\lambda)$, $n>2$ (more on this below). The $n=2$ term will produce the wave function renormalization.

Non-locality of RH

We must now see how the idea of controlling the large fields may be combined with a high temperature analysis of the integral (25).

The perturbation expansion (28) for V', e.g. for the $(\nabla\phi)^4$ model leads to

$$V'(\nabla\psi') = \sum_{n=0}^{\infty} \int dx_1 \ldots dx_n K_n(x_1 \ldots x_n) \nabla\psi'_{x_1} \ldots \nabla\psi'_{x_n} \tag{37}$$

with the kernels K_n given by connected grapths with lines Γ, and hence exhibiting due to (24) exponential falloff in $x_1 \ldots x_n$. According to the previous discussion, we would like to obtain such a result in the region where $\nabla\psi'_x$ is less than $K|\log\lambda|^{\nu}$, whereas a stability bound of the type (34) elsewhere. The problem is, how to separate the regions, or, to find a representation for e^{-V} exhibiting both properties. Such a representation is obtained by applying a high temperature (cluster) expansion to the $\exp\left[-\tfrac{1}{2}(Z,\Gamma^{-1}z)\right]_{-V}$ part of (25). The result is the following expression for e^{-V}.

$$e^{-V'(\nabla\psi')} = \sum_{\{X_j\}} \prod_j g^{D'}_{X_j}(\nabla\psi') \exp -\sum_{Y \cap X_j = \emptyset} V_Y(\nabla\psi') \tag{38}$$

Here $g^{D'}_{X_j}$ and V_Y are functions of $\nabla\psi'$, depending only on $\nabla\psi'|_{X_j}$ and $\nabla\psi'|_Y$ respectively. $g^{D'}_{X_j}$ represent the large field contribution: D' is the region where $\nabla\psi'_x$ is greater than $K|\log\lambda|^\nu$, and $\cup X_j \supset D$ with X_j disjoint. X_j represent exponential tails around the "islands" of large fields, coming from non-locality of Γ. V_Y are the small field potentials. They correspond to the perturbative multibody potentials as in (37). Let us now state the main result on the properties of this representation [8] .

Assumptions for v

Let v be even, $v(0)=0$ (for convenience) s.t.

(1) $v(\nabla\phi)$ is analytic in C^d in $|\mathrm{Im}\nabla_\mu\phi|<|\log\lambda_o|^\nu$ with

$$|\exp[-v(\nabla\phi)]| \leq \exp[\tfrac{\kappa}{2}|\nabla\phi|^2]$$

there.

(2) $|v(\nabla\phi) \leq \lambda_0$ for $|\nabla\phi| < |\log\lambda|^\nu$.

Then for $\kappa=\frac{1}{10}$ say, and λ_0 small

Result for $R^n_L H$

$R^n_L H$ is given by

$$\exp[-(R^n_L H)](\nabla\psi^n) = \text{const.}\ \exp[-\frac{Z_{n^{-1}}}{2}\int(\nabla\psi'_x)^2$$

$$\sum_{\{X_j\}} \prod_j g^{nD}_{X_j}(\nabla\psi^n)\exp[-\sum_{Y\cap X_j=\emptyset} V^n_Y(\nabla\psi^n)], \qquad (39)$$

(1n) $g^{nD}_{X_j}(\nabla\psi^n)$ are analytic on $|\mathrm{Im}\nabla\psi^n_x| < |\log\lambda_n|^\nu$, $x\in X_j$, and $|g^{nD}_{X_j}(\nabla\psi^n)| \leq \exp\kappa\int_{X_j}(\nabla\psi^n_x)^2 - \varepsilon\mathcal{L}(x_j)$

there,

(2n) $V_Y(\nabla\psi^n)$ are analytic on $|\nabla\psi^n_x| \leq |\log\lambda_n|^\nu$ with

$$|V_Y(\nabla\psi^n)| \leq \lambda_n\, e^{-\varepsilon\mathcal{L}(Y)} .$$

(3n) $|z_n - z_{n-1}| \leq \lambda_n$

<u>Remarks</u>

(a) Here $\lambda_n = \delta^n \lambda_0$, $\delta<1$, $\mathcal{L}(X_j)$ is a geometrical factor (length of the shortest tree graph on the points of X_j), and $\nabla\psi^n = \nabla A^n \phi$ lives on $L^{-n}Z^D$ with ϕ on Z^d (in [8] slightly stronger results are needed and proved).

λ_n plays the role of effective coupling on scale L^n.

(b) Clearly, in a strong sense $R^n H$ converges to a gaussian. This leads to the properties of $\langle - \rangle_H$, $\langle - \rangle_{R^n H}$ mentioned in (A).

(c) The analyticity of v in parameters (e.g. Z or λ in (10)) in regions s.t. (1) and (2) hold translates to corresponding analyticity of g^n, V^n and finally of $\langle - \rangle_H$. Thus we are able to prove analyticity of the dipole gas correlations in activity $|Z|<Z_0$ leading to <u>convergence</u> of perturbation theory (Mayer expansion) in Z, which is highly non-evident since the number of terms grows very rapidly (the best bound has been O(n!) estimate for the n:th order). For $\lambda(\nabla\phi)^4$ model in a hierarchical approximation <u>Borel-summability</u> follows [9].

<u>Iteration of the representation</u>

Let us finally sketch, how (39) is iterated. From the analog of (22) we have

$$\exp[-H^{n+1}(\nabla\psi^{n+1})] = \text{const.}\exp\left[-\frac{Z_n^{-1}}{2}\int(\nabla\psi^{n+1})\right] \times$$

$$\int d\mu_{\Gamma_n}(Z)\ \exp[-V(L^{-\frac{d}{2}}\nabla\psi^{n+1}_{\cdot/L} + Z)$$

$$= \text{const.}\ \exp\left[\frac{-z_n^{-1}}{2}\int(\nabla\psi^{n+1})^2\right]\int d\mu_{\Gamma_n}(Z)\sum_{\{X_j\}}$$

$$\Pi g^{nD}_{X_j}(L^{-\frac{d}{2}}\nabla\psi^{n+1}_{/L} + \nabla Z) \times$$

$$\times \exp\left[-\sum_{Y} V^n_Y(L^{-\frac{d}{2}}\nabla\psi^{n+1}_{/L} + \nabla Z)\right] \qquad (40)$$

(40) has two kinds of non-localities, the ones coming from $d\mu_\Gamma$ and the ones from $\exp[-\Sigma V_Y]$. Let us first Mayer expand the latter:

$$e^{-\Sigma V_Y} = \prod_Y (e^{-V_Y}-1+1) = \sum_{\{Y\}} \prod_Y (e^{-V_Y}-1) \quad . \tag{41}$$

Hence the integral in (40) reads

$$\int d\mu_{\Gamma_n} \sum_{\{X_j\},\{Y\}} \Pi g_{X_j}^{nD} \Pi (e^{-V_Y}-1) \equiv I \tag{42}$$

Now $d\mu_\Gamma$ still couples the Z_x, x in different X_j, Y: we pergorm a Mayer (cluster) expansion to it too. The result will then be

$$I = \sum_{\{X_\alpha\}} \prod_\alpha \rho_{X_\alpha} (\nabla\psi^{n+1}) \tag{43}$$

with the X_α disjoint sets and each ρ_X having an expression of the same type as (42) (with Mayer-factors from the expansion of $d\mu_{\Gamma_n}$).

The point of (43) is now, that if $|\nabla\psi_x^{n+1}| < |\log\lambda_{n+1}|$ in X_α, ρ_{X_α} will be small, being computable, essentially as in our simplified example, by perturbation theory for small Z and a non-perturbative contribution from large Z. Due to the Mayer-expansion ρ_{X_α} also carries an $\exp[-\varepsilon\mathcal{L}(X_N)]$ factor.

Thus we may exponentiate the polymers X_α in (43) not intersecting the large field region D:

$$I = \sum_{\{X_\alpha\}:X_\alpha \cap D \neq \emptyset} \prod_\alpha \rho_{X_\alpha} (\nabla\psi^{n+1}) \exp\Big[-\sum_{Y\cap X\alpha=\emptyset} \tilde{V}_Y(\nabla\psi^{n+1})\Big] \tag{44}$$

The last step will now be to extract the $\int(\nabla\psi^{n+1})^2$ part from the $\tilde{V}_Y$'s. The result will be (39) for n+1. The bounds of (1_n)-(3_n) carry over to n+1 by a rather straightforward analysis.

References

1. Aragao de Carvalho,C.,Caracciolo,S.,Fröchlich,J.: Nucl.Phys. B215,209(1983)
2. Brezin,E.,Zinn-Justin,J.:Phys.Rev.B14,3110(1976)
3. Domb,C.,Green,M.S,Phase Transitions and Critical Phenomena Vol.3, Academic Press,London 1974
4. Fröchlich,J.,Spencer,T.:J.Stat.Phys.24,617(1981)
5. Gawedzki,K.,Kupiainen,A.:Commun.Math.Phys.77,31(1980)
6. Gawedzki,K.,Kupiainen,A.:J.Stat.Phys.29,683(1982)
7. Gawedzki,K.,Kupiainen,A.:Contribution in this volume
8. Gawedzki,K.,Kupiainen,A.:Annals of Phys.147,198(1983)
9. Gawedzki,K.,Kupiainen,A.,Tirozzi,B.:J.Stat.Phys., to appear
10. Kadanoff,L.P.:Physics 2,263(1965)
11. Magnen,J.,Seneor,R.:Annals of Physics, to appear
12. Polyakov,A.:Phys.Lett.59B,79(1975)
13. Wilson,K.,Kogut,J.:Phys.Rep.12,75(1974)
14. Ma,S.:Phys.Lett.43A,475(1973)

ON A RELATION BETWEEN FINITE SIZE EFFECTS AND ELASTIC SCATTERING PROCESSES *

M. Lüscher

Deutsches Elektronen-Synchrotron DESY

Hamburg

ABSTRACT

Finite volume effects on the mass spectrum of massive quantum field theories (including QCD) are shown to be related to forward scattering amplitudes of the infinite volume theory. Using theoretical and experimental information on the latter, the finite size effects can be estimated. For the pion and the nucleon they are found to be small, provided $m_\pi L \geq 1$ (m_π: pion mass, L: box size).

1. INTRODUCTION

Consider, for example, an SU(2) gauge theory on a TxLxLxL periodic space-time lattice, where the time-like extent T is supposed to be big, i.e. ideally $T = \infty$. The mass gap M_0 of the theory can then be determined from the exponential fall-off of the plaquette two-point function C(t) at large times t:

---

* Lecture given at the Nato Advanced Study Institute in Cargèse (September 1983)

(1) $C(t) \propto e^{-M_o t}$

Here and below the lattice spacing is set equal to one so that L is an integer and M_o is dimensionless. The energy gap M_o not only depends on the bare coupling constant g_o, but also on the box size L. If one attempts to extract the infinite volume mass gap

(2) $$m_o = \lim_{L \to \infty} M_o \qquad (g_o \text{ fixed})$$

from a finite volume calculation [1)3)], the relative deviation

(3) $$\delta_o = (M_o - m_o)/m_o$$

must therefore be accounted for as a systematic error. The size dependence of the energy gap is not a lattice artefact. One rather expects that δ_o becomes a universal function of

(4) $$\zeta = m_o L$$

in the finite volume continuum limit, which is obtained by taking $g_o \to 0$ and simultaneously $L \to \infty$ while keeping ζ fixed.*

Of course, finite size effects on the mass spectrum are a common feature of Lagrangian quantum field theories including QCD and lattice spin models. This lecture is devoted to the question of how δ_o typically varies with respect to ζ and in particular of how big ζ must be in order that (say) $|\delta_o| < 0.1$. The general experience is

---

* ζ should not be confused with the parameter $z = M_o L$, which I have used earlier as a weak coupling expansion parameter [2)3)]. The two parameters are, however, related by $z = \zeta(1 + \delta_o)$ so that $z \simeq \zeta$ at large ζ.

that δ_0 is small for $\varphi \gtrsim 3$, but this statement is presumably not always true and in any case needs proof. Here I shall establish a formula for δ_0 valid at large φ by which δ_0 can be accurately estimated provided something is known about the forward elastic scattering amplitude of the infinite volume theory. Conversely, when δ_0 has already been computed by other means, the formula allows (at least in principle) to extract the complete forward elastic scattering amplitude.

2. ISING MODEL

To warm up, let us first discuss the soluble case of the 2-dimensional Ising model. Thus, pick a T x L square lattice Λ with periodic boundary conditions and attach to each site n of Λ an Ising spin $\sigma(n)$ taking values ± 1. The action S of a spin field at inverse temperature β is given by

$$S = -\beta \sum_{n\in\Lambda} \sum_{\mu=0,1} \sigma(n)\,\sigma(n+\hat{\mu}),$$

where $\hat{\mu}$ denotes the unit vector in the positive μ-direction ($\mu = 0$ for time, $\mu = 1$ for space). In what follows, we shall only consider the high temperature (small β) phase of the model. As in the gauge theory case, the energy gap $M_0(\beta,L)$ can then be read off from the exponential decay at large times of the spin correlation function, which is defined by

$$\langle \sigma(n)\,\sigma(0)\rangle = \lim_{T\to\infty} \sum_{\text{fields}} \sigma(n)\sigma(0)\, e^{-S} \Big/ \sum_{\text{fields}} e^{-S}$$

Alternatively, we may note that e^{-M_0} is equal to the ratio of the next to highest to highest eigenvalue of the transfer matrix. These eigenvalues have already been computed by Schultz, Mattis and Lieb[4]) almost twenty years ago so that here we only have to recall their beautiful exact results.

In the infinite volume limit ($L = \infty$), the mass gap is given by

(5) $$m_o = -2\beta - \ln \operatorname{tgh}\beta$$

As β rises, m_o is monotonically decreasing and eventually vanishes at the critical point

(6) $$\beta_c = \frac{1}{2} \ln (1 + \sqrt{2})$$

At $L = \infty$, there is actually a whole "mass shell" of 1-particle states with momentum q, $-\pi \leq q \leq \pi$, and energy ϵ_q given by

(7) $$\operatorname{ch} \epsilon_q = \operatorname{ch} m_o + 1 - \cos q, \quad \epsilon_q \geqslant m_o .$$

Near β_c and at small q, this formula reduces to the relativistic energy momentum relation

(8) $$\epsilon_q^2 = m_o^2 + q^2$$

The finite volume energy gap M_o, which has also been obtained by Schultz et al., reads

(9) $$M_o = m_o + \frac{1}{2} \sum_{\nu=0}^{L-1} \epsilon_{\frac{\pi}{L}(2\nu+1)} - \frac{1}{2} \sum_{\nu=0}^{L-1} \epsilon_{\frac{\pi}{L}2\nu}$$

The first sum here represents the zero point energy of the vacuum state and the second sum the zero point energy of the zero momentum 1-particle state. These frequency sums come with the "wrong" sign, because of the fermionic character of the harmonic oscillators in terms of which the diagonalization of the transfer matrix is achieved.

For further analysis, it is useful to rewrite eq. (9) as follows. First, using the Poisson summation formula, we have

$$M_o = m_o - L \sum_{\nu=-\infty}^{\infty} \int_{-\pi}^{\pi} \frac{dq}{2\pi} \; e^{iqL(2\nu+1)} \epsilon_q$$

Then, shifting the contour of integration to imaginary q and performing a number of substitutions, one arrives at

$$(10) \quad M_o = m_o + \sum_{\nu=0}^{\infty} \frac{2}{2\nu+1} \int_{-\pi}^{\pi} \frac{dq}{2\pi} \; e^{-(2\nu+1)L\epsilon_q}$$

It follows immediately that for fixed $\beta < \beta_c$ and growing L, the energy gap M_o is monotonically decreasing towards m_o. At large L, δ_o eventually vanishes exponentially fast:

$$(11) \quad \delta_o \underset{L\to\infty}{\propto} L^{-1/2} \, e^{-m_o L}$$

We shall see later that this exponential falling off is typical for massive quantum field theories, although the decay rate is not always equal to one.

The continuum limit of δ_o at β_c is also easily obtained from eq. (10). Thus, letting β approach β_c and sending L to infinity in such a way that $\zeta = m_o L$ remains fixed, leads to

$$(12) \quad \delta_o = \sum_{\nu=0}^{\infty} \frac{2}{2\nu+1} \int_{-\infty}^{\infty} \frac{dp}{2\pi} \; e^{-(2\nu+1)\zeta p_o} , \qquad p_o = \sqrt{1+p^2}$$

Evaluating this expression numerically, yields the table

ξ	δ_0
1	0.392
2	0.089
3	0.026
4	0.008

As expected, δ_0 quickly becomes small when ξ grows, so that beyond $\xi = 3$ the finite size effects shift the mass gap by at most a few percent.

Before we turn to the general case, let us briefly consider the d+1-dimensional Ising model. No exact solution is available here, of course, but we may instead rely on the high temperature cluster expansion. Recently I found that the graphs contributing to δ_0 can be arranged in such a way that to all orders

$$(13) \quad \delta_0 \underset{\xi \to \infty}{\sim} \xi^{-d/2} e^{-\xi} \sum_{\nu=0}^{\infty} c_\nu \xi^{-\nu}$$

with perturbatively computable coefficients $c_\nu(\beta)$ [5]. While this result is interesting, its proof is mainly a book-keeping task. The following two features are, however, remarkable:

- High temperature graphs contributing to δ_0 must wind around the world in such a way that they cannot be unwrapped. This fact alone already explains why δ_0 is exponentially small at large L, because every such graph has a weight proportional to β^L.
- The coefficients c_ν in eq. (13) derive from graphs, which also contribute to the four point vertex function at certain special momenta, thus suggesting a connection with the elastic scattering amplitude. This is the topic, to which we turn now.

3. A FORMULA FOR δ_0 IN TERMS OF THE ELASTIC SCATTERING AMPLITUDE

Let us consider a scalar field theory in d+1 dimensions describing a single self-interacting spinless particle of mass m_o. In the course of the discussion it will become clear how to generalize to more complicated situations. From now on I shall furthermore assume that the continuum limit has already been taken *, although I believe that, with appropriate modifications, the result obtained below (eq. (15)) also applies in lattice theories.

It is well known that the (infinite volume) forward scattering amplitude F extends to an analytic function of the "crossing variable" ν in the cut plane shown in Fig. 1 (F and ν are defined in the Appendix). In addition to the cuts starting at $\nu = \pm m_o$, $F(\nu)$ has simple poles at $\nu = \pm \frac{1}{2}m_o$, if one-particle exchange scattering processes are possible. In that case, an effective three particle coupling constant λ can be defined through

$$(14) \quad \lim_{\nu \to \pm \frac{1}{2} m_o} (\nu^2 - \tfrac{1}{4} m_o^2)\, F(\nu) = \tfrac{1}{2} \lambda^2$$

Note that, due to crossing symmetry, $F(-\nu) = F(\nu)$, and the two limits in eq. (14) coincide.

Suppose now that the theory is enclosed in a d-dimensional periodic space box of size L and that δ_0 is defined as before. The formula alluded to in the introduction then reads

---

* In particular, some physical units are henceforth used for dimensionful quantities like M_o, m_o and L.

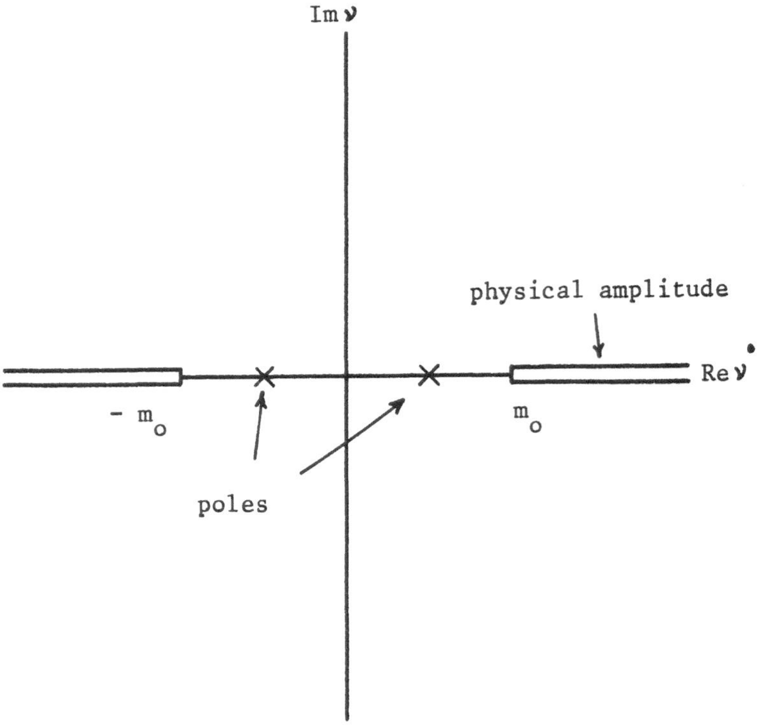

Fig. 1: Analyticity domain of F(ν).

$$\delta_0 = -\lambda^2 m_0^{d-5} \frac{d}{8\pi} \left(\frac{4\pi}{\sqrt{3}} \zeta\right)^{1-\frac{d}{2}} K_{\frac{d}{2}-1}\left(\frac{\sqrt{3}}{2}\zeta\right) - \frac{d}{2m_0^2} \int d\mu(q)\, e^{-q_0 L}\, F(iq_1) + O(e^{-\alpha\zeta}) \tag{15}$$

Here, K_1 denotes a modified Bessel function (Ref. 6, § 8.432) and, as usual, $\zeta = m_o L$. Furthermore,

$$d\mu(q) = \frac{d^d q}{(2\pi)^d\, 2q_0} , \qquad q_0 = \sqrt{m_0^2 + \vec{q}^{\,2}}$$

is the Lorentz invariant measure on the one-particle mass shell. Note that the forward scattering amplitude F is integrated over along the imaginary axis in the ν-plane, far away from singularities. The error term in eq. (15) is dependent on dynamical details of the theory, but α never becomes smaller than $\sqrt{\frac{3}{2}}$ and may, on the other hand, be as large as 3.

A most intriguing feature of eq. (15) is that it relates δ_o to an experimentally measurable quantity. Also, eq. (15) immediately implies that δ_o is exponentially decreasing at large ξ:

(16a) $$\delta_o \sim -\lambda^2 m_o^{d-5} \frac{d}{4\sqrt{3}} \left(\frac{4\pi}{\sqrt{3}}\xi\right)^{\frac{1}{2}(1-d)} e^{-\frac{\sqrt{3}}{2}\xi} \qquad (\text{if } \lambda \neq 0)$$

(16b) $$\delta_o \sim -F(0)\, m_o^{d-3} \frac{d}{4} (2\pi\xi)^{-\frac{d}{2}} e^{-\xi} \qquad (\text{if } \lambda = 0)$$

The actual size of δ_o at a given large ξ, however, depends on the strength of the interactions and may therefore vary considerably with respect to the parameters of the theory.

The best proof of eq. (15) I have so far been able to find goes as follows*. Let $\phi(x)$ be an interpolating scalar field so that at $L = \infty$

(17) $$\langle 0 | \phi(0) | p \rangle = 1$$

The finite L Euclidean 2-point function of ϕ is then written as

---

* When this lecture was delivered, I sketched a different argument, which did not make use of Feynman diagrams and was therefore universally applicable. However, I later found some of the steps involved difficult to justify rigorously and now prefer the reasoning outlined here.

$$\langle \phi(x)\phi(0)\rangle_L = \int \frac{dp_o}{2\pi} \; L^{-d} \sum_{\vec{p}} e^{ipx} \; G_L(p)$$

$$G_L(p)^{-1} = m_o^2 + p^2 - \Sigma_L(p)$$

(Euclidean scalar products are used here, e.g. $p^2 = p_o^2 + \vec{p}^2$). The normalization (17) and the requirement that m_o be the physical mass at $L = \infty$ amounts to

$$\Sigma_\infty(p) = \frac{\partial}{\partial p^2} \; \Sigma_\infty(p) = 0 \qquad \text{for } p = (im_o, \vec{0}).$$

On the other hand, the finite L energy gap M_o is implicitly determined by

$$G_L(p)^{-1} = 0 \text{ for } p = (iM_o, \vec{0}).$$

Writing $M_o = (1 + \delta_o) m_o$, it follows that

$$(18) \qquad \delta_o = -\frac{1}{2m_o^2} \left(\Sigma_L(\hat{p}) - \Sigma_\infty(\hat{p}) \right) + O(\delta_o^2)$$

where $\hat{p} = (im_o, \vec{0})$.

The second, more difficult step in the proof of eq. (15) consists in showing that

$$\Sigma_L(\hat{p}) - \Sigma_\infty(\hat{p}) =$$

(19)

$$\tfrac{1}{2}\left\{ \text{—} + \text{—} + \text{—} \right\} + O(e^{-\alpha \, \xi})$$

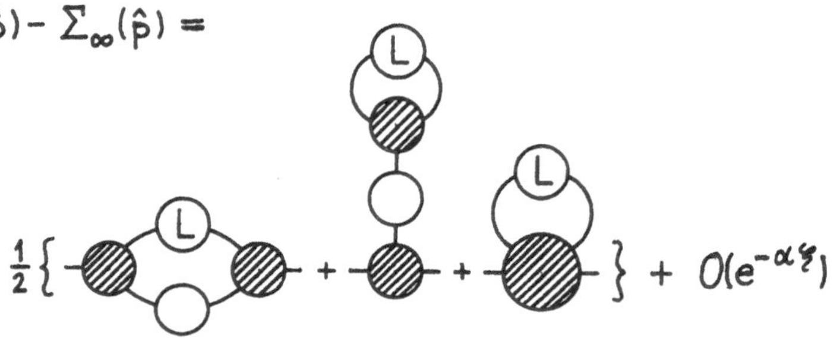

The shaded bubbles here represent the $L = \infty$ Euclidean vertex functions (full propagator amputated, 1-particle irreducible n-point functions of ϕ), and

$$-\bigcirc- = G_\infty(q)$$

$$-\textcircled{L}- = 2\sum_{i=1}^{d} \cos(q_i L)\; G_\infty(q) \tag{20}$$

The L-dependence of the rhs of eq. (19) stems entirely from the modified propagator (20). All other factors and loop momentum integrations are exactly as in the infinite volume theory.

Eq. (19) can be understood heuristically by noting that in position space the modified propagator (20) is equal to the old propagator with the argument shifted by a period L in any of the 2d space directions. The rhs of eq. (19) may then be interpreted as the sum of all self-energy diagrams where exactly one line is allowed to wind around the world. A mathematical proof of eq. (19) can be given, within the framework of Feynman diagrams, for any massive, local Lagrangian theory of the field ϕ [5]. The form of the self-interactions of ϕ is irrelevant for the proof, as long as a perturbative treatment is meaningful. In view of this generality, it seems reasonable to expect eq. (19) to be valid also beyond perturbation theory.

The proof of eq. (15) is now easily completed along the following lines. First note that, for symmetry reasons, the modified propagator (20) may be replaced by

$$2d\, e^{iq_1 L}\; G_\infty(q)$$

without changing the rhs of eq. (19). In each of the diagrams, we then substitute $q_1 \rightarrow q_1 + is$, $s > 0$, so that

$$e^{iq_1 L} \longrightarrow e^{-sL} e^{iq_1 L}$$

In the course of this shift of the q_1 integration contour, singularities are met. Those closest to the real axis are the poles of the propagators at $q^2 = -m_o^2$ and $(q + \hat{p})^2 = - m_o^2$ (in the first diagram). Other singularities are further off and need not be considered, if we allow for an error of order $e^{-\alpha \ell}$. The pole contributions, on the other hand, are easily evaluated by the residue theorem, which puts the momenta flowing through the propagators on the mass shell. We are then left with an amputated 4-point function with all legs on mass shell, i.e. with a scattering amplitude. In this way eq. (15) is obtained. In particular, the term proportional to λ^2 arises from a contribution of the first diagram on the rhs of eq. (19), where the momenta flowing through the propagators are both on the mass shell.*

4. APPLICATIONS

4.1 2-dimensional Ising model

This is one of the rare cases where δ_o is known exactly. Comparing the continuum limit result (12) with eq. (15), we can read off the scattering amplitude:

$$F(\nu) = - 8\, m_o \sqrt{m_o^2 - \nu^2} \tag{21}$$

In 1+1 dimensions, identical particles can only scatter forwardly so that F in fact determines the complete elastic scattering amplitude. Thus, the scattering operator $\mathbb{S}$ in the 2-particle sector can be calculated and comes out to be

---

* Note that $\lambda^2 = (\Gamma_3(k_1,k_2,k_3))^2$, where Γ_3 is the 3-point vertex function and $k_i^2 = - m_o^2$ for all i.

(22) $$\mathcal{S} = -1$$

In view of eq. (12), the striking simplicity of this result is perhaps not unexpected. Anyhow, eq. (22) has been known before [7)8)], and we may therefore consider the present calculation a non-perturbative check of the new relation (15).

4.2 2-dimensional O(n) non-linear σ-model

There has been increasing evidence that the particle spectrum of this model consists of a massive O(n) vector multiplet. With a few structural assumptions (most of which were later justified), the brothers Zamolodchikov have been able to calculate the scattering matrix exactly [9)]. In particular, the forward amplitude $F(\nu)$, as defined in the Appendix, is known and can be inserted into eq. (15), which is also valid in the present case. For n = 3, the formula for the mass shift of the O(n) vector multiplet obtained in this way, assumes the following simple form:

(23) $$\delta_o = 4\pi \int_{-\infty}^{\infty} dt\; e^{-\xi \operatorname{ch} t}\, \frac{\operatorname{ch} t}{t^2 + \frac{9}{4}\pi^2} + O(e^{-\alpha \xi})$$

Evaluating the integral numerically and neglecting the error term, yields the table

ξ	δ_o	
1	0.650	
2	0.155	(n = 3)
3	0.045	
4	0.014	

As in the Ising model, δ_o thus turns out to be small for $\xi \geq 3$.

Some time ago, I computed $m_o/\Lambda_{\overline{MS}}$ by a method, which was based on finite L perturbation theory [2] ($\Lambda_{\overline{MS}}$ denotes the Λ-parameter in the modified minimal subtraction scheme). A working hypothesis was that finite size effects on the mass gap are small for (say) $\zeta = 3...4$, an assumption, which is now seen to be entirely justified. Furthermore, the news on δ_o can be combined with the calculation of Ref. 2 to obtain the improved value

(24) $$m_o/\Lambda_{\overline{MS}} = 1.6 \qquad (n = 3)$$

I believe that this result is accurate to 10%, but a reliable error estimation has to wait for the 2-loop mass formula [15].

For n > 3, the situation is qualitatively the same as in the O(3) model, and at $n = \infty$ a further non-trivial check of eq. (15) is obtained, because δ_o can also be computed by other means in this case [2]. It is conceivable that eq. (23) is but the first term of a complete expansion of δ_o as in the Ising model (eq. (12)), the higher terms being related to many particle scattering processes. Such a magic formula would eventually lead to an exact determination of $m_o/\Lambda_{\overline{MS}}$.

4.3 4-dimensional pure SU(n) gauge theory

Little is known about the particle spectrum of these models, but all the recent investigations point to the existence of a lowest lying $J^{PC} = 0^{++}$ particle and of a perhaps nearby 2^{++} particle. As an interpolating field for the 0^{++} particle one may take, for example,

$$\phi(x) = F^a_{\mu\nu}(x)\, F^a_{\mu\nu}(x),$$

where $F^a_{\mu\nu}$ is the gauge field tensor. From perturbation theory one easily shows that the 3-point function of ϕ does not vanish.

Barring an accidental zero on the mass shell, it follows that the effective coupling λ does not vanish, too. We therefore conclude that

$$(25)\quad \delta_0 \underset{\zeta\to\infty}{\sim} -\left(\frac{\lambda}{m_0}\right)^2 \frac{3}{16\pi}\, \zeta^{-1}\, e^{-\frac{\sqrt{3}}{2}\zeta} ,$$

where m_o denotes the mass of the 0^{++} particle and δ_o the corresponding finite volume mass shift. Since λ must be real, eq. (25) implies that

$$(26)\quad \delta_o < 0 \text{ for large } \zeta .$$

On the other hand, $\delta_o \longrightarrow +\infty$ for $\zeta \longrightarrow 0$ [3)] so that there must be a value of ζ, where M_o/m_o is minimal. This behaviour is quite different from the situation in the non-linear σ-model (M_o is monotonically decreasing there) and is perhaps related to the small radius of convergence of the perturbation expansion of M_o derived in Ref. 3.

4.4 Realistic QCD *

Let us first consider the pion mass shift δ_π. Performing the integral over the momentum components q_2 and q_3 explicitly, eq. (15) reduces to

$$(27)\quad \delta_\pi = -\frac{3}{8\pi\zeta}\int_{-\infty}^{\infty}\frac{dp}{2\pi}\, e^{-\zeta\sqrt{1+p^2}}\, F_{\pi\pi}(i m_\pi p) + O(e^{-\alpha\zeta})$$

---

* Not the quenched approximation. Vacuum polarization is crucial for the validity of the formulae in this paragraph. Isospin breaking effects are neglected.

where m_π denotes the pion mass, $F_{\pi\pi}(\nu)$ the forward $\pi\pi$-scattering amplitude, and $\zeta = m_\pi L$. To estimate $F_{\pi\pi}(\nu)$ near $\nu = 0$, we have current algebra results and experimental data at our disposal. Thus, denoting isospin indices by $\alpha, \beta, \ldots$, the elastic $\pi\pi$-scattering amplitude can be written as

$$T_{\alpha'\beta',\alpha\beta} = \delta_{\alpha\beta}\delta_{\alpha'\beta'} A(s,t,u) + \delta_{\alpha\alpha'}\delta_{\beta\beta'} A(t,u,s) + \delta_{\alpha\beta'}\delta_{\beta\alpha'} A(u,s,t) \tag{28}$$

(see the Appendix for unexplained notation). To lowest order of the chiral low energy expansion, the invariant amplitude A is then given by Weinberg's formula [10]:

$$A(s,t,u) = \frac{1}{F_\pi^2}(s - m_\pi^2) \tag{29}$$

F_π = pion decay constant.

Recently, the next order has also been worked out [11], but, for simplicity, I shall stick to the lowest order expression. The forward amplitude is then easily calculated and eq. (27) becomes

$$\delta_\pi = \frac{3}{8\pi\zeta}\frac{m_\pi^2}{F_\pi^2}\int_{-\infty}^{\infty}\frac{dp}{2\pi}\, e^{-\zeta\sqrt{1+p^2}} + O(e^{-\alpha\zeta}) \tag{30}$$

Inserting the experimental values m_π = 139 MeV, F_π = 93 MeV, we finally obtain the table

ζ	δ_π
1.0	0.051
1.5	0.016
2.0	0.006
2.5	0.003

δ_π is thus roughly an order of magnitude smaller than the mass shift δ_0 in the two-dimensional non-linear σ-model.

A modified form of eq. (15) also applies to the finite size mass shift δ_p of the proton, which is defined by

$$\delta_p = (M_p - m_p)/m_p \tag{31}$$

(m_p is the physical proton mass and M_p the energy of the zero momentum proton state at finite L). The relevant scattering amplitude in this case is the pion-proton forward elastic scattering amplitude $F_{\pi p}(\nu)$, where, in the notation of the Appendix, the proton is the A-particle and the pion the B-particle. $F_{\pi p}(\nu)$ has poles at

$$\nu = \pm \nu_B \,, \qquad \nu_B = \frac{m_\pi^2}{2m_p} \,, \tag{32}$$

with residues

$$\lim_{\nu \to \pm\nu_B} (\nu^2 - \nu_B^2)\, F_{\pi p}(\nu) = -6\, g_{\pi N}^2\, \nu_B^2 \tag{33}$$

$g_{\pi N}$: pion-nucleon coupling constant.

These poles come from nucleon exchange diagrams, and they give rise to the leading term in the proton mass shift formula

$$\delta_p = \frac{9}{4}\left(\frac{m_\pi}{m_p}\right)^3 \frac{g_{\pi N}^2}{4\pi\xi}\, e^{-\xi\sqrt{1-\nu_B^2/m_\pi^2}}$$

$$- \frac{3}{8\pi\xi}\left(\frac{m_\pi}{m_p}\right)^2 \int_{-\infty}^{\infty} \frac{dp}{2\pi}\, e^{-\xi\sqrt{1+p^2}}\, F_{\pi p}(i m_\pi p) + O(e^{-\alpha\xi}) \tag{34}$$

(as before $\xi = m_\pi L$). Note, however, that for reasonable values of ξ, both terms are of the same order of magnitude, because $\nu_B^2/m_\pi^2 \ll 1$. The first term only dominates for academically large ξ.

The experimental pion-nucleon scattering data were fitted by Höhler et al. [12], and, using dispersion relations, they were able to determine the pion-nucleon coupling constant $g_{\pi N}$ and the forward scattering amplitude near $\nu = 0$ *. To display their results, we first subtract the pseudo-vector Born term from the scattering amplitude,

$$F_{\pi p}(\nu) = 6\, g_{\pi N}^2\, \nu_B^2 / (\nu_B^2 - \nu^2) + R(\nu), \tag{35}$$

and expand the remainder $R(\nu)$ in a convergent power series:

$$R(\nu) = \sum_{k=0}^{\infty} r_k\, (\nu/m_\pi)^{2k} \qquad (|\nu| < m_\pi). \tag{36}$$

From the analysis of Höhler et al., we then have

$$g_{\pi N}^2/4\pi = 14.3 \tag{37a}$$

$$r_0 = -60.7, \quad r_1 = 45.3, \quad r_2 = 8.1 \tag{37b}$$

(this order of magnitude of the coefficients r_i is also suggested from chiral perturbation theory [13]).

At first sight, $R(\nu)$ appears to be a rather large amplitude,

* $F_{\pi p}(\nu)$ is related to the amplitude $C^+(\nu)$ of Höhler et al. through $F_{\pi p}(\nu) = 6\, m_p\, C^+(\nu)$.

at least, when compared with $F_{\pi\pi}(0) \simeq -2$. In eq. (34), $R(\nu)$ is however multiplied by $(m_\pi/m_p)^2$ so that its contribution to the proton mass shift at say $\varsigma = 1$ comes out to be a few percent only. The same is true for the other two contributions, which, furthermore, come with opposite sign and almost cancel each other. The conclusion then is that

$$0 < \delta_p < 0.05 \text{ for } \varsigma = 1$$

$$0 < \delta_p < 0.01 \text{ for } \varsigma \geq 1.5$$

i.e. the proton mass shift is even a little smaller than the pion mass shift.

The smallness of the pion mass has always been a potential source of difficulties in hadron mass calculations. Here we have seen that it is balanced, to some extent, by the weakness of the interactions of the pion at low energies. This property is a consequence of spontaneous chiral symmetry breaking and current algebra. Because of the unphysical anomalies of the axial currents in lattice QCD, we must therefore be careful, when carrying over the results obtained above. The prospects for calculating the spectrum on small lattices are otherwise rather good, since the condition $m_\pi L \gtrsim 1$ is not very restrictive.

4.5 Low temperature quantum field theory

It is well-known that the canonical ensemble describing a given quantum field theory at temperature T can be realized in Euclidean space by letting the time coordinate become an angular variable with period $L = 1/k_B T$ (k_B: Boltzmann constant). From a mathematical point of view, the situation is then not very different from the one we have studied so far, except that time and space are interchanged and that now only one coordinate is compactified. Also, the physical meaning of M_o is not so much that of an energy, but rather of an inverse correlation length characterizing the falling off of correlation functions in space-like directions. In the present context, eq. (15) remains valid provided the rhs is divided by d. It represents the leading term of a low temperature expansion of M_o.

In most cases we have considered, δ_o is positive. This means that the correlation length decreases as the temperature rises, a behaviour, which is typical for systems where the low temperature phase is either unbounded or bounded by first order transitions. Especially, a phase transition in QCD in the region $0 \leq k_B T \leq m_\pi$ seems rather unlikely, because M_π and M_p are practically temperature independent there. In the pure Yang-Mills gauge theory, on the other hand, δ_o is negative and may eventually drop to -1 at some critical T. A phase transition is in fact known to exist in lattice gauge theories [14)], although the order of the transition has not yet been pinned down.

ACKNOWLEDGMENTS

I am indebted to J. Gasser for an introduction to pion-pion and pion-nucleon scattering. I have also benefitted from discussions with H. Lehmann on these matters and with G. Münster on the strong coupling expansion of the mass gap in lattice models. Finally, I would like to thank the organizers for their kind invitation to lecture at Cargèse.

APPENDIX: Scattering amplitude notations

My conventions concerning (d+1)-vectors are:

$$p^\mu = (p^o, \vec{p}) \ , \quad \vec{p} = (p^1, \dots, p^d)$$

$$p_o = p^o \, , \quad p_i = -p^i \quad (i = 1,\ldots,d)$$

$$p \cdot q = p_\mu q^\mu = p^o q^o - \vec{p} \cdot \vec{q}$$

One-particle states $|p\alpha\rangle$ with momentum p and quantum numbers α (spin, isospin) are always normalized such that

$$\langle p'\alpha' | p\alpha \rangle = \delta_{\alpha'\alpha} \, 2p^o \, (2\pi)^d \, \delta(\vec{p}\,' - \vec{p})$$

The T-matrix for elastic scattering A+B → A+B is defined by

$$\langle p'\alpha', q'\beta' \text{ out} | p\alpha, q\beta \text{ in} \rangle =$$

$$\delta_{fi} + i\,(2\pi)^{d+1} \delta(p'+q'-p-q) \, T(p'\alpha', q'\beta' | p\alpha, q\beta),$$

where $p\alpha$ and $p'\alpha'$ refer to particle A and

$$\delta_{fi} = \langle p'\alpha', q'\beta' \text{ in} | p\alpha, q\beta \text{ in} \rangle.$$

For spinless particles, we also write

$$T(p'\alpha', q'\beta' | p\alpha, q\beta) = T_{\alpha'\beta',\alpha\beta}(s,t,u)$$

$$s = (p+q)^2, \quad t = (q'-q)^2, \quad u = (q'-p)^2$$

The forward amplitude F, which enters the formula for δ_o, is defined through

$$F = \sum_\beta T(p\alpha, q\beta | p\alpha, q\beta).$$

In all cases considered here, F is a Lorentz scalar depending only on the "crossing variable"

$$\nu = \frac{s-u}{4m_A} \, , \quad m_A\text{: mass of particle A.}$$

REFERENCES

1) B. Berg: these proceedings
2) M. Lüscher: Phys. Lett. 118B (1982) 391
3) M. Lüscher: Nucl. Phys. B219 (1983) 233, and
M. Lüscher and G. Münster: Nucl. Phys. B232 (1984) 445, and
G. Münster: in Proceedings of the VII International Congress on Mathematical Physics, Boulder, Colorado (1983), to be published
4) T.D. Schultz, D.C. Mattis and E.H. Lieb: Rev. Mod. Phys. 36 (1964) 856
5) To be published elsewhere
6) I.S. Gradshteyn and I.M. Ryzhik: "Table of Integrals, Series and Products", Academic Press, New York (1965)
7) M. Sato, T. Miwa and M. Jimbo: "Field theory of the 2d Ising model in the scaling limit", Kyoto preprint IMS 207 (1976) (unpublished)
8) B. Berg, M. Karowski and P. Weisz: Phys. Rev. D19 (1979) 2477
9) A.B. Zamolodchikov and A.B. Zamolodchikov: Nucl. Phys. B133 (1978) 525
10) S. Weinberg: Phys. Rev. Lett. 17 (1966) 616
11) J. Gasser and H. Leutwyler: Bern preprint BUTP-83/5
12) G. Höhler, F. Kaiser, R. Koch and E. Pietarinen: "Handbook of Pion-Nucleon Scattering", Physics Data 12-1, Fachinformationzentrum Karlsruhe (1979)
13) H. Lehmann: private communication
14) C. Borgs and E. Seiler: Nucl. Phys. B215 [FS7] (1983) 125
15) D. Petcher and M. Floratos: in preparation

RENORMALIZATION GROUP AND MAYER EXPANSIONS

Gerhard Mack

II. Institut für Theoretische Physik
der Universität Hamburg

Mayer expansions promise to become a powerful tool in exact renormalization group calculations. Iterated Mayer expansions were sucessfully used in the rigorous analysis of 3-dimensional U(1) lattice gauge theory by Göpfert and the author [1,2] , and it is hoped that they will also be useful in the 2-dimensional nonlinear σ-model, and elsewhere.

SOME LATTICE MODELS

I will consider models that live on a d-dimensional hypercubic lattice Λ of lattice spacing a. Its sites are denoted by x, y, ... Standard lattice notation will be used, viz.

$$\int_x = a^d \sum_{x\in\Lambda} \quad , \quad (f,g) = \int_x f(x)g(x)$$

$$\nabla_\mu f(x) = a^{-1}[f(x+e_\mu)-f(x)] \quad , \quad -\Delta = \nabla_{-\mu}\nabla_\mu \quad . \text{ etc.} \tag{1.1}$$

e_μ is the lattice vector of length a in the μ-direction; $e_{-\mu} = -e_\mu$.

<u>O(N)-symmetric 2-dimensional Heisenberg ferromagnet = nonlinear σ-model</u>

The (random) variables of the model are N-dimensional unit vectors s(x) attached to the sites x of the lattice. The Hamiltonian is

$$\mathcal{H}(s) = \frac{N}{2f_0}\int_x [\nabla_\mu s(x)]^2 \tag{1.2}$$

Expectation values of observables O(s) are computed with Boltzmann factor $e^{-\mathcal{H}(s)}$, viz

$$\langle O(s)\rangle = Z^{-1}\int \prod_x d\vec{s}(x)\, e^{-\mathcal{H}(s)}\, O(s)$$
$$Z = \int \prod_x d\vec{s}(x)\, e^{-\mathcal{H}(s)} \qquad (1.3)$$

$d\vec{s}$ is the rotation invariant measure on the unit sphere S^{n-1}.

This model is of interest because it is renormalizable and asymptotically free in 2 dimensions in perturbation theory [3]. It is therefore expected that the model possesses a continuum limit when $a \to o$, $f_o \to \infty$ in the manner dictated by perturbation theory (and for suitably defined observables). Moreover, the model is expected to have a finite correlation length for all values of the coupling constant f_o if $N \geq 3$.

Discrete Gaussian model in 3 dimensions

The (random) variables of the model are again attached to the sites x of the lattice and take values $n(x) = 2\pi \cdot$ integer. So the field n has N = 1 component. The Hamiltonian is

$$\mathcal{H}(n) = \frac{1}{2\beta}\int_x [\nabla_\mu n(x)]^2 \quad , \qquad (1.4)$$

and expectation values are computed according to

$$\langle O(n)\rangle = Z^{-1}\sum_n e^{-\mathcal{H}(n)}\, O(n)$$
$$Z = \sum_n e^{-\mathcal{H}(n)} \qquad (1.5)$$

The model possesses a global symmetry

$$n(x) \to n(x) + 2\pi I \quad , \text{ I integer} \qquad (1.6)$$

This model is the Kramers Wannier dual transform of the 3-dimensional U(1) lattice gauge theory model with Villain action and el. charge squared $e^2 = 4\pi^2/\beta$. The correlation length of both models is the same, and the string tension α of the 3-dimensional U(1) lattice gauge theory equals the surface tension in the discrete Gaussian model. The global symmetry of the discrete Gaussian model is spontaneously broken if the expectation value $\langle n(x)\rangle$ exists at all (as it does in 3 dimensions), and the surface tension is defined as free energy/area of a domain wall between domains with different magnetization $n(x) = 0$ and 1. It was proven that the model possesses a finite correlation length ξ for large β, and asymptotically as $\beta \to \infty$ its inverse ξ^{-1} approaches the value m_D,

$$m_D^2 = (2\beta/a^3)e^{-\beta v_{Cb}(0)/2} \quad , \quad v_{Cb}(0) = 0.2527\, a^{-1} . \tag{1.7}$$

It was also shown that the string tension α is finite for all values of β, and obeys a bound

$$\alpha \geq c\, m_D \beta^{-1} \qquad \text{for large } \beta \tag{1.8}$$

c is a dimensionless constant. A (semi-)classical approximation starting from the effective action eq. (3.4) below (without the correction terms ...) gives

$$\alpha = 8\, m_D \beta^{-1} \qquad \text{for large } \beta \tag{1.9}$$

Existence of a continuum limit was also shown [2,4] . The model is unique in that it is much better understood by theoreticians than by computers. Recent Monte Carlo data are in agreement with the predictions (1.7), (1.9), however [5].

<u>A common formula</u>

Both models are similar in that they would be massless free field theories if it were not for the constraints which force s(x) to be a vector of unit length and n(x)/2π to be an integer, respectively. More generally, they are special cases of models with Boltzmann weights given by

$$\langle O \rangle = Z^{-1} \int d\mu_{\gamma v_{Cb}}(\varphi) \prod_x F(\varphi(x)) \tag{1.10}$$

$d\mu_{\gamma v_{Cb}}(\varphi)$ is the Gaussian measure which describes an N-component massless real free field φ with propagator γv_{Cb},

$$-\Delta_x v_{Cb}(x-y) = \delta(x-y)\cdot \mathbb{1} \tag{1.11}$$

$\mathbb{1}$ = N x N unit matrix. Explicitly

$$d\mu_{\gamma v_{Cb}}(\varphi) = Z_0^{-1} \exp\left\{ -\frac{1}{2\gamma} \int_x [\nabla_\mu \varphi(x)]^2 \right\} \prod_x d^N\varphi(x) \tag{1.12}$$

In the nonlinear σ-model

$$\gamma = f_0/N \quad , \quad F(\varphi) = \delta(\varphi^2 - 1) \quad , \tag{1.13a}$$

while in the discrete Gaussian model

$$\gamma = \beta \quad , \quad F(\varphi) = \sum_{n=0,\pm 1,..} \delta(2\pi n - \varphi) \tag{1.13b}$$

The class of models with Boltzmann weight (1.10) includes also

$\lambda\phi^4$-theory and other models studied by constructive field theorists [6] , the sine-Gordon theory, etc.

GENERAL STRATEGY FOR THEORIES WITH (POSSIBLY NONPERTURBATIVE) MASS

It is very helpful to divide the problem into two .

i) Compute an effective Hamiltonian $\mathcal{H}_{eff}(\Phi)$ with an UV-cutoff M of the order of the ultimate physical mass scale m, or a sequence of effective Hamiltonians with cutoffs M_1, M_2, ... $M_n = M = O(m)$. The cutoff may for instance be introduced in the form of a new lattice spacing $M^{-1} = (M^{-1}/a)a$

ii) Analyze the theory with this effective Hamiltonian.

We are interested in theories close to the continuum limit. In this case ma <<< 1 and therefore also Ma <<< 1. In an asymptotically free theory, such as the 2-dimensional nonlinear σ-model, the ultimate physical mass m is known a priori, as a function of a and bare coupling constant,apart from a dimensionless constant factor, if conventional wisdom is correct.

The point is that different problems arise in the two steps and different methods are applicable. For instance, spontaneous symmetry breaking, such as occurs in the discrete Gaussian model, affects only step ii). I advocate to use Mayer expansions in step i) to compute the effective actions $\mathcal{H}_{eff}$. In theories with a mass gap one can hope to carry through step ii) using standard methods of constructive field theory. For instance, Glimm Jaffe Spencer expansions around mean field theory [7] were used in case of the discrete Gaussian model. As a rule, semiclassical approximations become accurate when the UV cutoff M is low enough.

If we start from a theory on a lattice of lattice spacing a, then the effective Hamiltonian $\mathcal{H}_{eff}$ on the block lattice of lattice spacing M^{-1} will depend on a. If step i) can be mastered then one may examine the existence of a continuum limit by studying whether $\mathcal{H}_{eff}$ possesses a limit as $a \to o$, when the bare coupling constant in the original theory depends on a, and the field Φ is rescaled, in the appropriate way. To complete the proof of the existence of a continuum limit one would need to show in addition that an infinitesimal "local" perturbation of $\mathcal{H}_{eff}$ does not lead to a finite change in the correlation functions. To establish such a stability property can be a difficult problem which really belongs to step ii). In any case, however, only the presence of a first order phase transition could cause trouble of $\lim_{a\to o}\mathcal{H}_{eff}$ exists

BLOCK SPIN TRANSFORMATION

The effective Hamiltonians will depend on a new field Φ which is costumarily called the block spin. The simplest choice is $\Phi \propto$ block average of φ. One partitions the original lattice into blocks x' of side length La = M^{-1} and sets e.g.

$$\Phi(x') = L^{-2} \sum_{x \in x'} \varphi(x) \tag{3.1}$$

in 2 dimensions. The field Φ lives on a new "block" lattice of lattice spacing La, and the effective Hamiltonian is defined by

$$e^{-\mathcal{H}_{eff}(\Phi)} = \int \prod_x d^N\varphi(x)\, e^{-\mathcal{H}(\varphi)} \prod_{x'} \delta\Big(\Phi(x') - L^{-2} \sum_{x \in x'} \varphi(x)\Big) \tag{3.2}$$

One speaks in this case of a "real space renormalization group transformation" $\mathcal{H}(\varphi) \to \mathcal{H}_{eff}(\Phi)$ [0].

An alternative possibility is to introduce a Pauli Villars cutoff M in place of a block lattice. In this case the cutoff is buildt into the kinetic term in $\mathcal{H}_{eff}$, for instance

$$\text{kinetic term in } \mathcal{H}_{eff} = \frac{1}{2} \left(\nabla_\mu \Phi, \left[1 - \frac{\Delta}{M^2}\right] \nabla_\mu \Phi \right) \tag{3.3}$$

In the gas picture of quantum field theory, which will be described below, both possibilities amount to a split of the propagator γv_{Cb} of the original theory.

Example 1. The exact effective action for the discrete Gaussian model has the following form (for large enough β, i.e. close to the continuum limit) [2]. $\mathcal{H}_{eff}$ depends on a real field Φ and

$$\mathcal{H}_{eff}(\Phi) = \text{kinetic term} + V_{eff}(\Phi) \tag{3.4a}$$

If a Pauli Villars cutoff M is used then Φ lives on the original lattice, the kinetic term is given by expression (3.3) divided by β, and

$$-V_{eff}(\Phi) = \sum_{s=1,2,\ldots} \frac{1}{s!} \sum_{m_1 = \pm 1, \pm 2, \ldots} \cdots \sum_{m_s = \pm 1, \pm 2, \ldots} \int_{x_1} \cdots \int_{x_s} \tag{3.4b}$$

$$\rho_s(m_1, x_1, \ldots, m_s, x_s) \left[e^{i m_1 \Phi(x_1)} - 1\right] \cdots \left[e^{i m_s \Phi(x_s)} - 1\right]$$

$$= -\frac{m_D^2}{\beta} \int_{x \in \Lambda} \left[1 - \cos \Phi(x)\right] + \cdots \tag{3.4c}$$

$$\ldots = (\text{terms from } s = 1,\ |m_1| \geq 2) + \text{nonlocal terms.}$$

It is crucial for the success of any renormalization group strategy that the effective Hamiltionans are essentially local. It is inevitable that the exact effective Hamiltonian will have nonlocalities in the form of exponential tails, but these should decay over a distance of order M^{-1} for UV-cutoff M. This is satisfied in the example. The coefficients ρ_s $(m_1 x_1 \ldots m_s x_s)$ decay exponentially with distance between points $x_1 \ldots x_s$ with a decay rate govered by M. The precise statement of the bounds which incorporate the results about convergence of the infinite sums in (3.4a) and decay properties of the kernels ρ_s is a follows [2]. Given $\kappa < \infty$, $\mu < \infty$ arbitrarily large, and $C > o$, $\delta > o$ arbitrarily small and < 1, then the following bound is true for $s \geq 1$, uniformely in the size of the lattice Λ, provided $\lambda = m_D/M$ is sufficiently small (depending on κ, μ, C, δ) and β/a is sufficiently large (depending on κ, μ, C, δ, λ). Let $a_1 \ldots a_s$ some arbitrary subsets of Λ of volume $\mathrm{Vol}(a_i) = \int_{x \in a_i}$. Then

$$\sum_{m_1} \ldots \sum_{m_s} \int_{x_1 \in a_1} \ldots \int_{x_s \in a_s} |\rho_s(m_1, x_1, \ldots, m_s, x_s)| \, e^{2\kappa(-1+\Sigma |m_i|)}$$
$$\leq s! \, Vol(a_1) \, m_D^2 \beta^{-1} e^{-(1-\delta) M L(a_1, \ldots, a_s) - \mu(s-1)} (1-C)^{-s-10} \tag{3.5}$$

$L(a_1 \ldots a_s)$ is the "length of the shortest highway net that can be buildt to connect all the citites $a_1 \ldots a_s$", see figure 1. The length is defined using the distance $||x|| = 3^{-1/2} \sum_\mu |x_\mu|$.

The factor $e^{-\mu(s-1)}$ in these bounds shows that the correction terms ... in expression (3.4c) are very small so that the effective Hamiltonian is essentially that of a sine - Gordon theory. These results for the discrete Gaussian model are generalizations of earlier results of Brydges and Federbush [8] for very dilute Coulomb systems.

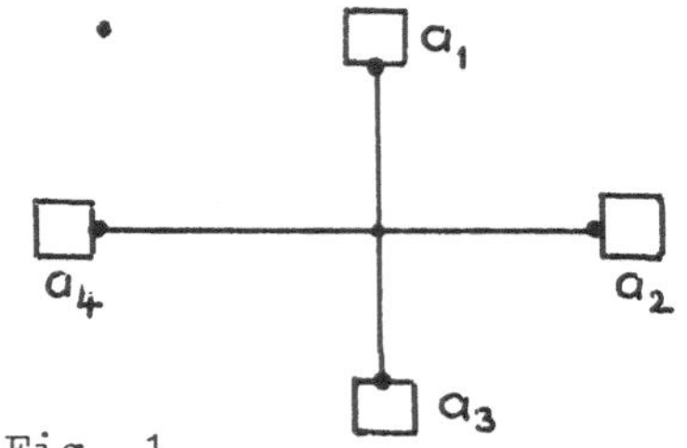

Fig. 1
The shortest highway net connecting cities $a_1 \ldots a_4$.

Let me finally emphasize that the block spin Φ need not be a field of the same kind as the original field φ. For instance in our examples

$$\text{unit N-vector } \varphi = s \quad \to \text{ arbitrary N-vector } \Phi \tag{3.6a}$$

$$\text{integer valued field } \varphi = n \to \text{ real field } \Phi \tag{3.6b}$$

It turns out that an attempt to use an integer valued block spin in the discrete Gaussian model leads to disaster. At some intermediate stage of the renormalization group procedure, when M is still very much larger than m_D (e.g. of order β^{-1}), the exponential tails in the effective Hamiltonian do not decay with a rate determined by M, but very much slower. That is, the effective Hamiltonian becomes very nonlocal.

Example 2. The effective Hamiltonian for the 2-dimensional nonlinear σ-model is not under control yet, but it is expected that it will be of the following form for large N, for choice (3.1) of block spin.

$$\mathcal{H}_{eff}(\Phi) = \text{kinetic term} + \int_x \mathcal{V}\left(|\Phi(x)|^2\right) \tag{3.7a}$$

$$+ \text{ small nonlocal terms involving } \nabla\nabla\Phi$$

$$e^{-\mathcal{V}(\xi^2)} \approx \int_{S^{N-1}} d\eta \exp\left\{-\frac{N}{2f_0 v(0)}(\xi-\eta)^2\right\} \tag{3.7b}$$

Integration of η is over the unit sphere S^{N-1} in N dimensions. The potential $\mathcal{V}$ has the form shown in figure 2.

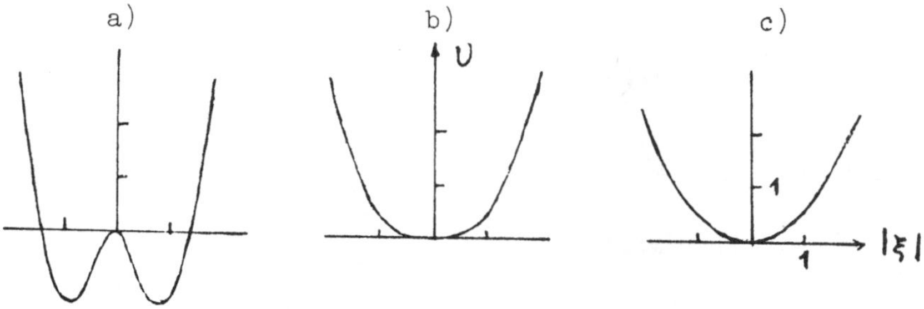

Figure 2 Potential $\mathcal{V}$ given by expression (3.7b).

a) for $f_0 v(0) < 1$ b) $f_0 v(0) = 1$ c) $f_0 v(0) > 1$

In the drawings N = 4, $f_0 v(0)$ = .4, 1.0, 2.0 respectively.

$v(0)$ is the propagator for the fluctuation part of the original field at zero distance. If a Pauli Villars cutoff were in use instead of a block lattice, this would be

$$v(0) = (-\Delta+M^2)^{-1}(x,x)$$

In any case,

$$v(0) \to \infty \quad \text{as} \quad M \to 0 \quad , \text{ viz } \quad v(0) \sim -\frac{1}{2\pi} \ln Ma + \text{const}$$

in 2 dimensions. Therefore, the situation of Figure 2c) is expected to be reached for low enough cutoff M. A semiclassical treatment of the resulting theory with Hamiltonian $\mathcal{H}_{eff}$ will then produce a finite correlation length (= mass gap). There exists a theorem of Brydges, Fröhlich and Spencer which confirms that a potential $\mathcal{V}(\xi^2)$ which is a monotone increasing function of ξ^2 assures a mass gap (assuming the nonlocal correction terms can be neglected or treated as a sufficiently small perturbation) [9].

Lesson of this school: Exact renormalization group calculations are not necessarily unmanageably complicated.

4. Gas picture

Mayer expansions are a familiar tool for the treatment of dilute gases [10]. Quantum field theory with Boltzmann weight given by eq. (1.10) admit a representation as a generalized kind of gas. It is obtained by Fourier transformation in the field.

The Fourier transform of a Gaussian measure is well known

$$\int d\mu_v(\varphi)\, e^{i(k,\varphi)} = e^{-\frac{1}{2}(k,vk)} \tag{4.1}$$

Let us set a = 1 in this section, and expand the factors F in the Boltzmann weight into a Fourier integral.

$$F(\xi) = \int dK\, \tilde{F}(K)\, e^{iK\xi} \tag{4.2}$$

Making use of (4.1) the partition function takes the form

$$Z = \int \Big[\prod_x dk(x)\, \tilde{F}(k(x))\Big] e^{-\gamma(k,v_{cb}k)/2} \tag{4.3}$$

This is the partition function of a "gas" of particles which may occupy the sites x of the lattice and which may carry arbitrary charge K = k(x). They interact through a Coulomb potential, and have a hard core which prevents any two of them from occupying the same site. If k(x) = 0 we may regard the site x as empty. The particles have a fugacity F(K) which depends on their charge K. It is not necessarily positive real (in contrast with a real gas).

Example: 3-dimensional discrete Gaussian model.
Here F is periodic and is therefore expanded in a Fourier-series

$$\sum_n \delta(2\pi n - \varphi) = \sum_{m=0,\pm1,\dots} e^{im\varphi} \tag{4.4}$$

As a result

$$\begin{aligned} Z &= \sum_{\{m(x)=0,\pm1,\dots\}} e^{-\beta(m,v_{cb}m)/2} \\ &= \sum_m z^{\Sigma m(x)^2} \exp\Big\{-\beta \sum_{x\neq y} m(x) v_{cb}(x,y) m(y)/2\Big\} \end{aligned} \tag{4.5}$$

This is the partition function of a Coulomb gas of particles with a hard core and arbitrary integer charge $m = \pm 1, \pm 2, \ldots$. They have a positive charge dependent fugacity z^{m^2},

$$z = e^{-\beta v_{Cb}(0)/2} \quad , \quad v_{Cb}(0) = 0.2527 \tag{4.6}$$

In this model, where the charge takes on only a discrete number of values, the partition function can also be written in grand canonical form. Let N be the number of occupied sites, and set

$$\xi = (m, x) \quad , \quad \int d\xi = \sum_{m=\pm 1, \pm 2, \ldots} \int_{x \in \Lambda}$$

$$v_{ij}(\xi_i, \xi_j) = \begin{cases} +\infty & \text{if } i \neq j \,,\, x_i = x_j \\ m_i v_{Cb}(x_i, x_j) m_j & \text{otherwise} \end{cases} \tag{4.7}$$

Then

$$Z = \sum_{N=0,1,2,\ldots} \frac{1}{N!} \int d\xi_1 \ldots d\xi_N \exp\left\{-\beta \sum_{i,j=1}^{N} v_{ij}(\xi_i, \xi_j)/2\right\} \tag{4.8}$$

BLOCK SPIN TRANSFORMATION VIA SPLIT OF THE POTENTIAL v_{Cb}.

Suppose for simplicity that the observables O depends on the original field φ only through the block spin Φ which may for instance be defined by eq. (3.1). Thus $O(\varphi) = O_1(\Phi)$. Then by (3.2),

$$\langle O \rangle = Z^{-1} \int O_1(\Phi) Z_1(\Phi) \prod_{x'} d\Phi(x')$$

$$Z_1(\Phi) = \exp\left[-\mathcal{H}_{eff}(\Phi)\right] \quad , \quad Z = \int Z_1(\Phi) \prod_{x'} d\Phi(x') . \tag{5.1}$$

(More generally this is true for a suitable O_1 which has to be found together with $\mathcal{H}_{eff}$.) I will now present a formula for $Z_1(\Phi)$ which generalizes the gas representation (4.3) of the full partition function. A similar formula was first used by Fröhlich [18]. It is obtained by splitting the potential v_{Cb}

$$v_{Cb} = v + u \tag{5.2}$$

The precise form of the split depends on the choice of the block spin, or, more generally, on the way in which the new UV cutoff M is introduced. But always $v(x,y)$ decays exponentially with distance $|x-y|$ with a rate of order M, while it approximates v_{Cb} well for distances considerably less than M. As a result, u has the same long distance behavior as v_{Cb} but is less singular (i.e. smaller) at short distances.

$$Z_1(\Phi)\prod_{x'} d\Phi(x') = d\mu_{\gamma u_1}(\Phi)\, e^{-V_{eff}(\Phi)} \equiv d\mu_{\gamma u_1}(\Phi) Z(\Phi) \quad (5.3)$$

$$e^{-V_{eff}(\Phi)} = \int\Big[\prod_x dq(x)\tilde{F}(k(x))\, e^{iq(x)\tilde{\Phi}(x)}\Big]\, e^{-\gamma(q,vq)/2} \quad (5.4)$$

$\tilde{\Phi}$ is a field on the original lattice which is determined by and linearly related to Φ, while $u_1(x',y')$ is related to u (s. below).

It is seen that $\exp(-V_{eff})$ is expressed as partition function of a "gas" of particles in the same way as Z was, but now these particles interact through a potential v which decays exponentially with distance. The fugacity is complex and x-dependent because of the factor $\exp iq(x)\tilde{\Phi}(x)$ which depends on Φ. The Gaussian measure $d\mu_{\gamma u}$ contributes a kinetic term to $\mathcal{H}_{eff}$, viz

$$\mathcal{H}_{eff} = \frac{1}{2\gamma}(\Phi, u_1^{-1}\Phi) + V_{eff} \quad (5.5)$$

In the case of a Pauli Villars cutoff, Φ lives on the original lattice, $\tilde{\Phi} = \Phi$ and $u_1 = u$, and

$$v = (-\Delta + M^2)^{-1}$$
$$u = (-\Delta)^{-1} - (-\Delta + M^2)^{-1} = \Big[-\Delta\Big(1 - \frac{\Delta}{M^2}\Big)\Big]^{-1} \quad (5.6)$$

For the choice (3.1) of block spin, the appropriate split of the potential v_{cb} is as follows. Let C be the map which takes square summable functions on the original lattice into squares summable functions on the block lattice through averaging over blocks

$$C\varphi(x') = (La)^{-d}\int_{x\in x'}\varphi(x) \quad (5.7)$$

so that $\Phi = C\varphi$, and let C^* be its adjoint. Let $u_1(x',y')$ be defined by

$$u_1 = C v_{cb} C^* \quad (5.8)$$

Then

$$u = \mathcal{A} u_1 \mathcal{A}^* \quad \text{and} \quad \tilde{\Phi} = \mathcal{A}\Phi \quad \text{with} \quad \mathcal{A} = v_{cb} C^* u_1^{-1} \quad (5.9)$$

As a result, $v = v_{Cb} - u$ takes the form

$$v = v_{cb} - v_{cb} C^* u_1^{-1} C v_{cb} = \lim_{\lambda\to\infty}(-\Delta + \lambda C^* C)^{-1} \quad (5.10)$$

It was shown by Kupiainen and Gawedzki [11] that the kernels $\mathcal{A}(y,x')$ and $v(x,y)$ decay exponentially with distance. Another instructive proof of the exponential deday of v was given by Bałaban [12]; it starts from the second formula for v. Let us verify that expression (5.3), (5.4) for $\mathcal{Z}_1(\Phi) = \exp(-\mathcal{H}_{eff}(\Phi))$ is in agreement with the definition (3.2), when v and u_1 are as defined in eqs. (5.7) ... (5.10).

The first step is to verify the following formula for the decomposition of the Gaussian measure $d\mu_{\gamma v_{Cb}}(\phi)$ into a Gaussian measure for the block spin Φ and a Gaussian measure for the fluctuation part ζ of $\varphi = \mathcal{A}\Phi + \zeta$. This formula is of independent interest and forms the starting point of the work of ref. 10.

$$\int d\mu_{\gamma v_{Cb}}(\varphi) f(\varphi) = \int d\mu_{\gamma u_1}(\Phi)\, d\mu_{\gamma v}(\zeta)\, f(A\Phi + \zeta) \tag{5.11}$$

for any f. If suffices to verify validity of this formula for $f(\varphi) = \exp i(p,\varphi)$. Formula (4.1) for the Fourier transform of a Gaussian measure gives

$$\int d\mu_{\gamma v_{Cb}}(\varphi) e^{i(p,\varphi)} = e^{-\gamma(p, v_{Cb} p)/2}$$

and

$$\int d\mu_{\gamma u_1}(\Phi)\, d\mu_{\gamma v}(\zeta) e^{i(p,\mathcal{A}\Phi)+i(p,\zeta)} = \int d\mu_{\gamma u_1}(\Phi) e^{i(\mathcal{A}^* p,\Phi) - \frac{\gamma}{2}(p,vp)}$$

$$= \exp\left\{-\frac{\gamma}{2}\left[(p,\mathcal{A}u_1\mathcal{A}^* p) + (p,vp)\right]\right\} = e^{-\gamma(p,v_{Cb}p)/2}$$

since $\mathcal{A}u_1\mathcal{A}^* + v = u + v = v_{Cb}$. This establishes (5.11).

Using the "common formula" of section 1, definition (3.2) of $\mathcal{H}_{eff}$ takes the following form for our models

$$\mathcal{Z}_1(\Phi) = \int d\mu_{\gamma v_{Cb}}(\varphi) \prod_x F(\varphi(x)) \prod_{x'} \delta(\Phi(x') - C\varphi(x')) \tag{5.12}$$

If follows from the definitions (5.8) ... (5.10) that

$$C\mathcal{A} = id \quad , \quad Cv = 0 \, .$$

As a result, the Gaussian measure $d\mu_{\gamma v}(\zeta)$ is concentrated on fields ζ with $C\zeta = 0$, i.e. vanishing block average (i.e. it is a Dirac δ-measure in some directions in field space). Therefore

$$\delta(\Phi(x') - C[\mathcal{A}\Psi + \zeta](x')) = \delta(\Phi(x') - \Psi(x'))$$

Rename the integration variable Φ of eq. (5.11) and insert (5.11) into (5.12). As a result

$$Z_1(\Phi)\prod_{x'} d\Phi(x') = \int d\mu_{\gamma u_1}(\Psi)\, d\mu_{\gamma v}(\zeta) \prod_x F\big((\mathcal{A}\Psi+\zeta)(x)\big)$$
$$\prod_{x'} \delta\big(\Phi(x') - \Psi(x')\big) d\Phi(x')$$
$$= d\mu_{\gamma u_1}(\Phi) \int d\mu_{\gamma v}(\zeta) \prod_x F\big(\mathcal{A}\Phi(x) - \zeta(x)\big)$$

Now one inserts the Fourier integral representation (4.2) of F and carries out the Gaussian integration over ζ, as in the derivation of eq. (4.3). The result is eq. (5.3), (5.4) with $\tilde{\Phi} = \mathcal{A}\Phi$.

MAYER EXPANSIONS

For a lattice gas there are in general several ways to set up a Mayer expansion [9]. The most familiar one starts from a grand canonical representation such as (4.8). This formalism is applicable to the discrete Gaussian model but it requires discrete values of the charges q. Alternatively one may proceed as follows.

Let us again set the original lattice spacing a = 1 and consider partition functions of the form

$$Z = \int \prod_x d\lambda_x(q(x)) \exp\Big[-\frac{\gamma}{2}\sum_{x\neq y} q(x)v(x,y)q(y)\Big] \qquad (6.1)$$

We wish to use expansions to compute $\mathcal{L}_{eff}(\Phi) = -\ln Z_1(\Phi)$ as given by eqs. (5.3), (5.4). In this case

$$d\lambda_x(q) = dq\, \tilde{F}(q)\, e^{iq\tilde{\Phi}(x)}\, e^{-\gamma q v(x,x) q/2} \qquad (6.2)$$

The last factor is present because we restricted the sum in the exponential in expression (6.1) to $x \neq y$.

One defines

$$f_{xy}(q) = e^{-\gamma q(x)v(x,y)q(y)} - 1 \qquad (6.3)$$

so that

$$Z = \int \Big[\prod_x d\lambda_x(q(x))\Big] \prod_{(xy)} \big[f_{xy}(q) + 1\big] \qquad (6.4)$$

The second product runs over unordered pairs (x,y) of distinct points.

In the sum in the exponential of (6.1) every such pair appears twice, this cancels the factor $\frac{1}{2}$.

Now one expands the product into products of f's. The terms in this expansion can be labelled by graphs $\mathcal{B}$. The vertices of these graphs are the sites x of the lattice Λ, and $\mathcal{B}$ specifies a set of (nondirected) links $(x,y) \in \mathcal{B}$, (x,y) an unordered pair of distinct points

$$Z = \sum_{\mathcal{B}} \int \Big[\prod_x d\lambda_x(q(x)) \Big] \prod_{(x,y)\in\mathcal{B}} f_{xy}(q) \tag{6.5}$$

Finally the graphs $\mathcal{B}$ are decomposed into connected graphs G_α with vertices $x \in \alpha \subset \Lambda$. The sum over $\mathcal{B}$ amounts to a sum over partitions $\{\alpha\}$ of Λ into subsets α, and sums over connected graphs G_α with prescribed set α of vertices. The integral factorizes so that

$$Z = \sum_{\{\alpha\}} \prod_{\alpha\in\{\alpha\}} A(\alpha) \tag{6.6a}$$

$$A(\alpha) = \sum_{G_\alpha} \int \prod_{x\in\alpha} d\lambda(q(x)) \prod_{(x,y)\in G_\alpha} f_{xy}(q) \tag{6.6b}$$

Expression (6.6) is called a <u>polymer representation</u> of the partition function. The polymers α are subsets of Λ in (6.6). In the more familiar example of a Mayer expansion which starts from a grand canonical representation $Z = \Sigma(N!)^{-1}Z_N$ such as (4.8), the polymer are subsets $\alpha = \{i_1 \ldots i_k\} \subseteq \{1 \ldots N\}$ of the set of particles $\{1 \ldots N\}$. In any case, an activity $A(\alpha)$ is assigned to each polymer, and a partition function Z resp Z_N is calculated by considereing all partitions of Λ resp. of the set of particles $\{1 \ldots N\}$ into polymers and adding the product of their activities.

To obtain the free energy $\ln Z$ (i.e. $\mathcal{H}_{eff}$ in renormalization group applications) one has to take the logarithm. In the familiar approach which starts from a grand canonical representation this is simple, viz.

$$Z = \sum_N \frac{1}{N!} Z_N = \exp \sum_\alpha \frac{1}{|\alpha|!} A(\alpha)$$

where $|\alpha| = k$ if $\alpha = \{i_1 \ldots i_k\}$. For a general polymer system (6.6a) one can use expansion methods as will be discussed in the next section. But in theories with a mass gap one is really only interested in an ultimate $\mathcal{H}_{eff}$ which has an UV cutoff that is as low as is feasible. It is then advantageous to avoid taking the logarithm at intermediate stages of a renormalization group procedure. Instead, one works with polymer representations of the partition functions. That is, to perform a renormalization group transformation that lowers the cutoff from M_k to M_{k+1} one computes new acticities $A'(\alpha')$ from old ones, and one established recursive bounds, which bound suitable norms $||A'(\alpha')||$ in terms of norms of old activities. In the end, the recursive estimates are used to establish bounds such as (3.5) on the terms in the expansion of the ultimate $\mathcal{H}_{eff}$.

In truly massless theories the situation is somewhat different. There one is interested in the renormalization group flow, i.e. in how $\mathcal{H}_{eff}$ depends on the cutoff M. A hybrid approach which takes the logarithm "partially" has been used [13] and is described in the lectures of Kupiainen and Imbrie [14].

POLYMER REPRESENTATION FOR $\mathcal{Z}(\Phi)$ AND LOCALIZATION OF $\mathcal{H}_{eff}(\Phi)$

Polymer representations for $\mathcal{Z}(\Phi) = \exp[-V_{eff}(\Phi)]$ involve activities that depend on Φ

$$\mathcal{Z}(\Phi) = \sum_{\{\alpha\}} \prod_{\alpha\in\{\alpha\}} A(\alpha|\Phi) \tag{7.1}$$

The activity $A(\alpha|\Phi)$ will only depend on $\Phi(x)$ for x a site that is occupied by a constituent of α. This is true in the simple Mayer expansions of the last section where $\alpha = \{x_1, \ldots, x_k\}$ is a set of sites, and also in the iterated Mayer expansions that will be described later on.

Given a polymer representation of $\mathcal{Z}(\Phi)$, $-V_{eff}(\Phi) = \ln\mathcal{Z}(\Phi)$ can be exhibited as a sum of terms that are localized in subsets X of Λ

$$-\ln \mathcal{Z}(\Phi) = \sum_{X\subseteq\Lambda} V_X(\Phi) \tag{7.2a}$$

This is achieved by a Möbius inversion [15]. Given subsets $W\subseteq Y\subseteq\Lambda$ one defines

$$\vartheta_{WY} = (-1)^{|Y|-|W|}, \quad |W| = \text{no of sites in } W .$$

Then

$$-V_X(\Phi) = \sum_{W\subseteq X} \vartheta_{WX} \ln\Big(\sum_{\substack{\{\alpha\}\\ \Sigma\alpha = X}} \prod_{\alpha\in\{\alpha\}} A(\alpha|\Phi)\Big) \tag{7.2b}$$

The inner sum runs over partitions of X into polymers. V_X is zero if X is not polymer connected, i.e. if X can be split into two non-empty pieces such that there exists no polymer α with constituents in both pieces. Validity of eqs. (7.2) follows from the identity

$$\sum_{\substack{Y \\ X \subseteq Y \subseteq X'}} \vartheta_{YX'} = \begin{cases} 1 & \text{if } X' = X \\ 0 & \text{otherwise} \end{cases} \tag{7.3}$$

Let us write $\alpha = x$, $X = x$ for polymers resp. subsets of Λ with only one site x, and define reduced activities

$$\bar{A}(\alpha|\Phi) = A(\alpha|\Phi) / \prod_{x \in \alpha} A(x|\Phi) \tag{7.4}$$

Then

$$V_x = -\ln A(x|\Phi) \tag{7.5a}$$

and

$$V_X = \sum_{Y} \vartheta_{YX} \ln\Big(1 + \sum_{\substack{\{\alpha\} \\ \Sigma\alpha \subseteq Y}}' \prod \bar{A}(\alpha|\Phi)\Big) \text{ for } X = \{x_1, \ldots x_s\},\ s \geq 2 \tag{7.5b}$$

Summation ofer $\{\alpha\}$ is now over partitions of subsets of Y into polymers with at least two occupied sites.

Expansion (7.2a) involves a finite sum on a finite lattice Λ. There is therefore no convergence problem.

In conclusion, a polymer representation of $\mathcal{Z}(\Phi)$ also contains information about localization properties of the effective action $\mathcal{H}_{eff}$. Roughly speaking, $\mathcal{H}_{eff}$ will have good localization properties when the activities decrease fast in magnitude with increasing size of the polymer (size = length of the shortest highway net that connects all constitutents, cp. section 3).

When <u>all</u> the reduced activities $\bar{A}(\alpha|\Phi)$ are small enough (for polymers α with more than one site), expression (7.5b) for V_X admits a power series expansion on the $\bar{A}$'s. As a result one obtains an expression of the form [16]

$$-V_X = \sum_{Q} a(Q)\, \bar{A}(Q|\Phi) \quad \text{with} \quad \bar{A}(Q|\Phi) = \prod_{\alpha \in Q} \bar{A}(\alpha|\Phi) \tag{7.6}$$

Summation is over sets Q of not necessarily distinct polymers $\alpha \subseteq X$ (polymer α_i may occur with multiplicity n_i). Every site x in X must be occupied by at least one constituent of some polymer α_i in Q. a(Q) are combinatorial coefficients independent of the activities. a(Q) = 1 if Q consists of a single polymer. The expansion converges if there exists a real number $\xi > 1$ auch that [16]

$$\xi^{-1}\left[1+\sup_{x}\prod_{\substack{\alpha\\ x\in\alpha}}\xi^{|\alpha|}\bar{A}(\alpha)\right]<1$$

Because of the restrictive conditions for convergence, use of expansion (7.6) should be avoided at intermediate stages of a renormalization group calculation, cp. the discussion in section 6.

THE LEADING TERM

Let us consider the simplest approximation where only the contribution from single sites X = x is retained in expansion (7.2a) for $V_{eff}(\Phi) = -\ln Z(\Phi)$. By eq. (7.5a) this gives

$$V_{eff}(\Phi) = -\sum_{x}\ln A(x|\Phi) + \cdots \tag{8.1}$$

The activities are given by eqs. (6.6b), with f_{xy} and $d\lambda$ from eqs. (6.3) and (6.2). There is only one connected graph with one vertex, it has no link. Therefore, in units a = 1,

$$A(x,\Phi) = \int d\lambda(q(x)) = \int dk\,\tilde{F}(k)e^{ik\tilde{\Phi}(x)}e^{-\gamma k v(x,x)k/2}$$

$\tilde{F}$ is the Fourier transform of the factor $F(\varphi(x))$ in the original Boltzmann weight (1.10). Under an inverse Fourier transformation, product goes into convolution. Therefore

$$A(x|\Phi) = \text{const}\int_{\mathbb{R}^N} d^N\eta\; F(\eta)\exp\left[-\frac{1}{2\gamma v(x,x)}(\eta-\tilde{\Phi}(x))^2\right] \tag{8.2}$$

For the discrete Gaussian model and a Pauli Villars cutoff (viz $\tilde{\Phi} = \Phi$) this gives

$$A(x,\Phi) = \sum_{k=0,\pm1,\ldots}\exp\left[-\frac{1}{2\beta v(x,x)}(\Phi(x)-2\pi k)^2\right] \tag{8.3}$$

For large β and small cutoff M, $v(x,x)\approx v_{Cb}(0)$ in 3 dimensions, and the logarithm of the periodized Gaussian can be approximated by a cosine. As a result, eq. (3.4c) for V_{eff} is obtained.

In the case of the nonlinear σ-model we obtain

$$-V_{eff}(\Phi) = \sum_{x}\ln\int_{S^{N-1}} d\eta\,\exp\left[-\frac{N}{2f_0 v(x,x)}(\tilde{\Phi}(x)-\eta)^2\right] \tag{8.4}$$

If we make the further approximation of replacing $\tilde{\Phi}(x) = \mathcal{A}\Phi(x)$ by $\Phi(x')$ for all x in the block x', we obtain eq. (3.7b) of section 3.

In this simple approximation the coefficient γ^{-1} of the kinetic term remains the same as in the original Hamiltonian. In the discrete Gaussian model this is a good approximation. The same is true for the nonlinear σ-model at large N. but for finite N, important contributions from later terms in the Mayer series are absorbed into a change $\gamma \rightarrow (Z/Z_1)\gamma$ of this coefficient (cp. sect. 10).

ITERATED MAYER EXPANSIONS

Exact expressions for effective actions come in the form of infinite series like (3.4b). In order to establish their convergence in the infinite volume limit and decay properties of the kernels one needs estimates on the polymer activities in the Mayer expansion. Conventional technology yields bounds which are much too weak to be useful for the 3-dimensional discrete Gaussian model at large β/a, for instance.

This problem is overcome by a renormalization group procedure. One does not compute $\mathcal{H}_{eff}$ for a cutoff M = O(m) through a simple Mayer expansion at once, but one lowers the UV-cutoff in several steps $M_1 > M_2 > \ldots > M_R = M$, and determine the associated effective actions $\mathcal{H}^{(\ell)}_{eff}$ = kinetic terms + $V^{(\ell)}_{eff}$, $\ell = 1 \ldots R$. $V^{(\ell)}_{eff}$ is determined through eqs. (7.2) by a polymer representation of $Z^{(\ell)}(\Phi) = \exp -V^{(\ell)}_{eff}$ It is not necessary to write down $V^{(\ell)}_{eff}$ explicitly because the polymer activities at stage ℓ are determined in terms of those at stage ℓ - 1, but it is worth noting that a formula to do it exists.

We saw already, in section 5, that a block spin transformation $\mathcal{H} \rightarrow \mathcal{H}_{eff}$ is determined by a split of the potential v_{Cb}, $v_{Cb} = u+v$. To decompose this transformation into several subsequent renormalization group transformations, one splits further

$$v = v^0 + v^1 + \ldots + v^{R-1} \tag{9.1}$$

This is done in such a way that the pieces v^ℓ have increasing range and decreasing strength at short distances as ℓ = 1 ... R. Then the pieces v^ℓ of the potential are treated one by one by using Mayer expansions, in the order ℓ = 1,2, ..., i.e. in the order of increasing range.

Let me explain why there is an advantage to proceeding in this way. Let us inspect the simplest nontrivial polymer $\alpha = (x_1,x_2)$ in a simple Mayer expansion for $\mathcal{Z}(\Phi)$. Consider the discrete Gaussian model, where charges take discrete values q = ±1, ±2, ... and a Pauli Villars cutoff. There is only one connected G_α with two vertices. Therefore

$$A(\alpha) = \sum_{q_1,q_2} \left(\exp\left[-\beta q_1 v(x_1,x_2) q_2\right] - 1\right) e^{-\beta v(0)(q_1^2+q_2^2)/2} \cdot \exp\left[i q_1 \phi(x_1) + i q_2 \phi(x_2)\right] \tag{9.2}$$

One needs bounds on $b_2 = \frac{1}{2}|\Lambda|^{-1}\int_{x_1}\int_{x_2} A(\alpha)$, for instance, in order to established estimates like (3.5). The decay properties of the kernels ρ_s are obtained from bounds on $\int_{x_1}\int_{x_2} e^{(1-\delta)M|x_1-x_2|}A(\alpha)$. v is a Yukawa potential, which is kernel of a positive operator, therefore

$$\sum_{1\le i<j\le N} q_i v(x_i,x_j) q_j \geq -B \sum_i q_i^2 , \qquad B = \frac{1}{2} v(0) \equiv \frac{1}{2} v(x,x) \tag{9.3}$$

One may write

$$|\Lambda| b_2 = -\frac{1}{2}\beta \sum_{q_1,q_2} e^{-\frac{\beta}{2}v(0)(q_1^2+q_2^2)} \int_{x_1,x_2} e^{i q_1\phi(x_1) + i q_2 \phi(x_2)} \cdot \int_0^1 ds\, q_1 v(x_1,x_2) q_2 \exp\left[-\beta s q_1 v(x_1,x_2) q_2\right]$$

Now one may estimate the exponential in the s-integrand by $\exp\left[+\frac{1}{2}\beta v(0)(q_1^2+q_2^2)\right]$ using inequality (9.2). Unfortunately this is not good enough to make the q-summations convergent, but this problem could be overcome by a slight improvement of (9.3). If we could restrict q_1, q_2 to ± 1, we would obtain a bound $|b_2| < 2\beta\tilde{v}(0)$, with $\tilde{v}(0) = \int_x |v(x,0)|$. Estimates on the activities for larger polymer α can be obtained in the same way by using the "tree formula" [17] for the sum over connected graphs in (6.6b). But this way of estimating is inefficient for interactions which are at the same time strong and possibly attractive at short distances, and have a rather long range. Bounding the exponential by its absolute maximum one ignores the fact that the interaction is strong only at short distances and not over its whole range. If one splits v as in (9.1) and then treats the individual pieces v^ℓ by Mayer expansion with estimates as above, the inefficiently is less severe because the pieces v^ℓ of the potential are either less strong at short distances, or they have a lesser range, or both.

In exact renormalization group calculations, efficient book keeping is a necessity. The formalism developed by Göpfert and the author [1] is designed to cope with this problem.

One builds up the ultimate polymers, to be called R-vertices in inductive steps, out of constituents = particles of the gas (or sites of the lattice). A O-vertex is a single particle i which is its own constituent. Associated with it is a variable $\xi_i = (q_i, x_i)$ which specifies its state and position, and a vertex function, or fugacity, σ°.

$$\sigma^\circ(\xi_i)\, dq_i = d\lambda(q_i) \quad ; \quad \int d\xi_i \equiv \sum_{q_i} \int_{x_i \in \Lambda} \tag{9.4}$$

Higher vertices are defined inductively. An ℓ-vertex α' is a finite collection $\{\alpha\}$ of $(\ell-1)$ vertices, no two of which share a constituent. There is an associated variable

$$\xi_{\alpha'} = (\{\xi_\alpha\}_{\alpha\in\alpha'}) \quad ; \quad d\xi_{\alpha'} = \prod_{\alpha\in\alpha'} d\xi_\alpha \tag{9.5}$$

A paricle is constituent of α' if it is constituent of one of the $(\ell-1)$-vertices $\alpha\in\alpha'$. We write $i\underline{\in}\alpha$ in this case. We use the symbol Σ for union of sets with vanishing mutual intersection.

It is convenient to introduce an abbreviation for interactions. If α,γ are two ℓ-vertices we write

$$v^\ell(\alpha\gamma) = \sum_{i\underline{\in}\alpha}\sum_{j\underline{\in}\gamma} v^\ell(\xi_i,\xi_j) \quad \text{with} \quad v^\ell(\xi_i,\xi_j) = q_i v^\ell(x_i,x_j) q_j \tag{9.6}$$

$v^\ell(\alpha\alpha)$ includes in particular the self interaction of all the constituents of the ℓ-cluster α that is due to the piece v^ℓ in the potential. We introduce the quantity $f^\ell_{\alpha\gamma}$ for distinct ℓ-vertices α,γ. It depends on ξ_α and ξ_γ and is given by

$$1 + f^\ell_{\alpha\gamma} = \exp[-\beta v^\ell(\alpha\gamma)] \tag{9.7}$$

Higher vertex functions are defined inductively as follows. We write $\{G_{\alpha'}\}$ for the set of all connected graphs with vertices $\alpha\in\alpha'$. Such a graph is specified by a set of links = unordered pairs (α,γ), with $\alpha \neq \gamma$ and $\alpha,\gamma\in\alpha'$. We define

$$\sigma^{\ell+1}_{\alpha'} = (\text{comb. factor}) \sum_{G_{\alpha'}} \left(\prod_{\alpha\in\alpha'} \sigma^\ell_\alpha e^{-\beta v^\ell(\alpha\alpha)/2}\right) \cdot \prod_{(\alpha\gamma)\in G_\alpha} f^\ell_{\alpha\gamma} \tag{9.8}$$

It depends on $\xi_{\alpha'}$. The comb. factor will be set = 1 for the following discussion, but in the articles [1,2] a different choice was made.

Mayer expansions are very flexible because one has the freedom of what to consider as the particles of the gas. In the $(\ell+1)$-st step of an interated Mayer expansion (which defines $\sigma^{\ell+1}$), the particles are ℓ-clusters α, and they have a fugacity $\sigma^\ell_\alpha e^{-\beta v^\ell(\alpha\alpha)/2}$ which incorporates interactions between the constituents of the ℓ-cluster α due to interactions v^r with $r < \ell$. It depends on the state ξ_α of the ℓ-cluster, i.e. on the charges and (relative) positions of its constituents.

The final polymer activity is obtained when all the interaction v is taken into account, viz.

$$A(\alpha) = \int dq_1 \ldots dq_n \; \sigma_\alpha^R(\xi_\alpha) \tag{9.9}$$

for an R-cluster α with n constituents. It depends on the positions $x_1 \ldots x_n$ of the constituents $i = 1 \ldots n$ of α, and on Φ through the Φ-dependent factors $\lambda(q)$, eq. (6.2). They determine σ° by (9.4), and enter therefore into all vertex functions σ^ℓ. Summation in the polymer representation (7.1) of $Z(\Phi) = \exp[-V_{eff}(\Phi)]$ must now be extended over all collections of R-clusters and sites of their constituents. No two constituents may occupy the same site of the lattice, and all of them together fill up the whole lattice. (Instead of considering a site x as empty when $q(x) = 0$ one considers it as occupied with a constituent of charge 0 which does not interact with the others.)

The leading term in an iterated Mayer expansion is the same as in the simple Mayer expansion (see section 8), it comes from R-vertices with a single constituent.

The vertex functions σ_α^ℓ with $\ell < R$ determine the effective Hamiltonian $\mathcal{H}_{eff}$ at intermediate stages of the renormalization group procedure. To see this note that we may make another split

$$\upsilon_{Cb} = v' + u' \tag{9.10a}$$

in place of $\upsilon_{Cb} = v + u$, with

$$v' = v^0 + \ldots + v^{\ell-1} \quad , \qquad u' = u + v^\ell + \ldots + v^{R-1} \tag{9.10b}$$

eqs. (5.1) ... (5.4) remain valid when u', v', V'_{eff} are substituted for u, v, V_{eff}, and $\mathcal{H}^{(\ell)}_{eff} = -\frac{1}{2\gamma}(\Phi, u'^{-1}\Phi) + V'_{eff}$ is effective Hamiltonian for a cutoff M^ℓ that determines the range of $v^{\ell-1}$, and therefore of v'. When σ^ℓ is computed, all of the interaction v' has been taken into account. Therefore σ^ℓ determines the polymer activities which in turn determine $Z'(\Phi) = \exp[-V'_{eff}]$

CONTINUUM LIMIT AND WAVE FUNCTION RENORMALIZATION

The simple choice (3.1) of the new field Φ produces what is sometimes called an "unrenormalized Block spin". If one wants to take a continuum limit in the nonlinear σ-model, for instance, it is necessary first to rescale the field Φ and to express the effective Hamiltonian and the observables in terms of the renormallized quantity

$$\Phi_r = Z^{-1/2}\Phi$$

Z is called the wave function renormalization constant. A continuum limit is expected to exist when one lets $a \to 0$, $f_o \to 0$ and $Z \to 0$ in the appropriate way dictated by perturbation theory.

To give an example, let us discuss the O(N)-symmetric nonlinear σ-model in the simple approximation of section 8 which becomes accurate when $N \to \infty$. For simplicity we choose a Pauli-Villars cutoff M as UV cutoff in the effective Hamiltonian $\mathcal{H}_{\text{eff}}$, so that

$$\mathcal{H}_{\text{eff}}(\Phi) = \frac{N}{2f_0}\int_x \nabla_\mu\Phi(x)\left(1-\frac{\Delta}{M^2}\right)\nabla_\mu\Phi(x) + a^{-2}\int_x \mathcal{V}(\Phi(x)) \quad (10.1)$$

with $\mathcal{V}$ given by

$$e^{-\mathcal{V}(\xi)} = \text{const}\int_{S^{N-1}} d\eta\, \exp\left[-\frac{N}{2f_0 v_M(0)}(\xi-\eta)^2\right] \quad (10.2)$$

$$v_M(0) \equiv (-\Delta+M^2)^{-1}(x,x) \sim -\frac{1}{2\pi}\ln Ma + \text{const} \quad (10.3)$$

In the large N limit, $\mathcal{V}(\xi)$ can be evaluated by a saddle point method.

$$\mathcal{V}(\xi) \sim \frac{1}{2\lambda^2}(\lambda-1)\xi^2 + \frac{1}{4\lambda^4}\xi^4 + \dots \qquad \text{with } \lambda = f_0 v_M(0). \quad (10.4)$$

The dots ... represent terms that are $o(f_0^2)$ when $\xi \equiv \Phi(x) = O(f_0^{\frac{1}{2}})$ as it shall turn out to be. One introduces a mass m by

$$f_0 v_m(0) = 1 \qquad \text{so that} \qquad m \sim \text{const}\; a^{-1} e^{-2\pi/f_0} \quad (10.5)$$

According to perturbation theory, m is the physical mass scale. We define a rescaled field Φ_r and an effective coupling constant f which depends on the UV-cutoff M in the effective Hamiltonian by

$$\Phi_r = Z^{-1/2}\Phi \quad , \quad f(M) = Z_1^{-1} f_0$$

with

$$Z = Z_1 = \frac{f_0}{2\pi}\ln M/m \qquad \text{for} \quad M/m \gg 1 . \quad (10.6)$$

We emphasize that this is an expression for Z in terms of the bare coupling constant f_0 which tends to zero in the continuum limit. One can deduce from perturbation theory [19] that $Z^{\frac{1}{2}} \sim f_0^{\alpha}$ modulo logs, as $f_0 \to 0$, with $\alpha = \frac{1}{2}(N-1)/(N-2)$ so that $\frac{1}{2} < \alpha < 1$ for $N > 3$, $\alpha = \frac{1}{2}$ for $N = \infty$ and $\alpha = 1$ for $N = 3$. The simple approximation which we have used here is only accurate for large N.

We reexpress $\mathcal{H}_{\text{eff}}$ in terms of Φ_r and see what happens when we take the continuum limit $a \to 0$, $f_0 \to 0$ with m fixed. According to the discussion in section 2 we are interested in values of the cutoff M which are a fixed multiple M/m of the physical mass scale m. Then

$$f(M) = \frac{1}{2\pi}\ln\frac{M}{m} \qquad \text{for} \quad M/m \gg 1$$

remains constant. The potential

$$V_{eff}(\Phi_r) \equiv a^{-2}\int_x v(\phi(x)) = \frac{a^{-2}f_o^2}{2f(M)^2}\int_x [-\phi_r(x)^2 + \phi_r(x)^4] + \cdots$$

acquires an infinitely sharp maximum at $\Phi_r(x)^2 = 1$ because $a^{-2}f_o^2 \to \infty$ in physical units, if $M/m > 1$. Thus, in the continuum limit the fluctuations on the length of Φ freeze again and we recover as our effective theory a nonlinear σ-model with Pauli-Villars modification of the interaction between spins

$$e^{-\mathcal{H}_{eff}(\phi_r)}\prod_x d^N\phi_r(x) = const \cdot \exp\left[-\frac{N}{2f(M)}\int_x \nabla_\mu \phi_r \left(1-\frac{\Delta}{M^2}\right)\nabla_\mu \phi_r\right]\prod_x d\vec{\phi}_r(x)$$

where $d\vec{\Phi} = d^N\Phi\delta(\Phi^2-1)$ is the uniform measure on the unit sphere.

ACKNOWLEDGEMENT

For many years Kurt Symanzik has always been ready to share his vast treasure of knowledge. With great patience and clarity he explained so many things to me. It was he who introduced Mayer expansions as a tool for studying Quantum Field Theory nearly 20 years ago (ref. 20), and they were a topic of our last conversation few days before his untimely death. With deep sorrow I dedicate these notes to his memory.

REFERENCES

0. K. Wilson, Phys. Rev. D2, 1473 (1970).
1. M. Göpfert and G. Mack, Commun. Math. Phys. 81, 97 (1981).
2. M. Göpfert and G. Mack, Commun. Math. Phys. 82, 545 (1982).
3. E. Brezin, J. Zinn Justin, J. C. LeGuillon, Phys. Rev. D14, 2615 (1976);
   A. M. Polyakov, Phys. Letters B59, 79 (1975).
4. L. Gross, Commun. Math. Phys. 92, 137 (1983).
5. M. Karliner and G. Mack, Nucl. Phys. B225 [FS9], 371 (1983).
   T. Sterling, J. Greensite, Nucl. Phys. B220 FS8 , 327 (1983).
6. J. Glimm and A. Jaffe, Quantum physics, Springer Verlag Heidelberg 1981, and references given there.
7. J. Glimm, A. Jaffe, T. Spencer, Ann. Phys. 101, 610, 631 (1975).
8. D. Brydges, Commun. Math. Phys. 58, 313 (1978);
   D. Brydges and P. Federbush, Commun. Math. Phys. 73, 197 (1980).
9. D. Brydges, J. Fröhlich, T. Spencer, Commun. Math. Phys. 83, 123 (1982).
10. D. Ruelle, Statistical mechanics, New York: Benjamin 1969.
11. K. Gawedzki and A. Kupiainen, Commun. Math. Phys. 77, 31 (1980).
12. T. Balaban, Commun. Math. Phys. 89, 571 (1983).
13. K. Gawedzki and A. Kupiainen, Ann. Phys. 147, 198 (1983).
14. A. Kupiainen, lectures presented at this school;
    J. Imbrie, lectures presented at this school.

15. G. S. Rushbrooke, J. Math. Phys. 5, 1106 (1964);
G. S. Rushbrooke, G. A. Baker and P. J. Woods, in: Phase transitions and critical phenomena,
C. Domb and M. Green eds., Academic press, New York, 1974, vol. 3.
C. Domb, ibid., p. 77.
16. Ch. Gruber, A. Kunz, Commun. Math. Phys. 22, 133 (1971).
17. D. Brydges, P. Federbush, J. Math. Phys. 19, 2064 (1978).
18. J. Fröhlich, Commun. Math. Phys. 47, 233 (1976).
19. E. Brezin, J. Zinn Justin, J. C. LeGuillon, Phys. Rev. D14, 2615 (1976).
20. K. Symanzik, J. Math. Phys. 7, 510 (1966).

GUAGE INVARIANT FREQUENCY SPLITTING IN

NON ABELIAN LATTICE GUAGE THEORY

P. K. Mitter

Laboratoire de Physique Théorique et Hautes Energies+
Université Pierre et Marie Curie (Paris VI)
Tour 16, 1er étage, 4 Place Jussieu
75230 Paris Cedex 05, France

I. INTRODUCTION

Lattice gauge theory, the problem of quark confinement and renormalization group (RG) methods have constituted a central theme of this school. As is well known, quark confinement, the existence of a mass gap, etc., can be proved in pure 4-dimensional Euclidean lattice gauge theory for a sufficiently large coupling constant / 1 /. On the other hand, perturbative RG studies show asymptotic freedom at short distances.++ As K.G. Wilson has forcefully advocated over the years, the modern RG / 2 / is essential to study the connection between the short and long distance behavior. Fortunately, the modern RG is slowly coming under mathematical control§, and it is to be hoped that in the not too distant future these methods will become powerful enough for the control of the most challenging and fascinating of renormalizable field theories: non-abelian gauge field theory in four dimensions.

The purpose of these notes§§ is to initiate a line of thinking which may turn out to be not unfruitful for a modern RG study of pure non-abelian gauge fields in four dimensions. The essential

+Laboratoire associé 280 au CNRS.

++See the lectures of C.P. Korthals-Altes, this volume, and references therein to the original literature.

§See the lectures of K. Gawedski and A. Kupiainen, J. Imbrie and G. Mack, this volume, together with the references therein. Given the modest scope of these notes, I do not attempt a complete listing of references. The basic work of G. Gallavotti et al /19/ is noted.

§§Expanded and modified version of lectures at the 1983 Cargèse School on "Progress in Gauge Field Theory," 1-15 Sept., 1983.

contribution is to give a theory of gauge-invariant high-low frequency splitting in Wilson's lattice gauge theory. Such a splitting for massless scalar field theory is well known. It corresponds to Pauli-Villars splits of the free massless propagator into soft and hard ones. The soft propagator is massless but with improved ultraviolet (UV) behavior, whereas the hard propagator is massive (Λ) but with standard UV behavior. Corresponding to such a split, the free field is the sum of two independent Gaussian random fields. One of them, the soft field, may be thought of as a "block field" which, on a lattice, is slowly varying over the scale $\Lambda^{-1} >> \epsilon$ (=lattice spacing), whereas the other, the hard field, is fluctuating around the block. A thorough-going perturbative renormalization theory for massless $(\phi^4)_4$ was developped in / 3 / using minimal powercounting and Λ as the renormalization scale, together with an appropriate perturbative RG equation. Extensions to continuum non-abelian gauge fields were made in /4 , 3 /. It should be noted that frequency splitting has been effectively used non-perturbatively in / 5 /.

Our main purpose then is to extend this circle of ideas gauge invariantly to Wilson's lattice gauge theory. We show that this theory can be rewritten in terms of two new fields U , V corresponding to frequency splitting. The gauge field U is super-renormalizable and (up to guage equivalence) slowly varying over a block size $\Lambda^{-1} >> \epsilon$. It is thus an effective block gauge field with Λ acting as an UV cutoff. It is our low frequency field. The high frequency field is massive (mass Λ), is renormalizable, and acts as the fluctuating field. U acts as a lattice gauge field, whereas V acts as a (vector) massive matter field transforming in the adjoint representation of the group SU(N). (There is some analogy to the so-called "background field method;" see, for example+, but should not be confused with it.) Both U,V are SU(N) valued. However, it is an important point that V does not require gauge fixing even in the continuum limit to have a well defined renormalizable propagator. Thus, the high frequency field V polarizes the medium in which the gauge field U lives; the low frequency gauge field U carries the long distance information of the theory.

Chapter II is a very brief introduction to local continuum differential geometric notations for non-abelian gauge fields. The sophisticated reader will, of course, dispense with it. In Chapter III we develop extensively an invariant lattice differential calculus for non-abelian gauge fields in the form that we need. Using the machinery of Ch. III, we develop the theory of gauge invariant frequency splitting in the Wilson theory in Chapter IV. Finally, in Chapter V, we make some tentative comments on the role of "algebraic topology" in quantum non-abelian gauge theory. Our main contention is that observable topological effects should "almost surely" be associated only with low frequency fields.

+C.P. Korthals-Altes, this volume

II. CONTINUUM LOCAL DIFFERENTIAL GEOMETRY: NOTATIONS

In this section we summarize for the reader's convenience the formalism of continuum local differential geometry / 6 / as extended to non-abelian gauge fields (see e.g. / 7 / and / 8 / where we borrow our notations). Closely related to this formalism is the lattice differential geometry (of non-abelian gauge fields) developped extensively in Chapter III.

G denotes a compact connected semi-simple Lie group and $\mathcal{G}$ its Lie algebra.+

$$P(\mathbb{R}^n, G) = \mathbb{R}^n \times G \tag{2.1}$$

is a trivial principal G bundle.++ $\Omega^p \otimes \mathcal{G}$ denotes the vector space of $\mathcal{G}$ valued C^∞ p-forms on $\mathbb{R}^n$. If $\omega \in \Omega^p \otimes \mathcal{G}$,

$$\omega = \sum_{1 \leq i_1 < \cdots < i_p \leq n} \omega_{i_1 \cdots i_p}(x)\, dx^{i_1} \wedge \cdots \wedge dx^{i_p} \tag{2.2}$$

where the coefficients have values in $\mathcal{G}$. Under a gauge transformation $P \to P$ (fibre preserving equivariant automorphism of P), we have:

$$\omega \mapsto ad_{g^{-1}}\, \omega = g \omega g^{-1} \tag{2.3}$$

where $g \in \mathcal{G}$ = Maps $(\mathbb{R}^n, G)$, the group of gauge transformations, and in (2.3) pointwise multiplication is implied.

We now introduce two bilinear forms:

$$\text{(a)} \quad (\Omega^p \otimes \mathcal{G}) \otimes (\Omega^q \otimes \mathcal{G}) \to \Omega^{p+q} \otimes \mathcal{G}$$
$$\omega \otimes \eta \mapsto [\omega, \eta] \tag{2.4}$$

where

$$[\omega, \eta] = \sum_{\substack{i_1 < \cdots < i_p \\ j_1 < \cdots < j_p}} \left[\omega_{i_1 \cdots i_p}(x), \omega_{j_1 \cdots j_q}(x)\right] dx^{i_1} \wedge \cdots \wedge dx^{i_p} \wedge dx^{j_1} \wedge \cdots \wedge dx^{j_q} \tag{2.5}$$

+We can (and will in Ch. III) choose, for simplicity G=SU(N). The reader may take this to be the case in this chapter also.

++We deal entirely with the local theory, which is all that is necessary to motivate Ch. III.

(b)
$$(\Omega^p \otimes \mathcal{G}) \otimes (\Omega^q \otimes \mathcal{G}) \to \Omega^{p+q} \equiv \Omega^{p+q} \otimes \mathbb{R}^1$$

$$\omega \otimes \eta \longmapsto \omega \wedge \eta \tag{2.6}$$

$$\omega \wedge \eta = \sum_{\substack{i_1 < \cdots < i_p \\ j_1 < \cdots < j_q}} \langle \omega_{i_1 \cdots i_p}(x), \; \eta_{j_1 \cdots j_q}(x) \rangle \, dx^{i_1} \wedge \cdots \wedge dx^{i_p} \wedge dx^{j_1} \wedge \cdots dx^{j_q} \tag{2.7}$$

Here $\langle , \rangle$ is a scalar product on $\mathcal{G}$, invariant under the adjoint G action. (More simply, if G=SU(N), $\mathcal{G}$ = Lie SU(N), then $\langle \xi, \eta \rangle = \mathrm{tr}\, \xi^* \eta$).

Choose a fixed orientation of $\mathbb{R}^n$, by specifying an ordered basis. Let

$$* : \quad \Omega^p \to \Omega^{n-p} \tag{2.8}$$

be the Hodge duality operator / 6 /. We have,

$$*^2 = (-1)^{p(n-p)} \quad : \quad \Omega^p \to \Omega^p \tag{2.9}$$

The Hodge scalar product on $\Omega^p \otimes \mathcal{G}$ is defined by:

$$(\omega, \eta) = \int_{\mathbb{R}^n} \omega \wedge * \eta = \int_{\mathbb{R}^n} d^n x \sum_{i_1 < \cdots < i_p} \langle \omega_{i_1 \cdots i_p}(x), \eta_{i_1 \cdots i_p}(x) \rangle$$

and

$$\| \omega \|^2 = (\omega, \omega) \tag{2.10}$$

We also use:

$$\langle \omega, \eta \rangle (x) = \omega \wedge * \eta (x) = \sum_{i_1 < \cdots < i_p} \langle \omega_{i_1 \cdots i_p}(x), \eta_{i_1 \cdots i_p}(x) \rangle$$

and

$$|\omega|^2 = \langle \omega, \omega \rangle \tag{2.11}$$

The exterior covariant derivative, acting on C^∞ p-forms,

$$d_A : \Omega^p \otimes \mathcal{G} \to \Omega^{p+1} \otimes \mathcal{G}$$

is given by

$$d_A = d + [A, \cdot\,] \tag{2.12}$$

and d_A^* is its formal adjoint: $\Omega^{p+1} \otimes \mathcal{G} \to \Omega^p \otimes \mathcal{G}$ in the scalar product $(,)$. It is minus the covariant divergence operator. We have:

$$d_A^* = (-1)^{n+np+1} * d_A * \tag{2.13}$$

on $\Omega^p \otimes \mathcal{G}$.

In the above,

$$A = \sum_{i=1}^{n} A_i(x)\, dx^i \tag{2.14}$$

is a C^∞, $\mathcal{G}$-valued connection 1-form on $\mathbb{R}^n$. Then

$$F(A) = dA + \tfrac{1}{2}[A, A] \tag{2.15}$$

is the curvature 2-form, an element of $\Omega^2 \otimes \mathcal{G}$. Under a gauge transformation $g \in \mathcal{G}$ we have:

$$A \mapsto A^g = gAg^{-1} + g\,dg^{-1}$$
$$F(A^g) = g\, F(A)\, g^{-1} \tag{2.16}$$

The classical Yang-Mills action functional S(A) is

$$S(A) = \tfrac{1}{2} \| F(A) \|^2 \tag{2.17}$$

Using

$$F(A + \tau) = F(A) + d_A \tau + \tfrac{1}{2}[\tau, \tau] \tag{2.18}$$

we derive from (2.16) the field equation

$$d_A^* F(A) = 0 \tag{2.19}$$

We recall the Bianchi identity

$$d_A F(A) = 0 \tag{2.20}$$

We recall briefly the notion of parallel transport. /7/
If $\omega \in \Omega^p \otimes \mathcal{G}$, we say ω is horizontal with respect to a connection A, if

$$d_A \omega = 0 \tag{2.21}$$

Let $\gamma = \{x(t) : 0 \leq t \leq 1,\ \gamma(0) = x, \gamma(1) = y\}$ be a C^1 curve in $\mathbb{R}^n$. $\gamma^{-1} = \{x(1-t) : 0 \leq t \leq 1\}$ is the reversed curve. Consider a partition $0 \leq t_1 \leq t_2 \leq \cdots \leq 1$ of $[0,1]$. Given a connection A, we say ω is parallel along γ if it is horizontal at every point x(t) of γ, i.e.

$$d_{A(x(t))}\, \omega(x(t)) = 0 \tag{2.22}$$

Given an initial point $x(t_1)$, and $\omega(x(t_1))$, the above system of ordinary differential equations has a unique solution $\omega(x(t)),\ t \geq t_1$. Thus we have a map $\omega(x(t_1)) \xrightarrow[U_\gamma]{} \omega(x(t_2))$ with

$$\omega(x(t_2)) = U_\gamma(x(t_1), x(t_2); A)^{-1}\, \omega(x(t_1))\, U_\gamma(x(t_1), x(t_2); A) \tag{2.23}$$

where $U_\gamma(\cdot,\cdot; A) \in G$ is the parallel transport operator. It is easy to verify that

$$U_\gamma(x(t_1), x(t_2); A)\, U_\gamma(x(t_2), x(t_3); A) = U_\gamma(x(t_1), x(t_3); A)$$

$$U_\gamma^{-1} = U_{\gamma^{-1}} \tag{2.24}$$

Under a gauge transformation $g \in \mathcal{G}$ we have

$$U_\gamma(x(t_1), x(t_2); A) \rightarrow U_\gamma(x(t_1), x(t_2); A^g)$$

$$= g(x(t_1))\, U_\gamma(x(t_1), x(t_2); A)\, g(x(t_2))^{-1} \tag{2.25}$$

If γ_x is a closed curve beginning and ending at x, we write:

$$U_{\gamma_x}(A) = U_\gamma(x(0), x(1); A) \tag{2.26}$$

It is an element of the so-called holonomy group (at x).

Given a connection A, we obtain a parallel transporter $U_\gamma(\cdot,\cdot;A)$. Conversely, given parallel transporters, we can recover connections. To this end let $\gamma_{xy} = (x,y)$ be the straight line path $x \to y$, and we write simply $U(x,y;A)$. Let $\{A_i(x)\}$ be the components of the connection 1-form A. Then,

$$A_i(x) = \lim_{\epsilon \to 0} \epsilon^{-1}\left(U(x, x+\epsilon e_i; A) - I\right) \tag{2.27}$$

where $\{e_i\}$ is an ordered basis of $\mathbb{R}^n$. Thus given parallel transporters $U_\gamma(x,y)$ we can use (2.27) to define the connection.

Let F_{ij} be the components of the curvature 2-form F(A). Since G is compact, there is a natural inclusion $G \subset M(N,\mathbb{C})$, for some N. $M(N,\mathbb{C})$ is the vector space of complex N x N matrices. Now define, following / 10 /,

$$\tilde{\mathcal{F}}^{(\epsilon)}_{ij}(U) = \epsilon^{-2}\Big\{U(x, x+\epsilon e_i; A)\, U(x+\epsilon e_i, x+\epsilon e_i+\epsilon e_j; A) - U(x, x+\epsilon e_j; A)\, U(x+\epsilon e_j, x+\epsilon e_i+\epsilon e_j; A)\Big\} \tag{2.28}$$

as an element of $M(N,\mathbb{C})$. Then

$$\lim_{\epsilon \to 0} \tilde{\mathcal{F}}^{(\epsilon)}_{ij}(U) = F_{ij}(A) \in \mathcal{G} \tag{2.29}$$

Finally let $\omega \in \Omega^p \otimes \mathcal{G}$, and $\omega_{j_1 \cdots j_p}$ its components. $d_A\omega \in \Omega^{p+1} \otimes \mathcal{G}$ and $D_i\omega_{j_1\cdots j_p}$ are its components. We have

$$D_i\omega_{j_1\cdots j_p}(x) = \lim_{\epsilon \to 0} \epsilon^{-1}\Big[U(x, x+\epsilon e_i)\,\omega_{j_1\cdots j_p}(x+\epsilon e_i)\, U(x+\epsilon e_i, x) - \omega_{j_1\cdots j_p}(x)\Big] \tag{2.30}$$

(2.27) - (2.30) record the well-known fact that connections, curvatures and covariant derivatives can be recovered given parallel transporters. This is of course at the root of Wilson's lattice gauge theory / 1 / which is completely geometric. We are going to play various games (with the renormalization group in sight) in lattice gauge theory. For this purpose we need an appropriate invariant lattice differential calculus for non-abelian gauge fields ("lattice differential geometry") to which we now turn. In particular we encounter $\mathcal{G}$ valued (or even M (N,C) valued) p-cochains which in the continuum are naturally associated to components of p-forms. An acquaintance with the equivalent of Ch. II, IV of / 6 / will be helpful.

III. LATTICE DIFFERENTIAL CALCULUS FOR NON-ABELIAN GAUGE FIELDS

The lattice differential geometry of Abelian gauge fields is well known (see e.g. / 9 /). In this chapter we will develop extensively the lattice differential geometry of non-abelian gauge fields suitable for our purposes. It is an elaboration on that to be found in / 10 / with additional ingredients.

We restrict ourselves to G = SU(N), $N \geq 2$, the group of N x N complex unitary matrices, identified with its fundamental representation. Its Lie algebra $\mathcal{G}$ is identified with the vector space of N x N complex, traceless, anti-hermitian matrices. M(N,C) denotes the vector space of complex N x N matrices. We have a natural inclusion

$$\mathcal{G} \subset M(N, \mathbb{C})$$

which induces

$$\Omega^p \otimes \mathcal{G} \subset \Omega^p \otimes M(N, \mathbb{C})$$

for p-forms on $\mathbb{R}^n$. If $\omega \in \Omega^p \otimes M(N, \mathbb{C})$ then

$$\omega = \sum_{i_1 < i_2 < \cdots < i_p} \omega_{i_1 \cdots i_p}(x)\, dx^{i_1} \wedge \cdots \wedge dx^{i_p} \tag{3.1}$$

where $\omega_{i_1 \cdots i_p} \in M(N, \mathbb{C})$

If, moreover, $\omega \in \Omega^p \otimes \mathcal{G}$, then we can expand

$$\omega_{i_1 \cdots i_p}(x) = \sum_{\alpha=1}^{N^2-1} \omega^{\alpha}_{i_1 \cdots i_p} \lambda_\alpha \tag{3.2}$$

where $\{\lambda_\alpha\}$ is a basis of $\mathcal{G}$. For both $\Omega^p \otimes \mathcal{G}$, $\Omega^p \otimes M(N, \mathbb{C})$ we define

$$\langle \omega_{i_1 \cdots i_p}(x), \eta_{j_1 \cdots j_p}(x) \rangle = \operatorname{tr} \omega^*_{i_1 \cdots i_p} \eta_{j_1 \cdots j_p}(x) \tag{3.3}$$

For both $\Omega^p \otimes \mathcal{G}$, $\Omega^p \otimes M(N, \mathbb{C})$ a gauge transformation $g \in \mathcal{G}$ induces the adjoint action:

$$\omega \mapsto \mathrm{ad}_{g^{-1}} \omega = g(x) \omega(x) g(x)^{-1} \tag{3.4}$$

Let now $\mathbb{Z}^n_\epsilon$ be the hypercubic n-dimensional lattice with equal lattice spacing ϵ in all directions. (In Chapter IV we choose n=4, but temporarily we do not fix n). The standard orthonormal basis in $\mathbb{R}^n$ induces a fixed orientation in $\mathbb{Z}^n_\epsilon$ which will be used throughout. Let $\Lambda_\epsilon \subset \mathbb{Z}^n_\epsilon$ be a finite subset (hypercube). Let $\mathcal{C}_p$ be a set of oriented p-cells whose vertices belong to Λ_ϵ. Let $\mathcal{C} = \{\mathcal{C}_p\}_{p=0}^n$ be a cell complex. Recall that 0-cells are points (vertices), oriented 1-cells are directed bonds, oriented 2-cells are oriented plaquettes, oriented 3-cells are oriented cubes, etc.

We shall keep in mind open complexes (Dirichlet b.c), closed complexes (free b.c.), or Λ_ϵ as an n-torus (periodic b.c.). Recall that a cell complex $\mathcal{C}$ is <u>open</u> if for any $c_p \in \mathcal{C}$ and any c_{p+1}, $c_p \in \partial c_{p+1} \Longrightarrow c_{p+1} \in \mathcal{C}$. It is <u>closed</u> if for each $c_{p+1} \in \mathcal{C}$, $c_p \in \partial c_{p+1} \Longrightarrow c_p \in \mathcal{C}$.

An (oriented) p-cell $c_p \in \mathcal{C}_p$ can be identified with a vertex ("lower left hand corner") x and edges $(x, x+\epsilon e_{i_k})$, $k=1,\cdots,p$, $i_1<\cdots<i_p$. Here e_{i_k} are basis vectors, and the orientation is that induced by the basis. We can write:

$$c_p = c_p(x; i_1, \cdots, i_p), \quad i_1 < \cdots < i_p \tag{3.5}$$

Let $V = \mathcal{G}$ or $M(N,\mathbb{C})$. $\Omega^p_\epsilon \otimes V$ is the vector space of V-valued p-cochains over the cell complex $\mathcal{C} = \{\mathcal{C}_p\}_{p=0}^n$. Thus if $\omega \in \Omega^p_\epsilon \otimes V$

$$\begin{aligned} \omega :\ & \mathcal{C}_p \longrightarrow V \\ & c_p \mapsto \omega(c_p) \equiv \omega_{i_1 \cdots i_p}(x), \quad i_1 < \cdots < i_p \end{aligned} \tag{3.6}$$

Under a gauge transformation $g \in \mathcal{G}$, $\Omega^p_\epsilon \otimes V \to \Omega^p_\epsilon \otimes V$ via

$$\omega \mapsto \omega^g, \quad \omega^g(c_p) = g(x)\, \omega(c_p)\, g(x)^{-1} \tag{3.7}$$

This property <u>is fundamental</u> for $\Omega^p \otimes V$. Later we introduce another space $\hat{\Omega} \otimes V$ with a different gauge transformation property.

We now introduce the Hodge $*$ operator:

$$* : \Omega^p_\epsilon \otimes V \longrightarrow \Omega^{n-p}_\epsilon \otimes V, \quad \omega \mapsto *\omega \tag{3.8}$$

defined as follows: Let $c_{n-p} \in \mathcal{C}_{n-p}$, so $c_{n-p} = c_{n-p}(x; j_1, \cdots, j_{n-p})$. Then

$$(*\omega)(c_{n-p}) = (*\omega)_{j_1\cdots j_{n-p}}(x) = \sum_{i_1<\cdots<i_p} \epsilon_{i_1\cdots i_p j_1\cdots j_{n-p}}\, \omega_{i_1\cdots i_p}(x) \tag{3.9}$$

where ϵ is the antisymmetric Kronecker pseudotensor. As usual,

$$*^2 = (-1)^{p(n-p)} \quad \text{on } p\text{-cochains} \tag{3.10}$$

We introduce bilinear forms in parallel to (2.4, 2.6).

(i)

$$(\Omega^p_\epsilon \otimes V) \otimes (\Omega^q_\epsilon \otimes V) \to \Omega^{p+q}_\epsilon \otimes V$$

$$\omega \otimes \eta \mapsto [\omega, \eta]$$

Let $c_{p+q} \in \mathcal{C}_{p+q}$. Then $c_{p+q} = c_{p+q}(x; k_1, \cdots k_{p+q})$

$k_1 < \cdots < k_{p+q}$.

$$[\omega,\eta](c_{p+q}) = [\omega,\eta]_{k_1\cdots k_{p+q}}(x)$$

$$= \sum_{\substack{i_1<\cdots<i_p \\ j_1<\cdots<j_q}} \delta^{i_1\cdots i_p j_1\cdots j_q}_{k_1\cdots\cdots\cdots k_{p+q}} [\omega_{i_1\cdots i_p}(x), \eta_{j_1\cdots j_q}(x)] \tag{3.11}$$

(ii)

$$(\Omega^p_\epsilon \otimes V) \otimes (\Omega^q_\epsilon \otimes V) \to \Omega^{p+q}$$

$$\omega \otimes \eta \mapsto \omega \wedge \eta$$

$$\omega\wedge\eta\,(c_{p+q}) = \sum_{\substack{i_1<\cdots<i_p \\ j_1<\cdots<j_q}} \delta^{i_1\cdots i_p j_1\cdots j_q}_{k_1\cdots\cdots k_{p+q}}\, \mathrm{tr}\, \omega^*_{i_1\cdots i_p}(x)\, \eta_{j_1\cdots j_q}(x) \tag{3.12}$$

Here, for $1 \le p \le n$, $\delta^{j_1\cdots j_p}_{i_1\cdots i_p}$ is completely antisymmetric in the j_k and in the i_k. For $i_1 < \cdots < i_p$, $j_1 < \cdots < j_p$

$$\delta^{j_1\cdots j_p}_{i_1\cdots i_p} = \begin{cases} 1 & \text{if } i_k = j_k,\ 1 \le k \le n \\ 0 & \text{if } i_k \ne j_k \text{ for at least one k} \end{cases} \tag{3.13}$$

For $\omega, \eta \in \Omega^p_\epsilon \otimes V$ we define first the pointwise G invariant scalar product $\langle\cdot,\cdot\rangle$:

$$\langle \omega(c_p), \eta(c_p)\rangle = \mathrm{tr}\, \omega^*(c_p)\, \eta(c_p) = \mathrm{tr}\, \omega^*_{i_1\cdots i_p}(x)\, \eta_{i_1\cdots i_p}(x)$$

$$|\omega(c_p)|^2 = \langle \omega(c_p), \omega(c_p)\rangle \tag{3.14a}$$

Then on $\Omega^p_\epsilon \otimes V$ we have the G invariant scalar product $(\ ,\ )$:

$$(\omega,\eta) = \epsilon^n \sum_{c_p \in \mathcal{C}_p} \langle \omega(c_p), \eta(c_p) \rangle = \epsilon^n \sum_x \sum_{i_1 < \cdots < i_p} \operatorname{tr} \omega^*_{i_1 \cdots i_p}(x)\, \eta_{i_1 \cdots i_p}(x)$$

$$\|\omega\|^2 = (\omega,\omega)$$

(3.14b)

Now let $U: \mathcal{C}_1 \to G$ be an elementary lattice parallel transporter (the name is justified via (3.16) below). Let $c_1 \in \mathcal{C}_1$, thus $c_1 = c_1(x,i)$. We write

$$U(c_1) = U_i(x) = U(x, x+\epsilon e_i)$$

$$U(-c_1) = U^*(c_1) = U_i^*(x) = U(x+\epsilon e_i, x)$$
$$= U_{-i}(x+\epsilon e_i) \quad (3.15)$$

Under a gauge transformation $g \in \mathcal{G}$

$$U \mapsto U^g, \qquad U^g(c_1) = g(x)\, U(c_1)\, g(x+\epsilon e_i)^{-1}$$
$$U^{*g}(c_1) = g(x+\epsilon e_i)\, U^*(c_1)\, g(x)^{-1} \quad (3.16)$$

where $c = c_1(x,i) \in \mathcal{C}_1$. Define:

$$A_i^\epsilon(x) \doteq \epsilon^{-1}\{U(x, x+\epsilon e_i) - I\}$$

(3.16a)

Then we identify the continuum connection $A = \sum A_i dx^i$ via a limit $A_i^\epsilon(x) \xrightarrow[\epsilon \to 0]{} A_i(x)$ to be taken in an appropriate topology. Corresponding to (2.5), we introduce a lattice covariant exterior derivative,

$$d_U : \Omega^p_\epsilon \otimes V \to \Omega^{p+1}_\epsilon \otimes V, \quad \omega \mapsto d_U \omega$$

(3.17)

where if

$$c_{p+1} = c_{p+1}(x, k_1, \ldots, k_{p+1}) \in \mathcal{C}_{p+1}$$
$$k_1 < \cdots < k_{p+1}$$

then

$$d_U \omega(c_{p+1}) = (d_U\omega)_{k_1 \cdots k_{p+1}}(x) = \sum_{\substack{j_1 < \cdots < j_p \\ i}} \delta^{i j_1 \cdots j_p}_{k_1 \cdots k_{p+1}}\, \mathbb{D}_i(U)\, \omega_{j_1 \cdots j_p}(x)$$

(3.17a)

and

$$\mathbb{D}_i(U)\omega_{j_1\cdots j_p}(x) = \epsilon^{-1}\left(U_i(x)\omega_{j_1\cdots j_p}(x+\epsilon e_i)U_i^*(x) - \omega_{j_1\cdots j_p}(x)\right) \tag{3.17b}$$

Let $d_U^* : \Omega_\epsilon^p \otimes \mathcal{V} \to \Omega_\epsilon^{p-1} \otimes \mathcal{V}$

be the adjoint of d_U with respect to the scalar product (3.14). It is straightforward to verify:

$$\begin{aligned} d_U^*\omega(c_{p-1}) &= (d_U^*\omega)_{k_1\cdots k_{p-1}}(x) \\ &= \sum_i \mathbb{D}_{-i}(U)\omega_{i\,k_1\cdots k_{p-1}}(x) \end{aligned} \tag{3.18}$$

where $\mathbb{D}_{-i}$ is given by replacing in (3.17b), $x+\epsilon e_i \to x+\epsilon e_{-i} = x-\epsilon e_i$ and $U_i(x) \to U_{-i}(x) = U_i^*(x-\epsilon e_i) = U(x, x-\epsilon e_i)$, see (3.15). Note that under d_U or d_U^* the gauge transformation property (3.7) remains intact, so that these maps are properly defined.

The space of $\mathcal{V}$ ($= \mathcal{G}, M(N,\mathbb{C})$) valued p-cochains that we have been studying so far have as their gauge transformation property (3.7). It is convenient to introduce another vector space of $M(N,\mathbb{C})$ valued p-cochains $\tilde{\Omega}_\epsilon^p \otimes M(N,\mathbb{C})$ with a slightly different gauge transformation property. Let $\tilde{\omega} \in \tilde{\Omega}_\epsilon^p \otimes M(N,\mathbb{C})$, and let $c_p = c_p(x, e_{i_1} \cdots e_{i_p})$. Then under a gauge transformation g, $\omega \mapsto \omega^g$ where

$$\tilde{\omega}^g(c_p) = g(x)\,\tilde{\omega}(c_p)\,g(x+\epsilon e_{i_1}+\cdots+\epsilon e_{i_p})^{-1} \tag{3.19}$$

and $\tilde{\omega}(c_p) = \tilde{\omega}_{i_1\cdots i_p}(x)$. Such a space was introduced in / 10 /. Note that a parallel transporter $U : \mathcal{C}_1 \to G$ belongs to $\tilde{\Omega}_\epsilon^1 \otimes M(N,\mathbb{C})$ because of (3.16). Following / 10 /, we introduce again the lattice exterior covariant derivative d_U (we use the same symbol as before) as a map

$$d_U : \tilde{\Omega}_\epsilon^p \otimes M(N,\mathbb{C}) \to \tilde{\Omega}_\epsilon^{p+1} \otimes M(N,\mathbb{C})$$

where if

$$\mathcal{C}_{p+1} \ni c_{p+1} = c_{p+1}(x; e_{k_1} \cdots e_{k_{p+1}}),\ k_1 < \cdots < k_{p+1}$$

then

$$d_U \tilde{\omega}\,(c_{p+1}) = (d_U \tilde{\omega})_{\kappa_1 \cdots \kappa_{p+1}}(x)$$

$$= \sum_{\substack{j_1 < \cdots < j_p \\ i}} \delta^{i j_1 \cdots j_p}_{\kappa_1 \cdots \kappa_{p+1}} \, \mathbb{D}_i(U)\, \tilde{\omega}_{j_1 \cdots j_p}(x) \tag{3.20}$$

and

$$\mathbb{D}_i(U)\,\tilde{\omega}_{j_1 \cdots j_p}(x) = \epsilon^{-1}\left\{ U_i(x)\,\tilde{\omega}_{j_1 \cdots j_p}(x+\epsilon e_i) - \tilde{\omega}_{j_1 \cdots j_p}(x)\, U_i(x+\epsilon e_{j_1}+\cdots+\epsilon e_{j_p}) \right\} \tag{3.21}$$

We noted earlier that parallel transporters $U \in \tilde{\Omega}^1_\epsilon \otimes M(N,\mathbb{C})$
We define following / 10 / the lattice curvature 2-cochain $\tilde{\mathcal{F}}(U)$ by:

$$\tilde{\mathcal{F}}(U) = (2\epsilon)^{-1}\, d_U U \tag{3.22}$$

which is an element of $\tilde{\Omega}^{(2)}_\epsilon \otimes M(N,\mathbb{C})$ Under d_U the gauge transformation property (3.19) is preserved so that the map is properly defined. We equip $\tilde{\Omega}^p \otimes M(N,\mathbb{C})$ with the same scalar product (3.14) used earlier. The scalar product is then invariant under the action (3.19) of $\mathcal{G}_1$.

Let

$$d_U^* : \tilde{\Omega}^p_\epsilon \otimes M(N,\mathbb{C}) \rightarrow \tilde{\Omega}^{p-1}_\epsilon \otimes M(N,\mathbb{C}) \tag{3.24}$$

be the adjoint of d_U with respect to the scalar product (3.14).
Then for $\mathcal{C}_{p-1} \ni c_{p-1} = c_{p-1}(x;\kappa_1 \cdots \kappa_{p-1})$, $\kappa_1 < \cdots < \kappa_{p-1}$, we have (exactly as in (3.18)

$$d_U^* \tilde{\omega}\,(c_{p-1}) = (d_U^* \tilde{\omega})_{\kappa_1 \cdots \kappa_{p-1}}(x)$$

$$= \sum_i \mathbb{D}_{-i}(U)\, \tilde{\omega}_{i \kappa_1 \cdots \kappa_{p-1}}(x) \tag{3.25}$$

with $\mathbb{D}_{-i}$ given by (3.21) with the replacement

$$x + \epsilon e_i \longrightarrow x + \epsilon e_{-i} = x - \epsilon e_i$$

$$U_i(x) \longrightarrow U_{-i}(x) = U_i^*(x - \epsilon e_i) = U(x, x - \epsilon e_i)$$

We define the covariant Laplace-Beltrami operator

$$\Delta_U = d_U^* d_U + d_U d_U^* \tag{3.26}$$

as a map $\Omega^p_\epsilon \otimes M(N,\mathbb{C}) \to \Omega^p_\epsilon \otimes M(N,\mathbb{C})$ or $\tilde{\Omega}^p_\epsilon \otimes M(N,\mathbb{C}) \to \tilde{\Omega}^p_\epsilon \otimes M(N,\mathbb{C})$ using the relevant definitions of d_U. Other variants of Δ_U will also be encountered (see later).

We now introduce a map,

$$\tau : \tilde{\Omega}^p_\epsilon \otimes M(N,\mathbb{C}) \to \Omega^p_\epsilon \otimes M(N,\mathbb{C}), \; \tilde{\omega} \mapsto \tilde{\omega}^{(\tau)}$$

defined as follows: Let $c_p = c_p(x, i_1, \ldots, i_p)$, $i_1 < \cdots < i_p \in \mathcal{C}_p$

Then,

$$\begin{aligned} \tilde{\omega}^{(\tau)}(c_p) &= \tilde{\omega}^{(\tau)}_{i_1 \cdots i_p}(x) \\ &= \tilde{\omega}_{i_1 \cdots i_p}(x) \prod_{k=0}^{p-1} U^*_{i_{p-k}}(x + \epsilon e_{i_1} + \cdots + \epsilon e_{i_{p-k-1}}) \end{aligned} \tag{3.27}$$

We have

$$(\tilde{\omega}^{(\tau)}, \tilde{\omega}^{(\tau)}) = (\omega, \omega) \tag{3.28}$$

so that τ is an isometry. Let τ^* be the adjoint of τ with respect to the scalar product $(\cdot, \cdot)$. Then

$$\tau^* : \Omega^p_\epsilon \otimes M(N,\mathbb{C}) \to \tilde{\Omega}^p_\epsilon \otimes M(N,\mathbb{C})$$

and

$$(\tilde{\omega}^{(\tau)}, \omega) = (\tilde{\omega}, \omega^{(\tau^*)}) \tag{3.29}$$

<u>Example</u>: Let $\tilde{\omega} \in \tilde{\Omega}^1_\epsilon \otimes M(N,\mathbb{C})$

Then $\tilde{\omega}^{(\tau)}_i(x) = \tilde{\omega}_i(x)\, U_i^*(x)$

If $\omega \in \Omega^1_\epsilon \otimes M(N,\mathbb{C})$, then

$$\omega^{(\tau^*)}_i(x) = \omega_i(x)\, U_i(x) \tag{3.30}$$

<u>We now write two useful formulae involving the map τ.</u>

Let $\phi \in \Omega^1_\epsilon \otimes \mathcal{G}$. Then $\phi^{(\tau^*)} \in \tilde{\Omega}^1_\epsilon \otimes M(N,\mathbb{C})$
Define, as an element of $\tilde{\Omega}^1_\epsilon \otimes M(N,\mathbb{C})$,

$$\tilde{U} = U + \epsilon\, \phi^{(\tau^*)} \tag{3.31}$$

Then, using (3.17b), we obtain for $\omega \in \Omega^p_\epsilon \otimes M(N,\mathbb{C})$

$$\mathbb{D}_i(\tilde{U})\,\omega_{j_1\cdots j_p}(x)$$

$$= \mathbb{D}_i(U)\,\omega_{j_1\cdots j_p}(x) + \left[\phi_i(x), U_i(x)\,\omega_{j_1\cdots j_p}(x+\epsilon e_i)\,U_i^*(x)\right] - \epsilon\,\phi_i(x)\,U_i(x)\,\omega_{j_1\cdots j_p}(x+\epsilon e_i)\,U_i^*(x)\,\phi_i(x) \tag{3.32}$$

or, using (3.17a),

$$d_{\tilde{U}} = d_U + [\phi, U(\cdot)U^*] - \epsilon\,\phi U(\cdot)U^*\phi$$

Similarly, using (3.21) and (3.22), we obtain:

$$\tilde{\mathcal{F}}(\tilde{U}) = \tilde{\mathcal{F}}(U) + d_U\,\phi^{(\tau^*)} + [\phi^{(\tau^*)}, \phi^{(\tau^*)}] \tag{3.33}$$

where

$$[\phi^{(\tau^*)}, \phi^{(\tau^*)}](c_2(x;i,j))$$

$$= \phi_i^{(\tau^*)}(x)\,\phi_j^{(\tau^*)}(x+\epsilon e_i) - \phi_j^{(\tau^*)}(x)\,\phi_i^{(\tau^*)}(x+\epsilon e_j) \tag{3.34}$$

It is useful for later purposes to give some additional formulae related to (3.33). We have:

$$\mathbb{D}_i(U)\,\phi_j^{(\tau^*)}(x) = \mathbb{D}_i(U)\,\phi_j(x)\,U_j(x)\,U_i(x+\epsilon e_j) + \epsilon\,U_i(x)\,\phi_j(x)\,U_i^*(x)\,\tilde{\mathcal{F}}_{ij}(U)(x) \tag{3.35}$$

whence

$$\| d_U\,\phi^{(\tau^*)}\|^2 = \| d_U\,\phi\|^2 + O(\epsilon) \tag{3.36}$$

where $O(\epsilon)$ represents gauge invariant "anomalous" vertices vanishing in the (naive) continuum limit.

We also have:

$$\left[\phi^{(\tau^*)}, \phi^{(\tau^*)}\right](c_2(x;i,j))$$
$$= [\phi_i(x), \phi_j(x)]\, U_j(x)\, U_i(x+\epsilon e_j) - \epsilon^2 \phi_j(x)\phi_i(x)\, \tilde{\mathcal{F}}_{ij}(U)(x)$$
$$+ \epsilon\left(\mathbb{D}_i(U)\, \phi_j^{(\tau^*)}(x) - \mathbb{D}_j(U)\, \phi_i^{(\tau^*)}(x)\right) \qquad (3.37)$$

whence

$$\| [\phi^{(\tau^*)}, \phi^{(\tau^*)}] \|^2 = \| [\phi,\phi] \|^2 + O(\epsilon) \qquad (3.38)$$

where, once again, $O(\epsilon)$ represents gauge invariant "anomalous" vertices vanishing in the (naive) continuum limit. (3.39)

Weitzenböck decomposition of Δ_U

Consider the Laplace-Beltrami operator:

$$\Delta_U : \Omega^1_\epsilon \otimes \mathcal{G} \longrightarrow \Omega^1_\epsilon \otimes \mathcal{G}$$

defined by (3.26). We also define the Bochner lattice Laplacian $\Delta_U^{(B)} : \Omega^1_\epsilon \otimes \mathcal{G} \to \Omega^1_\epsilon \otimes \mathcal{G}$, by

$$(\Delta_U^{(B)}\omega)_K(x) \doteq \epsilon^{-2} \sum_i \Big[2\omega_K(x) - U_i^*(x-\epsilon e_i)\,\omega_K(x-\epsilon e_i)\, U_i(x-\epsilon e_i) - U_i(x)\,\omega_K(x+\epsilon e_i)\, U_i^*(x) \Big] \qquad (3.40)$$

Define

$$\mathrm{ad}\, \tilde{\mathcal{F}}_{iK}(x) \doteq \epsilon^{-2}\left[\mathrm{ad}\, U_i(x)\, \mathrm{ad}\, U_K(x+\epsilon e_i) - \mathrm{ad}\, U_K(x)\, \mathrm{ad}\, U_i(x+\epsilon e_K)\right]$$

$$\mathrm{ad}\, \tilde{\mathcal{F}}_{-iK}(x) \doteq \epsilon^{-2}\left[\mathrm{ad}\, U_i^*(x-\epsilon e_i)\, \mathrm{ad}\, U_K(x-\epsilon e_i) - \mathrm{ad}\, U_K(x)\, \mathrm{ad}\, U_i^*(x+\epsilon e_K - \epsilon e_i)\right] \qquad (3.41)$$

Then:

$$(\Delta_U \omega)_K(x) = (\Delta_U^B \omega)_K(x) + \sum_i \mathrm{ad}\,\tilde{\mathfrak{F}}_{-iK}(x)\,\omega_i(x-\epsilon e_i+\epsilon e_K) \tag{3.42}$$

which is a Weitzenböck decomposition.

Examples

We wish to note some examples of the formalism elaborated so far.

1 $\|\tilde{\mathfrak{F}}(U)\|^2$ is nothing but the standard Wilson action, as noted in /10/.

2 Consider

$$\|d_U^* \tilde{\mathfrak{F}}(U)\|^2 = \epsilon^4 \sum_{i,j,K} \mathrm{tr}\,\left(\mathbb{D}_{-j}\tilde{\mathfrak{F}}_{jK}(x)\right)^*\left(\mathbb{D}_{-i}\tilde{\mathfrak{F}}_{iK}(x)\right) \tag{3.43}$$

It is easy to check that

$$\epsilon^3\,\mathbb{D}_{-i}\tilde{\mathfrak{F}}_{iK}(x) = 2U_K(x) - U_i^*(x-\epsilon e_i)\,U_K(x-\epsilon e_i)\,U_i(x-\epsilon e_i+\epsilon e_K) - U_i(x)\,U_K(x+\epsilon e_i)\,U_i^*(x+\epsilon e_K)$$

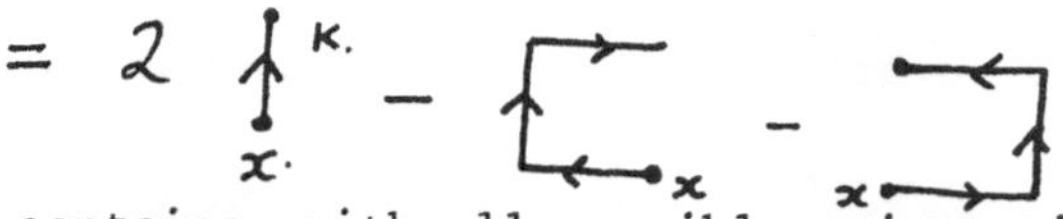

(3.44)

Hence (3.43) contains, with all possible orientations,

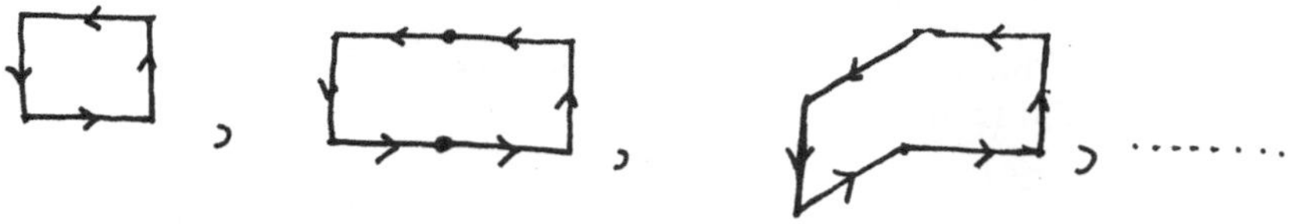

3 The Wilson lattice gauge action can be written as:

$$S(U) = \frac{1}{2}\|\tilde{\mathfrak{F}}(U)\|^2$$

$$S(U+\epsilon\tilde{\Phi}) = \frac{1}{2}\|\tilde{\mathfrak{F}}(U+\epsilon\tilde{\Phi})\|^2$$

$$= S(U) + (d_U^*\tilde{\mathfrak{F}}(U), \tilde{\Phi}) + O(\tilde{\Phi}^2)$$

where we have used (3.31) and (3.33), with $\tilde{\phi} \in \tilde{\Omega}^1_\epsilon \otimes M(N,\mathbb{C})$. Hence the "classical field equation" in Wilson's lattice gauge theory is:

$$d_U^* \tilde{\mathcal{F}}(U) = 0 \tag{3.45}$$

4 Under a gauge transformation

$$U_i^g(x) = g(x)\, U_i(x)\, g_i(x+\epsilon e_i)^{-1}$$

Let $g(x) = I + \xi(x)$ be an infinitesimal gauge transformation. Then,

$$\delta U = -\epsilon\, d_U \xi$$

with $\xi \in \tilde{\Omega}^{(0)}_\epsilon \otimes \mathcal{G}$ and d_U defined via (3.20)-(3.21). Then we have, using (3.31), (3.33)

$$\begin{aligned} \tilde{\mathcal{F}}(U+\delta U) &= \tilde{\mathcal{F}}(U - \epsilon d_U \xi) \\ &= \tilde{\mathcal{F}}(U) - d_U d_U \xi + [d_U \xi, d_U \xi] \end{aligned}$$

Hence,

$$S(U+\delta U) = S(U) - (d_U^* d_U^* \tilde{\mathcal{F}}(U), \xi) + O(\xi^2)$$

From the gauge invariance of S(U) we have the identity:

$$d_U^* d_U^* \tilde{\mathcal{F}}(U) = 0 \tag{3.46}$$

which is very useful in practice!

5 (3.46) will be used in the next chapter. To employ it effectively we note the following:

Let $\tilde{\omega} \in \tilde{\Omega}^1_\epsilon \otimes M(N,\mathbb{C})$. Then via the isometry $\mathcal{T}$, (3.27), we obtain $\tilde{\omega}^{(\mathcal{T})} \in \Omega^1_\epsilon \otimes M(N,\mathbb{C})$. We have the following useful identity:

$$d_U^* (\tilde{\omega}^{(\mathcal{T})}) = d_U^* \tilde{\omega} \tag{3.47}$$

where, of course, on the LHS we employ (3.18), (3.17), whereas on the RHS we employ (3.25), (3.21).

As a particular application of (3.47) we have:

$$d_U^* \left((d_U^* \mathcal{F}(U))^{(\tau)} \right) = d_U^* d_U^* \mathcal{F}(U)$$
$$= 0 \tag{3.48}$$

by virtue of (3.46). Note that for 0-cochains (M(N,C) valued functions)

$$\tilde{\Omega}_\epsilon^{(0)} \otimes M(N,\mathbb{C}) = \Omega_\epsilon^{(0)} \otimes M(N,\mathbb{C})$$

Depending on the choice of d_U , (3.20),(3.21) or (3.17), we have

$$\text{0-cochains} \begin{array}{l} \xrightarrow{d_U} \tilde{\Omega}_\epsilon^1 \otimes M(N,\mathbb{C}) \\ \xrightarrow{d_U} \Omega_\epsilon^1 \otimes M(N,\mathbb{C}) \end{array}$$

The top arrow has been used in the derivation of (3.46).

The lattice differential geometric formalism of this chapter goes over in the continuum limit ($\epsilon \to 0$), via the identification (3.16), to the continuum differential geometric formalism employed in Chapter II. In particular p-cochains in $\Omega_\epsilon^p \otimes \mathcal{V}$ go over to differential forms in $\Omega^p \otimes \mathcal{V}$. As $\epsilon \to 0$, the spaces $\Omega_\epsilon^p \otimes \mathcal{V}$ and $\tilde{\Omega}_\epsilon^p \otimes \mathcal{V}$ coincide, and τ becomes the identity map. The invariant scalar products, and covariant derivatives become continuum ones. Furthermore, the lattice curvature 2-cochain $\tilde{\mathcal{F}}(U)$ goes over to the continuum curvature 2-form F(A).

IV. FREQUENCY SPLITTING IN THE WILSON LATTICE GAUGE THEORY (4 DIMENSIONS

In this section we fix n = 4, the case of physical interest. The partition function Z of Wilson's lattice gauge theory / 1 / can be written as / 10 /,

$$Z_{\Lambda_\epsilon} = \int_{G^{|\mathcal{C}_1|}} \left(\prod_{b \in \mathcal{C}_1} dU(b) \right) e^{-\frac{1}{2g_0^2} \| \tilde{\mathcal{F}}(U) \|^2} \tag{4.1}$$

where $\tilde{\mathcal{F}}$ is the lattice curvature tensor of Ch. III, (3.22). $dU(b)$, for fixed b, is the normalized Haar measure on G. Let $\Theta(U)$ be a gauge invariant function. Then its Euclidean expectation is:

$$\langle \mathcal{O} \rangle = \frac{1}{Z_{\Lambda_\epsilon}} \int_{G^{|\mathcal{C}_1|}} \Big(\prod_{b \in \mathcal{C}_1} dU(b) \Big) e^{-\frac{1}{2g_0^2}\| \mathcal{F}(U) \|^2} \mathcal{O}(U) \tag{4.2}$$

Using the invariance of the Haar measure,

$$Z_{\Lambda_\epsilon} = \int_{G^{|\mathcal{C}_1|} \otimes G^{|\mathcal{C}_1|}} \prod_{b \in \mathcal{C}_1} \big(dU(b)\, dV(b) \big) e^{-\frac{1}{2g_0^2}\| \tilde{\mathcal{F}}(VU) \|^2}$$

$$\langle \mathcal{O} \rangle = Z_{\Lambda_\epsilon}^{-1} \int_{G^{|\mathcal{C}_1|} \otimes G^{|\mathcal{C}_1|}} \prod_{b \in \mathcal{C}_1} \big(dU(b)\, dV(b) \big) e^{-\frac{1}{2g_0^2}\| \tilde{\mathcal{F}}(VU) \|^2} \mathcal{O}(VU) \tag{4.3}$$

Here, V is a G-valued element of $\Omega^1_\epsilon \otimes M(N,\mathbb{C})$ whereas we recall that U is a G-valued element of $\tilde{\Omega}^1_\epsilon \otimes M(N,\mathbb{C})$. In addition to the usual local gauge invariance corresponding to:

$$V_i(x) \longrightarrow g(x)\, V_i(x)\, g(x)^{-1}$$

$$U_i(x) \longrightarrow g(x)\, U_i(x)\, g(x+\epsilon e_i)^{-1} \tag{4.4}$$

(4.3) has an additional local invariance: "hard-soft gauge invariance," corresponding to

$$V_i(x) \longrightarrow V_i(x)\, g(x)^{-1}$$

$$U_i(x) \longrightarrow g(x)\, U_i(x) \tag{4.5}$$

We shall "fix the hard-soft gauge" using the Faddeev-Popov method. To this end let $\psi(U,V) \in \Omega^1_\epsilon \otimes M(N,\mathbb{C})$ be a hard-soft gauge fixing function. Then, in the Faddev-Popov formalism, (4.3) is equivalent to:

$$Z_{\Lambda_\epsilon} = \int_{G^{|\mathcal{C}_1|} \otimes G^{|\mathcal{C}_1|}} \prod_{b \in \mathcal{C}_1} \big(dU(b)\, dV(b) \big) \det \Big(\frac{\delta \psi(gU, Vg^{-1})}{\delta g} \Big) \cdot e^{-\frac{1}{2g_0^2}\| \tilde{\mathcal{F}}(VU) \|^2 - \frac{\Lambda^2}{2g_0^2} \| \psi(U,V) \|^2} \tag{4.6}$$

where Λ, the hard-soft gauge fixing parameter, is on arbitrary mass scale. We choose $\Lambda \ll \epsilon^{-1}$.

Define

$$A_i^{h,\epsilon}(x) \doteq \epsilon^{-1}\left(V_i(x) - I\right) \in \Omega^1_\epsilon \otimes M(N,\mathbb{C})$$

$$A_i^{s,\epsilon}(x) = \epsilon^{-1}\left(U_i(x) - I\right) \tag{4.7}$$

In the putative continuum limit $\epsilon \to 0$, to be taken in some topology, $\sum A_i^h(x)dx^i \in \Omega^1 \otimes \mathcal{G}$ whereas $\sum A_i^s dx^i$ is a $\mathcal{G}$-valued connection 1-form.

There is a considerable amount of freedom+ in the choice of $\psi(U,V)$. Guided by continuum insight we choose <u>for simplicity</u> $\psi \in \tilde{\Omega}^1_\epsilon \otimes M(N,\mathbb{C})$, with

$$\psi(U,V) = A^{h,\epsilon} - \frac{1}{\Lambda^2}\left(d_U^* \mathcal{F}(U)\right)^{(\tau)} + \frac{\alpha}{\Lambda^2} d_U d_U^* A^{h,\epsilon} \tag{4.8}$$

with $A^{h,\epsilon} \in \Omega^1_\epsilon \otimes M(N,\mathbb{C})$, using (4.7). By virtue of (4.7),

$$V = I + \epsilon A^{h,\epsilon}$$

$$\begin{aligned} VU = U + \epsilon A^{h,\epsilon} U &= U + \epsilon \left(A^{h,\epsilon}\right)^{(\tau^*)} \\ &\in \tilde{\Omega}^1_\epsilon \otimes M(N,\mathbb{C}) \end{aligned} \tag{4.9}$$

where, in the last line we have used the map τ^* of (3.29). From (4.9), (3.31) and (3.33) we have:

$$\tilde{\mathcal{F}}(VU) = \tilde{\mathcal{F}}(U) + d_U\, A^{h,\epsilon\,(\tau^*)} + \left[A^{h,\epsilon\,(\tau^*)}, A^{h,\epsilon\,(\tau^*)}\right] \tag{4.10}$$

Squaring (4.10) we have:

+For a discussion of the hard-soft gauge fixing ambiguity, see later.

$$\frac{1}{2}\|\tilde{\mathcal{F}}(VU)\|^2 = \frac{1}{2}\|\tilde{\mathcal{F}}(U)\|^2 + \frac{1}{2}\|d_U (A^{h,\epsilon})^{\tau^*}\|^2 +$$
$$+ \left(d_U A^{h,\epsilon(\tau^*)}, [A^{h,\epsilon(\tau^*)}, A^{h,\epsilon(\tau^*)}]\right) +$$
$$+ \frac{1}{2}\|[A^{h,\epsilon(\tau^*)}, A^{h,\epsilon(\tau^*)}]\|^2 + \left(\tilde{\mathcal{F}}(U), d_U (A^{h,\epsilon(\tau^*)})\right)$$
$$+ \left(\tilde{\mathcal{F}}(U), [A^{h,\epsilon(\tau^*)}, A^{h,\epsilon(\tau^*)}]\right) \tag{4.11}$$

Squaring (4.8) we have:

$$\frac{\Lambda^2}{2}\|\psi(U,V)\|^2 = \frac{\Lambda^2}{2}\|A^{h,\epsilon}\|^2 + \frac{1}{2\Lambda^2}\|(d_U^* \tilde{\mathcal{F}}(U))^{\tau}\|^2 +$$
$$+ \frac{\alpha^2}{2\Lambda^2}\|d_U d_U^* A^{h,\epsilon}\|^2 + \alpha\|d_U^* A^{h,\epsilon}\|^2$$
$$- \left(A^{h,\epsilon}, (d_U^* \tilde{\mathcal{F}}(U))^{\tau}\right) \tag{4.12}$$

where we have used (see 3.46 - 3.48)

$$d_U^* \left((d_U^* \tilde{\mathcal{F}}(U))^{\tau}\right) = d_U^* d_U^* \tilde{\mathcal{F}}(U) = 0$$

Now note that

$$\left(\tilde{\mathcal{F}}(U), d_U (A^{h,\epsilon(\tau^*)})\right) = \left((d_U^* \mathcal{F}(U))^{(\tau)}, A^{h,\epsilon}\right) \tag{4.13}$$

Adding (4.11) and (4.12) we obtain

$$\frac{1}{2}\|\tilde{\mathcal{F}}(VU)\|^2 + \frac{\Lambda^2}{2}\|\psi(U,V)\|^2$$
$$= S^{(1)}(U) + \tilde{S}_U^{(1)}(A^{h,\epsilon}) + V_U(A^{h,\epsilon}) \tag{4.14}$$

where,

$$S^1(U) = \frac{1}{2}\|\tilde{\mathcal{F}}(U)\|^2 + \frac{1}{2\Lambda^2}\|d_U^* \tilde{\mathcal{F}}(U)\|^2 \tag{4.15}$$

where we have used the isometry of $\mathcal{T}$,

$$\tilde{S}^{(1)}_U(A^{h,\epsilon}) = \frac{1}{2}\|d_U A^{h,\epsilon(\mathcal{T}^*)}\|^2 + \alpha\|d_U^* A^{h,\epsilon}\|^2 + \frac{\alpha^2}{2\Lambda^2}\|d_U d_U^* A^{h,\epsilon}\|^2 + \frac{\Lambda^2}{2}\|A^{h,\epsilon}\|^2 \tag{4.16}$$

$$= \frac{1}{2}\|d_U A^{h,\epsilon}\|^2 + \alpha\|d_U^* A^{h,\epsilon}\|^2 + \frac{\alpha^2}{2\Lambda^2}\|d_U d_U^* A^{h,\epsilon}\|^2 + \frac{\Lambda^2}{2}\|A^{h,\epsilon}\|^2 + O(\epsilon) \tag{4.17}$$

where $O(\epsilon)$ represents interaction terms vanishing in the naive continuum limit.

$$V_U(A^{h,\epsilon}) = \left(d_U A^{h,\epsilon(\mathcal{T}^*)}, [A^{h,\epsilon(\mathcal{T}^*)}, A^{h,\epsilon(\mathcal{T}^*)}]\right) + \frac{1}{2}\|[A^{h,\epsilon(\mathcal{T}^*)}, A^{h,\epsilon(\mathcal{T}^*)}]\|^2 + \left(\tilde{\mathcal{F}}(U), [A^{h,\epsilon(\mathcal{T}^*)}, A^{h,\epsilon(\mathcal{T}^*)}]\right) \tag{4.18}$$

Finally the operator

$$K_U(A^{h,\epsilon}) = \frac{\delta\psi}{\delta g}(gU, Vg^{-1})$$

can be computed. It is

$$K_U(A^{h,\epsilon}) = \Lambda^2 + \alpha\, d_U d_U^* + d_U^* d_U - *[\cdot, *\,\tilde{\mathcal{F}}(U)^{(\mathcal{T})}] - \alpha[\cdot, d_U^* A^{h,\epsilon}] + \alpha\, d_U *[\cdot, *A^{h,\epsilon}] + O(\epsilon) \tag{4.19}$$

where $O(\epsilon)$ represents terms vanishing in the naive continuum limit. $K_U(A^{h,\epsilon})$ acts on $L^2(\Omega^1_\epsilon \otimes M(N,\mathbb{C}))$. As a result (4.6) reads

$$Z_{\Lambda_\epsilon} = \int_{G^{|\mathcal{E}_1|}} (\prod_{b\in\mathcal{E}_1} dU(b)) e^{-\frac{1}{g_0^2} S^1(U)} \int_{G^{|\mathcal{E}_1|}} (\prod_{b\in\mathcal{E}_1} dV(b)) \det(K_U(A^{h,\epsilon})) \cdot e^{-\frac{1}{g_0^2} \tilde{S}_U^{(1)}(A^{h,\epsilon}) - \frac{1}{g_0^2} V_U(A^{h,\epsilon})} \tag{4.20}$$

with (4.7), (4.15) - (4.19). The pictorial representation of $S^1(U)$ can be seen from (3.43) plus the standard plaquette term.

We shall now discuss some crucial properties of the representation (4.20). Define

$$d\mu_U^{h,\epsilon} = \prod_{b\in\mathcal{E}_1} dV(b)\, e^{-\frac{1}{g_0^2} \tilde{S}_U^{(1)}(A^{h,\epsilon})} \tag{4.21}$$

Introducing canonical coordinates $A^h \in \Omega^1_\epsilon \otimes \mathcal{G}$ for V, i.e.

$$V = e^{\epsilon A^h} \tag{4.22}$$

we have from (4.7),

$$A^{h,\epsilon} = A^h + O(\epsilon)$$

and (4.21) now reads:

$$d\mu_U^{h,\epsilon} = d\mu_U^h(A^h)\, \chi(|A^h| \le C_\epsilon)\, e^{-O(\epsilon)} [\det(\Delta_U^h + \Lambda^2)]^{-\frac{1}{2}} \tag{4.23}$$

where $C_\epsilon \to \infty$ as $\epsilon \to 0$, $O(\epsilon)$ represents interactions vanishing in the naive continuum limit and $d\mu_U^h(A^h)$ is an equivariant normalized Gaussian measure on $\Omega'_\epsilon \otimes \mathcal{G}$ with covariance

$$C_U^h = g_0^2 (\Delta_U^h + \Lambda^2)^{-1} \tag{4.24}$$

where

$$\Delta^h_U = d_U^* d_U + \alpha d_U d_U^* + \frac{\alpha^2}{2\Lambda^2} d_U d_U^* d_U d_U^* \tag{4.25}$$

$\Delta^h_U + \Lambda^2$ is an invertible operator.

In the continuum limit (set U = I), (4.24) gives (the coupling constant g_0 can be put back into interactions via rescalings) the hard propagators in momentum space:

$$\tilde{C}^h(k) = \left(\delta_{\mu\nu} - \frac{k_\mu k_\nu}{k^2}\right)\frac{1}{k^2+\Lambda^2} + \frac{1}{\alpha}\frac{k_\mu k_\nu}{k^2} \cdot \frac{1}{2\alpha k^2 + \frac{\alpha^2}{\Lambda^2}(k^2)^2 + \Lambda^2} \tag{4.26}$$

which is massive and automatically gauge fixed. Calling $A^h_{(t)}$, $A^h_{(\ell)}$, the transverse and longitudinal components, we can read off the ultraviolet (UV) dimensions from (4.26):

$$\dim_{UV} A^h_{(t)} = 1, \quad \dim_{UV} A^h_{(\ell)} = 0 \tag{4.27}$$

The reason for calling A^h a hard field is precisely (4.27) plus the fact that it is massive (mass Λ).

Next we turn to the measure $d\mu^s$ defined by:

$$d\mu^s = \left(\prod_{b \in \mathcal{C}_1} dU(b)\right) e^{-\frac{1}{g_0^2} S^1(U)} \tag{4.28}$$

Introducing canonical coordinates

$$U = e^{\epsilon A^s}$$

we have in the naive continuum limit

$$d\mu^s \longrightarrow \mathcal{D}A^s \, e^{-\frac{1}{2g_0^2}\left(A^s,\, \Delta(I + \frac{\Delta}{\Lambda^2})A^s\right) + \text{interactions}}$$
$$\Delta = d^* d \tag{4.29}$$

Because $\Delta(I+\frac{\Delta}{\Lambda^2})$ is not invertible, (4.29) is not well defined; the Faddeev-Popov "covariant" guage fixing formalism (which can be rigorously justified on the lattice / 11 / after suitable precautions) can be employed in (4.28). Suppose then that we have fixed a gauge. Then, apart from the Faddeev-Popov determinant, the quadratic form in the exponent of (4.29) should be replaced by:

$$\frac{1}{2}\left(A^s, \Delta\left(I+\frac{\Delta}{\Lambda^2}\right)A^s\right) + \alpha'\left(d^* A^s, \left(I+\frac{\Delta}{\Lambda^2}\right)d^* A^s\right) \tag{4.30}$$

where α' is a gauge fixing parameter. From (4.30) we obtain the momentum space propagator:

$$\tilde{C}^{(s)}(k) = \left(\delta_{\mu\nu} - \frac{k_\mu k_\nu}{k^2}\right)\frac{\Lambda^2}{k^2(k^2+\Lambda^2)} + \frac{1}{\alpha'}\frac{k_\mu k_\nu}{k^2}\frac{\Lambda^2}{k^2(k^2+\Lambda^2)} \tag{4.31}$$

From (4.29) we see

(i) the (free) A^s field is massless

(ii) $\dim_{UV} A^s = 0$ i.e. improved UV behavior (4.32)

This justifies the name "soft field" for A^s.

Let us now turn to the problem of the "hard-soft gauge fixing" ambiguity. The entire formalism developped so far depended on the local hard-soft transformations (4.5), together with the Faddeev-Popov formalism employed in (4.6) together with the choice (4.8). It has been shown+ in / 11 /, following earlier work /12 /, that with suitable precautions in the actual lattice representation of gauge fixing, the Faddeev-Popov formalism can be rigorously justified on the lattice in spite of gauge fixing ambiguities (for "covariant gauges"). The effect of the finite number of Gribov copies is a multiplicative finite constant in the partition function which cancels out for normalized Schwinger functions, a truly remarkable fact.

The traditional lattice representations of the Landau gauge are well known. One can exploit the freedom of the lattice gauge representation to choose a form (which may differ from somebody else's by $O(\epsilon)$ terms vanishing in the naive continuum limit) such that one fits into the mathematical framework of / 11 /. Exactly

---

+ I thank Erhard Seiler for bringing this to my attention and for stimulating discussions.

the same remarks are valid for us. One should so define the hard-soft gauge fixing that (i), one fits into the mathematical framework of / 11 /, and (ii), one recovers $\frac{1}{2}\Lambda^2 \|\psi(U,V)\|^2$ with ψ as in (4.8) up to $O(\epsilon)$ terms vanishing in the naive continuum limit. Since the qualitative aspects of our discussion do not change, we will not enter into this complication here.

Renormalization group aspects

The continuum action corresponding to $S^{(1)}(U)$ is

$$\frac{1}{2g_0^2}\left(F(A^s),\ \Delta_{A^s}\left(I+\frac{\Delta_{A^s}}{\Lambda^2}\right)F(A^s)\right) \tag{4.33}$$

Using the propagator (4.31), the interactions generated by (4.33) and the Faddeev-Popov determinant, it can be seen, using the power-counting analysis of Appendix D of / 13 /, that the pure U theory is super-renormalizable, i.e. UV divergences stop at a finite order of perturbation theory in g_0. The strictly renormalizable character of (4.20) is due to the hard V integration which, as we have seen, requires no explicit gauge fixing for the continuum limit.

Note that U is gauge equivalent to some U' which, in probability, is nearly piecewise constant on a scale $\Lambda^{-1} \gg \epsilon$. It is thus an "approximate block field" with block size Λ^{-1} whereas the massive high frequency V field (mass Λ) is a "fluctuating field" which (in the gaussian approximation) is orthogonal to the block field. In fact the physical picture is that of a slowly varying (scale Λ^{-1}) super-renormalizable gauge field (with Λ acting as an UV cutoff) coupled to a gauge covariant massive high frequency matter field which polarizes the medium in which U operates.+

The RG analysis of the high frequency V integration, which reflects the strictly renormalizable character of the theory, is the hardest part of the full RG analysis (given that U is super- renormalizable). We must chop up V into a sequence of frequency scales which must be integrated step by step (starting from the highest frequency and integrating down). To do this we must form at each step a block field and integrate out the fluctuating field around it. To form a V block field we may use the fact that U is already an "approximate block field" (block size Λ^{-1}). Let us now show how a gauge covariant V block field may be formed.

Let B_x be a block of size L ($\epsilon \ll L \leq \Lambda^{-1}$) centered at x. Let γ_{xy} denote the shortest path in $B_x \subset \Lambda_\epsilon$ from x to $y \in B_x$.

+This interpretation is akin to that of G. Mack (/ 10 / and his seminar at this school).

Let $\{x_i\}_{0\le i\le m}$, $x_0 = x$ and $x_m = y$ where x_i are vertices of $\gamma_{xy} \cap B_x$. We form the parallel transport operator along γ_{xy} from x to y, namely

$$U(\gamma_{xy}) = \prod_{i=0}^{m-1} U_\gamma(x_i, x_{i+1}) \tag{4.34}$$

where $U_\gamma(\cdot,\cdot)$ are the elementary parallel transport operators. We then define the V-block field, $V_i^{Bl}(x)$ by

$$V_i^{Bl}(x) = L^{-4}\epsilon^4 \sum_{y\in B_x} U(\gamma_{xy}) V_i(y) U(\gamma_{xy})^{-1} \tag{4.35}$$

which has the correct gauge transformation property. Using (4.7), we obtain

$$A_i^{h,\epsilon\,(Bl)}(x) = L^{-4}\epsilon^4 \sum_{y\in B_x} U(\gamma_{xy}) A_i^{h,\epsilon}(y) U(\gamma_{xy})^{-1} \tag{4.36}$$

Using the method of / 14 / and the measure (4.21),(4.23), the fluctuating field around the block can be calculated and the measure (4.21) can be rewritten in terms of a product measure (plus interactions). The RG transformation then consists in integrating out the fluctuating field.

<u>A remark on "long-distance observables"</u>

As is clear from (4.3), (4.6) and (4.20), the observables, whose Euclidean expectations we have to calculate, are gauge invariant functions of VU. However for long distance effects (e.g. the investigation of quark confinement) it is sufficient to integrate gauge invariant functions of the low frequency U field alone. Because of the improved ultraviolet behavior of the U field (with Λ acting as an effective cutoff), renormalized Wilson loops involving these fields will be much easier to define. From a practical point of view, these are the only Wilson loops that matter.

V. LOW FREQUENCY GAUGE FIELDS AND TOPOLOGY

In this section, we wish to point out that global topological field configurations, if they indeed play a role in quantum gauge field theory, should be naturally associated to low frequency fields. Let us return to the frequency split representation (4.20) for the partition function $\mathcal{Z}$ of lattice regularized gauge field theory. For Feynman graph continuum limit calculations it is convenient to rescale fields by the bare coupling constant g_0. This is done by redefining (4.7) via

$$A_i^{h,\epsilon}(x) = (g_0\epsilon)^{-1}(V_i(x) - I)$$
$$A_i^{s,\epsilon}(x) = (g_0\epsilon)^{-1}(U_i(x) - I) \tag{4.7'}$$

so that we get rid of the explicit factors g_0^{-2} in (4.20). With this change, define the gauge invariant effective potential $\mathcal{V}_{eff}^{(1)}(U)$ via

$$e^{-\mathcal{V}_{eff}^{(1)}(U)} = \int_{G^{|\mathcal{C}_1|}} \prod_{b\in\mathcal{C}_1} dV(b) \det(K_U(A^{h,\epsilon}))\, e^{-\tilde{S}_U^{(1)}(A^{h,\epsilon})} \cdot e^{-V_U(A^{h,\epsilon})} \tag{5.1}$$

$\mathcal{V}_{eff}^{(1)}(U)$ is thus generated, via the cumulant expansion, as a sum of connected Feynman graphs with A^h internal lines and U external lines. As noted earlier, even in the continuum limit no additional gauge fixing is required for the A^h integrations. We thus obtain an effective low frequency action:

$$S_{eff}^{(1)}(U) = S^1(U) + \mathcal{V}_{eff}^{(1)}(U) \tag{5.2}$$

which governs the long distance effects in the theory. From the analysis in Ch. IV we know that almost all (i.e. all but a finite number of) continuum limit UV divergent Feynman graphs are those with A^h internal lines. We shall now take a shortcut which is not entirely in the spirit of constructive RG theory but is sufficient for the point of this section.

Let us imagine that appropriate gauge invariant subtractions have been made for the A^h integration (e.g. in perturbation theory using dimensional renormalization+ with Λ as the renormalization scale /3,4/). The resulting low frequency effective action:

$$S_{eff}^{(1),R}(U) = S^1(U) + \mathcal{V}_{eff}^{(1),R}(U, g, \Lambda, \epsilon) \tag{5.3}$$

describes a super-renormalizable low frequency field theory. Up to gauge equivalence, the gauge field U lives effectively on blocks of size $\Lambda^{-1} >> \epsilon$ (see Ch. IV).

For the point we wish to make eventually it may be necessary to make one further iteration of the high low frequency split of Ch. IV. In other words, the low frequency field U of (5.3) is itself divided up into another soft lattice gauge field U and a hard field V. We do not give the details which are a straightforward generalization of the method of Ch. IV. Then repeating the previous steps we arrive at a new low frequency gauge invariant effective action:

$$S_{eff}^{(2),R}(U) = S^{(2)}(U) + \mathcal{V}_{eff}^{(2),R}(U, g, \Lambda, \epsilon) \tag{5.4}$$

where

$$S^{(2)}(U) = \frac{1}{2}\left(\tilde{\mathcal{F}}(U), \left(I + \frac{\Delta_U}{\Lambda^2}\right)^2 \tilde{\mathcal{F}}(U)\right) \tag{5.5}$$

with

$$\Delta_U = d_U d_U^*$$

(5.4) too describes a super-renormalizable theory, divergent Feynman graphs now stopping at the 1-loop order.

To pass to the continuum limit it may be necessary to fix a gauge (say the Landau gauge) via the Faddeev-Popov procedure. Then expanding out (5.5) using

$$U_i(x) = e^{\epsilon g A_i^{ss}(x)}$$

we find (in the Landau gauge) that the corresponding continuum free

+For lattice dimensional renormalization see /15/, the graphwise subtraction scheme being that of /16/.

field propagator for is:

$$\left(\delta_{ij} - \frac{k_i k_j}{k^2}\right)\frac{1}{k^2\left(1+\frac{k^2}{\Lambda^2}\right)^2} \tag{5.6}$$

The corresponding continuum Gaussian measure (with covariance (5.6) is supported on $1-\delta, \delta>0,$ differentiable path with Höhler continuity of order $\alpha, 0<\alpha<1/2$. This function space (a Banach space) is denoted $C_{1-\delta,\alpha}(\Lambda)$ where Λ is a periodic box in $\mathbb{R}^4$. It is a nice "classical" function space. The fact that we have fixed a gauge means that we are (at least locally) on the space of gauge inequivalent potentials with the above topology.

We shall now indulge in a bit of continuum global analysis. Recall that in the continuum the gauge transformation group $\mathcal{G}$ is (essentially) Maps $(\Lambda, SU(N))$. Let $\mathcal{G}^0$ = Maps$_0$ $(\Lambda, SU(N))$ (the subset of maps such that $g(x_0) = e$, where x_0 is a point on the face of the box). $\mathcal{G}^0$ is a normal subgroup of $\mathcal{G}$ and $\mathcal{G}/\mathcal{G}^0 \approx SU(N)$ Let us assign the $C_{2-\delta,\alpha}(\Lambda)$ topology to these objects and call them $\mathcal{G}_{2-\delta,\alpha}$, $\mathcal{G}^0_{2-\delta,\alpha}$. Then by the methods of /17,18/ one can prove that these are well defined Banach Lie groups and $\mathcal{G}^0_{2-\delta,\alpha}$ is a closed subgroup of $\mathcal{G}_{2-\delta,\alpha}$.

Let $\mathcal{A}_{1-\delta,\alpha}$ be the space of continuum connections on Λ with the topology of the function space on which the Gaussian measure with covariance (5.2) is supported. Using the methods of /17,18/ one can prove that $\mathcal{A}_{1-\delta,\alpha}$ is a smooth nontrivial $\mathcal{G}^0_{2-\delta,\alpha}$ principal bundle on the orbit space $\mathcal{A}_{1-\delta,\alpha}/\mathcal{G}_{2-\delta,\alpha}$. This orbit space is a smooth paracompact Banach manifold. It has a very complicated topology. For example, its deRham cohomology groups $H^p(\mathcal{A}/\mathcal{G}^0)$ can be computed by the methods of /8/ (see also /18/). For simplicity let N = 2 and recall Λ is a 4-torus. Then the Poincaré series

$$P_t\left(\mathcal{A}/\mathcal{G}^0\right) = \sum_{p=0}^{\infty} t^p \dim H^p\left(\mathcal{A}/\mathcal{G}^0\right)$$

is given by

$$P_t\left(\mathcal{A}/\mathcal{G}^0\right) = \left(1+t^3\right)^4\left(1-t^2\right)^{-6}\left(1+t\right)^4 \tag{5.7}$$

which gives a very rich cohomology.+

Is this topology of any relevance++ to quantum continuum gauge field theory looked at from a global non-perturbative viewpoint? If it is, then the previous analysis shows that it should be associated with an effective "low frequency continuum gauge theory" with improved ultraviolet behavior. As a consequence the function spaces on which the effective "low frequency" measure is expected to be supported tend to become "classical" so that global analysis (infinite dimensional geometry and topology) might become applicable.

ACKNOWLEDGEMENTS

I thank James Glimm for stimulating conversations and collaboration which motivated me to return to the circle of ideas in /3,4 /. I thank Erhard Seiler for participation in some aspects of the project presented in these notes, although of course he is not responsible for any errors on my part. The interest of Krzysztof Gawedski and Gerhard Mack is gratefully acknowledged. Leonard Gross stimulated my interest in lattice differential geometry.

Over the years I profitted from the scientific advice and friendship of Kurt Symanzik. It is with great sadness that I dedicate these notes to his memory.

REFERENCES

1. K.G. Wilson, Phys. Rev. D10, 2445 (1974).
   __________, in:"New Developments in Quantum Field Theory and Statistical Mechanics," M. Lévy and P.K. Mitter, eds., Plenum Press, N.Y. (1977). K. Osterwalder, ibid.
   J. Drouffe and C. Itzykson, Phys. Rep. 38C (1978), 133.
   K. Osterwalder and E. Seiler, Ann. Phys. 110, 440 (1978).
   E. Seiler: Lecture notes in Physics No. 159 , Springer (1982).
2. K.G. Wilson and J. Kogut, Phys. Rep. C, 12 (1974), 75
   K.G. Wilson, Rev. Mod. Phys. 47 (1975), 773
   __________, Rev. Mod. Phys. 55 (1983), 583
   __________, in: "Recent Developments in Gauge Theories," G. 'tHooft et al, eds., Plenum Press, N.Y. (1980).
3. J.H. Lowenstein and P.K. Mitter, Ann. of Physics 105 (1977), 138.
   G. Valent, Thèse d'Etat (Université Paris VII, 1979).
4. P.K. Mitter, G. Valent, Phys. Lett. 70B (1977), 65.

+Construction of cocycles is due to I.M. Singer (see R. Stora, this volume) and M. Asorey (private communication) who has generalized the method of /18/ to higher cohomology groups.

++See R. Stora (this volume) and references therein (in particular to the work of Witten and Ramadas).

5. M. Göpfert and G. Mack, Comm. Math. Phys. 82 (1982), 545.
J. Frölich, Comm. Math. Phys. 47 (1976), 233.
D. Brydges, Comm. Math. Phys. 58 (1978), 313.
6. G. de Rham, "Variétés différentiables," Hermann, Paris (1960).
7. J.L. Koszul, Lectures on Fibre Bundles and Differential Geometry, Tata Institute of Fundamental Research, Bombay (1960).
S.S. Chern, "Complex Manifolds Without Potential Theory," D. Van Nostrand, Princeton, N.J. (1967).
8. M. Atiyah and R. Bott, Yang-Mills fields on Riemann surfaces (to be published).
9. J. Glimm and A. Jaffe, Comm. Math. Phys 56 (1977), 195.
L. Gross, Convergence of $U(1)_3$ lattice gauge theory to its continuum limit (to be published).
A. Guth, Phys. Rev. D21 (1980), 2291.
10. G. Mack, Dielectric lattice gauge theory (DESY 83-052), to appear in Nuclear Physics B.
11. B. Sharpe, Gribov copies and the Faddeev-Popov formula in lattice gauge theories (to be published in Nuclear Physics B).
12. P. Hirschfeld, Nucl. Phys. B 157 (1978), 37.
13. B.W. Lee and J. Zinn-Justin, Phys. Rev. D5 (1972), 3121.
14. K. Gawedski and A. Kupiainen, Comm. Math. Phys. 77 (1980), 31.
15. K. Symanzik, in: "Recent Developments in Gauge Theories," op.cit.
16. P. Breitenlohner and D. Maison, Comm. Math. Phys. 52 (1977), 11, 39, 55.
17. I.M. Singer, Comm. Math. Phys. 60 (1978), 7.
M.S. Narasimhan and T.R. Ramadas, Comm. Math. Phys. 67 (1979), 21.
P.K. Mitter and C.M. Viallet, ibid., 79 (1981), 457.
18. M. Asorey and P.K. Mitter, CERN/TH 3424.
19. G. Gallavotti et al, Comm. Math. Phys. 59 (1978), 143; 71 (1980), 95.

PROLEGOMENA TO ANY FUTURE COMPUTER EVALUATION OF THE QCD MASS SPECTRUM

Giorgio Parisi

Universita' di Roma II "Tor Vergata" 00173 Roma (Italy)
and
INFN-Laboratori Nazionali Frascati 00040 Frascati
(Italy)

I. INTRODUCTION

In recent years we have seen many computer based evaluations of the QCD mass spectrum[1]. At the present moment a reliable control of the systematic errors is not yet achieved; as far as the main sources of systematic errors are the non zero values of the lattice spacing and the finite size of the box, in which the hadrons are confined, we need to do extensive computations on lattices of different shapes in order to be able to extrapolate to zero lattice spacing and to infinite box. While it is necessary to go to larger lattices, we also need efficient algorithms in order to minimize the statistical and systematic errors and to decrease the CPU time (and the memory) used in the computation.

In these lectures the reader will find a review of the most common algorithms (with the exclusion of the application to gauge theories of the hopping parameter expansion in the form I have proposed[2]: it can be found in Montvay's contribution to this school); the weak points of the various algorithms are discussed and, when possible, the way to improve them is suggested.

For reader convenience the basic formulae are recalled in the second section; in section third we find a discussion of finite volume effects, while the effects of a finite lattice spacing are discussed in section fourth; some techniques for fighting against the statistical errors and the critical slowing down are found in section fifth and sixth respectively. Finally the conclusions are in sections seventh.

Although these last four years have seen a great improvement of the techniques, there is still a lot to do: it would be very nice if our collegues, which have never used a computer, would start to study these problems, not with the aim of doing simulations themselves but for finding theoretically the best way for performing simulations: the need of better algorithms cannot be overstimated.

II. BASIC FORMULAE

In this Section the basic formulae of Euclidean field theory are recalled.[3] Let us consider a four dimensional box of sides L^3xT (with periodic boundary conditions or any other kind of homogeneous boundary condition); in most of the cases we suppose $L \gg T$ and T so large that it can be practically considered to be infinite).

If only bosons are present, there is a probability measure $d\mu[A]$ (A being the generic Bosonic field) which is proportional to $\exp\{-S[A]\} d[A]$, $S[A]$ being the Euclidean action: it can be written as the integral of a local function.

If our Euclidean theory satisfies the Osterwalder Schrader axioms (which imply the existence of a corresponding Wightman type field theory in Minkowski space), we have that:

$$O_i(t)\ O_j(o) \;=\; \Sigma_n\ C_i^{(n)}\ C_j^{(n)}\ \exp\left(-E^{(n)}t\right) \qquad (1)$$

where the operators $O_i(t)$ are functionals depending only on the A field at time t, the E_n's are the energies of the states at rest (which are supposed to be discreate) in the Minkowski space in a box of side L^3. For most of the physical application we are interested to compute the E_n's in the limit $L \to \infty$, although interesting informations on the low energy hadron-hadron interaction may be obtained if we study the L dependence.

In presence of fermionic field ($S_F = \int d^Dx\, \bar{\psi}\Delta[A]\psi$) if C invariance is not violated, after the elimination of the Fermionic field by Gaussian integration, we remain with an effective probability distribution for the Bosonic field:

$$d\mu_F[A] = d\mu[A]\ \det[\Delta] \equiv d[A]\ \det[\Delta] \exp\left\{-S[A]\right\} \qquad (2)$$

The correlation functions of the fermionic field can be evaluated easyly using formulae like[3]:

$$\begin{aligned} &< \bar{\psi}(x)\gamma_5 \psi(x)\ \bar{\psi}(o)\gamma_5\ \psi(o)> = \int d\mu_F[A]\, |G(x,o|A)|^2 \\ &\Delta[A]\ G(x,y|A) = \delta(x-y) \end{aligned} \qquad (3)$$

Eq. 3 holds only if the bilinear operator $\bar{\psi}\gamma_5\psi$ is not a flavour singlet, otherwise a slightly more complex formula holds.

Although the "natural" formulation of fermions is done using anticommuting variables, only commuting quantities enter in Eqs. (2) and (3): it is possible to generate the bosonic field according to the probability distribution (2) by using a modified Monte Carlo method[4] and the Green function $G(x,y|A)$ can be analytically computed using a fast method for solving elliptic differential equations. In the so called quenched approximation the determinant in (3) is removed: this correspond to neglect virtual quarks loops.

This program may be implemented only by introducing in the space time a mesh of size a (i.e. we consider lattice field theory): on the top of the statistical errors common to any probability based computation we have systematic errors due to the non vanishing of L^{-1} and a. Although at the end we need to do an extrapolation by considering different values of a and L, it is convenient to use algorithms which have the smallest possible systematic errors; they will be described in the next section.

III. FINITE VOLUME EFFECTS

Le us discuss firstly finite volume effects in an SU(N) theory in the limit $N \to \infty$. In this case in a box of size $L^3 \times \infty$ two phases are possible:

a) $< P > = 0$
b) $< P > \neq 0$

where P is the trace of the Wilson loop of lenght L winding the box. It has been argued that for $L > L_c$ $< P > = 0$ and in this phase no physical quantity depends on L[5]. This result is confirmed by the explicite formulae for finite volume correction written in terms of the S matrix, if we use the conjecture that the S-matrix is the identity for an SU(∞) theory.

For infinite N, $L > L_c$ would be enough for killing all the finite volume corrections. For finite N we cannot have phase transitions in a finite box and L_c is not sharply defined, however we can speak of two different regimes a) and b). In regime a) the effect of finite volume corrections may be systematically evaluated by considering the effect of virtual particles winding through the box[6,7]; these effects are rather small for all the virtual particles but the pion: they are proportional (roughly speaking) to[6]

$$\exp(-mL) \tag{4}$$

Moreover the corrections to the meson spectrum are Zweig suppressed and have a small prefactor, unfortunately the corrections to the barion spectrum are not Zweig suppressed and may be relatively high.

For boxes of reasonable size (i.e. 2 Fermi) the only effects on the masses may come from virtual pion exchange; fortunately these exchange is suppresed due to the Goldstone nature of the pion (Adler zeros); the decoupling of the pion in this limit may be checked by computing the zero energy scattering lenght following Ref. 4.

At the present moment the most fashionable method for decreasing finite volume effects due to meson exchange consists in changing the masses of the quarks and to perform extrapolations to small masses[F1]. A more fancy possibility for reducing finite volume effects consists in playing with the boundary conditions.

In the rich men version we introduce an addictional U(9) smell group: quarks transform under the fundamental representation of the group and are fermion of parastatistic 1/9; physical objects are singlect under this group; we can now impose twisted boundary conditions in the U(9) directions[9], strongly reducing the finite volume effects. Obviously in the infinite volume limit we recover the original theory.

In the poor men version we introduce only an U(3) smell group and the boundary conditions are imposed using the following matrix diagonal in smell space[10]:

$$\begin{vmatrix} 1 & 0 & 0 \\ 1 & \exp(i2\pi/3) & 0 \\ 0 & 0 & \exp(i4\pi/3) \end{vmatrix} \qquad (5)$$

Different quarks with different smell get different phases at the boundary: for an SU(2) theory the same prescription correspond of imposing periodic and antiperiodic boundary conditions for the two smells respectively.

The poor men version kills only the leading terms $(\exp(-m_M L))$ leaving subleading terms $\exp(-m_M \sqrt{2}L)$, while the rich version kills all terms up to $\exp(-3m_M L)$ but costs more in CPU time and memory.

The poor man version is recommended in all cases where memory and not CPU is the limiting factor while the rich man version is compulsory for studing more subtle effects like the ρ width.

VI. THE LATTICE SPACING

In order to perform a simulation it is necessary to introduce a lattice spacing (let us call it a).

In a pure gauge theory it was proven by Symanzik[11] that the finite lattice spacing corrections are proportional to a^2 and it is possible to find out an action on the lattice such that these corrections are absent: in an asymptotically free theory the action may be computed in perturbation theory in the bare coupling constant. When fermions are present the corrections are proportional to a and are much more serious.

The advantages of using an improved action have been carefully investigated by Monte Carlo in the case of the two dimensional σ-models[12] and start to be investigated in the case of the lattice gauge theories.

In my opinion the effects of the improvement would be small for pure gauge theories (provided that we stay far away from the critical point in the β-fundamental β-adjoint plane[12]) while the improvement seem to be absolutely necessary for fermions where theeffects are much strongers; careful studied in this direction would be very interesting: the field is rapidly developing and it is difficult to provided an updated list of references.

A related subject which is not yet studied is how to improve the results of a simulation done in the Langevin approach[13] by trying to compensate the finite time step effects (in this context time is the fith dimensional computer time) with the finite lattice spacing effects. Let us consider a trivial case: massless free field theory in one dimension. In the continuum the Langevin equation is

$$\dot{\varphi} = \Delta\varphi + \eta \qquad \overline{\eta(x_1 t)\,\eta(y_1 t)} = 2\,\delta(x-y)\,\delta(t-t') \tag{6}$$

When we introduce a lattice spacing a and a time spacing ε the Langevin equation can be written as:

$$\begin{aligned} \tilde{\varphi}_n(i) &= \varphi_n(i) + R_n(i)\,\varepsilon^{\frac{1}{2}} \qquad \overline{R_n(i)R_m(j)} = \frac{2}{a}\,\delta_{i,j}\,\delta_{n,m} \\ \varphi_{n+1}(i) &= \tilde{\varphi}_n(i) + \varepsilon\left[\tilde{\varphi}_n(i+1) + \tilde{\varphi}_n(i-1) - 2\,\tilde{\varphi}_n(i)\right]/(2\,a^2) \end{aligned} \tag{7}$$

An easy computation in momentum space tell us that:

$$G(P) = \varphi(P)\varphi(-P) = \frac{1}{P^2 + (\frac{\varepsilon}{2} - \frac{a^2}{12})P^4 + O(P^6)} \tag{8}$$

at the magic value $\varepsilon = a^2/6$ the effects of order a^2 cancels with those of order ε. Independently of the possibility of cancelling errors of different origines, it is clear that if the continuum limit is done at a^2/ε constant, the final errors remain of order a^2. A more careful investigation of these problems would be welcome, also given the relevance of the Langevin equation for introducing Fermion loops.

V. FIGHTING AGAINST STATISTICAL ERRORS

Everyone would agree that if we want to measure something (e.g. $\langle O \rangle$ near the continuum limit) it is better to consider a quantity which has a definite probability distribution in the continuum limit, i.e. if (P(z) is the probability that O=Z, we would like that in the continuum limit the limiting probability $P_c(Z)$ exists and it is such that:

$$\int_{-\infty}^{+\infty} P_c(z)\, dz = 1 \tag{9}$$

As far as the statistical errors are proportional to $< 0^2 > - < 0 >^2$ we would like that the noise to signal ratio

$$R = \frac{N^2}{S^2} = \frac{< 0^2 >}{< 0 >^2} - 1 \tag{10}$$

remains finite in the continuum limit.

Unfortunately it is well known that if 0 is a local operator in more than one dimensions:

$$P_c(z) = o \tag{11}$$

Fields are not functions but distributions: only observable constructed with smeared fields have a well defined probability distribution.

If we use local operators to compute the masses the ratio noise to signal diverges in the continuum limit; e.g. if we measure the glueball mass by looking to the plaquette plaquette correlation function the noise to signal ratio diverges like a^{-16}, making the computation impossible for small a.

Generally speaking in order to compute masses it is convenient to consider observables which are functional of the field smeared in space and not in the time in order to preserve Eq. (1). This can be trivially done in a non gauge theory. In a gauge theory we have two options:

a) we fix the gauge (i.e. the Coulomb gauge[14]) and we smear the gauge variables in this gauge;

b) we use a gauge invariant construction for the block gauge field(e.g. the one proposed by Wilson restricted on a space slice). A very simple and efficient procedure is discussed in Ref. 15, as far as the computation of the glueball is concerned.

We notice en passant that all the informations on the low energy spectrum are conteined in the block fields and going from the original lattice to a new lattice with twice lattice spacing we loose only informations on the high energy part of the spectrum, which is not so important; however the number of variables will decrease of a factor 8-16. The variational approach for computing the eigenvalus can be done by starting directly from the block field configurations (reducing the amount of work needed if we consider many operators); moreover we could save on a tape only the block field configurations and the block quarks propagator, strongly reducing the input-output problem.

It is also possible to decrease the statistical error[16] by a careful use of the DLR equations in order to find a new observable O' such that $< O' > = < O >$ and $< (O')^2 > \ll < O^2 >$.

This procedure has been used for computing the string tension with rather good results. The most efficient way I can think to measure the sting tension using the DLR identities is the following: we consider a lattice L^3xT where all gauge fields at t=0 are equal to zero (U=1). The expectation value of the Poliakoff loop P in the x direction decrease for large L as

$$< P(t) > \simeq \exp(-KLt) \tag{12}$$

The appropriate DLR identity is:

$$< \Pi\, U_x > = \overline{\Pi \quad < U_x >_x} \tag{13}$$

where $< U_x >_x$ denotes the average over the links in the x direction and the bar the average over the other links. The implementation of the identity is rather simple: we start from independent gauge field configurations and we upgrade only the field in the x direction with the other fields quenched; in this way we compute an approximate expectation value of each link U_x and the Poliakoff loop is computed as the product of all the approximated expectation values. At the end we average the configurations of the gauge fields. This procedure may be also used to compute the correlation function of two Poliakoff loops at distance larger than 1 if the slice at t=0 is not cold (i.e. A=0) but it is at thermal equilibium and we quench also the values of the U_x's on the two slices (i.e. at t=0 and t=T/2). This procedure should be used only if one is interested to compute the potential, not only the string tension; indeed it is well known that unless special techniques are used the best way for computing the exponential decay of correlation functions consists in measuring the responce function[17].

VI. FIGHTING AGAINST THE CRITICAL SLOWING DOWN

From the lattice point view the continuum limit is a second order phase transition at which the coherence lengh becomes infinite. For a theory whose Lagrangian is quadratic in the momentum it is well known that the dynamics of the low frequency modes slows down of a factor a^2, i.e. the number of Monte Carlo interaction needed to produce independent configurations increase as a^{-2}, neglecting logarithms. The same result follows in the Langevin approach where the time step must be of order a^2. Pictorial we could say that the information diffuses on the lattice (or makes a random walk) so that we need a time proportional to ℓ^2 in order to change the block variable associated to a region, of size ℓ, in lattice spacing units.

This slowing down should be avoided as far as possible: at my knowledge there are two methods which solve this problem, the FFT preconditioned Langevin equation and the Multi Grid Monte Carlo.

Let us first describe the Langevin approach for a scalar field theory with action

$$S\,[\varphi] = \int d^d x\ \frac{1}{2}\left[(\partial_\mu\ \varphi\ )^2 + V(\varphi)\right] \tag{14}$$

As we have already mentioned it is well known that:

$$< 0\,[\varphi] > = \overline{0\,[\varphi]} \tag{15}$$

where as usual the bar denotes the average over the noise and φ is the solution of the Langevin equation:

$$\dot{\varphi} = \Delta\varphi - V'(\ \varphi) + \eta \quad \eta(x\ t)\ \eta(y\ t') = 2\delta(t-t')\delta(x-y) \tag{16}$$

Now we would like to increase the speed of the slow variables at low momenta; a very interesting theorem tell us that equation (15) still holds if we consider the generalized Langevin equation:

$$\begin{aligned} \dot{\varphi}(x) &= \int dy\ Q(x-y)\ \left[\Delta(y) - V'(\varphi(y))\right] + \eta_Q(x) \\ &\eta_Q(x\ t)\ \eta_Q(y\ t') = 2\delta\ (t-t')\ Q(x-y) \end{aligned} \tag{17}$$

where the kernel Q has been chosen in such a way to have a fast speed up the low momentum region: e.g. $Q(x) = dk\int \exp(ikx)\ (k^2+m^2)^{-1}$.

The computation of the convolution can be done in momentum space and the Fourier transform can be done using the Fast Fourier Transform algorithm which is quite efficient (the slowing down is now proportional to ln a). It looks like that the best choice of Q(x) consists in taking

$$Q(x) \simeq < \varphi(x)\,\varphi(0) > \tag{18}$$

however this question should be investigated in a more detailed way.

The same technique may be used for gauge theories (in the Landau gauge it is likely that Q(x) should behaves like $\exp(-x^2)$ at large distance) if we remove the constraint that the U variables belong to the gauge group and we replace it with an appropriate weight in the action; on the other hand this method could be well used for Fermions both in the pseudofermions approach and in the computation of the Green function, using the Gauss Seidel or the conjugate gradient methods.

The basic idea of the multigrad Monte Carlo approach consists in introducing a varaible for each field and also for each block field. In lattice of size $L=2^n$ we can write[18]

$$\varphi(i) = \sum_{0}^{n} {}_k \quad \sum_{j} c_{i,j}^{(k)} \quad \varphi^{(k)} \ (j),$$

where each component of the the index j runs from 1 to 2^{n-k} and the $c_i^{(k)}$ are appropriate constant.

We can now perform a Monte Carlo simulation on all the variables together; also in absence of tricks for fastening the computation, the total time for a multigrad Monte Carlo cycle will be at worse proportional to n, i.e. the algorithms is slow only of a factor ln a. The benefit of the multigrad algorithm is that the efficiency for changing the low momentum variables should be quite high and no critical slowing down should be present. This may happens if the constants C(k) are chosen in the appropriate way: simple minded arguments suggest that for an action with two derivatives the c's must be linear in j while for the action with only one derivative the c's can be constant inside each block.

The method should be particularity efficient for quadratic actions: (the CPU time for a multigrid sweep is similar to the one for an usual sweep) expecially for the computation of the Fermionic propagator which is a first order differential equation. On the contrary the application of the multigrid method to the gauge field sector seems to be particularily paintfull but it may be rewarding.

It could be very interesting to see how these methods work in a concrete case.

VII. CONCLUSIONS

I believe that in the future the methods described in the two last sections will lead to a strong reduction of the CPU time needed to perform a computation. The use of block fields for constructing the observable decreases the statistical errors; moreover as far as the block fields should excite from the vacuum only low energy particles Eq. (1) will be dominated by the lowest energy state also at moderate values of t, allowing therefore an unbiased determination of the masses of the lowing lying states. On the other hand the multigrid method should strongly fasten the slow part of the computation, i.e. the treatment of the Fermions.

It seems that we have at our disposal all ingredients for performing a successful computation of the hadronic spectrum keeping under control the systematic errors.

At the present moment, on most of the super computers the main limitation seems to be lack of memory space; the trend is reversing: new generation supercomputers will have a reasonable amount of memory (may be not enough). With the advent of 256 kb chips also dedicated computers may be equipped with a sufficient large amount of memory. A serious problem that slows down the progress in this field is the difficulty in writing down the computer code which implements the various algorithms, i.e. the software problem. This implements the various algorithms, i.e. the software problem. This problem becomes stronger if we want to write down efficient codes for pipelined or parallel machines. A possible way to overcome these difficulties is discussed in Wilson's contribution to this school.

Of course the final solution for decreasing the memory space (may be not the CPU time) is to push Symansik's improvement program or/and Wilson's renormalization group approch. This may be crucial expecially for the properties of the pion in a realistic simulation in order of not upsetting the cancellation of virtual pion exchange due to the quasi Goldstone nature of the pion.

ACKNOWLEDGMENTS

It is a pleasure to thank M. Falcioni, E. Marinari, G. Martinelli, M. Paciello, R. Petronzio, F. Rapuano, B. Taglienti, Y.C. Zhang for many interesting discussions.

I am also grateful to H. Hamber and K.G. Wilson for useful discussions and suggestions concerning the multigrid method.

REFERENCES

1. See for example H. Hamber, Proc. of the Intern. Conf. on "Matematical Physics", Boulder (1983), to be published; G. Parisi, Proc. Summer School Cargese (1983), to be published.
2. G. Parisi, Nuclear Phys. B205, /FS 5/, 337 (1982).
3. H. Hamber and G. Parisi, Phys. Rev. Letters 47, 1972 (1981); E. Marinari, G. Parisi and C. Rebbi, Phys. Rev. Letters 47, 1975 (1981).
4. F. Fucito, E. Marinari, G. Parisi and C. Rebbi, Nuclear Phys. B180, /FS 2/, 369 (1981); H. Hamber, E. Marinari, G. Parisi and C. Rebbi, Nuclear Phys. to be published and references therein.
5. T. Eguchi and H. Kawai, Phys. Rev. Letters 48, 1063 (1982); G. Bhanot, U. Heller and H. Neuberger, Phys. Letters 113B, 47 (1982); G. Parisi, Phys. Letters 112B, 463 (1982); G. Parisi and Y.C. Zhang, Nuclear Phys. 215, /FS 4/, 182 (1983).
6. H. Hamber and G. Parisi, Phys. Rev. D27, 3215 (1982).
7. See H. Luscher contribution to this school.
8. For a recent review see J. Gasser and H. Leutvyler, Phys. Reports, in press.
9. T. Eguchi, J. Jurkiewicz and C.P. Korthals Altes, in Proc. Workshop Word Scientific (1982).
10. G. Martinelli, G. Parisi, R. Petronzio and F. Rapuano, Phys. Letters 122B, 283 (1983).
11. K. Symanzik in "Methematical Problems in Theoretical Physics" eds. R. Schrader et al. Lecture Notes in Physics 153, (Springer, Berlin, 1982).
12. M. Falcioni, G. Martinelli, M.L. Paciello, G. Parisi, B. Taglienti, Nuclear Phys. B225, /FS 9/, 313 (1983); B. Berg, S. Meyer, I. Montvay and K. Symanzik, Phys. Letters 126B, 467 (1983).

13. G. Parisi and Y.S. Wu, Scientia Sinica 24, 483 (1981); M. Falcioni, E. Marinari, M.L. Paciello, G. Parisi, B. Taglienti and Y.S. Zhang, Nuclear Phys. B215, /FS 7/, 265 (1983).
14. G. Parisi, R. Petronzio, R. Rapuano, in preparation.
15. M. Falcioni, G. Parisi, M. Paciello and B. Taglienti, in preparation.
16. G. Parisi, R. Petronzio and F. Rapuano, Phys. Letters 128B, 418 (1983).
17. For a discussion of this point see G. Parisi in the Proc. of the Workshop "Word Scientific", Trieste (1982).
18. For a review of the multigrid method see A. Brandt and N. Dinar in "Numerical Methods for Partial Differential Equations" ed. by S.V. Parter, (Academic Press 1979).

FOOTNOTES

F1. It is well know that in reality the masses of the barions and the mesons contain non analytic terms like $m_q^{3/2}$ and $m_q^2 \ln m_q$ (m_q being the quark mass) with computable coefficients; moreover there are indications from the experimental value of the sigma term[8] in pion proton scattering that strong non linearities in the barionic masses are present when the quark mass changes from the up to the strange mass. These terms should be practically absent in the quenched approximation: they are Zweigh suppressed: similar arguments suggest that the nucleon mass may be about 1100 MeV in the quenched approximation, i.e. it should be similar to the purified mass of Ref. 8.

F2. The observed "scaling" behaviour of the gluball mass near 5.6 implies that these data are useless to conclude something on the continuum limit: indeed both the string tension and the deconfinement critical temperature do not scale in this region and the scalar glueball mass is the most sensitive quantity to the existence of the nearby critical point: it seems likely the observed "scaling" of the glueball is simple a coincidence without any deep meaning.

ALGEBRAIC STRUCTURE AND TOPOLOGICAL ORIGIN OF ANOMALIES

Raymond Stora

L.A.P.P.
B.P. 909
74019 Annecy-le-Vieux Cedex, France

INTRODUCTION

Although the subject of this seminar is not central in the present activity in gauge theories, mostly concerned with non perturbative aspects, a recent peak in the published literature on this old subject makes it worthwhile to review some of the methods which have emerged slowly since 1976.

Whereas the abelian anomaly was first discovered by J. Steinberger, J. Schwinger, S. Adler, J.S. Bell, R. Jackiw[1], we shall be mostly concerned with the non abelian Adler Bardeen[2] anomaly and the related Wess Zumino[3] Lagrangian, but the same techniques apply to the gravitational anomalies recently discovered by L. Alvarez Gaumé and E. Witten[4].

The whole machinery relies on two items: one has to do with an extensive use of the Faddeev Popov ghost whose interpretation as the Maurer Cartan form of the gauge group is both natural and effective[5]. The other item is what we shall call the Cartan[6] homotopy formula. These items take care of the algebra which was elaborated in collaboration with B. Zumino[7]. An exposition of this can already be found in B. Zumino's lectures in Les Houches, this summer[8].

The topological origin of anomolies has been investigated by I.M. Singer[9], M.F. Atiyah and I.M. Singer[9], following work by M.F, Atiyah and R. Bott[9] on two dimensional guage theories. The conclusion is:

Seminar given at the Cargese Institute of theoretical physics on "Progress in gauge field theory", Sept. 1 - Sept. 15 1983.

the Adler Bardeen anomaly is a reflection of the non vanishing of the second cohomology of orbit space. On the other hand, T.R. Ramadas[9] has found another way to spot the topological origin of this anomaly and shown how to relate it to the second cohomology of the guage group in three dimensions (fixed time).

These notes are organized as follows. Section II is devoted to the introduction of Faddeev Popov ghosts and the Slavnov operation in the expression of the Wess Zumino consistency condition. The "Russian formula" is then established and shown to provide solutions of the desired type - all solutions in four dimensional renormalizable theories[10]-.

Section III is devoted to the derivation of the Cartan homotopy formula[6] in the general non commutative form given by B. Zumino[8]. The Wess Zumino lagrangian is then exhibited by a simple application of the formula.

Section IV provides a description of the Alvarez Gaumé - Witten[4] gravitational anomalies as an application of the above mentioned formulae which was carried out by T. Schucker, F. Langouche and myself[11] after the school was over.

Section V gives a sketchy view of Ramadas and Atiyah-Singer's constructions.

Acknowledgements

It is a pleasure to thank I.M. Singer and B. Zumino for communicating their unpublished work through intense correspondence to which these notes owe must of the material; T.R. Ramadas for informing P.K. Mitter and myself of his results; A. Haefliger, S.S. Chern I.M. Singer for kindly answering my inquiry on the Cartan homotopy formula. I also wish to thank L. Alvarez Gaumé for a discussion on the gravitational anomaly and for keeping me informed on this work, as well as E. Witten for correspondence on this subject. I have found it difficult to quote properly all contributors to this subject since 1976: it is my impression that some of the crucial remarks have been made independently and almost simultaneously by several authors or groups of authors either in published or in unpublished form. I therefore wish to apologize to the many authors whom I have not properly quoted. I would also like to thank in chronological order J. Dixon, J. Thierry Mieg, P. Cotta Ramusino, M. Tonin, L. Bonora, L. Baulieu with whom I had the pleasure to discuss some of the ideas involved here, at various stages of their developments[12].

II. SOLUTIONS OF THE WESS ZUMINO CONSISTENCY CONDITIONS IN TERMS OF FADDEEV POPOV GHOSTS

Consider, to be specific, some renormalizable σ-type model, with eventual symmetry breaking of an internal compact group G, assumed to be simple until otherwise specified. Current algebra can be defined by coupling the theory to an external Yang Mills field

$$a = \sum_{\alpha,\mu} a_\mu^\alpha e_\alpha dx^\mu \tag{1}$$

The corresponding Ward identity reads

$$W(\omega)\Gamma(\phi,a) = \Delta(\omega,a) = \int \mathcal{O}\!\mathcal{L}(\omega,a) \tag{2}$$

where Γ is the vertex functional involving the matter fields and symmetry breaking fields collectively denoted ϕ; $W(\omega)$ is the functional differential operator realizing the infinitesimal gauge transformation parametrized by $\omega(x) = \Sigma\omega^\alpha(x)e_\alpha$; $\Delta(\omega,a)$ is the Adler Bardeen anomaly, the integral of a 4-form $\mathcal{O}\!\mathcal{L}(\omega,a)$ linear in ω, local in a and ω, defined up to d (local form).

The Wess Zumino consistency condition

$$W(\omega)\Delta(\omega',a) - W(\omega')\Delta(\omega,a) = \Delta([\omega,\omega'],a) \tag{3}$$

is a mere consequence of the commutation relations

$$[W(\omega), W(\omega')] \equiv W(\omega)W(\omega') - W(\omega')W(\omega) = W([\omega,\omega']) \tag{4}$$

where

$$[\omega,\omega']^\alpha(x) = f^\alpha{}_{\beta\gamma}\,\omega^\beta(x)\,\omega^\gamma(x)$$

where the f's are the structure constants of Lie G, the Lie algebra of G.

It is convenient to introduce the Faddeev Popov ghost, which we shall still denote ω and the Slavnov operation s defined by

$$\begin{aligned} sa &= -D\omega \equiv -(d\omega + [a,\omega]) \\ s\omega &= -\tfrac{1}{2}[\omega,\omega] \end{aligned} \tag{5}$$

so that the W.Z. consistency condition can be rewritten as

$$s\,\Delta(\omega,a) = 0 \tag{6}$$

[s is defined as an antiderivation anticommuting with the exterior differential d and odd differential forms].

From the algebraic Poincaré lemma[13], it follows that

$$s\mathcal{A}(\omega,a) = -dQ_3^2 \tag{7}$$

where Q_3^2 is a 3-form with $\phi\pi$ ghost number 2. Solutions of the W.Z. consistency condition are now easy to obtain: let $\mathcal{I}_3(.,.,.)$ be a third rank symmetric tensor on (Lie G), invariant under Ad G; the classical "transgression formula"[14]

$$\mathcal{I}_3(F,F,F) = 3d\int_0^1 dt\, \mathcal{I}_3(a, F_t, F_t) \tag{8}$$

where

$$F = da + \frac{1}{2}[a,a] = \frac{1}{2}\sum_\alpha F_{\mu\nu}^\alpha e_\alpha\, dx^\mu \wedge dx^\nu \tag{9}$$

$$F_t = t\,da + \frac{t^2}{2}[a,a] = tF + \frac{t^2-t}{2}[a,a] \tag{10}$$

can be used as follows: Let

$$A = a+\omega \quad , \quad D = d+s \tag{11}$$

It is easy to see that, as a consequence of the definition of s

$$F \equiv DA + \frac{1}{2}[A,A] \tag{12}$$

so that

$$\mathcal{I}_3(F,F,F) = 3D\int_0^1 dt\, \mathcal{I}_3(A,\mathcal{F}_t,\mathcal{F}_t) \equiv DQ_5 \tag{13}$$

where

$$\mathcal{F}_t = t\,DA + \frac{t^2}{2}[A,A] \tag{14}$$

(The algebraic properties needed to prove the formula will be exhibited in section III). Expanding the 5-form Q_5 according to the lower degree referring to space time and the upper degree, which indicates the $\phi\pi$ ghost number (degree in ω):

$$Q_5 = Q_5^0 + Q_4^1 + Q_3^2 + Q_2^3 + Q_1^4 + Q_0^5 \tag{15}$$

we get:

$$\begin{gathered} \mathcal{I}_3(F,F,F) = dQ_5^0 \\ sQ_5^0 + dQ_4^1 = 0 \;,\; sQ_4^1 + dQ_3^2 = 0\,,\ldots \end{gathered} \tag{16}$$

The last equation shows that Q_4^1 solves the W.Z. consistency condition. It provides the answer in the case of four dimensional renormalizable theories[10,13], up to a numerical factor which can be evaluated by computing a convergent Feynman diagram and is well known to be non vanishing only in the case where chiral fermions are coupled to a. Of course, this statement needs to be qualified since the solution is always ambiguous up to terms of the form $s\,Q_4^0$, where Q_4^0 is a local four form which may be added to or substracted from the Lagrangian density, at will.

For higher dimensional theories, in even dimension $d = 2n-2$, the same method works provided there is a symmetric invariant polynomial of degree n on (Lie G) $\mathfrak{I}_n$, but, in this case, there is no known uniqueness theorem.

The formula is then

$$\mathfrak{I}_n(F,F,F) = n\,D\int_0^1 dt\, \mathfrak{I}_n(A, \mathcal{F}_t, \ldots \mathcal{F}_t) = D\,Q_{2n-1} \qquad (17)$$

and the anomaly is Q^1_{2n-2}, in self understood notation.

Remarks

i) Q_5 has been called a five form, following the interpretation of ω as the Maurer Cartan form of the gauge group $(x \to g(x) \in G)$, an interpretation which has slowly become more and more accurate. In fact[8] an alternative version of Eq. 11 is obtained by replacing $A = a+\omega$ by $A_g = {}^g a + g^{-1}\delta g$ where ${}^g a = g^{-1}ag + g^{-1}dg$ is the gauge transformed of a by g, δ the differential on the gauge group, and interpreting s as δ: the algebra is the same. This was first suggested by using the version of (11) "on the bundle", in the Neeman Regge Thierry Mieg scheme[15] (replace a by $g^{-1}ag + g^{-1}dg$ where (x,g) are the bundle coordinates) in spite of the difficulties attached to this scheme. A superspace version of Eq. 11 was given by the Padua group[16], leading to analogous results but I believe that "anticommuting" parameters are not essential to deal with differential forms.

ii) Formula (13) will be called the "Russian formula" because it seems to exist in the russian literature where I was however unable to spot it.

III. THE CARTAN HOMOTOPY FORMULA AND THE WESS ZUMINO LAGRANGIAN

Formula (8) can be generalized as follows: Let P be a possibly non commutative polynomial in a, F (with values in the enveloping algebra of Lie G); a_1, a_2 two Yang Mills fields. The Cartan formula is

$$P(a_1; F_1) - P(a_2, F_2) = (k_{12} d + d k_{12}) P \tag{18}$$

where k_{12} is the homotopy operator - an antiderivation of degree -1,

$$k_{12} P = \sum_F (-)^{\sigma(F)} \int_0^1 dt\, P(a_t; F \to a_1 - a_2, F) \tag{19}$$

where

$$a_t = t a_1 + (1-t) a_2 \quad , \quad F_t = d a_t + \frac{1}{2} [a_t, a_t]$$

and the symbol Σ indicates that one should sum over all possibilities to replace one F by $a_1 - a_2$ and simultaneously all others by F_t, whereas the sign $(-)^{\sigma(F)}$ takes care of the antiderivation character of k_{12}. The extremely simple proof given by B. Zumino[8] goes as follows: Pick an interpolating field

$$a(t) \; , \; 0 \leq t \leq 1, \quad \text{with} \quad a(0) = a_1 \; , \; a(1) = a_2 \tag{20}$$

e.g.

$$a(t) = t a_1 + (1-t) a_2 \tag{21}$$

Define on the algebra generated by a_t, F_t the antiderivation k_t by:

$$k_t a_t = 0 \qquad\qquad k_t F_t = \frac{d a(t)}{dt} \tag{22}$$

Then it is trivial to verify that

$$\frac{d}{dt} = k_t d + d k_t \tag{23}$$

using the formula for F_t, and the Bianchi identity

$$d F_t + [a_t, F_t] = 0 \tag{24}$$

which follows from

$$d^2 = 0 \tag{25}$$

The final formula follows by integration over t from 0 to 1.

It is now easy to construct the Wess Zumino Lagrangian, namely the local action $\Gamma_{W.Z.}(g,a)$ where g is a chiral field with values in G (i.e, an element of the guage group), such that

$$W(\omega)\,\Gamma_{W.Z.}(g,a) = \Delta(\omega,a) \tag{26}$$

for the gauge group action

$$\begin{aligned} a &\to {}^{\gamma}a = \gamma^{-1} a \gamma + \gamma^{-1} d\gamma \\ g &\to \gamma^{-1} g \end{aligned} \tag{27}$$

The trick is that used by E. Witten[17]: g maps space time, which we take as euclidean, compact, e.g. S_4 into G. Let $g(S_5^+)$ be a half sphere, with boundary $g(S_4)$, in G (see Fig. 1).

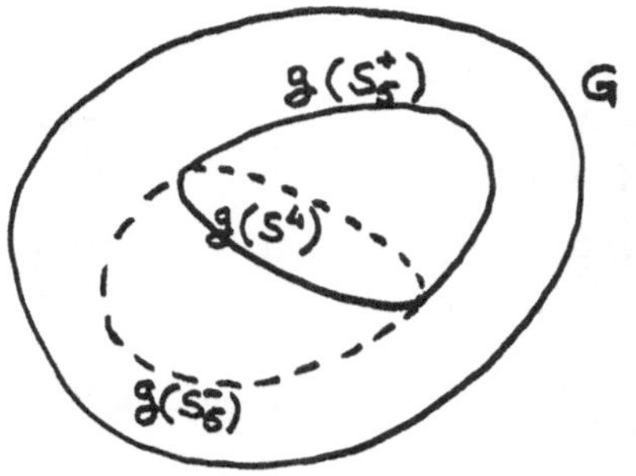

Fig. 1

(g is continuable to $g(S_5^+)$ if $\pi_4(G) = 0$). Then, we claim that

$$\Gamma_{W.Z.}(g,a) = \int_{g(S_5^+)} \left[Q_5^0({}^{g}a) - Q_5^0(a) \right] \tag{28}$$

Indeed, the first term is gauge invariant, in view of Eq. 27. Next we know

$$W(\omega)\, Q_5^0(a) \equiv s\, Q_5^0(a) = -d\, Q_4^1(\omega,a) \tag{29}$$

Hence

$$s\,\Gamma_{W.Z.}(g,a) = s\int_{g(S_5^+)} \left(-Q_5^0(a)\right) = \int_{g(S^4)} Q_4^1(\omega,a) = \Delta(\omega,a) \tag{30}$$

There remains to transform $\Gamma_{W.Z.}$ in terms of four dimensional integrals, which is done by means of the Cartan formula:

$$Q_5^0(a_1;F_1) - Q_5^0(a_2;F_2) = (k_{12}d + dk_{12})Q_5^0$$
$$Q_5^0({}^g a_1;{}^g F_1) - Q_5^0({}^g a_2;{}^g F_2) = ({}^g k_{12}d + d\,{}^g k_{12})Q_5^0 \tag{31}$$

Now,

$$dQ_5^0 = \mathcal{J}_3(F,F,F) \tag{32}$$

and, by writing out k_{12}, ${}^g k_{12}$ and using the invariance of $\mathcal{J}_3$ one can see that

$$k_{12}\mathcal{J}_3 = {}^g k_{12}\mathcal{J}_3 \tag{33}$$

Hence

$$Q_5^0({}^g a_1;{}^g F_1) - Q_5^0(a_1;F_1) = Q_5^0({}^g a_2;{}^g F_2) - Q_5^0(a_2;F_2) + d({}^g k_{12}Q_5^0 - k_{12}Q_5^0) \tag{34}$$

Now put $a_1 = a$, $a_2 = 0$. We get[17,18]

$$T_{W.Z.}(g,a) = \int_{g(S_5^+)} Q_5^0(g^{-1}dg,0) + \int_{g(S_4)} ({}^g k_{12} - k_{12})Q_5^0 \tag{35}$$

The first term, which describes pure meson couplings is the Witten[17] Lagrangian:

$$Q_5^0(g^{-1}dg,0) \propto \mathcal{J}_3(g^{-1}dg, d(g^{-1}dg), d(g^{-1}dg)) \tag{36}$$

$dQ_5^0 = 0$ from the Jacobi identity and the invariance of $\mathcal{J}_3$. Hence one may write locally (e.g. on $g(S_5^+)$ which is contractible to 0)

$$Q_5^0(g^{-1}dg,0) = d\chi_4 \tag{37}$$

according to the usual Poincaré lemma on $g(S_5^+)$. On the other hand the second term of eq.(35) is polynomial in a and g and describes the couplings of mesons and external gauge fields.

So, finally

$$\begin{aligned}\Gamma_{W.Z.}(g,a) &= \int_{S_4} \chi_4 + \int_{S_4} ({}^g k_{12} - k_{12})\, Q_5^0 \\ &= \Gamma_{\text{Witten}}(g) + \tilde{\Gamma}_{W.Z.}(g,a)\end{aligned} \tag{38}$$

with

$$\begin{aligned}\tilde{\Gamma}_{W.Z.}(g,0) &= 0 \\ \Gamma_{W.Z.}(g,0) &= \Gamma_{\text{Witten}}(g)\end{aligned} \tag{39}$$

There remains now to extend this analysis to the case of a compact, not necessarily simple group G (a product of simple factors and U(1)'s). The various components of the anomaly are still parametrized by third rank symmetric invariant tensors on the full Lie algebra. Note that factorizable anomalies may occur in the case where U(1) subgroups are involved e.g. the product of a 0-dimensional anomaly $\omega^{ab.}$ (coming from $F^{ab.}$) with a four dimensional invariant belonging to a simple factor (Killing (F,F)). Let K be the subgroup of G along which the anomaly vanishes i.e. is the gauge variation of a local functional. This is the situation in the chiral case $G = G_R \times G_L$ with vanishing anomaly on the diagonal subgroup $G_{diag} = \{g_L = g_R = g\}$.

First note that a given invariant polynomial on Lie G, even if it is factorizable gives rise to one and only one anomaly. Indeed, if

$$\begin{aligned}&\mathcal{J}(F,F,F) = (d+s)Q_5 = (d+s)Q'_5\,, \\ &d(Q_5^0 - Q_5'^0) = 0 \;\rightarrow\; Q_5^0 - Q_5'^0 = d\chi \\ &d(Q_4^1 - Q_4'^1) = -s(Q_5^0 - Q_5'^0) = -s\,d\chi \\ &\Rightarrow\; Q_4^1 - Q_4'^1 = s\chi + d\lambda\end{aligned} \tag{40}$$

where χ and λ are local expressions in a,ω. We may thus choose Q_5 in its canonical form (Eq. 13). Let us then decompose a along Lie K and an orthogonal complement stable under K :

$$a = a_{|K} + a_{\perp K} \tag{41}$$

(e.g. $a_{\substack{R\\L}} = V \pm A$, in the chiral case).

Then we have

$$\mathcal{Q}_5\left(a+\omega;\, F(a)\right) = \mathcal{Q}_5\left(a+\omega;\, \mathcal{F}(a+\omega)\right) - \mathcal{Q}_5\left(a_{|K}+\omega_{|K};\, \mathcal{F}(a_{|K}+\omega_{|K})\right) + \mathcal{Q}_5\left(a_{|K}+\omega_{|K};\, \mathcal{F}(a_{|K}+\omega_{|K})\right) \tag{42}$$

Now

$$\mathcal{Q}_5\left(a+\omega;\mathcal{F}(a+\omega)\right) - \mathcal{Q}_5\left(a_{|K}+\omega_{|K};\, \mathcal{F}(a_{|K}+\omega_{|K})\right) = \left[k_K\,(d+s) + (d+s)\,k_K \right] \mathcal{Q}_5 \tag{43}$$

Since

$$(d+s)\,\mathcal{Q}_5 = \mathcal{J} \tag{44}$$

and

$$k_K \mathcal{J} = 3\int_0^1 dt\; \mathcal{J}\left(a_{\perp K}+\omega_{\perp K},\, \mathcal{F}_t \cdot\cdot\right) \tag{45}$$

does not contain $\omega_{|K}$, introducing into the action the counter term

$$\int \Delta_1 \mathcal{L} = \int_{S_4} k_K \mathcal{Q}_5 \tag{46}$$

transforms the anomaly into

$$k_K \mathcal{J} + \mathcal{Q}_5\left(a_{|K}+\omega_{|K},\, \mathcal{F}(a_{|K}+\omega_{|K})\right) \tag{47}$$

If we now assume that the invariant $\mathcal{J}$ is such that

$$\mathcal{J}\left(F(a_{|K})\right) = 0 = \mathcal{Q}_5\left(a_{|K}+\omega_{|K},\, F(a_{|K})\right) \tag{48}$$

the anomaly vanishes along K and its component $\perp$ K is given by Eq. (47).

Once the anomaly has been made to vanish along K, the WZW

action is invariant under the corresponding gauge transformations and one may choose a gauge to parametrize G/K and restrict the chiral fields g_α to representatives of the coset space (e.g., in the previous example, choose the Wess Zumino gauge $g_L = g_R^{-1} = \dot{g}$).

We have envisaged one situation where anomalies cancel. It would actually be nice to know necessary and sufficient conditions for the vanishing of anomalies[13].

Before ending this sketchy description of the situation, let us recall again that, in general, anomalies may arise from all invariants of Lie G, including factorizable ones, i.e. the anomaly may come out as a product of some invariant with a lower dimensional anomaly (for instance if $G = U(1) \times SU2$ or $G = U(1) \times U(1)$, the $U(1)$ anomaly is of the form $\omega_{U(1)}(\mathrm{tr})FF$ where ω is the 0-dimensional form derived from $F_{U(1)}$ by the general procedure).

IV GRAVITATIONAL ANOMALIES[18]

Consider e.g. chiral fermions in d dimensions coupled to an external gravitational field represented by a vielbein form $\{e^a\}$ $a = 1,\dots,d$ and a spin connection form $\omega^a{}_b$. The base manifold is supposed to be compact and the internal metric $\{\eta_{ab}\}$ to be positive definite. The classical action is invariant under local orthogonal transformations as well as field transformations induced by diffeomorphisms of the base manifold. The connected vacuum functional may however suffer from anomalies

$$\begin{aligned} W(\Omega)\Gamma(e,\omega) &= \Delta(\Omega;e,\omega) = \int \alpha(\Omega;e,\omega) \\ W(\xi)\Gamma(e,\omega) &= \Delta(\xi;e,\omega) = \int \alpha(\xi;e,\omega) \end{aligned} \tag{49}$$

with

$$\begin{aligned} W(\Omega) &= \int \delta_\Omega e \frac{\delta}{\delta e} + \delta_\Omega \omega \frac{\delta}{\delta \omega} \\ W(\xi) &= \int \delta_\xi e \frac{\delta}{\delta e} + \delta_\xi \omega \frac{\delta}{\delta \omega} \end{aligned} \tag{50}$$

where

$$\begin{aligned} \delta_\Omega e &= -\Omega e \\ \delta_\Omega \omega &= d\Omega + [\omega, \Omega] = D\Omega \end{aligned} \tag{51}$$

represent the orthogonal gauge Lie algebra.

$$\begin{aligned} \delta_\xi e &= -L_\xi e = -[i(\xi)d + d\,i(\xi)]e \\ \delta_\xi \omega &= -L_\xi \omega = -[i(\xi)d + d\,i(\xi)]\omega \end{aligned} \tag{52}$$

are a possible choice of the action of infinitesimal diffeormorphisms of the base (vector fields $\xi = \xi^\mu\, \partial/\partial x^\mu$) through the action of the Lie derivative L_ξ ($i(\xi)$ denotes the innerproduct of a form with - or evaluation of a form on - the vector field ξ). The action of diffeomorphisms is defined up to gauge transformation e.g. we could have defined $\delta_\xi^{cov} = \delta_\xi + [i(\xi)\omega]$ where $[i(\xi)\omega]$ is an infinitesimal gauge transformation of parameter $\Omega = i(\xi)\omega$, but this choice which leads to nicer formulae has also some disadvantages[11] on which we shall not insist here, and we shall stick to the ordinary Lie derivative.

The corresponding commutation relations are easily derived:

$$\begin{aligned} [W(\Omega), W(\Omega')] &= W([\Omega, \Omega']) \\ [W(\xi), W(\Omega)] &= -W(L_\xi \Omega) \\ [W(\xi), W(\xi')] &= W([\xi, \xi']) \end{aligned} \tag{53}$$

where $[\xi, \xi']$ denotes the Lie bracket. The Wess Zumino consistency relations follow from application to Γ:

$$\begin{aligned} W(\Omega)\alpha(\Omega') - W(\Omega')\alpha(\Omega) - \alpha([\Omega, \Omega']) &= d(\text{local}) \\ W(\xi)\alpha(\Omega) - W(\Omega)\alpha(\xi) + \alpha(L_\xi \Omega) &= \quad '' \\ W(\xi)\alpha(\xi') - W(\xi')\alpha(\xi) - \alpha([\xi, \xi']) &= \quad '' \end{aligned} \tag{54}$$

Particular solutions can be found as follows: first consider the non-abelian anomaly associated with an invariant polynomial $\mathcal{J}_{2k}$ on (Lie O_d) necessarily of even degree (e.g. $\mathrm{Tr}\, X^{2k}$ or products thereof, where $X = -X^T \in \text{Lie } O_d$). Make the substitution $X \to R$, which yields a $4k$ form $\mathcal{J}_{2k}(R \ldots R)$; compute the corresponding

$$\mathcal{Q}_{4k-1}(\omega, R) = 2k \int_0^1 dt\, \mathcal{J}_{2k}(\omega, R_t, \ldots R_t) \tag{55}$$

and the corresponding anomaly $\mathcal{A}(\Omega) = Q^1_{4k-2}$ which therefore only occurs in dimensions d of the form 4k-2.

Now we have

$$\begin{aligned} I_\xi \mathcal{A}(\Omega) &= W(\xi)\mathcal{A}(\Omega) + \mathcal{A}(I_\xi \Omega) \\ &= \big(i(\xi)d + d\, i(\xi)\big)\mathcal{A}(\Omega) \\ &= i(\xi)\, W(\Omega)\, \mathcal{Q}_{4k-1} + d\,(\text{local form}) \\ &= W(\Omega)\, i(\xi)\, \mathcal{Q}_{4k-1} + \quad '' \end{aligned} \tag{56}$$

So, a candidate for $\mathcal{A}(\xi)$ is $i(\xi)Q$, which is identically zero in 4k-2 dimensions because Q_{4k-1} is a 4k-1 form.

This was to be expected because the usual non abelian internal gauge anomaly is consistent with a vanishing gravitational anomaly and the consistency check is identical with the above calculation.

Other candidates for gravitational anomalies would be solutions of the homogeneous system obtained from Eq.(54) by putting $\mathcal{A}(\Omega) = 0$:

$$\begin{aligned} &W(\Omega)\mathcal{A}(\xi) = 0 \quad \text{mod } d\,(\text{local form}) \\ &W(\xi)\mathcal{A}(\xi') - W(\xi')\mathcal{A}(\xi) - \mathcal{A}([\xi,\xi']) = 0 \quad '' \end{aligned} \tag{57}$$

It is to be noted that (57) is independent from the definition of $W(\xi)$ up to a gauge transformation. The gravitational anomalies found so far[18] seem to be rigidly coupled to the non abelian orthogonal gauge anomalies, and it is not known whether homogeneous gravitational anomalies exist or not.

V TOPOLOGICAL INTERPRETATIONS OF ANOMALIES[9]

We shall conclude this report by mentioning two beautiful connections of the existence of anomalies with the topology of gauge groups and of the orbit space of gauge fields.

a) The Atiyah-Singer Constructions[9]

One considers gauge fields over compactified euclidean four space M, e.g. S^4; we shall assume the instanton number to be 0 or 1 because it is almost irrelevant to the discussion. Let $\mathcal{A}$ be the (affine) space of such gauge fields (no topology, contractible). Let $\mathcal{G}$ be the gauge group (e.g. Maps $M \overset{g}{\rightarrow} G$, pointed, i.e. such that g(north pole) = identity of G). Let $\mathcal{A}/\mathcal{G}$ be the corresponding orbit space. Because $\mathcal{G}$ has topology (homotopy, cohomology can be computed) $\mathcal{A}/\mathcal{G}$ has topology[9,19]. In particular, it has a non vanishing second cohomology for which one can find representatives as the lowest Chern class of suitable bundles over $\mathcal{A}/\mathcal{G}$. Among the bundles which have been considered so far there are two which are natural and interesting: first of all $\mathcal{A}$ itself; secondly an "index bundle" associated with the Dirac-Weyl operator $\not{\partial}_A$:"ker $\not{\partial}_A$ - ker $\not{\partial}_A^{+}$" (with dimension the integer Ind $\not{\partial}_A$). Then one constructs a connection on some such bundle, which requiresthe definition of a Green's function, either of a covariant Laplacian or of the Dirac operator. The bundle and its connection are then pulled back on $M \times \mathcal{A}/\mathcal{G}$, the fiber over (x, a) consisting of pairs (f(x), f) where f(x) is the value of f at x. The characteristic classes on $\mathcal{A}/\mathcal{G}$ are then expressed by integrating higher characteristic classes over $M \times \mathcal{A}/\mathcal{G}$ on cycles of M (e.g. all of M). Such classes, when lifted to $\mathcal{A}$ are exact. The lowest Chern class thus becomes the differential of a one form on $\mathcal{A}$ which, restricted to a fiber $\mathcal{G}_a$ is a one form on $\mathcal{G}$ depending on a, and turns out to be the anomaly, with again the $\phi\pi$ ghost identified with the Maurer Cartan form of $\mathcal{G}$. I.M. Singer's calculation is sketched in the Appendix.

b) The Ramadas Interpretation[9]

The Witten Lagrangian[17] Eq.(39) leads to an action defined up to an additive integer multiple of 2π (if suitably normalized), because another choice of S_5^+ e.g. S_5^- yields an answer differing from the former by $\int_{g(S_5)} \mathcal{J}_3(\omega, [\omega,\omega], [\omega,\omega])$, an integer $\times 2\pi$ because of the non triviality of the integer cohomology class $\mathcal{J}_3(\omega, [\omega,\omega]\ [\omega,\omega])$ of G.

This suggests that, if the dynamics of the chiral field g is defined via a symplectic form on phase space[21], the Witten equation of motion[17] will be appropriately recovered. Fixed time phase space is provided by e.g.[20] Maps $S^3 \rightarrow G$, with (north pole $\rightarrow$ identity), and canonically conjugate momentum $g^{-1}\partial_0 g$. We thus look for a closed two form on configuration space, i.e. a cohomology class of the gauge group over S^3 [19]. Then, apply Eq.(13) with $A = g^{-1}dg + g^{-1}\delta g$, $s=\delta$ we find that

$$\delta Q_3^2 = - dQ_2^3$$

Hence

$$\int_{S^3} \mathcal{Q}_3^2$$

provides a closed 2-form on $[S_3 \to G]$. Its contribution to the equations of motion is obtained by polarizing it with respect to $g^{-1}\delta g$, look at the coefficient of one of the variables and replace there the other variable by $g^{-1}\partial_0 g$. One can check than one recovers this way the equationsof motion found by E. Witten[17].

CONCLUSION

We have tried to summarize some of the recent progress on what could be called the kinematics of anomalies. It involves algebra and correspondingly the topology which emerges whenever the field theory under study is confined to a space or euclidean space time box. Only locality has been used except in once case where uniqueness was obtained by superimposing power counting. In each specific case, one has to compute one or finitely many coefficients which may have further remarkable properties[18].

REFERENCES

1. J. Steinberger, P.R. 76:1180 (1949);
   J. Schwinger, P.R. 82:664 (1951);
   S. Adler, P.R. 177:2426 (1969);
   J.S. Bell, R. Jackiw, N.C. 60A:47 (1969).
2. S. Adler, W.A. Bardeen, P.R. 182:1517 (1969);
   W.A. Bardeen, P.R. 184:1848 (1969);
   S. Adler, in Lectures on elementary particles and quantum field theory, 1970 Brandeis lectures M.I.T. Press, S. Deser, M. Grisaru, H. Pendleton Ed.
3. J. Wess, B. Zumino, P.L. 37B:95 (1971).
4. L. Alvarez Gaumé, E. Witten, Nucl. Phys. to appear.
5. This interpretation was already alluded to in: C. Becchi, A. Rouet, R. Stora, in "Renormalization Theory", Erice 1975, G. Velo & A.S. Wightman Ed., NATO ASI Series Vol. C23 Reidel (1976). It is strengthened by the possibility to repeat the formal Faddeev Popov argument and express the "volume" of the gauge group (very likely not to exist, in the continuum), in terms of the Maurer Cartan form of that group (cf. J.M. Leinaas, K. Olaussen, P.L. 108B:199 (1982)). It has been spelled out most accurately by L. Bonora, P. Cotta-Ramusino, C.M.P. 87:589 (1983).
   An alternative interpretation: a basis of generators for the dual (Lie $\mathcal{G}$)* of (Lie $\mathcal{G}$), cf. D. Sullivan, IHES Pub. n° 47, p. 269 (1977).

6. H. Cartan in Colloque de Topologie (Espaces fibrés) Bruxelles (1950).
7. R. Stora, B. Zumino, in preparation.
8. B. Zumino, Les Houches Lectures, August (1983);
B. Zumino, Wu Yong Shi, A. Zee, to appear in Nucl. Phys.;
L. Bonora, P. Pasti, Padua preprint IFPD 20/83;
P.K. Townsend, G. Sierra, LPTENS 83/12 (1983);
Feynman graph calculations are well represented in several papers by P.H. Frampton, T.W. Kephart, P.R. D28:1010 (1983); P.L. 50:1347 (1983); and a recent preprint. See also refs 4, 18, for the gravitational case.
Supersymmetric techniques are given by L. Alvarez Gaumé, this volume; D. Friedan and P. Windey in proceedings of Zakopane's school (1983), CERN preprint; also ref. 4.
9. I.M. Singer, seminar notes, Santa Barbara seminars (1982);
M.F. Atiyah, R. Bott, "The Yang Mills equations over Riemann surfaces", preprint;
M.F. Atiyah, I.M. Singer, in preparation;
M.F. Atiyah, K. theory, Benjamin (1969);
T.R. Ramadas, C.M.P., to appear.
For the computation of the (co)homology of $\mathcal{G}$, $\mathcal{A}/\mathcal{G}$, see R. Thom in Colloque de Topologie Algébrique, Louvain 1956, Masson 1957.
10. C. Becchi, A. Rouet, R. Stora in Field theory, Quantization and Statistical Physics, E. Tirapegui Ed. Reidel (1981) pp. 3-32.
11. F. Langouche, T. Schucker, R. Stora, in preparation. The direct Faddeev Popov treatment will be found there.
12. Although "BRS" originally made use of an anticommuting parameter generating the Slavnov transformation s, this parameter was quickly gotten rid of, (cf. e.g. ref. 5), and the Slavnov operation s was there defined as an antiderivation anticipating that the whole machinery was merely involving differential forms. Anticommuting parameters have nevertheless been systematically used in the superspace formalism advocated by the Tel Aviv - Paris group[15] (Y. Nee'man, L. Baulieu, J. Thierry Mieg) by the Padua-Milano group[5,16] (L. Bonora, P. Cotta Ramusino, P. Pasti, M. Tonin) and early introduced by G. Curci, R. Ferrari, P.L. 63B:91 (1976) and O. Piguet, M. Schweda.
13. cf. e.g. R. Stora in "New developments in quantum field theory and Statistical Mechanics", Cargèse 1976, M. Levy, P. Mitter Ed., NATO ASI Series B26, Plenum 1977;
J. Dixon, unpublished.
14. S. Kobayashi, K. Nomizu, Foundations of differential geometry, Interscience (1963);
S.S. Chern, Complex manifolds without potential theory, Springer (1979);
See also ref. 13.

15. Y. Nee'man, T. Regge, J. Thierry-Mieg in Proc. XIXth International Conference on high energy physics, Tokyo (1978); This is the first article of a long list which I apologize not to be able to include here. Recent contributions to this programme can for instance be found in L. Baulieu, J. Thierry-Mieg, LPTHE 83/25, to appear in Nucl. Phys.
16. L. Bonora, P. Cotta Ramusino, P.L. 107B:87 (1981); Other references to the PaduaMilano scheme can be found here.
17. E. Witten, N.P. B223:422 (1983);
A.P. Balachandran, V.P. Nair, C.G. Trahern, P.R. D27:1369 (1983);
Chou Kuang chao, Guo Han ying, Wu Ke, Song Xing chang, AS ITP 83-027, 83-033.
18. L. Alvarez Gaumé, this volume;
L. Alvarez Gaumé, E. Witten, Nucl. Phys. to appear[4];
See also ref. 11.
19. See ref. 9. Also M. Asorey, P.K. Mitter, CERN TH-3424 to be published in CMP.
20. fixed-time 3-space is compactified to S^3.
21. E. Cartan, Systèmes différentiels extérieurs, Hermann, Paris;
J.M. Souriau, Systèmes dynamiques, Dunod Paris;
R. Abraham, J. Marsden, Foundations of mechanics, Benjamin (1978).

APPENDIX

We sketch out here the calculation of I.M. Singer indicated in section V a). (Santa Barbara seminar 1982). We can locally parametrize $\mathfrak{A}$ by

$$A = g^{-1} a g + g^{-1} dg \tag{A1}$$

where a represents an orbit.
Then

$$\delta A = g^{-1} \delta a\, g - D_A\, g^{-1} \delta g \tag{A2}$$

So, we can introduce on $\mathfrak{A}$ the connection

$$\mathcal{A} = - G_A D_A^{+} \delta A \qquad \text{where} \qquad G_A = (D_A^{+} D_A)^{-1} \tag{A3}$$

such that

$$\mathcal{A}\big|_{\text{fiber}} = g^{-1} \delta g \tag{A4}$$

The total connection over $M \times \mathfrak{A}/\mathcal{G}$ at $(g(x), g(.); x, A)$ is

$$A + \mathcal{A} = A - G_A D_A^{+} \delta A \tag{A5}$$

The corresponding curvature is

$$\begin{aligned} \mathcal{F} &= (d+\delta)(A+\mathcal{A}) + \tfrac{1}{2}\left[A+\mathcal{A}, A+\mathcal{A}\right] \\ &= F + (1 - D_A G_A D_A^{+})\delta A - \delta(G_A D_A^{+})\delta A \\ &\quad + \tfrac{1}{2}\left[G_A D_A^{+} \delta A, G_A D_A^{+} \delta A\right] \end{aligned} \tag{A6}$$

The third rank invariant $\mathcal{J}_3$ gives rise to $\mathcal{J}_3(\mathcal{F}, \mathcal{F}, \mathcal{F})$, which, integrated over M yields a 2-form on $\mathfrak{A}$

$$\begin{aligned} \Omega = 3 \int_M \Big\{ & \mathcal{J}_3\big((1 - D_A G_A D_A^{+})\delta A, (1 - D_A G_A D_A^{+})\delta A, F\big) \\ & + \mathcal{J}_3\big(-\delta(G_A D_A^{+})\delta A + \tfrac{1}{2}\left[G_A D_A^{+}\delta A, G_A D_A^{+}\delta A\right], F, F\big)\Big\} \end{aligned} \tag{A7}$$

Since

$$(d+\delta)\,\mathcal{J}_3(\mathcal{F},\mathcal{F},\mathcal{F}) = 0 \tag{A8}$$

it follows that

$$\delta\,\Omega = 0 \tag{A9}$$

Furthermore,

$$\begin{aligned}\Omega &= \int_M 3(d+\delta)\int_0^1 dt\, \mathcal{J}_3\Big(A - G_A D_A^{+}\delta A,\ t\mathcal{F} + \frac{t^2-t}{2}\big[A - G_A D_A^{+}\delta A, \\ &\qquad A - G_A D_A^{+}\delta A\big],\ t\mathcal{F} + \frac{t^2-t}{2}\big[A - G_A D_A^{+}\delta A,\ A - G_A D_A^{+}\delta A\big]\Big) \\ &= 3\,\delta \int_M \int_0^1 dt\,\Big\{ \mathcal{J}_3\Big(G_A D_A^{+}\delta A,\ tF + \frac{t^2-t}{2}[A,A],\ \text{i.e.}\Big) \\ &\qquad + \mathcal{J}_3\Big(A,\ t(1 - D_A G_A D_A^{+})\delta A + t^2 - t\,\big[G_A D_A^{+}\delta A, A\big], \ldots \\ &\qquad\qquad \ldots\ tF + \frac{t^2-t}{2}[A,A]\Big)\Big\} \\ &= 3\,\delta\, Q_4^1\end{aligned} \tag{A10}$$

Restricting Q_4^1 to the fiber through A:

$$\delta A = -D_A\, g^{-1}\delta g \tag{A11}$$

yields

$$\begin{aligned} Q_4^1\Big|_{\substack{\text{fiber}\\ \text{of } A}} = -\int_M \int_0^1 dt\,\Big\{ &\mathcal{J}_3\Big(g^{-1}\delta g,\ tF + \frac{t^2-t}{2}[A,A],\ \text{i.e.}\Big) \\ &+ \mathcal{J}_3\Big(A,\ (t^2-t)\big[g^{-1}\delta g, A\big],\ tF + \frac{t^2-t}{2}[A,A]\Big)\Big\}\end{aligned} \tag{A12}$$

which is the russian formula for the anomaly with ω identified with $g^{-1}\delta g$.

There remains to prove that Ω, defined on $\mathcal{A}$ defines a form on $\mathcal{A}/\mathcal{G}$, (which can then be written by choosing the gauge $D_A^+ \delta A = 0$).

For that purpose, it is enough to check that the term $-D_A g^{-1}\delta g$ in δA does not contribute to the result (up to d of some form if we look at the non integrated $[\mathcal{J}_3(\mathcal{F},\mathcal{F},\mathcal{F})]_4^2$.

The first term in (A7) yields 0, each factor being left unaltered. In the second term, the part quadratic in $g^{-1}\delta g$ vanishes by the Maurer Cartan structure equation. The linear part can be written as follows

$$\mathcal{J}_3(-\delta G_A D_A^+ \delta A, F, F) = -\delta \mathcal{J}_3(G_A D_A^+ \delta A, F, F) - 2\mathcal{J}_3(G_A D_A^+ \delta A, D_A \delta A, F) \quad \text{(A13)}$$

In the polarization process, the total differential reduces to

$$\delta_A \mathcal{J}_3(\omega, F, F) = -2\mathcal{J}_3(\omega, D_A \delta A, F) \quad \text{(A14)}$$

because of gauge invariance,
the second term of (A13) gives

$$2\mathcal{J}_3(\omega, D_A \delta A, F) - 2\mathcal{J}_3(G_A D_A^+ \delta A, [F,\omega], F) \quad \text{(A15)}$$

the first part cancels (A14) and the second part cancels the polarized contribution to

$$\mathcal{J}_3\left(\tfrac{1}{2}[G_A D_A^+ \delta A, G_A D_A^+ \delta A], F, F\right) \quad \text{(A16)}$$

MORSE THEORY AND MONOPOLES: TOPOLOGY IN LONG RANGE FORCES†

Clifford Henry Taubes*

Department of Mathematics
University of California, Berkeley, CA 94720

OUTLINE OF LECTURES

I. Forces Between Monopoles

- A. Forces Between Monopoles
- B. Are They [Solutions] Really Present?
- C. Topology of $\mathcal{C}/\mathcal{G}$ [The Orbit Space]
- D. Application of min-max to $\mathcal{C}/\mathcal{G}$
- E. New Results

II. The Newly Discovered, Fake $\mathbb{R}^4$

- F. The Fake $\mathbb{R}^4$
- G. Smooth Structures
- H. Smooth Structures on Manifolds
- I. Properties of $\mathbb{R}^4_\Sigma$
- J. The Existence Proof

I. FORCES BETWEEN MONOPOLES

Consider the t'Hooft-Polyakov monopole [1]. This is a static, finite energy solution to the evolution equations of a non-Abelian Higgs model with adjoint Higgs. For simplicity, take $G = SU(2)$, but one could do a similar analysis for any group.

---

*NSF Postdoctoral Fellow in Mathematics

†The text of a lecture delivered at the 1983 Cargese NATO Summer School: Progress in Gauge Theories.

Remember that the energy of a configuration (A,Φ) of static gauge field, $A = (A_i)_{i=1}^{3} = (A_i^a \tfrac{1}{2}\tau^a)_{i=1}^{3}$, and Higgs field $\Phi = \Phi^a \tfrac{1}{2}\tau^a$, (here, $\{\tau^a\}_{i=1}$ are the anti-hermitian Pauli matrices, $\tau^a\tau^b = -\delta^{ab} - \varepsilon^{abc}\tau^c$) is given by the expression

$$\mathcal{A}(A,\Phi) = \tfrac{1}{2}\int_{\mathbb{R}} \{|B_A|^2 + |\nabla_A\Phi|^2 + \lambda V(\Phi)\} \quad . \tag{1.1}$$

Here, B_A is the magnetic field,

$$B_A = \left(B_{A_i} = \tfrac{1}{2}\varepsilon^{ijk}(\partial_j A_k - \partial_k A_j + [A_k, A_j])\right)_{i=1}^{3} \quad ;$$

while $\nabla_A\Phi$ is the minimally coupled, covariant derivative of Φ,

$$\nabla_A\Phi = (\nabla_{A_i}\Phi = \partial_i\Phi + [A_i,\Phi])_{i=1}^{3} \quad .$$

The Higgs potential is $V(\Phi) = (1 - |\Phi|^2)^2$, with $\lambda \geq 0$. Finally, the norms here are $|a|^2 = 2\,\mathrm{trace}(a^+a)$.

The t'Hooft-Polyakov monopole satisfies the variational equations of $\mathcal{A}$,

$$\begin{aligned} \varepsilon^{ijk}\nabla_{A_i}B_{A_j} + [\Phi, \nabla_{A_k}\Phi] &= 0 \quad , \\ \nabla_{A_i}\nabla_{A_i}\Phi + \lambda(1 + |\Phi|^2)\Phi &= 0 \quad . \end{aligned} \tag{1.2}$$

(With $\lambda = 0$, this is the BPS [2] limit.) The t'Hooft-Polyakov monopole is characterized by its being the minimum energy solution to Eq. (1.2) with 1-unit of magnetic charge,

$$Q(A,\Phi) = \frac{1}{4\pi}\int_{S^2_\infty} \mathrm{tr}(\Phi B_{A_k})d^2\Sigma_k = 1$$

The t'Hooft-Polyakov anti-monopole comes from the monopole by changing Φ to $-\Phi$; it has $Q = -1$.

> QUESTION: Can there be charged, or uncharged solutions to Eq. (1.2) with higher energy? Do monopoles and anti-monopoles form (unstable) bound states?

Let us look first at the forces between monopole and anti-monopole. The non-abelian Higgs model is designed to describe the spontaneous symmetry breakdown from the group SU(2) to the group U(1). At large distances, $|x| >> 1$ for a monopole at the origin, the fields are abelian. The abelian direction at each $x \in \mathbb{R}^3$ is determined by $\phi(x) \in$ Lie Alg. SU(2). So a monopole and anti-monopole interact via a long range, *attractive* Coulomb force. (Remember: For vector forces, like charges attract.)

$$\mathcal{E}_{\text{Coul}}(r) = -\frac{1}{r} + \mathcal{O}(e^{-r}) \quad .$$

The $\mathcal{O}(e^{-r})$ is due to a massive, intermediate vector boson; the mass is 1 due to the Higgs effect.

There is also a long range scalar field, the Higgs field, which is attractive whether between like charges or opposite charges:

$$\mathcal{E}_{\text{Higgs}} = -\frac{e^{-\sqrt{\lambda r}}}{r} + \mathcal{O}(e^{-r}) \quad .$$

So, the net long range potential energy between a monopole and an anti-monopole $(m-\bar{m})$ is

$$\mathcal{E}_{\ell.r.} = -\frac{1}{r} - \frac{e^{-\sqrt{\lambda r}}}{r} + \mathcal{O}(e^{-r}) \quad . \qquad (1.4)$$

Note that the monopole-monopole potential (m–m) is

$$\mathcal{E}_{\ell.r.} = +\frac{1}{r} - \frac{e^{-\sqrt{\lambda r}}}{r} \quad .$$

The conclusion from Eq. (1.4) is that whatever the value of λ, monopoles attract anti-monopoles at large separation. Note that except for $\lambda = 0$ (the BPS limit), monopoles repel monopoles at large r [3].

The long range attraction of $m-\bar{m}$ is good if bound pairs are to be found, but a mechanism which keeps them from annihilating is still required.

The intermediate vector bosons, W_+ and W_- also mediate vector forces. And again, for vector forces, likes attract but opposites repel. So the monopole, considering (e^{-r}) corrections, looks like

Here, e.m. is the $1/r$ magnetic coulomb force, and W_+, W_-, at right angles to it and each other in isospin space, have e^{-r} decay. The arrows signify three orthogonal directions in isospin space, with ⊙ meaning "points out of the paper" and ⊕ meaning "points into the paper". The anti-monopole is

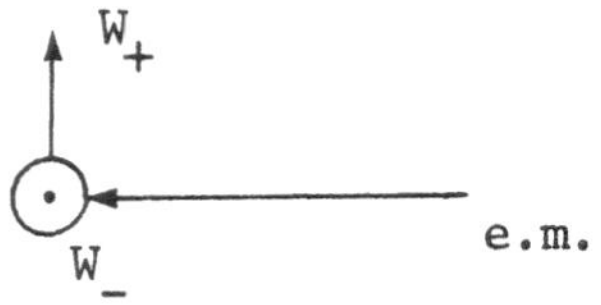

Note that the direction of (W_+,W_-) in the plane in isospin space orthogonal to e.m. is not well defined, as a U(1) gauge transformation causes them to rotate: A gauge transformation $g = \exp(i\pi\Phi(x))$ causes the change

But, when you have a monopole with an anti-monopole, the relative orientation of the pairs (W_+^m, W_-^m) to $(W_+^{\bar{m}}, W_-^{\bar{m}})$ *is* gauge invariant: the U(1) gauge transformations rotate both pairs equally. Thus, our m-$\bar{m}$ configuration space has an extra parameter, namely the relative angle $\theta \in [-\pi,\pi]$.

The potential energy as a function of θ is

$$\mathcal{E}_{\text{Gauge}}(r,\theta) = -\frac{1}{r} + \frac{2}{r} e^{-r} \cos\theta \quad ,$$

which is the correction to $\mathcal{E}_{\text{coul}}(r)$. The $\theta = 0$ case looks like

$$= -\frac{1}{r} + \frac{2}{r} e^{-r} \quad .$$

The $\theta = \pi$ case looks like

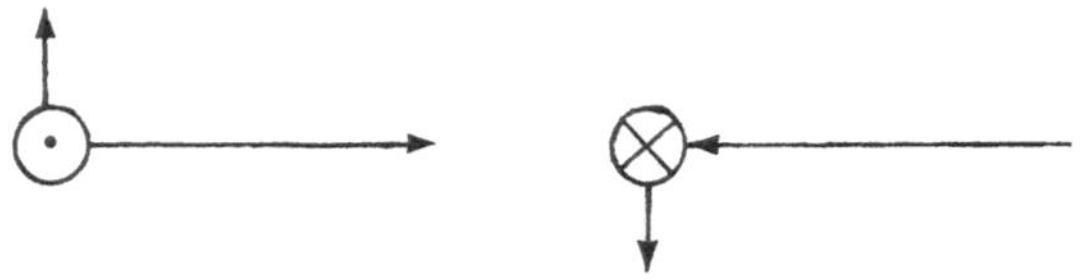

$$= -\frac{1}{r} - \frac{2}{r} e^{-r} \; .$$

One must also take into account the short-distance behavior of the Higgs force. At short range, symmetry is restored, as all three forces, e.m., W_+, W_-, have the same strength. The Higgs field, as it is supposed to be an order parameter, vanishes at the monopole center. So, schematically,

$$\mathcal{E}_{\text{Higgs}}(r,\theta) = -\frac{1}{r} e^{-\sqrt{\lambda r}} (1 - e^{-r})$$

The total potential energy is easiest to visualize in the BPS limit where $\lambda = 0$:

$$\mathcal{E}_T^{\text{BPS}}(r,\theta) = -\frac{2}{r} (1 - e^{-r}(\tfrac{1}{2} + \cos\theta))$$

A plot of $\mathcal{E}_T$ versus (r,θ) yields

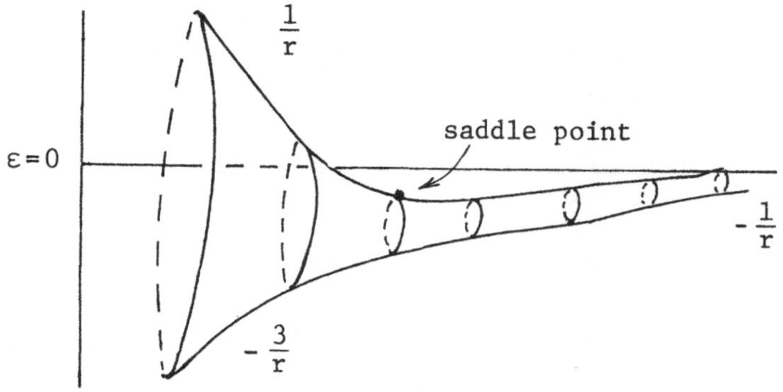

When $\lambda>0$, the saddle point is pushed out to larger radius, but except for providing the spontaneous symmetry breaking, the Higgs force is not relevant to the discussion. Indeed,

$$\mathcal{E}_{\text{Gauge}} = -\frac{1}{r} + \frac{2}{r} e^{-r} \cos\theta$$

contains all the essential terms.

The important fact is that a saddle point exists at $\theta = 0$ and $r \sim \mathcal{O}(1)$ = the boson mass.

The conclusion here is that unstable solutions of higher energy should not be unexpected.

II. ARE THEY REALLY PRESENT ?

The heading of this section asks the important question: Can one rigorously establish the existence of these solutions? The proof for the $\lambda = 0$ case was developed by the author [4] and $\lambda > 0$ was proved by D.Groisser [5].

The strategy is to translate the problem of finding solutions to a partial differential equation into a problem in the calculus of variations. By this, one means the following: $c = (A,\Phi)$ is a solution to the Yang-Mills-Higgs equations if and only if the energy, $\mathcal{A}$, is stationary with respect to all infinitesimal (but compactly supported on $\mathbb{R}^3$) variations of the fields. That is, if and only if

$$\frac{d}{dt}\mathcal{A}(c + t\psi)\Big|_{t=0} = 0 \quad ,$$

for all $\psi = (a_i,\phi)$ of compact support.

One should think of $\mathcal{A}$ as a functional on the configuration space; the space of gauge orbits $\mathcal{C}/\mathcal{G}$. Here,

$$\mathcal{C} = \left\{\text{smooth } c = (A,\Phi) : \mathcal{A}(c) < \infty \text{ and } \lim_{|x|\to\infty} |\Phi| = 1\right\} \quad ,$$

and

$$\mathcal{G} = \left\{\text{smooth maps from } \mathbb{R}^3 \text{ to } SU(2)\right\} \quad . \tag{2.1}$$

Then $c = (A,\Phi) \in \mathcal{C}$ is a solution to Eq. (1.2) if and only if $[c] \in \mathcal{C}/\mathcal{G}$ is a critical point of $\mathcal{A}$.

Morse theory proposes to relate the critical points of $\mathcal{A}$ to the topology of $\mathcal{C}/\mathcal{G}$. If $\mathcal{A}$ is nice (in a technical sense [6]) then $\mathcal{C}$ being topologically convoluted will imply that $\mathcal{A}$ has many critical points — solutions to Eq. (1.1).

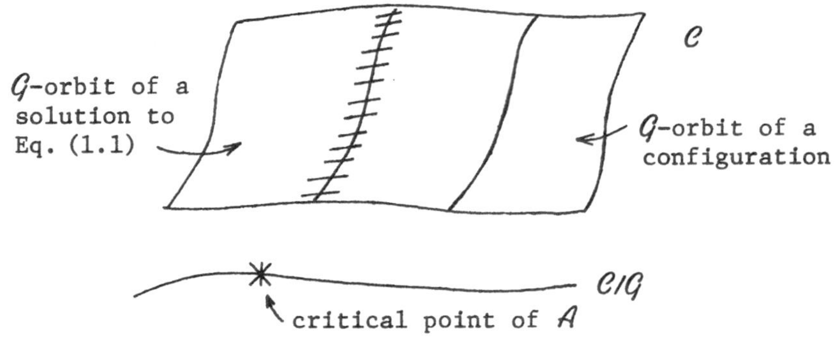

It is perhaps useful to look at a finite dimensional example first. Let T = torus in $\mathbb{R}^3$, and let $f: T \to [0,1]$ be a

C^2-function on T. Look for critical points of f; points $p \in T$ where

$$\vec{\nabla} f|_p = \left\{\frac{\partial f}{\partial x^i}\right\}_{i=1}^{2} = 0 \quad .$$

f :

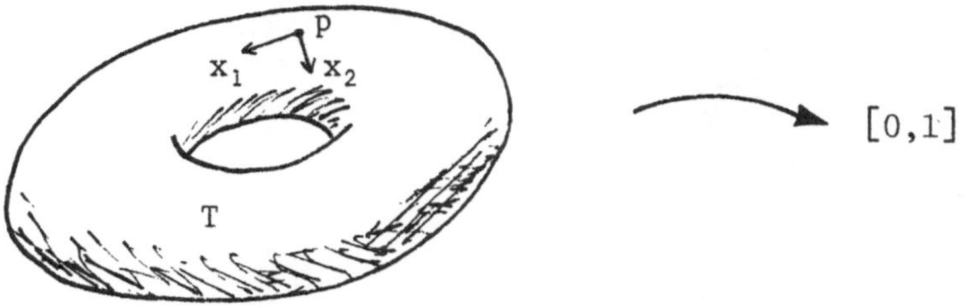

Here T corresponds to $\mathcal{C}/\mathcal{G}$, f corresponds to $\mathcal{A}$, and a point $p \in T$ where $\vec{\nabla f}|_p = 0$ corresponds to a solution of Eq. (1.1).

CLAIM: Any such f has at least *three* critical points.

First, be convinced that the maxima and minima of f are critical points. For example, if p is a minima, then one can Taylor expand f about p to obtain for x near p,

$$f(x) = f(p) + \vec{\nabla} f|_p \cdot (\vec{x} - \vec{p}) + \mathcal{O}(|x-p|^2) \quad .$$

If $\vec{\nabla} f|_p \neq 0$, then consider $x = \vec{p} - \varepsilon \vec{\nabla} f|_p$ for ε small. Then the formula above says that $f(x) < f(p)$ which contradicts the fact that p is minimizing.

Notice that if the minimum (or maximum) for f is not obtained at a unique point, then an extra minima gives already a third critical point as required. So one can now assume with no loss of generality that $f^{-1}(0) = \{p\}$ is just one point.

To find the third critical point in this case, one can use the noncontractible loops on T. That T has such loops is well known, e.g.

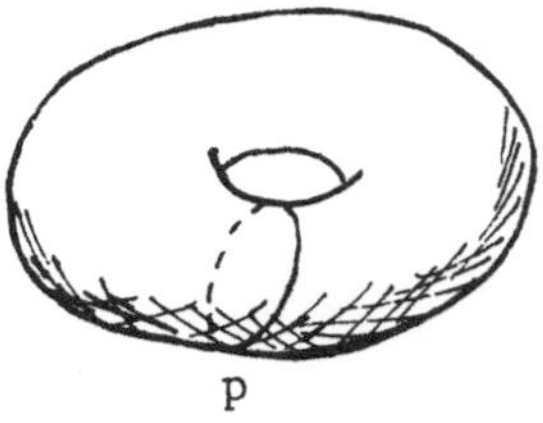

or

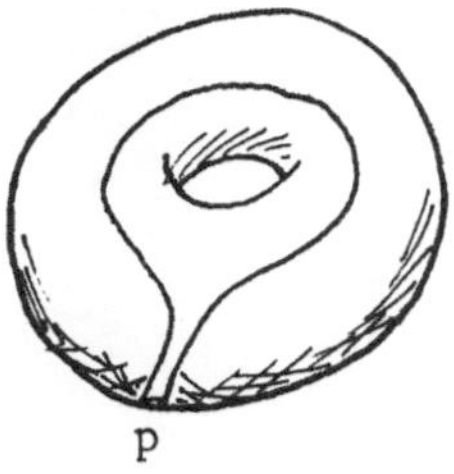

Let $\Lambda = \{$continuous loops on T, starting and ending at p and not contractible on $T\}$.

The idea is to do min-max over Λ. This is a non-linear version of the Rayleigh-Ritz variational method to find eigenvalues of a quadratic form.

To each $\ell \in \Lambda$, associate a point $\bar{\ell} \in \ell$ where f is maximal, i.e. $f(\bar{\ell}) \geqslant f(\ell(t))$ for all $t \in [0,1]$. If f is maximal at more than one point on ℓ, choose $\bar{\ell}$ for which $|\nabla f_{\bar{\ell}}|$ is smallest.

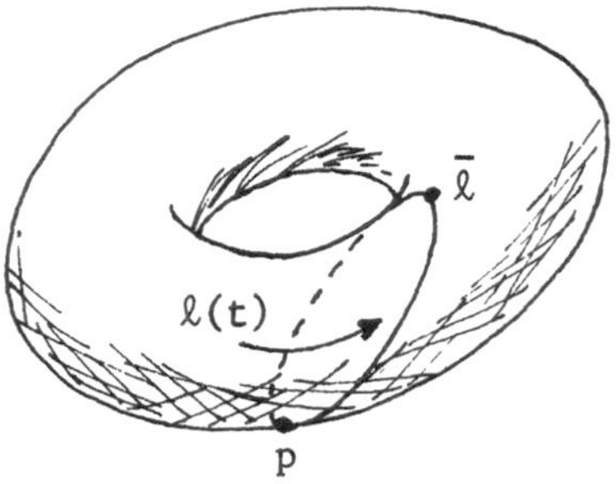

Associate to Λ the number $f_\Lambda = \inf_{\ell \in \Lambda} f(\bar{\ell})$. Notice that $f_\Lambda > 0$. Indeed, if $f_\Lambda = 0$, then since f is a C^2-function and p is an isolated minimum, there are loops in Λ which lie completely in small neighborhoods of p. But such a loop would surely be contractible since the small neighborhoods of p look like discs.

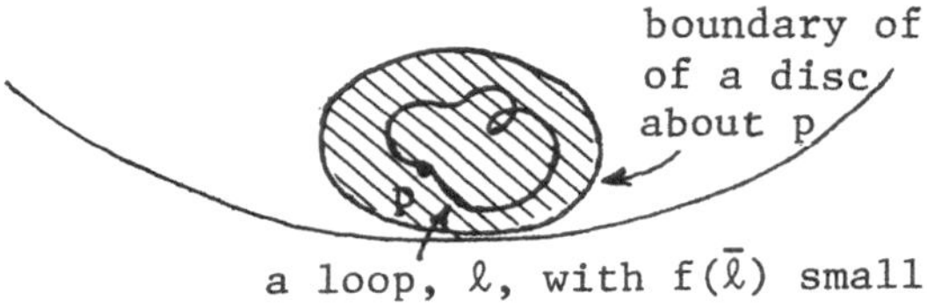

Now, the claim is that there exists $x \in T$ with $f(x) = f_\Lambda$ and $\nabla f|_x = 0$. It can be found by looking at sequences of loops, $\{\ell_i\} \subset \Lambda$ with $f(\bar{\ell}_i) \geqslant f(\bar{\ell}_{i+1}) \to f_\Lambda$.

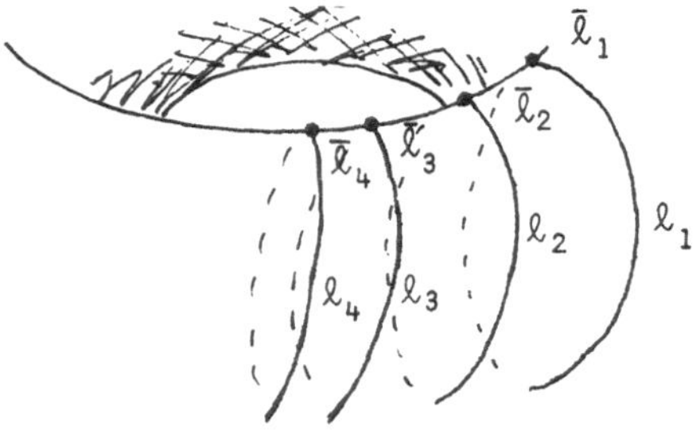

As T *is compact*, the sequence $\{\bar{\ell}_i\}$ has a limit point, $x \in T$ and $f(x) = f_\Lambda$. If $|\nabla f_x| > \delta$, then for all i sufficiently large, $|\nabla f_{\bar{\ell}_i}| > \delta$ too. Then the following picture emerges:

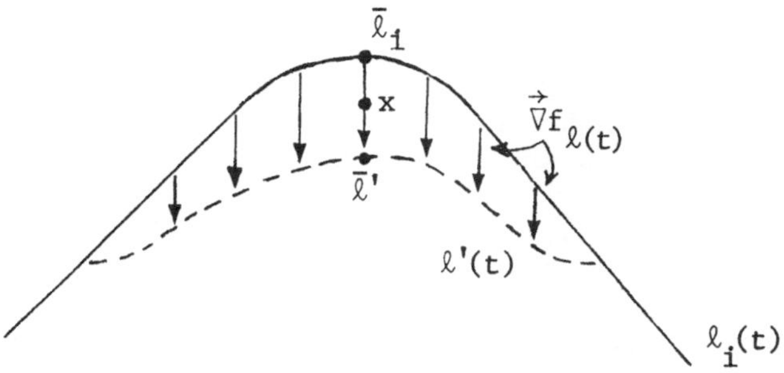

Let $\ell'(t) = \ell_i(t) - \varepsilon \vec{\nabla} f|_{\ell_i(t)}$. Then $\ell'(t)$ is a continuous deformation of $\ell(t)$, so it is in Λ. If $\vec{\nabla} f_x \neq 0$, then for i large and ε small, one would have $f(\bar{\ell}') < f_\Lambda$ which is not possible (by definition).

III. TOPOLOGY OF $\mathcal{C}/\mathcal{G}$

The relevant lessons from the torus example are as follows: First, one requires non-trivial topology to have a good apriori reason to have extra, non-minimal critical points. Second, one requires a compactness to deduce that the min-max sequences actually converge.

As a remark, be aware that this min-max technique, known as Ljusternik-Šnirelman (L-S) theory [7],[8] works in much more generality. Given a space X, and a space Λ of non-contractible maps of another space, Y, into X, one can use L-S theory to *try* to prove that a function $f: X \to [0,\infty)$ has a critical point, $x \in X$ with critical value

$$f(x) = f_\Lambda = \inf_{Y \in \Lambda} \sup_{y \in Y} f(Y(y)) \quad .$$

For the monopoles, the configuration space is $\mathcal{C}/\mathcal{G}$. It is topologically complicated. First, $\mathcal{C}$ has a countable number of path components, $\mathcal{C} = \cup_{n=-\infty}^{\infty} \mathcal{C}_n$. Indeed, n is the well known monopole number or topological charge in the model. Remember that it is defined because each $(A,\Phi) \in \mathcal{C}$ has $|\Phi| \to 1$ as $|x| \to \infty$. Therefore there is for each Φ, a radius $R(\Phi)$ so that $|\Phi|(x) > \frac{1}{2}$

if $|x| > R(\Phi)$. Then, on a large sphere, $S_{|x|=r}$ for $r > R(\Phi)$, the map

$$\left.\frac{\Phi}{|\Phi|}\right|_{|x|=r} : S_r^2 \longrightarrow S^2 = \text{unit sphere in Lie algebra SU(2)}$$

has a well defined winding number. The winding number is independent of $r > R(\Phi)$ and one puts $(A,\Phi) \in \mathcal{C}_n$ if and only if $\Phi/|\Phi|$ has winding number n on sufficiently large spheres.

This number is $\mathcal{G}$ invariant, so $\mathcal{C}/\mathcal{G} = \cup_{n=-\infty}^{\infty} \mathcal{C}_n/\mathcal{G}$.

Alternately, one can compute n from a surface integral at $|x| = \infty$ since D. Groisser has proved that for any finite action (A,Φ),

$$\lim_{r\to\infty} \frac{1}{4\pi}\int_{|x|=r} \text{trace}(\Phi B_{A_i})d\Sigma^i$$

is an integer [9].

In fact, the asymptotic values of Φ give $\mathcal{C}_n/\mathcal{G}$ richer structure than one might think [4]. Consider a 1-parameter family, $(A^t,\Phi^t) \in \mathcal{C}_0$, That is, at each $t \in [0,2\pi]$, this is a configuration of gauge field plus Higgs field. Assume that the dependence on t is continuous, and that

$$(A^0(x), \Phi^0(x)) = (A^{2\pi}(x), \Phi^{2\pi}(x)) = (0, \tfrac{1}{2}\tau^3) .$$

Thus, (A^t,Φ^t) is a loop in $\mathcal{C}_0$ which is based at the trivial configuation.

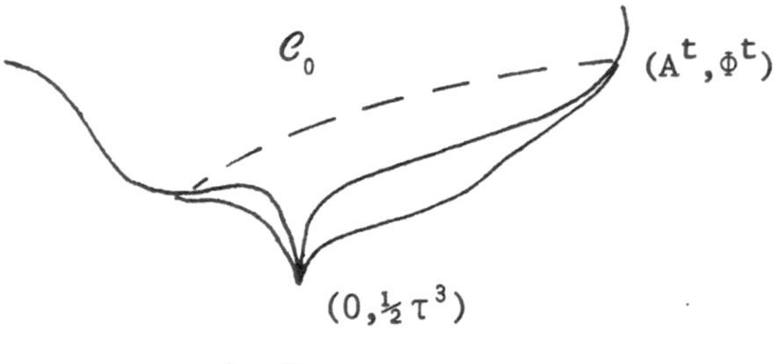

QUESTION: When is (A^t,Φ^t) not contractible? The answer is found by looking at the asymptotic values of $\Phi^t(x)$: Define $\hat{\Phi}^t(\theta,\phi)$ as an S^1-parametrized map from S^2 to S^2 by

$$\hat{\Phi}^t(\theta,\phi) = \lim_{|x|\to\infty} \Phi^t(|x|,\theta,\phi) .$$

Alternately, $\hat{\Phi}^t(\theta,\phi)$ is a map from S^1 into $M_0(S^2;S^2)$ — the space of degree zero maps from S^2 to S^2.

Suppose that $\hat{\Phi}^t(\theta,\phi)$ is not contractible to the constant map of S^1 into $M_0(S^2;S^2)$; the map sending all $t \in S^1$ to $\frac{1}{2}\tau^3$ Then (A^t,Φ^t) cannot be contractible in $\mathcal{C}_0$ to the constant map of S^1 into $\mathcal{C}_0$ which sends all $t \in S^1$ to $(0,\frac{1}{2}\tau^3)$. The converse here is also true.

The homotopy groups of $M_0(S^2;S^2)$ are known in principal, and one can look up in a book to find $\pi_1(M_0(S^2;S^2)) \simeq \mathbb{Z}$. The generator can be drawn schematically: It is

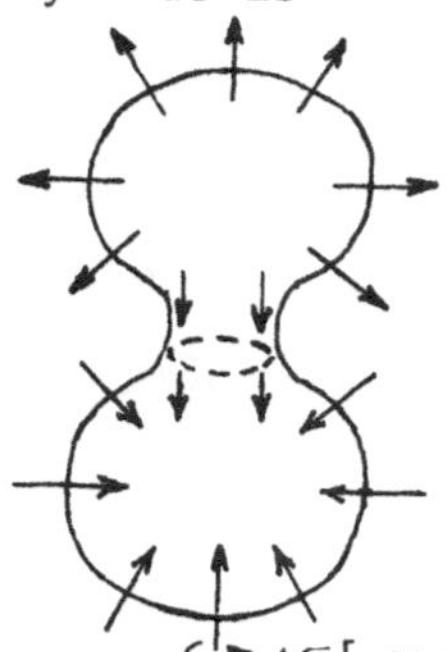

(3.1)

The distorted sphere above is the domain S^2. Picture it being surrounded by a larger, round sphere. Follow an arrow from its start on the distorted sphere, out to the large, surrounding sphere. This tells you where to map the inner sphere onto the outer. As t goes through $[-\pi,\pi]$, rotate counterclockwise the lower hemisphere (with the arrows) of the inner sphere. This one parameter family is not contractible.

After taking $\mathcal{G}$ into account, one finds that $\pi_1(\mathcal{C}_0/\mathcal{G}) \simeq \mathbb{Z}$, $\pi_1(\mathcal{C}_n/\mathcal{G}) \simeq \mathbb{Z}_{|n|}$ if $n \neq 0$, and $\pi_\ell(\mathcal{C}_n/\mathcal{G}) \simeq \pi_{\ell+2}(S^2)$. For example, $\pi_2(\mathcal{C}_0/\mathcal{G}) \simeq \mathbb{Z}_2$, $\pi_3(\mathcal{C}_0/\mathcal{G}) \simeq \mathbb{Z}_2$ and $\pi_4(\mathcal{C}_0/\mathcal{G}) \simeq \mathbb{Z}_{12}$. There is plenty of topology here to do Ljusternik-Snirelman theory.

IV. APPLICATION OF MIN-MAX TO $\mathcal{C}/\mathcal{G}$ (Day 2)

Recall now what has just been established: $\mathcal{C}_0/\mathcal{G}$ is not simply connected, $\pi_1(\mathcal{C}_0/\mathcal{G}) \simeq \mathbb{Z}$. Further, $\mathcal{A}^{-1}(0)$ is a single point in $\mathcal{C}_0/\mathcal{G}$, the $\mathcal{G}$ orbit of $(0,\frac{1}{2}\tau^3)$. The generator of π_1 is some $\ell(t) = (A^t,\Phi^t)$ with $(A^0,\Phi^0) = (A^{2\pi},\Phi^{2\pi}) = (0,\frac{1}{2}\tau^3)$.

To implement the min-max procedure, one mimics the finite dimensional example. Define

$$\Lambda = \{\text{continuous maps from } S^1 \text{ into } \mathcal{C}^0/\mathcal{G} \text{ which are not contractible and which begin and end at the } \mathcal{G} \text{ orbit of } (0,\tfrac{1}{2}\tau^3).\}$$

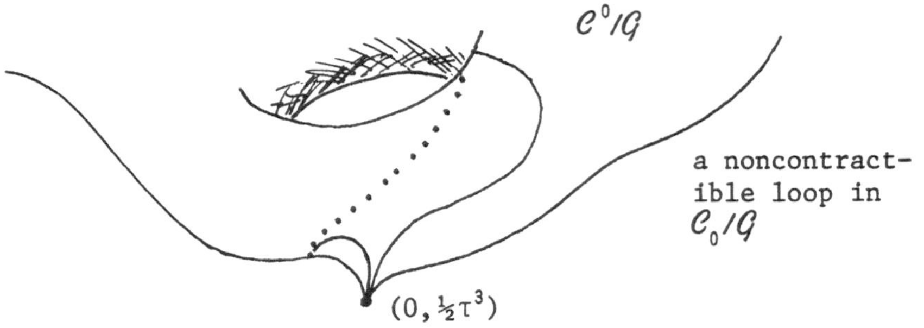

To each $\ell(t) = (A^t,\Phi^t) \in \Lambda$, associate the configuration, $\bar{\ell} = (A,\Phi) \in \ell([0,2\pi])$ where $\mathcal{A}$ is maximal and such that $\|\nabla\mathcal{A}_{\bar{\ell}}\|_*$ is the smallest over all the maxima on $\ell([0,2\pi])$. Here,

$$\|\nabla\mathcal{A}_{\bar{\ell}}\|_* = \sup_{\substack{\psi \text{ of compact} \\ \text{support in } \mathbb{R}^3}} \frac{\left|\frac{d}{dt}\,\mathcal{A}(\bar{\ell}+t\psi)\right|_{t=0}}{\int(|\nabla_A\psi|^2 + |[\Phi,\psi]|^2)} .$$

As in the torus example, define

$$\mathcal{A}_\Lambda = \inf_{\ell\in\Lambda}\ \sup_{t\in[0,2\pi]} \mathcal{A}(\ell(t)) ,$$

and choose a min-max sequence, $\{\ell_i(\cdot)\}_{i=1}^{\infty} \in \Lambda$ with $\mathcal{A}(\bar{\ell}_i) \geq \mathcal{A}(\bar{\ell}_{i+1}) \searrow \mathcal{A}_\Lambda$:

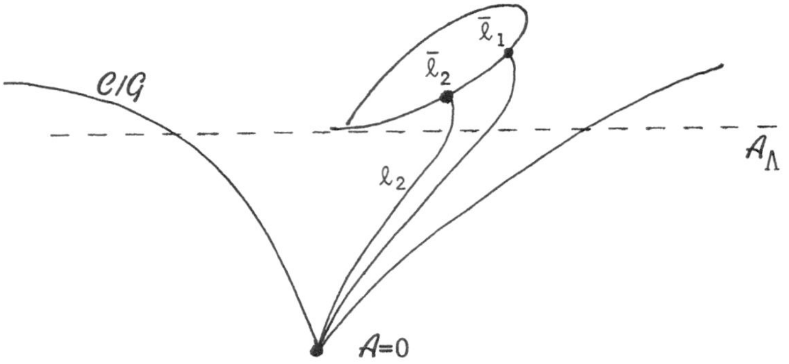

One must now demonstrate that the sequence, $\{\ell_i = (A^i,\Phi^i)\}$ of loop maxima converges to a nontrivial solution of Eq. (1.1) which is in $\mathcal{C}_0$.

Here, one is distinguishing between two pictures for $\mathcal{C}_0/\mathcal{G}$:

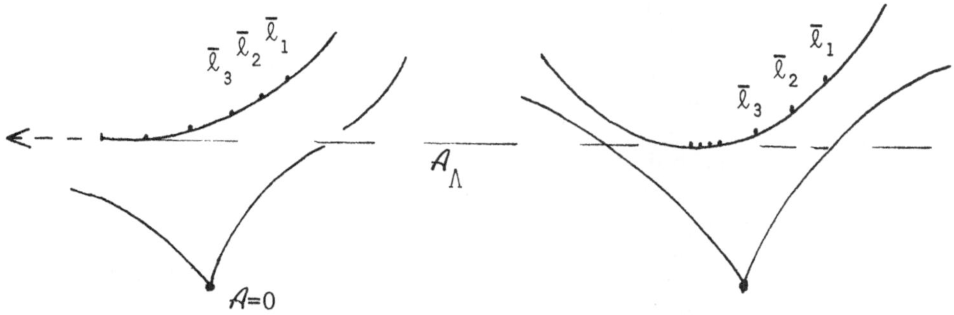

The question is one of convergence. Convergence is tricky. Most sequences don't converge, statistically.

On the 2-dimensional torus, every sequence of points has a convergent subsequence. This is a well known but yet miraculous result which is true for any compact space. This may be verified readily by contracting measles, chicken pox or poison ivy. The fact can be proved as follows: Let $\{x_i\}$ be the given sequence. Divide T in half into sets (T_1, T_2). An infinite subsequence, $\{x_{i1}\} \subset \{x_i\}$ lands in one of T_1 or T_2. For argument's sake, suppose T_1. Now divide T_1 in half into (T_{11}, T_{12}) and repeat .... an infinite number of times.

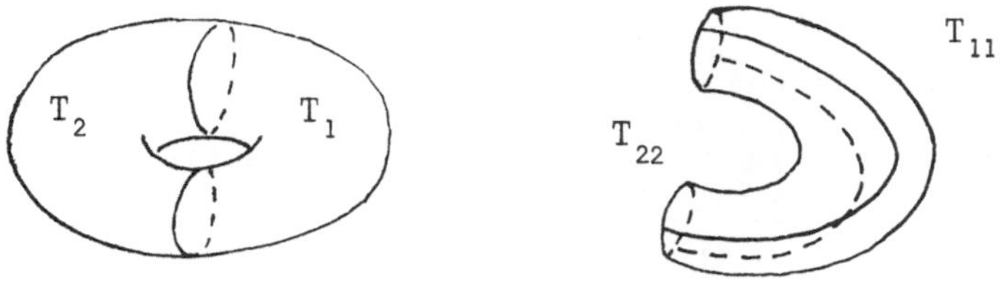

By continually subdividing, one constructs sets $\{T_1, ...\}$ whose intersection is a single point, $x \in T$, which has the property that every neighborhood of x contains points of $\{x_i\}$. This x is a limit point.

There is a fundamental difference between T and $\mathcal{C}_0/\mathcal{G}$. T is compact and $\mathcal{C}_0/\mathcal{G}$ is not. Consider the convergence of a sequence, $\{x_i\}$, of points on $\mathbb{R}^2$. By mimicking the analysis for T, one obtains the statement: Either $\{x_i\}$ has a convergent subsequence with limit $x \in \mathbb{R}^2$, or given $R < \infty$, there exists $i_R < \infty$ such that $|x_i| > R$ when $i > i_R$. That is, there is either convergence, or else the points wander off to ∞.

FUNDAMENTAL LESSON: To prove that a sequence on a non-compact space converges, it is necessary to provide a reason for it to stay in a bounded region.

The question one might ask is: In infinite dimensions, what

are the relevant bounded regions? The simplest example here is the Hilbert space $L^2(\mathbb{R}^3)$ of square integrable functions on $\mathbb{R}^3$. It has a countable basis, $\{v_\alpha\}_{\alpha=1}^{\infty}$, which can be the 3-dimensional, Hermite polynomials (the eigenfunctions of $-\Delta + |x|^2$).

A classical, but semi-miraculous theorem is that balls in $L^2(\mathbb{R}^3)$ are weakly compact. (The difference between weak and strong compactness will be described shortly.) The proof of this result is easy: Let $\{h_i\}$ be the given sequence with $\|h_i\| < B$ for all i. Then the sequence of *numbers* $\{\langle h_i, v_1\rangle\} \subset [-B,B] \subset \mathbb{R}$ has a convergent subsequence because $[-B,B]$ is compact; i.e. a subsequence $\{h_{i_1}\}$ has $\langle h_{i_1}, v_1\rangle \to \beta_1 \in [-B,B]$. Next consider the sequence of numbers $\{\langle h_{i_1}, v_2\rangle\} \subset [-B,B]$. It has a convergent subsequence, and so a subsequence $\{h_{i_2}\} \subset \{h_{i_1}\}$ has $\langle h_{i_2}, v_1\rangle \to \beta_1$ *and* $\langle h_{i_2}, v_2\rangle \to \beta_2$. Repeat this operation (called diagonalization) an infinite number of times to construct a subsequence $\{h_{i_\infty}\} \subset \{h_i\}$ and numbers $\{\beta_\alpha\}_{\alpha=1}^{\infty}$ which are the limits of $\langle h_{i_\infty}, v_\alpha\rangle$. The weak limit of $\{h_{i_\infty}\}$ is by definition $h = \sum_\alpha \beta_\alpha v_\alpha$. One can easily check that

1) $\|h\| = \sqrt{\sum_\alpha \beta_\alpha \beta_\alpha} < \lim_{i\to\infty} \|h_i\| < B$.

2) For any *fixed* $v \in L^2$, $\lim_{i\to\infty} \langle h_{i_\infty}, v\rangle = \langle h, v\rangle$.

Now turn to the min-max sequence, $\{(A^i,\Phi^i)\} \subset \mathcal{C}_0$. Since $\mathcal{A}(A^i,\Phi^i) < B$, the sequences $\{(B_A)^i\}$, $\{(\nabla_A\Phi)^i\}$ and $\{1 - |\Phi^i|^2\}$ all lie in balls in $L^2(\mathbb{R}^3)$, so they converge weakly for some subsequence of $\{(A^i,\Phi^i)\}$.

A very important theorem of K.Uhlenbeck [10] states that if, for a sequence $\{(A^i,\Phi^i)\}$, the curvatures and covariant derivatives $\{(B_A)^i\}$ and $\{(\nabla_A\Phi^i)\}$ converge in L^2, then there is a sequence $\{g^i\} \subset \mathcal{G}$ of gauge transformations such that the gauge transformed sequence $\{g^i(A^i,\Phi^i)\}$ has a subsequence which converges *weakly* to (A,Φ). *And*, if

$$\|\nabla\mathcal{A}_{(A^i,\Phi^i)}\|_* \searrow 0 ,$$

then $(A,\Phi) \in \mathcal{C}$ and it is a solution to Eq. (1.1) [ ].

For the min-max sequences here, it is always true that $\|\nabla\mathcal{A}\|_* \searrow 0$. Therefore, min-max always produces finite action solutions to Eq. (1.1). The hard part is to prove that the solutions that are so generated are not

1) the trivial solution, $(0, \tfrac{1}{2}\tau^3)$; and

2) the t'Hooft-Polyakov monopole, anti-monopole; and

for the Bogomol'nyi-Prasad-Sommerfield limit, not any of the self-dual monopoles which solve Bogomol'nyi's equation.

It is a fact that for the min-max over $\pi_1(\mathcal{C}_0/\mathcal{G})$, the number $\mathcal{A}_\Lambda > 0$. It is essentially for the same reason as in the finite dimensional example.

Scenarios 1) and 2) above illustrate the pitfalls of the notion of weak convergence. Consider $\{g^i(A^i,\Phi^i)\}$ as a sequence of configurations on $\mathbb{R}^3$ and also as a sequence of snap-shots (time-lapse photography) of an evolving configuration. It is useful to consider the sequence of energy densities,

$$s_i(x) = |(B_A)^i|^2(x) + |(\nabla_A\Phi)^i|^2(x) + \frac{\lambda}{2}(1-|\Phi^i|^2(x))^2 \quad .$$

Remember that for all i, $\int s_i(x) \geq \mathcal{A}_\Lambda > 0$.

The limit, (A,Φ), would be trivial (Scenario 1) above) if either

1a) The action densities diffuse away; as $i\to\infty$, $\sup_{x\in\mathbb{R}^3} s_i(x) \to 0$. Or,

1b) For all i, $\sup_x s_i(x) > \delta$, but this sup is taken on at points $\{x_i\}$ which have, themselves, no limit on $\mathbb{R}^3$.

These two possibilities are illustrated below:

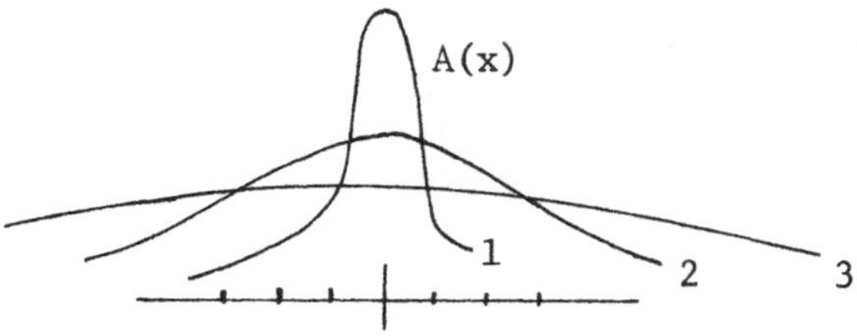

Case (1a)

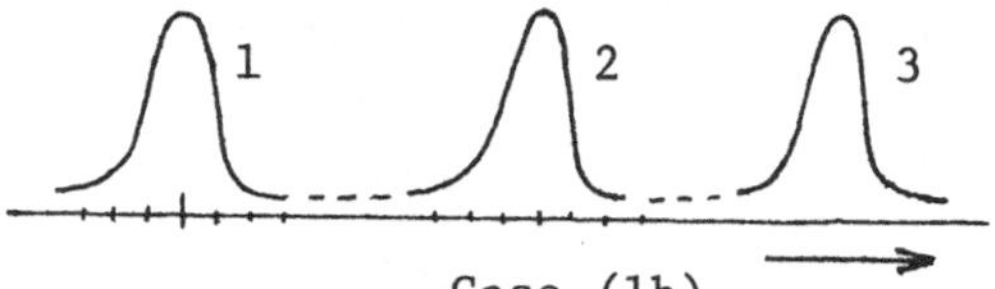

Case (1b)

The diffusion, case (1a), cannot occur. Physically, if $\sup s_i(x)$ is small enough, then the loop lies within a perturbation expansion's convergence radius of the minimum, $(0,\frac{1}{2}\tau^3)$. As in the torus case, this would contradict that the loops were not contractible. In fact, the conditions $\mathcal{A}_\Lambda > 0$, $\lim \|\nabla\mathcal{A}_i\|_* = 0$ and

$\lim(\sup_x s_i(x)) = 0$ are mutually incompatible.

If case (1b), above, were to occur, then one could use translational invariance to center as the origin in $\mathbb{R}^3$, for each i, a point where $s_i(\cdot)$ is maximal. But, this wouldn't stop other concentrations of $s_i(\cdot)$ from moving off to ∞ as $i \to \infty$. Nevertheless, by centering the maxima of s_i, one obtains in the limit a non-trivial solution.

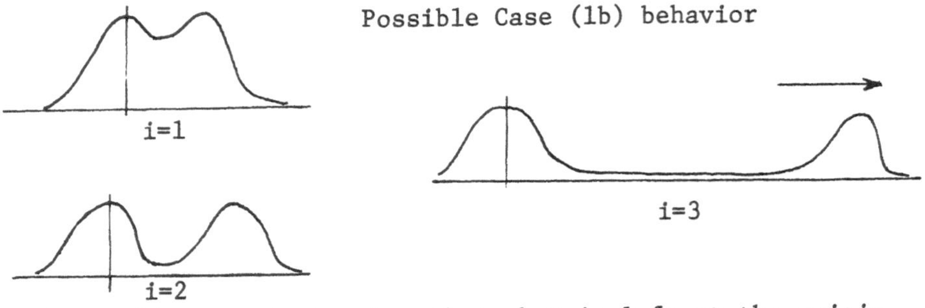

Possible Case (1b) behavior

The problem now is to show that what is left at the origin is not the 't Hooft-Polyakov monopole (or monopoles of higher charge [11]). Indeed, the blob of density that remains *in the limit as* $i \to \infty$ could have charge $\neq 0$ if, for large, but finite i, blobs of local charge + and − had formed and were separating as a function of i. This is Scenario (2).

Notice that such a scenario would be time lapse photography of a monopole-anti-monopole pair being created (plus possibly more) and separating out to ∞. But from this point of view, it is clear that for Scenario (2) to occur, it must be true that

$$\mathcal{A}_\Lambda \geq 2 \times (\text{one monopole mass}) .$$

Also, because m and $\bar{m}$ *attract* at large distances, and by construction, the action of (A^i, Φ^i) is *decreasing* to $\mathcal{A}_\Lambda$, one does not expect Scenario (2) to arise: pulling m and $\bar{m}$ apart should produce sequences with *increasing* action.

To prove that Scenario (2) cannot occur, one need only demonstrate that $\mathcal{A}_\Lambda < 2 \times (\text{one monopole mass}, M_m)$ by constructing an explicit, trial loop, $\ell(t) \in \Lambda$ with $\mathcal{A}(\bar{\ell}) < 2M_m$. Here, one is distinguishing:

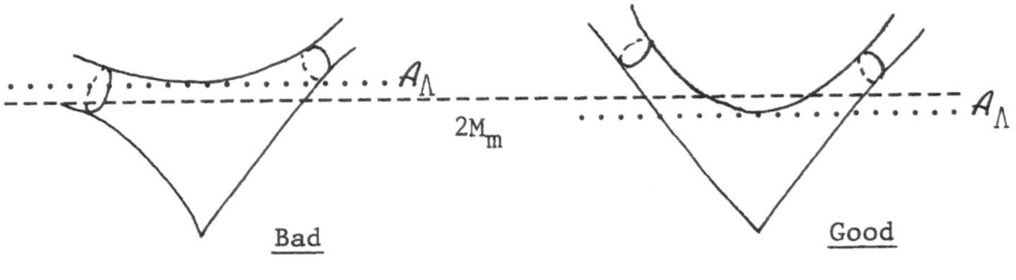

PROBLEM: How to select a useful trial loop in $\mathcal{C}_0$? Two qualities are important:

1) It is not contractible.
2) The action is easy to estimate.

To obtain a trial loop, (A^t, Φ^t), recall that Φ^t at $|x| = \infty$ must trace out the 1-parameter family of maps from S^2 to S^2 that is drawn in Eq. (3.1). Now, the 't Hooft-Polyakov monopole has for Φ at $|x| = \infty$ the hedge-hog configuration

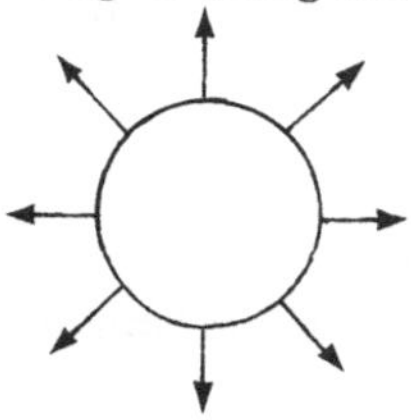

(4.1)

monopole

while the anti-monopole has the reversed picture:

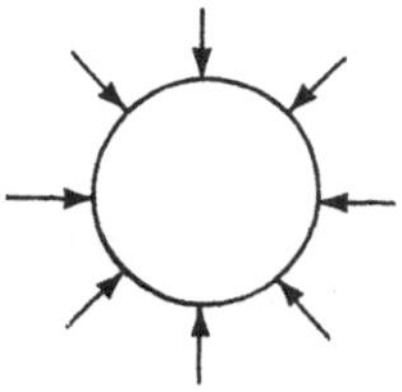

(4.2)

anti-monopole

So a loop $\ell(t) = (A^t, \Phi^t)$ which is not contractible comes from rotating an anti-monopole while holding a nearby monopole fixed. Compare Eq. (3.1) with Eqs. (4.1) and (4.2). Physically, one creates an m-$\bar{m}$ pair from the vacuum, $(0, \frac{1}{2}\tau^3)$. The relative coordinates to describe the m-$\bar{m}$ system are the (r,θ) coordinates of §1 (separation, and relative angle in isospin). Separate the m-$\bar{m}$ from $(r=0, \theta=-\pi)$ to $(r=d, \theta=-\pi)$ for some large $d \gg 1$. Then rotate the $\bar{m}$ relative to the m; i.e. let r stay fixed at $r=d$, and θ go from $-\pi$ to π. When θ is back to π, bring the m and m back together. This schematically describes the trial loop:

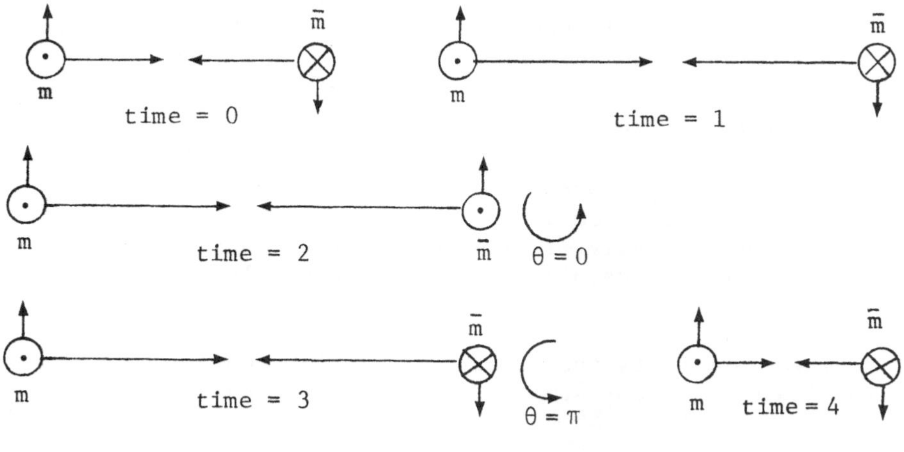

An explicit construction of this one parameter loop can be done with the explicit, 't Hooft-Polyakov monopole and anti-monopole described in Section 1 [4,5]. The energy as a function of r and θ, for large r, is independent of θ to first order: one finds

$$\mathcal{A}(r,\theta) = 2M_m - \frac{1}{r} + \mathcal{O}(r^{-2}) \quad .$$

In particular, if d is the maximum separation,

$$\mathcal{A}_{max} \leq 2M_m - \frac{1}{d} + \mathcal{O}(d^{-2})$$

for this loop.

The conclusion then is that $\mathcal{A}_\Lambda < 2M_m$ and there exists a non-minimal solution to the SU(2)-Yang-Mills equations with monopole number zero.

V. NEW RESULTS

By using these and similar long-range force estimates with the calculus of variations, new, stronger results have been obtained for this problem, and in 4-dimensional Yang-Mills theory. Two recent theorems are

THEOREM 5.1 (C.H Taubes [12, 13]: There exists in each monopole sector, an infinite number of guage inequivalent, smooth, finite action solutions to the SU(2) Yang-Mills-Higgs equations in the BPS limit. In each monopole sector, these solutions can be found with arbitrarily large action. Except for the solutions to the Bogomol'nyi equations (the self-dual monopoles), they are all unstable.

THEOREM 5.2 (C.H.Taubes [14]): For groups SU(2) and SU(3), the moduli spaces of instanton solutions (self-dual connections) of the Yang-Mills equations on S^4 are path connected manifolds.

VI. THE FAKE $\mathbb{R}^4$ (Day 3)

The general purpose of my lecture has been to present some mathematical facts which are peripheral to real physics (whatever that is). Specifically, I have concentrated on facts which are provable, knowing something about inter-monopole or, in today's case, inter-instanton forces.

I was asked, by the organizers of this conference, to report on the discovery, last summer, of a new, 4-dimensional manifold.

It is called a "fake" or "exotic $\mathbb{R}^4$" [15]. At least three fake $\mathbb{R}^4$'s [16] have been found and possibly there are an infinite number of distinct ones.

This fake $\mathbb{R}^4$ was completely unexpected — indeed, its discovery would be equivalent to a headline tomorrow in *Le Figaro*: "Free Quarks Found at CERN".

These fake $\mathbb{R}^4$'s are wierd things and my suspicion is that real ingenuity will be required to put them into physics. However, perhaps philosophical questions are raised by their existence. Perhaps not.

My involvement in this subject was peripheral and apostiori. If you want to know more about the fake $\mathbb{R}^4$, send at least $7 (U.S. dollars) to: Math Sciences Research Institute (MSRI), 2223 Fulton Street, Berkeley, CA 94720 (USA). You will receive a large and self-contained preprint by D.Freed, M.Freedman, and K.Uhlenbeck called "Gauge theories and 4-manifolds" [15]. This is to become a book later. (And $7 is certainly less than the future retail price.)

The existence of this fake $\mathbb{R}^4$ was proved by topologists M.Freedman at U.C. Santa Barbara and A. Casson at the University of Texas at Austin. The proof, a proof by contradiction, follows logically from a remarkable theorem of Simon Donaldson [17].

What I am going to say in the next hour comprises mostly plagiarism from two sources. First, the aforementioned notes of Freed, Freedman and Uhlenbeck, and second, a handwritten summary of the ideas involved due to Robion Kirby at U.C. Berkeley [18].

A fake $\mathbb{R}^4$, $\mathbb{R}^4_\Sigma$, is a smooth, four-dimensional manifold which is homeomorphic to $\mathbb{R}^4 = \mathrm{span}(e_1, e_2, e_3, e_4)$

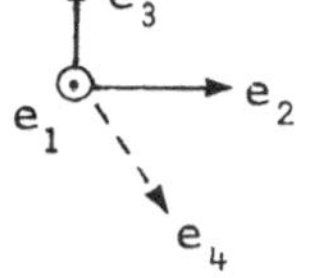

The true $\mathbb{R}^4$ is the vector space. But $\mathbb{R}^4_\Sigma$ is *not* diffeomorphic to the vector space $\mathbb{R}^4$. This means that there exists a continuous map, $f\colon \mathbb{R}^4_\Sigma \to \mathbb{R}^4$ which is one to one, and its inverse, $f^{-1}\colon \mathbb{R}^4 \to \mathbb{R}^4_\Sigma$ is also continuous. But no smooth map exists from $\mathbb{R}^4_\Sigma \to \mathbb{R}^4$ which is one to one with *smooth* inverse. In fact,

Property 1: No smooth one to one map exists between $\mathbb{R}^4_\Sigma$ and $\mathbb{R}^4$.

A creature who cannot detect *smooth* structures (he can't differentiate between derivatives) does not see $\mathbb{R}^4_\Sigma$ as distinct from $\mathbb{R}^4$. But a creature who does take derivatives sees two quite different objects.

Philosophical Question: Do humans really take derivatives? Can they tell the difference?

VII. SMOOTH STRUCTURES

This is a subject which rarely concerns the average man (cf. the philosophical question).

Consider the line, $\mathbb{R}$ with a parameter t. A function, $f: \mathbb{R} \to \mathbb{R}$ is continuous if

$$\lim_{\varepsilon \to 0} |f(t+\varepsilon) - f(t)| \searrow 0$$

for every $t \in \mathbb{R}$. The function is differentiable if

$$f'(t) \equiv \lim_{\varepsilon \to 0} \frac{1}{\varepsilon} (f(t+\varepsilon) - f(t))$$

is continuous for every $t \in (0,1)$.

Now define the space of once differentiable functions on (0,1) to be the set of continuous functions which are differentiable.

The function, f, is infinitely differentiable if all its derivatives are differentiable (defined inductively, $f'' = (f')'$ ... etc.).

A diffeomorphism $f: \mathbb{R} \to \mathbb{R}$ is a one to one, infinitely differentiable map whose inverse is also infinitely differentiable.

A smooth structure, Σ, on $\mathbb{R}$ is: (1) a cover of $\mathbb{R}$ by open sets $\{(-\frac{1}{2},t_0),(t_1',t_1),(t_2',t_2)\ldots$ and $(t_{-1}',\frac{1}{2}),(t_{-2}',t_{-2}),\ldots\}$ with $0 < t_1' < t_0 < t_1 \ldots$ and $0 > t_{-2} > t_{-1}' > t_{-2} \ldots$. (2) Smooth overlap functions,

$$f_{i,i+1} : (t_{i+1}',t_i) \to (t_{i+1}',t_i) \qquad \text{for } i \geq 0 ,$$

$$f_{-1,0} : (-\tfrac{1}{2},\tfrac{1}{2}) \to (-\tfrac{1}{2},\tfrac{1}{2}) ,$$

$$f_{-i-1,-i} : (t_{-i}',t_{-i-1}) \to (t_{-i}',t_{-i-1}) \qquad \text{for } i < 0 ,$$

which are one to one with smooth inverses and which take the end points to themselves.

Note that $f_{i,i+1}$, being 1 to 1 and invertible means that $f'_{i,i+1} \neq 0$ and hence $f_{i,i+1}$ is a *monotone* increasing function.

One labels $\mathbb{R}$ with this smooth structure Σ by $\mathbb{R}_\Sigma$. The manifold $\mathbb{R}_\Sigma$ would be diffeomorphic to $\mathbb{R}$ should there exist a smooth $g: \mathbb{R}_\Sigma \to \mathbb{R}$ which is 1 to 1, with smooth inverse.

In physics, one doesn't distinguish between diffeomorphic smooth structures — they are just coordinate transformations.

Such a map, g is represented by a set of functions, $\{g_i: (t'_i, t_i) \to (t'_i, t_i)\}$ with each g_i a diffeomorphism, i.e. $\partial g_i/\partial t > 0$. (Here g is assumed to be orientation preserving.) In addition, it is required that on (t'_{i+1}, t_i), $g_{i+1}(f_{i,i+1}(g_i^{-1}(t))) = t$. This is the condition for the definition to be consistent.

For example, if each $f_{i,i+1}(t) = t$, then $\mathbb{R}_\Sigma = \mathbb{R}$ and an orientation preserving diffeomorphism of $\mathbb{R}$ is just a monotonic function on $\mathbb{R}$. Consider $t \longrightarrow e^{+t}$.

> THE FUNDAMENTAL THEOREM IN ONE DIMENSION: Any orientation preserving smooth structure on $\mathbb{R}$ is diffeomorphic to the standard one. There is only one smooth line.

The proof goes loosely in the following way: Remember that $f_{i.i+j}$ on (t'_{i+1}, t_i) is a strictly monotone, smooth function which takes t'_{i+1} to t'_{i+1} and t_i to t_i.

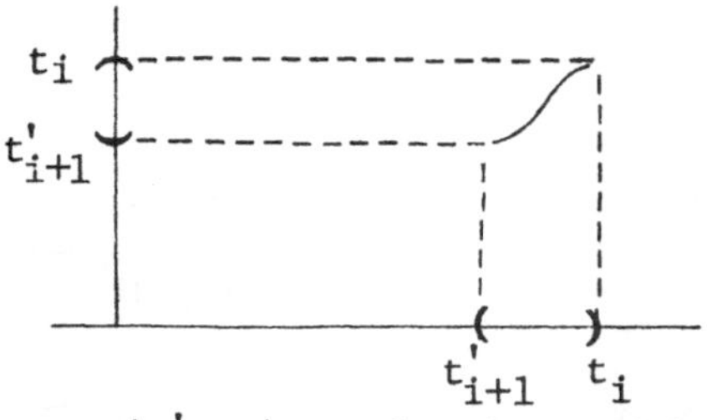

Graph of $f_{i,i+1}$

Now, define $g_i(t)$ on (t'_i, t_i) to be $f_{i,i+1}(t)$ when $t \in (t'_{i+1}, t_i)$, define $g_i(t)$ to be t when $t \in (t'_i, t_{i-1})$ and between t_{i-1} and t'_{i+1} take $g_i(t)$ to be any smooth, monotone interpolation between t and $f_{i,i+1}(t)$. The existence of such an interpolation can be seen best by drawing the picture shown on the next page. Thus, each $g_i(\cdot): (t'_i, t_i) \to (t'_i, t_i)$ is now smooth, monotone increasing, and by construction $g_{i+1}(f_{i,i+1}(g_i^{-1}(t))) = t$ when $t \in (t'_{i+1}, t_i)$. This completes the proof.

As an exercise, prove that S^1 has only one smooth structure up to diffeomorphism.

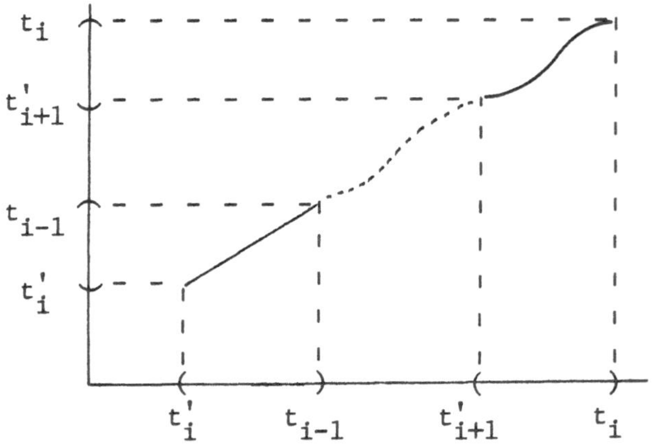

VIII. SMOOTH STRUCTURES ON MANIFOLDS

A smooth N-manifold is defined by the following data. First, there is the basic topological space, M_{top}, which is a point set together with an assignment of open subsets. Second, there is a cover of M_{top} by open sets $\{U_\alpha\}$ and homeomorphisms, $h_\alpha\colon U_\alpha \to \mathbb{R}^N$. Here $N \in [0,1,\ldots,\infty]$. Finally, there are *smooth* transition functions $g_{\alpha\beta} = h_\alpha(h_\beta^{-1}(\cdot))$ from $h_\beta(U_\alpha \cap U_\beta)$ to $h_\alpha(U_\alpha \cap U_\beta)$ as subsets of $\mathbb{R}^N$. These $g_{\alpha\beta}$ are required to be diffeomorphisms.

The given M_{top} is a *topological* N-manifold if the $g_{\alpha\beta}$'s are only continuous, i.e. they are homeomorphisms from $h_\alpha(U_\alpha \cap U_\beta)$ to $h_\beta(U_\alpha \cap U_\beta)$.

For example, the $\mathbb{R}_\Sigma$ of the last section is a smooth 1-manifold.

THE FUNDAMENTAL THEOREM ON $\mathbb{R}^N$, $N = 1,2,3,5,6,\ldots,\infty$. Any orientation preserving smooth manifold structure on $\mathbb{R}^N$ is diffeomorphic to the standard $\mathbb{R}^N$ as a $\times_N \mathbb{R}^1$.

The proof of the fundamental theorem on $\mathbb{R}^N$ for $N \neq 1$ is difficult, cf. [19].

Now it is known that the fundamental theorem for $\mathbb{R}^4$ is *false*. The exotic $\mathbb{R}^4$, $\mathbb{R}^4_\Sigma$ (to distinguish it from $\mathbb{R}^4 = \times_4 \mathbb{R}^1$) is defined by a locally finite cover of $\mathbb{R}^4$ by open balls $\{U_\alpha\}$, *homeomorphisms* $\{h_\alpha\colon U_\alpha \to U_\alpha\}$ such that

1) $\{g_{\alpha\beta} = h_\alpha(h_\beta^{-1}(\cdot))\colon U_\alpha \cap U_\beta \to U_\alpha \cap U_\beta\}$ are *diffeomorphisms* and

2) No set of *diffeomorphisms*, $\{\eta_\alpha\colon U_\alpha \to U_\alpha\}$ exist with $\{g_{\alpha\beta} = \eta_\alpha(\eta_\beta^{-1}(\cdot))\}$.

For *continuous* data, $\mathbb{R}^4_\Sigma$ and $\mathbb{R}^4$ are the same. But for smooth data, they are wildly different.

IX. PROPERTIES OF $\mathbb{R}^4_\Sigma$ [18]

At this time, there is no explicit construction of $\mathbb{R}^4_\Sigma$. The properties that $\mathbb{R}^4_\Sigma$ is known to have were deduced via a reductio absurdum argument; without them, a contradiction would exist with established fact.

1) There exists a compact set, $K \subset \mathbb{R}^4_\Sigma$, such that K cannot be contained in the interior of a smoothly imbedded 3-sphere. As $\mathbb{R}^4_\Sigma$ is homeomorphic to $\mathbb{R}^4$, there are certainly 3 spheres surrounding K which are topologically imbedded, i.e. the spheres $x_1^2 + x_2^2 + x_3^2 + x_4^2 = R$ for R sufficiently large. Here $\{x_i\}_{i=1}^4$ are the coordinate functions on the true $\mathbb{R}^4$. Indeed, each x_i is a continuous function on $\mathbb{R}^4_\Sigma$ — but not a smooth one.

2) $\mathbb{R}^4_\Sigma$ does not *smoothly* imbed into the standard 4-sphere. But, since it is homeomorphic to $\mathbb{R}^4$, it imbeds *continuously* in S^4 (as $S^4\setminus$point).

3) $\mathbb{R}^4_\Sigma$ does imbed in $S^2 \times S^2$ smoothly. Therefore, it has a complex analytic structure. There exists a non-standard $\mathbb{C}^2$ also.

4) There is no homeomorphism $\alpha\colon \mathbb{R}^4_\Sigma \to \mathbb{R}^4$ which is smooth, or has smooth inverse.

5) There is a smooth structure on $\mathbb{R}^4 \times [0,1]$ which restricts to the standard $\mathbb{R}^4$ on $\mathbb{R}^4 \times \{0\}$ and which restricts to $\mathbb{R}^4_\Sigma$ on $\mathbb{R}^4 \times \{1\}$.

6) This comment is related to 1). There is no smooth metric on $\mathbb{R}^4_\Sigma$ which is asymptotically flat. Does it have a complete Einstein metric?

7) Also related to 1), there is no 3-manifold which is smoothly imbedded near ∞ in $\mathbb{R}^4$ and $\mathbb{R}^4_\Sigma$ in a homotopically similar way.

Some questions that would be worthwhile solving are:

1) Is there a useful differential invariant of $\mathbb{R}^4_\Sigma$ which can distinguish it from $\mathbb{R}^4$; i.e. the index of some operator?

2) Is there a reasonably direct construction of $\mathbb{R}^4_\Sigma$?

For physicists, perhaps one should address: How are distinct smooth structures to be treated in quantum gravity? Indeed, there are 4-manifolds with *no* smooth structures at all [20],[17]. What does quantum gravity do with them?

X. THE EXISTENCE PROOF

Simon Donaldson [17] proved that except for S^4, there are no *smooth*, compact, oriented, simply connected, spin 4-manifold whose second Betti-number was equal, up to sign, to its Hirzebruch signature. That is, there is no *smooth*, oriented, ... 4-manifold which has only self-dual or only anti-self-dual harmonic 2-forms in its DeRham cohomology. He proved this by exhibiting a contradiction. The contradiction was obtained by examining the parameter space of 1-instanton solutions to the SU(2) Yang-Mills equations on a hypothetical manifold satisfying the constraints above (after giving it a Riemannian metric).

On the other hand, Mike Freedman [21] showed that there exist *topological* 4-manifolds, which, were they to admit a *smooth structure*, would satisfy the constraints above. Therefore, Freedman's manifolds do not have smooth structures. (Freedman had earlier proved the existence of topological 4-manifolds which can't be smoothed [20].)

Freedman and Casson went over Freedman's theorem for the existence of the *topological* manifolds. The purpose was to find exactly where it was that differentiability disappeared. They concluded that only a fake $\mathbb{R}^4$ with all of the listed properties in Section 9 would rule out differentiability of Freedman's manifolds.

REFERENCES

1. G.'t-Hooft, Nucl. Phys. B79: 276 (1974), and A.M. Polyakov, JETP Lett. 20: 194 (1974).
2. E.B.Bogomol'nyi, Sov. J. Nucl. Phys. 24: 449 (1976).
3. N.S.Manton, Nucl. Phys. B135: 319 (1978).
4. C.H.Taubes, Comm. Math. Phys. 86: 257 (1982); and Comm. Math. Phys. 86: 299 (1982).
5. D.Groisser, "SU(2) Yang-Mills-Higgs theory on $\mathbb{R}^3$", Harvard University preprint.
6. R.Palais, Proc. Symp. Pure Math. 15: 185 (1970); Amer. Math. Soc., Providence, RI.
7. R.Palais, Topology 5: 115 (1966).
8. M.Berger, Nonlinearity and Functional Analysis, Academic Press, New York (1977); and L.Ljusternik, Topology of the Calculus of Variations in the Large, (Amer. Math. Soc. Transl. 16), Amer. Math. Soc., Providence, RI (1966).

9. D.Groisser, "Integrability of the monopole number in SU(2) Yang-Mills-Higgs theory on $\mathbb{R}^3$", MSRI Berkeley preprint, (1983).
10. K.K.Uhlenbeck, Comm. Math. Phys. 83: 31 (1981).
11. A.Jaffe and C.H.Taubes, Vortices and Monopoles, Birkhauser, Boston (1980).
12. C.H.Taubes, "On the Yang-Mills-Higgs equations," Bull. AMS, to appear.
13. C.H.Taubes, in preparation.
14. C.H.Taubes, "Path connected Yang-Mills moduli spaces," Univ. of Calif., Berkeley, preprint (1983).
15. D.Freed, M.Freedman and K.K.Uhlenbeck, "Gauge theories and four manifolds," MSRI, Berkeley preprint, (1983).
16. R.Gompf, J. Diff. Geo. 18: 317 (1983).
17. S.K.Donaldson, J. Diff. Geo. 18: 279 (1983).
18. R.Kirby, private communication.
19. E.Moise, Ann. of Math. 54: 506 (1951); 55: 172 (1952); 55: 203 (1952); 55, 215 (1952); 56: 96 (1952). And J.Stallings, Proc. Camb. Phil. soc. 58: 481 (1962).
20. M.Freedman, Ann. of Math. 110: 177 (1979).
21. M.Freedman, J. Diff. Geo. 17: 357 (1982).

MONTE CARLO RENORMALIZATION GROUP AND THE THREE DIMENSIONAL ISING MODEL

Kenneth G. Wilson

Newman Laboratory of Nuclear Studies
Cornell University
Ithaca, NY 14853

ABSTRACT

Precise computations of critical properties of the three dimensional Ising model are available from computations on the ICL DAP (Edinburgh) and the Santa Barbara Ising processor. The computations however are not complete because they do not include reliable information on the correction to scaling exponent. The Monte Carlo Renormalization group is used in the Scottish computations; finite size scaling is used at Santa Barbara. The ideas behind the Monte Carlo Renormalization Group are explained, along with an analysis of the errors involved, to the extent they are presently known.

*****

In this report I shall discuss the current work on the Monte Carlo Renormalization Group applied to the critical behavior of the 3 dimensional Ising model. I will particularly emphasize the work reported in a paper by Pawley, Swendsen, Wallace, and myself.[1]

The critical point of the 3-dimensional Ising model is being explored with extreme care, in order to determine the effectiveness of the Monte Carlo Renormalization Group method. My own ultimate interest is in other problems, such as the lattice gauge theory. However, it is much easier to obtain high statistics for an Ising model computation than for a gauge theory computation; thus I have concentrated more on the Ising model case, and will continue to work on it until I am satisfied that the computation is understood.

The Ising model Monte Carlo Renormalization Group computations

I will discuss have been carried out on lattices of sizes ranging from 4^3 to 64^3, using the ICL DAP at Edinburgh[1] and other systems. The Santa Barbara group[2] has data from a 128^3 lattice using their special purpose processor. There is very substantial evidence for convergence as the lattice size increases, especially for the critical coupling of the Ising model. In contrast, the gauge theory computations have only extended to 16^4 lattices and there is very little reliable evidence yet for convergence for most qualities of interest.

In the case of the Ising model it should now be possible to obtain error estimates on the computations. In this report I shall try to emphasize the sources of error, the extent to which these sources are understood, and how they affect the numerical results. Unfortunately, the errors are still far from being mastered (in my opinion), especially errors associated with corrections to scaling near the critical point. Until these errors are understood, the Ising model computation will not be totally satisfactory, regardless of how spectacular some of the numerical results appear to be.

I will also try to explain the basic theory of the Monte Carlo Renormalization Group.

In this report I will discuss in Section II some aspects of renormalization group theory for a lattice of infinite extent. Then procedures for working on a finite lattice will be discussed in Section III. I will not say much about actual numerical results, since these are well reported in the literature.[1,2] Included in this report are a number of open questions for which further computations are required.

I. SOME RENORMALIZATION GROUP THEORY

To start with I will give a general discussion of real space renormalization group theory; this is needed before the Monte Carlo Renormalization Group ideas can be formulated.

Let $\mathcal{H}_o$ be a Hamiltonian, but with the Boltzmann factor $-\beta$ folded in:

$$\mathcal{H}_o = -\beta H \tag{1}$$

where H is the Hamiltonian itself. Define a sequence of effective interactions $\mathcal{H}_i$ by the rule

$$\exp\{\mathcal{H}_{i+1}[t]\} = \sum_{\{s\}} \Pi_{\vec{m}}\, K(t_{\vec{m}}, \{s\}_{\vec{m}})\, \exp\{\mathcal{H}_i[s]\} \tag{2}$$

where the set $\{s\}$ refers to a set of variables $s_{\vec{n}}$ on an infinite lattice labelled by the lattice vector $\vec{n}$. For the Ising model, $s_{\vec{n}}$ is restricted to be +1 or -1; real space renormalization group transformations can be defined for much more general configuration spaces. Each interaction $\mathcal{H}_i$ depends on a set of variables $\{s\}$. The partition function is given by

$$Z = \sum_{\{s\}} \exp \mathcal{H}_i[s] \tag{3}$$

where the sum is a sum over all configurations. The real space transformation (2) will be defined to ensure that all the effective interactions $\mathcal{H}_i$ yield the same partition function Z when Z is computed by (3).

The kernels K define "block variables" $t_{\vec{m}}$ on a block lattice labelled by vectors $\vec{m}$. In the computations described here, the block lattice is formed by dividing the original lattice into 2×2×2 blocks and assigning a block lattice location to each 2×2×2 block on the original lattice. For each block variable $t_{\vec{m}}$ there is a kernel $K(t_{\vec{m}},\{s\}_{\vec{m}})$ which involves $t_{\vec{m}}$ itself and variables $s_{\vec{n}}$ from the original lattice that lie within or in the neighborhood of the block $\vec{m}$. The kernel must satisfy the Kadanoff constraint[4]:

$$\Sigma_t \; K(t, \{s\}_{\vec{m}}) = 1 \tag{4}$$

where Σ_t means a sum over all possible values of a single t variable. The Kadanoff constraint ensures that the same partition function is obtained from the configuration sum (3) for every interaction $\mathcal{H}_i$ (i.e., for all block levels i).

The Kadanoff constraint allows a very large class of kernels K, and one way to reduce errors in actual computations is to adjust the choice of K to minimize errors. This possibility has not been explored yet in the context of the Monte Carlo Renormalization Group (to my knowledge); it should be.

In this report the kernel K is defined as follows. Firstly, the variables $t_{\vec{m}}$ are required to be Ising-like ($t_{\vec{m}} = \pm 1$ only). (This is not necessary — the Kadanoff constraint could be satisfied even if $t_{\vec{m}}$ has many values, or is continuous; but for practical convenience we work here with the Ising-like case.) Secondly, aside from $t_{\vec{m}}$ itself, we consider the kernel to depend on only one other variable, namely the sum $S_{\vec{m}}$ of the eight old spins $s_{\vec{n}}$ which lie in the 2×2×2 block $\vec{m}$. Since $S_{\vec{m}}$ is the sum of eight spins, each of which can be ±1, the sum $S_{\vec{m}}$ has nine possible values, namely -8, -6, -4, -2, 0, 2, 4, 6, or 8. We define K(t,S) as follows. If t and S have the same sign then K(t,S) is 1. If t and S have opposite signs then $K(t,S)$ is 0. For the case $S=0$ we set $K(t,0) = \frac{1}{2}$ for both choices $t = \pm 1$ for t. Another way of saying this is that if S is not 0, then the kernel forces t to be $S/|S|$. If S is 0 then t is given the values +1 or -1 with equal probability ½.

There is a wide range of possibilities for K. Even if one lumps the old spins into a single sum S, it is possible to choose S to be a weighted sum of spins outside as well as inside the block, or, for example, one can choose a weighted sum of spins centered about a single old spin rather than centered about a block.

In the calculations reported here the initial interaction was taken to be the nearest neighbor Ising model on a simple cubic lattice, with a single coupling constant K (not to be confused with the kernel function K)

$$\mathcal{H}_0 = K \sum_{\vec{n},i} s_{\vec{n}} s_{\vec{n}+\hat{i}} \tag{5}$$

where $\hat{i}$ is a unit vector and the sum over $\hat{i}$ is a sum over the three positive unit vectors on a cubic lattice.

The effective interactions $\mathcal{H}_i[t]\,(i > 0)$ contain all possible interactions involving all possible products of the block spins $t_{\vec{m}}$ that are consistent with the symmetries of the problem. In particular, 2 spin, 4 spin, 6 spin, etc. interactions of arbitrary range are present. However, it is hoped that the magnitude of the interactions drops rapidly as the separation of the block spins increases. In particular, it is hoped that there is an effective correlation length ξ_i governing the fall-off of interactions in $\mathcal{H}_i$ versus separation. That is, if an interaction involves the product of two block spins $t_{\vec{m}}$ and $t_{\vec{m}'}$ with separation $r = |\vec{m}-\vec{m}'|$ then the magnitude of the coupling constant for this interaction should include a factor $\exp\{-r/\xi_i\}$. It is hoped that ξ_i is finite, hopefully relatively small, even when the initial Hamiltonian is at its critical point and has a true correlation length of ∞. The true correlation length appears when correlation functions are computed for the original Hamiltonian without the block spin kernels appearing in (2). The block spin kernels are supposed to "freeze" the long range correlations present in the original Hamiltonian, thereby allowing the interactions in the block interaction $\mathcal{H}_i$ to be short-ranged. I say it is hoped that ξ_i is finite and relatively small because there is no proof of this except in simple cases (such as the Gaussian model) where the effective interactions can be determined analytically.[5] For each block spin computation one has to verify, usually indirectly, that the effective interactions $\mathcal{H}_i$ have short range. In the case of the block spin computations reported here the evidence for short range is indirect but strong.[1]

The initial Ising interaction has a critical point at a critical value K_c of K. K_c is known to be about .22169 from high temperature series analysis.[6] When K is at K_c it is expected that the sequence of effective interactions $\{\mathcal{H}_i\}$ approaches a fixed point interaction $\mathcal{H}^*$ as $i \to \infty$, with $\mathcal{H}^*$ also having short range. When i is large and $\mathcal{H}_i$ is near $\mathcal{H}^*$, it is expected that $\mathcal{H}_i$ will have an expansion about $\mathcal{H}^*$ in a set of "eigenoperators" O_j of the renormalization group transformation. These eigenoperators are for the linearized form of the transformation (2), linearized about the fixed point $\mathcal{H}^*$. The eigenoperators O_j have eigenvalues λ_j. The linearized transformation is not a symmetric transformation, so that the eigenvalues λ_j may be complex or in some cases might not be complete but in examples I have studied carefully to date one has only had to deal with simple real eigenvalues.

The eigenoperators have two classifications. First of all, eigenoperators with $|\lambda_j| > 1$ are "relevant", that is, the corresponding eigenoperators, if present at all in $\mathcal{H}_i$, grow in strength (as $(\lambda_j)^i$) as i increases, carrying $\mathcal{H}_i$ away from the fixed point $\mathcal{H}^*$. There are two important relevant operators for the three dimensional Ising model. One is an even eigenoperator (even in the sense that it contains only even products of spins) and is associated with changing K away from criticality. The eigenvalue λ_t for this eigenoperator is called the "thermal eigenvalue". Its value for the three dimensional Ising model is about 3.[1] There is also a relevant odd eigenoperator with an eigenvalue (called the magnetic eigenvalue) about 5.5.[1] Eigenoperators with eigenvalues less than 1 are called "irrelevant"; they govern the rate at which sequence $\{\mathcal{H}_i\}$ approach $\mathcal{H}^*$ when all relevant operators are absent. The even irrelevant operator with the largest eigenvalue controls corrections to the scaling theory of critical behavior. From other analyses of these corrections (high temperature series and the "ε" expansion[7]) it is expected that the leading irrelevant even eigenvalue will be about .7.

A separate classification due to Wegner[8] is into redundant and non-redundant operators. Redundant operators do not affect the physics of the model; in field theoretic language they are operators which vanish as a consequence of the field equations. However, redundant operators do appear in the effective interactions and can cause considerable confusion. The eigenvalues of redundant operators can change when the block spin kernel function K is changed, and if one is unfortunate, one or more of the redundant operators can become relevant. When this happens the $\mathcal{H}_i$ will not approach $\mathcal{H}^*$ when the initial Hamiltonian is critical, unless extra parameters are added to the initial Hamiltonian and adjusted precisely to eliminate the relevant, redundant operators from the sequence $\mathcal{H}_i$. Even when all the redundant operators are irrelevant they can still mask the irrelevant operator that governs corrections to scaling. The three dimensional Ising computations have turned up a second relevant odd eigenoperator which is suspected to be redundant, but there has been no verification of this suspicion yet. Since this operator is odd, it does not appear in the expansion of even interactions $\mathcal{H}_i$ and therefore is relatively harmless. Its eigenvalue is around 1.8,[9] well below the leading odd eigenvalue of 5.5.

A direct computation of the effective interactions $\mathcal{H}_i$ by Monte Carlo methods (as originally proposed by S.-K. Ma[10]) looks at first sight pretty hopeless. A separate Monte Carlo computation is required for each configuration of the block spins $\{t_{\vec{m}}\}$. As many configurations need to be looked at as there are parameters in $\mathcal{H}_i[t]$ to be determined. There might be thousands or even millions of parameters in $\mathcal{H}_i[t]$ that are important. I have carried out a direct computation of the interactions $\mathcal{H}_i$ for the two dimensional Ising model,[11] determining about 217 parameters for each i, but the number of possible interactions of a given range greatly increases as the dimension increases. However, there remains a possibility that the number of important parameters in $\mathcal{H}_i$ might be held within reasonable bounds and that direct computations of the $\mathcal{H}_i$ might someday be feasible. By studying various choices for the block spin kernel K, one might find a choice for which (a) the correlation length ξ_i is very small, making all but nearest neighbor and diagonal nearest neighbor couplings negligible, and (b) 4 spin interactions and higher spin interactions are negligible. In the 2-d computation, the four spin interactions were relatively small and 6-spin and higher were probably negligible, although I never carefully tested the effect of neglecting them. An interaction containing only short range two spin interactions contains only a small number of parameters and might be calculable with the current computer power available for Ising computations.

The current Monte Carlo Renormalization Group computations involve an indirect study of the effective interactions $\mathcal{H}_i$. Namely one studies expectation values of block spins with respect to the $\mathcal{H}_i$. The basic theorem used is that these expectation values can be reformulated as expectation values with respect to the original interaction. As a result, a single Monte Carlo computation can be used to generate all the expectation values of interest.

The basic theorem will be illustrated by discussing an expectation value with respect to the first block spin interaction $\mathcal{H}_1[t]$. Suppose one wants the expectation value of a function f(t): for example, f(t) might be $t_{\vec{m}}\, t_{\vec{o}}$ for some vector $\vec{m}$. Then one wants to compute

$$\langle f(t)\rangle = Z^{-1} \sum_{\{t\}} f(t)\, e^{\mathcal{H}_1[t]} \tag{6}$$

where Z is given by Eq. (3). Substituting the definition of $\mathcal{H}_1$, one obtains

$$\langle f(t)\rangle = Z^{-1} \sum_{\{t\}} \sum_{\{s\}} f(t) \, \Pi_{\vec{m}} \, K(t_{\vec{m}}, \{s\}_{\vec{m}}) \, e^{\mathcal{H}_0[s]} \tag{7}$$

The Kadanoff condition on the kernel K ensures that Z can be computed using $\mathcal{H}_0$ instead of $\mathcal{H}_1$ in (3). A Monte Carlo procedure applied to $\mathcal{H}_0$ (the Metropolis method, the heat bath method, or whatever[12]) generates sequences of configurations of old spins {s} distributed with probability $Z^{-1} \exp\{\mathcal{H}_0(s)\}$. To compute the expectation value of f(t), the Monte Carlo procedure is extended to generate configurations of block spins {t}. The Kadanoff condition (4) allows the kernel $K(t_{\vec{m}}, \{s\}_{\vec{m}})$ to be used itself as a probability distribution for $t_{\vec{m}}$ (provided that the kernel is always positive or zero for any value of $t_{\vec{m}}$). Thus Monte Carlo procedures can be used to generate configurations of {t} distributed with probability $\Pi_{\vec{m}} K(t_{\vec{m}}, \{s\}_{\vec{m}})$ for each given configuration {s} of old spins. The expectation value of f(t) can be determined in the usual way by averaging f(t) over the configurations {t}. In the computations reported on here, the Monte Carlo generation of configurations {t} was very simple since either $t_{\vec{m}}$ was determined uniquely by the block spin sum $S_{\vec{m}}$ or, if $S_{\vec{m}}$ was 0, $t_{\vec{m}}$ was set to +1 or -1 with equal probability 1/2 with the help of a computer-generated random number.

When computing expectation values of subsequent interactions $\mathcal{H}_2$, etc., all that is required is to build several levels of block spin configurations by the same Monte Carlo procedure.

The next step is to make some predictions about the behavior of block spin expectation values near the critical point. These predictions (once finite size effects have been studied) can be checked by Monte Carlo computations and are used to determine the critical temperature and critical exponents of the Ising model.

First of all, consider the expectation values of a set of simple block spin products, such as the nearest neighbor spin product $t_{\vec{o}}\, t_{\hat{x}}$, a diagonal nearest neighbor spin product $t_{\vec{o}}\, t_{\hat{x}+\hat{y}}$, perhaps a four spin product, etc. Denote these products for the ith block levels as $P_1{}^i$, $P_2{}^i$, etc., and their expectation values as $p_1{}^i$, $p_2{}^i$ etc. $p_1{}^i$ refers to a nearest neighbor product, $p_2{}^i$ to a diagonal nearest neighbor product, etc. Note that as the level i is increased, the t's refer to larger size blocks so that a nearest neighbor block spin product corresponds to a larger and larger physical separation as i increases.

If the sequence $\mathcal{H}_i$ approaches a limit $\mathcal{H}^*$ as i increases, then the expectation values $p_n{}^i$ should also approach limits p_n^* as $i\to\infty$. Conversely, if the initial interaction $\mathcal{H}_0$ is slightly off critical, due to the coupling K not being precisely K_c, then the $\mathcal{H}_i$ for large i have departures from $\mathcal{H}^*$ behaving as λ_t^i where λ_t is the thermal eigenvalue. Correspondingly, the p_n^i should have corrections away from p_n^* behaving as λ_t^i. Thus the expectation values p_n^i for large i should vary very rapidly as K varies; the critical value $K = K_c$ should be easily recognized as the unique value of K for which the p_n^i approach a value independent of i, instead of changing rapidly with i.

The next prediction involves derivatives of expectation values with respect to various coupling terms that can be added to block interactions. Coupling terms involve the same set of products P_n^i that are used for expectation values, except that in a coupling term these products must be summed over all translations and rotations on the lattice. Derivatives with respect to these coupling terms reduce to expectation values involving the connected part of products of two of the P_n^i. To be precise, we denote these derivatives by $d_{n,m}^{i,\ell}$, referring to the derivative of the expectation value p_n^i with respect to the coefficient of P_m^ℓ added to the block interaction at level ℓ. The formula for $d_{n,m}^{i,\ell}$ is

$$d_{n,m}^{i,\ell} = \sum_{\substack{\text{(translations} \\ \text{(rotations}}} \text{of } P_m^\ell) \left(<P_n^i \, P_m^\ell> - p_n^i \, p_m^\ell \right) \tag{8}$$

To make a prediction regarding these derivatives, it is convenient to consider the expansion of the P_m^ℓ in terms of eigenoperators O_j (more precisely, the expansion sums of the P_m^ℓ over all translations and rotations). We assume that $\mathcal{H}_\ell$ is approximately at the fixed point $\mathcal{H}^*$. Then each P_m^ℓ (summed over translations and rotations) can be written as a sum $c_{mj}{}^\ell \, O_j$ of the eigenoperators; the coefficients $c_{mj}{}^\ell$ are actually independent of ℓ.
But an eigenoperator O_j added to the interaction $\mathcal{H}_i$ with coefficient c_{mj} then contributes to a subsequent interaction $\mathcal{H}_i$ with coefficient $\lambda_j^{(i-\ell)} c_{mj}$. Thus the derivative of an expectation value on level i behaves as $\lambda_j^{(i-\ell)} c_{mj}$. To be precise, let $e_{n,j}{}^i$ be the derivative of the expectation value $<p_n^i>$ with respect to the coefficient of O_j added to the interaction $\mathcal{H}_i$.
Then

$$d_{n,m}^{i,\ell} = \Sigma_j \; c_{mj} \; \lambda_j^{(i-\ell)} \; e_{n,j}^{\;\;i} \tag{9}$$

When the interaction $\mathcal{H}_\ell$ is not close to $\mathcal{H}^*$ (for example, when $\ell=0$) but subsequent interactions do approach $\mathcal{H}^*$, then it is still true that a small addition of the interaction P_m^ℓ to $\mathcal{H}_\ell$ is equivalent to adding a linear combination $c_{mj}^\ell \; \lambda_j^{(i-\ell)} \; O_j$ to $\mathcal{H}_i$, but now the coefficient $c_{mj}{}^\ell$ depend on ℓ. Then

$$d_{n,m}^{\;i,\ell} = \Sigma_j \; c_{mj}^{\;\ell} \; \lambda_j^{\;(i-\ell)} \; e_{n,j}^{\;i} \tag{10}$$

The thermal eigenvalue is expected to be about four times larger than the next leading eigenvalue. This, when i-ℓ is large, the dominant term in this equation should be

$$d_{n,m}^{\;i,\ell} \simeq c_{mt}^{\;\ell} \; \lambda_t^{\;(i-\ell)} \; e_{n,t}^{\;i} \tag{11}$$

with an error that decreases by a factor of 4 (relative to the dominant term) each time i increases by 1. This formula says that $d_{n,m}^{i,\ell}$ thought of as a matrix in the indices n and m, is separable: a product of a factor depending only on m and ℓ times another factor depending only on n and i. This separability has been very obvious in the actual Ising computations.

If the leading irrelevant eigenvalue is much larger than any remaining eigenvalue, (including eigenvalues of redundant operators) then a more accurate formula should be

$$d_{n,m}^{\;i,\ell} = c_{mt}^{\;\ell} \; \lambda_t^{\;(i-\ell)} \; e_{n,t}^{\;i} + c_{m,2}^{\;\ell} \; \lambda_2^{\;(i-\ell)} \; e_{n,2}^{\;i} \tag{12}$$

where λ_2 is the leading irrelevant eigenvalue. This formula says that the matrix $d_{n,m}^{i,\ell}$ is the sum of two separable terms, to much greater accuracy than just fitting $d_{n,m}^{i,\ell}$ to a single separable expression. Up until now statistical errors have prevented any evidence developing that the two eigenvalue fit to $d_{n,m}^{i,\ell}$ is better than a single eigenvalue fit; see below for further discussion.

When $\mathcal{H}_i$ is approximately $\mathcal{H}^*$ then the derivatives $e_{n,j}^{\;i}$ are also independent of i. This means that there are two ways that the eigenvalue λ_t, can be determined from computations of $d_{n,m}^{i,\ell}$. The first way is to compute the ratio of $d_{n,m}^{i,\ell}$ for two adjacent values of i, both considerably larger than ℓ. As long as $\mathcal{H}_i$ and $\mathcal{H}_{i+1}$ are both close to $\mathcal{H}^*$, one predicts that

$$d_{n,m}^{\;i+1,\ell}/d_{n,m}^{\;i,\ell} \simeq \lambda_t \tag{13}$$

with an error that behaves as $(\lambda_2/\lambda_t)^{i-\ell}$. The prediction is that the same ratio is obtained for any choice of n and m. Conversely if $\mathcal{H}_\ell$ and $\mathcal{H}_{\ell+1}$ are both close to $\mathcal{H}^*$ then one predicts that the factors $c_{mj}^{\;\ell}$ are independent of ℓ, so that one can use ratios of $d_{n,m}^{\;i,\ell}$ for two adjacent values of ℓ:

$$d_{n,m}^{\;i,\ell}/d_{n,m}^{\;i,\ell+1} = \lambda_t \tag{14}$$

with an error of order $(\lambda_2/\lambda_t)^{(i-\ell+1)}$.

In these formulae, the method to reduce the error on the determination of λ_t is to increase the separation $i-\ell$. In the first ratio (13), one can accomplish this by setting $\ell=0$ and using i as large as possible. Assuming the initial interaction is precisely critical, $\mathcal{H}_i$ gets closer to $\mathcal{H}^*$ as i increases so that the errors due to $\mathcal{H}_i$ not being $\mathcal{H}^*$ decrease as well as the errors due to the λ_2 term in the ratio. In the second form of the ratio, it is necessary to increase ℓ as well as $i-\ell$ to improve accuracy since there are errors in the formula (14) due to $\mathcal{H}_\ell$ not being $\mathcal{H}^*$ and these errors should be reduced as well as the explicit effects of the λ_2 term in the ratio.

Swendsen, who first advocated the use of the derivative matrices $d_{n,m}{}^{i,\ell}$ in order to compute exponents, has used a different approach which might give a number of exponents, not just λ_t. Let $K_m{}^\ell$ denote the coupling coefficient for the spin product $p_m{}^\ell$ in the interaction $\mathcal{H}_\ell$; then

$$d_{n,m}{}^{i,\ell} = \partial p_n{}^i \,/\, \partial K_m{}^\ell \tag{15}$$

Swendsen's approach is to use the chain rule for differentiation to derive a formula for the derivatives of $K_m{}^{\ell+1}$ with respect to $K_m{}^\ell$: since

$$\partial K_{m'}{}^{\ell+1} / \partial K_m{}^\ell = \Sigma_n \frac{\partial K_{m'}{}^{\ell+1}}{\partial p_n{}^i} \frac{\partial p_n{}^i}{\partial K_m{}^\ell} \tag{16}$$

If we define the matrix $d^{i,\ell}$ to be the matrix with elements $d_{n,m}{}^{i,\ell}$, and we define the matrix C^ℓ as the matrix with elements

$$\partial K_{m'}{}^{\ell+1} \,/\, \partial K_m{}^\ell = C_{m'm}{}^\ell \tag{17}$$

then the formula (16) can be written

$$C^\ell = (d^{i,\ell+1})^{-1} \, d^{i,\ell} \tag{18}$$

The matrix C^ℓ defines the linearized renormalization group transformation whose eigenvalues are λ_j and eigenoperators are O_j, provided that $\mathcal{H}_\ell$ and $\mathcal{H}_{\ell+1}$ are close enough to $\mathcal{H}^*$. Swendsen's approach is to compute the matrix C^ℓ from Eq. (18) and then diagonalize C^ℓ to determine the eigenvalues λ_j. The problem with Swendsen's method is that the matrix to be inverted in Eq. (18) is an infinite matrix, but in any practical calculation only a finite number of its matrix elements are determined. I have found no way to estimate the error that is made when a truncated form of the matrix is inverted instead of the full matrix. What I mean is that I know of no way of determining the error introduced by truncation that I would stake any part of my professional reputation on, at least for the case Swendsen deals with.

Swendsen generally works with the case $i=\ell+1$ in Eq. (18). If one allows i to be much larger than ℓ, so that the contributions of different eigenvalues λ_j in Eq. (9) have very different sizes, then one can easily work out the effect of the truncation of the matrices in Swendsen's formula to matrices of size L×L. Namely the matrix C^ℓ is truncated to include the effects of the first L eigenvalues only of the full infinite matrix. Unfortunately, when i is much larger than ℓ the statistical errors of the Monte Carlo computations tend to be larger than the contributions of sub-dominant eigenvalues to the matrices $d^{i,\ell+1}$ and $d^{i,\ell}$. Thus one may have to live with Swendsen's method when trying to extract subdominant eigenvalues.

II FINITE SIZE EFFECTS

So far the whole discussion has assumed that the lattices involved are of infinite extent. Actual Monte Carlo computations must be made on lattices of finite size. The effects of finite size will now be examined.

Suppose the original interaction is confined to a lattice of size N^3, with periodic boundary conditions. When this lattice is divided into 2×2×2 blocks one is left with a periodic lattice of blocks of size $(N/2)^3$; when this lattice is blocked in turn, a new lattice of size $(N/4)^3$ is generated, etc. Thus the effective interaction $\mathcal{H}_1$ defined by Eq. 2 lives on an $(N/2)^3$ lattice; the effective interaction $\mathcal{H}_2$ lives on an $(N/4)^3$ lattice, etc.

Placing the effective interactions $\mathcal{H}_i$ on a finite lattice distorts the interaction itself since terms in $\mathcal{H}_i$ whose range is longer than the lattice size can no longer be present. However, if the effective correlation length ξ_i for $\mathcal{H}_i$ is much smaller than the lattice size for $\mathcal{H}_i$ this distortion should be negligible; in particular this distortion decreases exponentially as the effective lattice size increases, whereas all the other finite size effects to be discussed fall off as powers of the lattice size. In practice the finite size distortions of the effective interactions appear to be appreciable in the Ising model calculations only on effective block lattices of size 2^3; they are small for 4^3 lattices and have yet to be detected for lattices of size 8^3 or larger. See ref. 1.

The finite size effects become much more pronounced when one looks at expectation values with respect to block interactions. The finite size effects in the p_n^i are quite noticeable out to large lattices. Finite size effects are also present in the $d_{n,m}^{i,\ell}$ matrices, but in the factorized form of these matrices

(Eqs. 9,10) only the factors e_{nj}^{i} explicitly refer to expectation values of the effective interactions and carry large finite size effects. The other factors, namely the coefficients C_{mj}^{ℓ} and the eigenvalues λ_j, are related to properties of the effective interactions and the transformation (2) which defines the effective interactions, and these factors are expected to have only exponentially decreasing finite size effects, apart from irrelevant operators in $\mathcal{H}_i$ itself.

Due to finite size effects, a table of p_n^{i} versus i no longer shows convergence to a fixed point when $K = K_c$. The values of p_1^{i} versus i are shown, for example, in the left hand column of Table 1 for a 64^3 lattice: one sees almost a doubling of p_1^{i}

TABLE I. Expectation values of nearest neighbor product of block spins (summed over three axes) versus block level i and original lattice size N^3. The K value is .221663. Diagonals down and to the left correspond to constant finite size for the block lattice; agreement is expected when $\mathcal{H}_i$ are all $\mathcal{H}^*$. Reading across shows the finite size effects for a given block level. Statistical errors are shown where they exceed .01. (Data courtesty of S. Pawley, Edinburgh, prepared by interpolation from runs at K = .22161 and .22169).

i/N^3	64^3	32^3	16^3	8^3
0	.997	1.007	1.035	1.107
1	.962	.994	1.078	1.299
2	1.01	1.097	1.33	1.945
3	1.11±.01	1.35±.01	1.97	
4	1.36±.02	1.98±.02		
5	1.97±.04			

for large i due to finite size effects. However there is a fix for the problem of finite size effects. The next column of Table 1 shows the p_1^{i} for the case of a 32^3 lattice. For large i, $\mathcal{H}_i$ should be $\mathcal{H}^*$ independent of i (assuming K is K_c) and the only reason p_n^{i} varies with i is due to the size of the lattice that $\mathcal{H}_i$ resides on. Now notice that $\mathcal{H}_{i+1}$ starting from a 64^3 lattice and $\mathcal{H}_i$ starting from a 32^3 lattice reside on the same size block lattice. For example, $\mathcal{H}_3$ starting from a 64^3 lattice resides on an 8^3 lattice of blocks, as does $\mathcal{H}_2$ starting from a 32^3 lattice. Thus if $\mathcal{H}_2$ and $\mathcal{H}_3$ are both close to $\mathcal{H}^*$ then there is no distinction left between $\mathcal{H}_3$ from the 64^3 lattice and $\mathcal{H}_2$ from the 32^3 lattice: they are both close to $\mathcal{H}^*$ and both live on an 8^3 lattice. Hence the expectation values for these interactions should also be the same. This is clearly visible in Table I. If one compares p_1^{i+1}

on the left with p_n^i in the next column (i.e. compare entries along a diagonal left and down) one sees that the entries get closer to each other as i increases. As i increases, the block lattice size decreases and the finite size effects increase, but the finite size effects remain identical for the comparison of p_n^{i+1} from the 64^3 lattice with p_n^i for the 32^3 lattice; the only discrepancy left for this comparison comes from irrelevant operators which cause $\mathcal{H}_{i+1}$ and $\mathcal{H}_i$ to be distinct. The strength of the irrelevant operators decreases as i increases, hence the matching should improve as i increases.

In Table I, the same initial interaction $\mathcal{H}_o$ with the same value of K was used for both the 64^3 lattice computation and the 32^3 lattice computation. If K is somewhat different from K_c then $\mathcal{H}_i$ and $\mathcal{H}_{i+1}$ will both contain relevant operator contributions, with the contribution to $\mathcal{H}_{i+1}$ being about three times larger than the contribution to $\mathcal{H}_i$ (since the relevant thermal eigenvalue is about 3 for the three dimensional Ising model). In this case p_n^{i+1} will differ from p_n^1 and the diagonal matching will break down, in fact becoming three times worse for each unit increase in i. This means K must be extremely close to K_c to achieve the degree of matching shown in Table I for large i; in fact an error of only .00003 in K will spoil the matching shown, despite statistical uncertainties.

The computation of the critical value of K from this matching procedure shows very rapid convergence as the lattice size increases. Each time the lattice size is doubled, one more level of blocking becomes possible and thus the sensitivity of the matching to changes in K increases by about a factor of 3. At the same time more precise matching can be sought because the effects of irrelevant operators (which prevent a perfect match for all n values in p_n^i simultaneously) decrease. In practice this has meant about a factor 6 decrease in the finite size error in the determination of K_c for each doubling of the lattice size N.

In principle, a similar procedure is the best way to compute eigenvalues of the linearized renormalization group transformation. Namely one uses the formula (13) to determine, λ_t, with $\ell=0$ and i as large as possible; to eliminate finite size effects one computes $d_{n,m}^{i+1,0}$ starting from a lattice of size N^3 while $d_{n,m}^{i,0}$ is computed starting from a lattice of size $(N/2)^3$. The dominant error, for large i, is likely to come from irrelevant operators still present in $\mathcal{H}_i$ despite the large i value. Note that the error due to the leading irrelevant operator in $\mathcal{H}_i$ decreases only as λ_2^i whereas the error due to irrelevant operators explicitly in Eq. 10 decreases much faster, namely about $(\lambda_2/3)^i$.

Unfortunately, using the formula (10) with data from two different lattices results in large statistical errors. The best such results to date come from the Santa Barbara group.[2] Their results are based on conventional finite size scaling computations, but there is a close analogy between their formulae and Eq. 10. The Santa Barbara group has been limited to a few percent accuracy on exponent computations due to statistical errors.

In contrast, computations based on Swendsen's original approach using Eq. (18) can be carried out using results from a single finite lattice computation. In Eq. 18, the largest block level (i) is the same for both d factors, and therefore both should be computed for the same original lattice. One set of errors associated with the use of block levels ℓ and $\ell+1$ come from distortions of the effective interactions $\mathcal{H}_\ell$ and $\mathcal{H}_{\ell+1}$ due to finite block lattice sizes, but these appear to decrease[1] rapidly for block lattice sizes greater than 4^3. There are also errors due to irrelevant operators in $\mathcal{H}_\ell$ and $\mathcal{H}_{\ell+1}$. The block level ℓ cannot be as large as the level i of the two lattice method, so that the irrelevant operator correction is larger than in the two lattice comparison method but these errors appear to be roughly under control.[1] Unfortunately there is still the unknown error due to matrix truncation in Eq. 10. This truncation error can be reduced by taking $i-\ell$ large in Eq. 10, but then the level ℓ has to be reduced, bringing in a larger irrelevant operator correction.

The great advantage of the Swendsen approach is that it gives a spectacular improvement in statistical error compared to any comparison method involving two different lattices. The improvement in statistical accuracy comes about through a factor of about 10 cancellation of statistical errors in the ratio of the two truncated matrices of Eq. 10. Preliminary studies by me indicate that the origin of the cancellation of the statistical error is that there is no critical slowing down in the ratio, despite the presence of critical slowing down in each of the matrices entering the ratio. This means that it is beneficial to compute expectation values very frequently during a Monte Carlo run, if Swendsen's method of computing exponents is to be used. The expectation value computations on the ICL DAP were carried out every fourth Monte Carlo pass, which is much too frequent if critical slowing down is important, but was probably not too frequent for Swendsen's method of determining exponents.

Thus for determining the leading exponents (derived from the leading thermal and magnetic eigenvalues) the two lattice comparison methods are (in my view) preferred because their systematic errors are smaller and more easily understood, despite the greater cost in computing time to achieve adequate statistical accuracy. However, for non leading exponents, no good method has been found yet to determine them, and Swendsen's method will have to be pursued further. The big problem with Swendsen's method is that the matrix truncation error has been studied by considering matrices of size 1×1 up to 7×7 only, whereas there could easily be thousands of important couplings, meaning that the minimum sensible truncation size might be, say, 10000×10000. For example, there are probably five hundred to one thousand distinct four spin products that can fit into a 3×3×3 sublattice, none of which are very long range products. Even if one or two of these couplings (the ones that have been studied so far) are negligible no one can claim that the cumulative effect of these thousand different couplings is negligible. Thus in Ref. 1, the error quoted for ν (which has been computed by Swendsen's method) has an extra unknown term due to this aspect of the truncation error.

The quantitative results that have been obtained in references 1 and 2 are spectacular but even so, further statistical accuracy and formal study is needed to isolate the leading irrelevant operator and verify its dominance in $\mathcal{H}_i$ for large i when K is at K_c.

In the discussion of errors I have not discussed practical issues, namely the questions of statistical errors and questions of randomness of computer-generated random numbers. With respect to statistical errors, the unfortunate fact is that few practitioners of Monte Carlo computations compute their errors reliably or conservatively enough. The result is to reduce the credibility of all Monte Carlo computations, even good ones. To determine the error of a high precision computation, where the number of Monte Carlo passes vastly exceeds the correlation time, I find that the Monte Carlo run should be broken into at least thirty (30) subruns and the standard deviation determined for the thirty subruns. This gives a reasonably secure estimate for the statistical error. It is wishful thinking and dangerous to use much less than thirty subruns. It costs nothing to divide a run into thirty parts instead of two or four or eight.

With respect to the randomness of the random numbers, I can only urge that several important computations be repeated with several different forms of random number generators, and that "microcanonical" methods[13] also be used as a check whenever possible. The Scottish and Santa Barbara computations use distinct generators.

ACKNOWLEDGEMENTS

Part of my contribution to this work was carried out during a six month stay at the IBM Zurich Laboratory, in collaboration with R. Swendsen. I have used both the Array Processor at Cornell and IBM facilities for computations. I join other authors of this volume in dedicating this paper to the memory of

Kurt Symanzik and his lifetime contribution to the broad area of quantum field theory and statistical mechanics.

This work was supported in part by the National Science Foundation.

REFERENCES

1. G. S. Pawley, R. H. Swendsen, D. J. Wallace, and K. G. Wilson, Edinburgh preprint 83/238, submitted to Phys. Rev. B.

2. M. N. Barber, R. B. Pearson, D. Toussaint, and J. L. Richardson, Santa Barbara Institute for Theoretical Physics preprint NSF-ITP-83-144.

3. For reviews see T. W. Burkhardt and J. M. J. Van Leeuwen, eds. Real Space Renormalization, (Springer-Verlag, 1982).

4. L. P. Kadanoff, Phys. Rev. Lett. 34, 1005 (1975).

5. T. L. Bell, and K. G. Wilson, Phys. Rev. B11, 3431 (1975).

6. See Refs. 11-13 of Ref. 1.

7. See Ref. 1.

8. F. J. Wegner, in Phase Transitions and Critical Phenomena, C. Domb and M. S. Green, Eds., Vol. VI (Academic Press, 1976) p.34.

9. See Ref. 1.

10. S.-K. Ma, Phys. Rev. Lett. 37, 471 (1976).

11. K. G. Wilson, Revs. Mod. Phys. 47, 773 (1975).

12. See, e.g., K. Binder, ed., Monte Carlo Methods in Statistical Physics, Springer, Berlin, 1979.

13. D. Callaway and A. Rahman, Phys. Rev. Lett. 49, 613 (1982); M. Creutz, ibid. 50, 1411 (1983); G. Bhanot, Institute for Advanced Studies preprint.

INDEX

Anomaly, 71-75, 469, 543-562
  axial, 1, 2, 268
  gravitational, 1, 9-20, 543, 544, 553-555
  non-abelian, 2, 9, 554, 555
Anomalous dimension, 226, 239
Area law, 214
Asymptotic freedom, 79, 93, 197, 201-204, 271, 273, 280, 317, 319, 322, 328-330, 403, 404, 435, 438, 474, 476, 497, 534

Background field method, 418-422, 498
Beta-function, 110, 111, 173, 186, 188, 200, 238, 312, 329, 439
Block spins, 79-83, 86, 109, 156, 238, 239, 241, 435-438, 477, 478, 481-483, 489, 492, 592-595, 600
Boltzmann factor, 425, 473, 590
Boundary conditions, 5, 8, 60-64, 67-69, 74, 118, 176, 423, 425, 453, 532, 534, 599

Callan-Symanzik Equation, 200, 412-414, 423 (*see also* Renormalization group equation)
Canonical ensemble, 155, 156, 160, 162, 165, 469
Chiral symmetry, 48, 264
  breaking, 1, 202, 469
Cluster expansion, 82, 86, 93, 96, 102, 445, 448, 456
Confinement, 79, 145, 169, 171, 172, 174, 202, 204, 206, 207, 213, 214, 224, 383, 384, 392, 393, 400, 403, 404, 497, 524
Continuum Limit, 80, 106, 107, 111, 114, 115, 127, 130, 144, 148, 171, 173, 179-181, 184, 196, 200, 204, 209, 224, 233, 235, 253, 257, 266, 267, 455, 457, 462, 474-476, 492-494, 498, 511, 512, 515, 517, 519-526, 535-537, 541
Correlation
  functions, 80, 124, 131, 150, 156, 176, 191, 192, 453, 470, 476, 532, 536, 537, 592
  length, 111, 116, 176, 221, 436, 470, 474, 479, 592, 594, 599
Coulomb gas, 384, 481
Critical behaviour, 170, 589, 593
Critical dimension, 202, 206, 223, 226, 367, 424
Critical exponent, 170, 177, 178, 181, 221, 233, 595
Critical observation, 341
Critical phenomena, 169, 173, 174, 181, 435, 436, 440
Critical point, 72, 170, 177, 180-183, 201, 214, 221, 224, 435, 439, 454, 535, 589-595

Critical temperature, 389, 541, 595
Cross-over, 147
Cut-off, 301, 337, 404, 405, 410, 431, 440, 476-479, 485, 486, 489, 492, 493, 498

Debye screening, 384, 392
Dirac-Kähler equation, 247-269, 363
Dislocation, 375-380, 384, 387-390
Duality, 355, 357, 358

Faddeev-Popov method, 407, 418, 427, 516, 522, 526, 543-545
Fake R^4, 563, 580-586
Fibre bundle, 23, 24, 31-36, 499
Finite size effects, 117, 123, 133, 157, 424, 451, 452, 456, 464, 595, 599-603
Finite volume, 106, 136, 148, 204, 236, 238, 452, 454, 531, 533, 534
Fixed point, 173, 186, 201, 233, 237, 244, 245, 437, 438, 440, 593, 596
Flux, 207
  electric and magnetic, 27, 383
Functional integral, 6, 82, 173, 238

Gauge
  fixing, 89, 99, 317, 418, 427, 498, 522, 523, 525
  invariance, 1, 83, 89, 96, 201, 514, 516
  tranformation, 9, 28, 33, 66, 84, 87, 89, 94, 99, 100, 262, 263, 266, 405, 418, 427, 502-509, 514, 524, 527, 545, 553-555, 566, 576
Ginsburg-Landau theory, 387
Glueball, 107, 108, 111, 114, 117, 121, 124, 129, 148, 170, 205, 224, 258
  mass, 106, 112, 115, 120, 148, 201, 221, 536, 541
Goldstone boson, 36, 267
Gravity, 2, 11, 23, 24, 36, 171, 172, 228
Green's functions, 111, 169, 175, 178, 196, 263, 276, 282-286, 301-307, 311, 312, 315-318, 325-328, 370, 533, 538, 556

Heat bath, 595
Higgs mechanism, 68, 80, 81, 83, 85, 100, 329, 331

Improved action, 109, 110, 116, 128, 131, 135, 144, 149, 543
Index theorem, 2-8, 30, 31
Infrared, 437, 438
  divergence, 273, 312
  problems, 80, 81
  properties, 202
Instanton, 49, 71, 72, 272, 273, 328, 556, 580
  gravitational, 38
Irrelevant operator, 593, 601-603
Ising model, 144, 155, 157, 159, 162, 188, 204, 207, 223, 453, 456, 462-464, 589-603

Kaluza-Klein theories, 18, 23, 24, 34-37, 40, 51

Lattice approximation, 80, 83, 179, 247, 248, 261, 265, 266, 435
Lattice gauge theory, 105-150, 170, 171, 174, 201, 202, 205, 206, 219, 223, 271, 373, 400, 403-431, 470, 473, 474, 497-528, 535, 589
Localization, 228, 338, 367, 369

Manton's action, 109, 110, 115, 126, 130
Mass gap, 80, 82, 96, 108, 111-122, 132, 133, 146, 149, 193, 204, 440, 451-456, 464, 476, 479, 485, 497

Mayer expansion, 447, 448, 473, 476, 480, 484-486, 489-492
Mean field theory, 170, 197, 205, 220, 223-227, 233, 393, 400, 476
Metropilis algorithm, 165, 595
Microcanonical ensemble, 155-162, 165
Monopole, 51-76, 383, 384, 390-392, 563-580
  Dirac-, 23-50, 64, 67
  Kaluza-Klein-, 23, 24, 39-48
  't Hooft-Polyakov-, 42, 52, 63-65, 563, 564, 576, 578-580
Monte Carlo, 105-150, 156, 165, 197, 272, 405, 410, 413, 424, 430, 431, 475, 533-539, 589-591, 594, 595, 599, 602, 603
Morse theory, 563, 568

Non-Linear Sigma-model, 106, 193, 202, 203, 235, 439, 463, 465, 467, 473, 475, 476
1/N-expansion, 171, 188, 193, 202, 205, 223, 235, 271, 423, 440

Order parameter, 156, 373, 566
Osterwalder-Schrader-Positivity, 173, 205, 214, 225, 227

Partition function, 5, 80, 84, 85, 91, 92, 108, 159, 176, 384, 392-394, 406, 416, 420, 481, 484-486, 522, 525, 591
Perturbation
  expansion, 86, 98, 99, 233, 244, 271, 272, 283, 305, 319, 322, 326, 334, 445, 465, 577
  theory, 81, 93, 96, 100, 110, 171, 185, 186, 192, 193, 200-204, 272, 408, 409, 424, 442-444, 447, 448, 461, 464, 468, 474, 492, 493, 523, 525, 534
Phase
  diagram, 393
  structure, 92
  transition, 109, 177, 328, 373, 393, 400, 470, 476, 537
Planar diagrams, 271-331, 428-431
Positivity, 150
  of the action, 91

Quantum Chromodynamics, 105, 107, 171, 172, 201, 202, 261, 265, 268, 272, 273, 280, 328, 403, 404, 406, 451, 452, 465, 469, 470, 531

Random Lattice, 337-370
Random surface, 169-171, 174, 175, 201, 202, 207, 214, 215, 217, 219, 222, 223, 228
  planar, 170, 171, 215, 217-219, 224, 233
Random walk, 169, 171, 172, 174, 181, 183-185, 188-193, 198, 201, 202, 222, 245
Regularization, 105, 109, 111, 214, 219, 408, 409, 411, 414
  dimensional, 408, 411
  Pauli-Villars, 10, 13, 477, 479, 482, 488, 489, 493
Relevant operators, 201, 593, 594, 601
Renormalization, 82, 197, 282, 284, 329, 332, 333, 420, 439, 492, 526
  group, 79, 82, 83, 105, 116, 156, 173, 197, 201, 206, 235, 271, 322, 333, 435, 436, 439, 473, 477, 478, 480, 485-492, 497, 523, 524, 540, 589-599
  group equation, 184, 188, 271, 302, 312-316

Scale invariance, 36, 197, 437
Scaling, 106, 108, 111, 116, 122, 127, 130-133, 137, 139, 177, 226, 410, 431, 439, 440, 541

Scaling (continued)
  asymptotic, 107, 111, 112, 117, 118, 120, 127, 137, 140-142, 146-149
  region, 111, 125, 133, 413, 414
Schwinger model, 71, 76
Spin system, 116, 174, 175, 179-181, 189, 435, 452
Statistical mechanics, 169-173, 207, 219, 222, 225, 235, 337, 435, 436
String, 170, 204
  Dirac-, 28
  model, 19
  tension, 105, 107, 112, 119, 121, 148, 204, 205, 221, 410-413, 422, 430, 474, 475, 537, 541
Strong coupling expansion, 105-107, 117, 120, 121, 130, 132, 135, 136, 139, 142-144, 147, 148, 150, 171, 202, 204, 224, 337, 409, 416
Superfluid, 373, 378, 384, 385
Supergravity, 11, 18, 19, 49
Supersymmetry, 172, 260, 261
Supersymmetric quantum mechanics, 3
Susskind reduction, 252, 253, 258, 263, 269

Symmetry breaking, 1, 72, 171, 177, 222, 476, 544, 564, 567

Transfer matrix, 112, 115, 181, 194, 225, 453, 454

Ultraviolet, 438
  behavior, 79, 99, 498, 522, 524, 528
  divergence, 98, 281, 282, 523
  problems, 81, 100, 171, 244, 326, 439
  stability, 80, 83, 100
Universiality, 108, 109, 115, 120, 127, 136, 149, 219

Vortex, 81, 83, 373-375, 379, 384-387

Wilson
  action, 83, 100, 108, 110, 128, 131, 149, 205, 207, 406, 407, 409, 413, 414, 422, 513
  loop, 112-114, 122, 123, 132, 135, 137, 207, 209, 215, 406-409, 423, 424, 524, 533

Yang-Mills
  action, 108, 501
  equation, 71, 568, 580
  field, 545, 547
  theory, 83, 92, 470, 580